Amphiphilic Block Copolymers

Self-Assembly and Applications

Amphiphilic Block Copolymers

Self-Assembly and Applications

Edited by

Paschalis Alexandridis
Department of Chemical Engineering
State University of New York at Buffalo
Buffalo, NY 14260-4200, USA

Björn Lindman
Physical Chemistry 1,
Center for Chemistry and Chemical Engineering,
Lund University,
P.O. Box 124, S-221 00 Lund, Sweden

2000

ELSEVIER

Amsterdam - Lausanne - New York - Oxford
Shannon - Singapore - Tokyo

ELSEVIER SCIENCE B.V.
Sara Burgerhartstraat 25
P.O. Box 211, 1000 AE Amsterdam, The Netherlands

First edition 2000

Library of Congress Cataloging in Publication Data
A catalog record from the Library of Congress has been applied for.

ISBN: 0-444-82441-3

∞ The paper used in this publication meets the requirements of ANSI/NISO Z39.48-1992 (Permanence of Paper).
Printed in The Netherlands.

Contents

Techniques for the study of self-assembly structure and dynamics

Applications of amphiphilic copolymers

List of Contributors

V.Yu. Alakhov
Supratek Pharma Inc., Institute Armand-Frappier, Immunology Research Center, 531 boul. des Prairies, building 18, Laval, Québec, Canada H7B 1B7

P. Alexandridis
Physical Chemistry 1, Center for Chemistry and Chemical Engineering, and Center for Amphiphilic Polymers from Renewable Resources, Lund University, P.O. Box 124, S-221 00 Lund, Sweden, and Department of Chemical Engineering, State University of New York at Buffalo, Buffalo, NY 14260-4200, USA

T. Annable
ZENECA Resins, The Heath, Runcorn, Cheshire, UK

C. Booth
Manchester Polymer Centre, Department of Chemistry, University of Manchester, Manchester M13 9PL, UK

R. Buscall
ICI Technology Centre, Wilton, Middlesborough, Cleveland, UK

B. Chu
Department of Chemistry, State University of New York at Stony Brook, Stony Brook, NY 11790-3400, USA

A. Eisenberg
Department of Chemistry, McGill University, 801 Sherbrooke Street West, Montréal, Québec, Canada H3A 2K6

R. Ettelaie
ICI Technology Centre, Wilton, Middlesborough, Cleveland, UK

J.P.A. Fairclough
Manchester Materials Science Centre, UMIST, Grosvenor Street, Manchester M1 7HS, UK

M. Goldraich
Department of Chemical Engineering, Technion – Israel Institute of Technology, Haifa 32000, Israel

I.W. Hamley
School of Chemistry, University of Leeds, Leeds LS2 9JT, UK

K. Holmberg
Institute for Surface Chemistry, P.O. Box 5607, S-114 86 Stockholm, Sweden

H.-O. Johansson
Department of Biochemistry, Center for Chemistry and Chemical Engineering, Lund University, P.O. Box 124, S-221 00 Lund, Sweden

A.V. Kabanov
Department of Pharmaceutical Sciences, College of Pharmacy, University of Nebraska Medical Center, 600 South 42nd Street, Omaha, NE 68198-6025, USA

G. Karlström
Department of Theoretical Chemistry, Lund University, P.O. Box 124, S-22 100 Lund, Sweden

K. Khougaz
Department of Chemistry, McGill University, 801 Sherbrooke Street West, Montréal, Québec, Canada H3A 2K6

B. Lindman
Physical Chemistry 1, Center for Chemistry and Chemical Engineering, and Center for Amphiphilic Polymers from Renewable Resources, Lund University, P.O. Box 124, S-221 00 Lund, Sweden

P. Linse
Physical Chemistry 1, Center for Chemistry and Chemical Engineering, Lund University, P.O. Box 124, S-221 00 Lund, Sweden

L.-Z. Liu
Department of Chemistry, State University of New York at Stony Brook, Stony Brook, NY 11790-3400, USA

T. Liu
Department of Chemistry, State University of New York at Stony Brook, Stony Brook, NY 11790-3400, USA

S.-M. Mai
Department of Chemistry, University of Manchester, Manchester M13 9PL, UK

M. Malmsten
Institute for Surface Chemistry, P.O. Box 5607, S-114 86 Stockholm, Sweden

M. Moffitt
Department of Chemistry, McGill University, 801 Sherbrooke Street West, Montréal, Québec, Canada H3A 2K6

K. Mortensen
Condensed Matter Physics and Chemistry Department, Risø National Laboratory, DK-4000 Roskilde, Denmark

V.M. Nace
The Dow Chemical Company, Texas Operations, Research and Development, Freeport, TX 77541, USA

U. Olsson
Physical Chemistry 1, Center for Chemistry and Chemical Engineering, Lund University, P.O. Box 124, S-221 00 Lund, Sweden

R. Pons
Dep. Tecnologia de Tensioactius, C.I.D., C.S.I.C., c/ Jordi Girona 18-26, 08034 Barcelona, Spain

A.J. Ryan
Manchester Materials Science Centre, UMIST, Grosvenor Street, Manchester M1 7HS, and CLRC Daresbury Laboratory, Warrington WA4 4AD, UK

M. Svensson
Physical Chemistry 1, Center for Chemistry and Chemical Engineering, Lund University, P.O. Box 124, S-221 00 Lund, Sweden

Y. Talmon
Department of Chemical Engineering, Technion – Israel Institute of Technology, Haifa 32000, Israel

F. Tjerneld
Department of Biochemistry, Center for Chemistry and Chemical Engineering, Lund University, P.O. Box 124, S-221 00 Lund, Sweden

G.-E. Yu
Manchester Polymer Centre, Department of Chemistry, University of Manchester, Manchester M13 9PL, UK

R. Zana
Institut C. Sadron, CNRS-ULP, 6 rue Boussingault, F-67000 Strasbourg, France

L. Zhang
Department of Chemistry, McGill University, 801 Sherbrooke Street West, Montréal, Québec, Canada H3A 2K6

Preface

Amphiphilic (αμφιφιλος) are molecules which have an affinity (φιλος) for two different (αμφι) types of environments. This dual affiliation is built into the molecules by the covalent joining of parts (blocks) of different chemical character and solution properties. Given the opportunity, the two (or more) different parts strive to minimize their contact, and can thus coerce the amphiphilic molecules to attain a preferential orientation. As a result, amphiphilic molecules self-organize at interfaces and in solution and, in doing so, modify to a great extent interfacial properties and enhance dramatically compatibility or partitioning.

The first molecules that usually come to mind as being amphiphilic belong to the classes of surfactants and lipids. These molecules typically consist of an aliphatic hydrocarbon chain joined to a hydrophilic (often charged) head group, and they "thrive" (self-organize) in aqueous media, in mixtures of water and oil, at air-water interfaces, and at hydrophobic surfaces. Surfactants and lipids, some of them natural products, have attracted the scientific attention since the beginning of the 20th century. Over the last 3 decades, following up on the advent of synthetic polymers during WWII, a new class of amphiphilic molecules has emerged, that of block and graft copolymers. Perhaps the best known (but not always recognized as such) reflection of the self-organization properties of block copolymers is thermoplastic elastomers. These are rubber-like materials with the capability to melt and be processed and/or recycled in this state (note that rubber itself cannot melt as the mountains of used car tires can attest). Thermoplastic elastomers are triblock copolymers with a middle rubber-like (e.g., polyisoprene, PI) block and end-blocks made of polystyrene (PS). At a certain temperature range above the glass transition of PS, the PS end-blocks of different block copolymer molecules associate to form PS-rich domains. When the temperature is quenched below the PS glass transition, PS turns glassy and immobile, and the PS-rich domains act as the connecting links which give the desired elastomer properties.

Systems containing amphiphiles (surfactants or block copolymers) are notable in that no special "machining", other than mixing the "right" components at the "right" conditions, is required in order to achieve self-assembled morphologies of very well-defined structural order and nanometer scale dimensions. The challenge here is to control and tune the morphology, i.e., to determine the appropriate components and conditions leading to the desired microstructure. This is where the importance of self-assembly becomes paramount.

It is the belief of the editors of this book that the recognition of block copolymers as being amphiphilic molecules and sharing common features with other well-studied amphiphiles will prove beneficial to both the surfactant and the polymer communities. An aim of this book is to bridge the two communities and cross-fertilise the different fields. To this end, leading researchers in the field of amphiphilic block copolymer self-assembly, some having a background in surfactant chemistry, and others with polymer physics roots, have agreed to join forces and contribute to this book, titled "Amphiphilic Block Copolymers: Self-Assembly and Applications".

The book consists of four entities. The first part discusses theoretical considerations behind the block copolymer self-assembly in solution and in the melt. The second part provides case studies of self-assembly in different classes of block copolymers (e.g., polyethers, polyelectrolytes) and in different environments (e.g., in water, in non-aqueous solvents, or in the absence of solvents). The third part presents experimental tools, ranging from static (e.g., small angle neutron scattering) to dynamic (e.g., rheology), which can prove valuable in the characterization of block copolymer self-assemblies. The fourth part offers a sampling of current applications of block copolymers in, e.g., formulations, pharmaceutics, and separations, applications which are based on the unique self-assembly properties of block copolymers.

It is notable that in the late 70's, at the time when the surfactant solution behavior was becoming a very active field of investigation and a fundamental understanding was starting to emerge, but at a time when the phase behavior knowledge of block copolymers was at an embryonic stage, Pierre-Gilles de Gennes, Nobel Prize Laureate (Physics, 1991), stated that "block copolymers can give us the best model of amphiphilic behavior" [Solid State Phys. Suppl. 14 (1978) 1]. The present "Amphiphilic Block Copolymers: Self-Assembly and Applications" book is a testament to this foresight. The picture that emerges from this book is that amphiphilic block copolymers are very flexible in attaining a great variety of micro-structures in the absence or in the presence of solvents and/or other additives. Furthermore, the macromolecular nature of amphiphilic block copolymers allows access to a wide range of length scales and time scales. Amphiphilic block copolymers have a potential of great contributions to diverse applications such as synthesis of functional nanoporous materials or formulations of coatings, paints, pharmaceuticals, and personal care products.

A number of people and organizations have contributed directly or indirectly to this book. First and foremost we wish to thank each and all the authors. Each chapter is an excellent contribution. At the same time we feel that the chapters of this book provide a nice example of synergism (often observed in systems containing amphiphiles), where the sum (book) is even greater than the individual components (chapters) taken together. The Swedish Natural Science Research Council (NFR) has provided funding that supported significant part of the editors' research on Amphiphilic Block Copolymers in Lund. The Center for Amphiphilic Polymers from Renewable Recourses (CAP) at Lund University has embraced this work and has provided, through numerous meetings and workshops, a fertile forum for exchange of ideas and knowledge. Last, but not least, many thanks are due to the Elsevier Science B. V. editors, Dr. Kostas Marinakis for his support and encouragement at the inception stages of this book, and Dr. Huub Manten-Werker for patiently shepherding this work and bringing it to a fruitful completion.

PA & BL

Amphiphilic molecules: small and large

Björn Lindman and Paschalis Alexandridis

Physical Chemistry 1, Center for Chemistry and Chemical Engineering, and Center for Amphiphilic Polymers from Renewable Resources, Lund University, P.O. Box 124, Lund S-22100, Sweden.

Large amphiphilic molecules represent a new frontier in the field of surfactant science and technology. Their macromolecular nature affords a range of architectures, length scales, time scales, and levels of interactions much wider than those offered by small amphiphilic molecules. At the same time, such diversity poses great challenges in the characterization and understanding of the solution and surface properties of large amphiphiles. An obvious starting point is to utilize the arsenal accumulated over the last decades on the behavior of small amphiphiles. The evolution in self-assembly properties from small to large amphiphiles is outlined here, and the two classes of molecules are placed in perspective.

1. INTRODUCTION

Amphiphilic is an attribute which means "loving both" (or "having an affinity for both"). In the case of the molecules of particular interest here, this dual "affection" is usually expressed toward water and toward oil, but, in general, amphiphilicity can be expressed toward any two solvents which are incompatible with each other. For most purposes we consider as amphiphilic molecules those consisting of hydrophilic (water-loving) parts and hydrophobic (water-hating) parts, but in the present treatise we will include also molecules with two different parts being both hydrophilic. Typical amphiphilic molecules have molecular weights on the order of 500 and we would call them "small". Amphiphiles also come in "large" sizes, 10 - 1000 times bigger than the "small". These large amphiphiles are often block copolymers, where block(s) of one type of homopolymer are attached sequentially to block(s) of another type. There are two natural starting points for considering amphiphilic block copolymers in solvent environments, low molecular weight amphiphilic molecules in water and neat (solvent-free) block copolymers. In this introductory chapter we will present amphiphilic copolymers in such a broader context, and in doing so we will also compare amphiphilic block copolymers with other types of amphiphilic copolymers.

2. SELF-ASSEMBLY OF SMALL AMPHIPHILIC MOLECULES

Surfactants (surface-active agents) and (polar) lipids are representative examples of low molecular weight amphiphiles with distinct hydrophilic and hydrophobic parts. These classes, with broad technical applications and biological functions, have been extensively studied during the whole of the 20th century and a thorough understanding has emerged [1-6]. In the various applications, constant efforts are made to design new surfactants with an improved performance. It has been known since long time that the more amphiphilic a surfactant the more efficient it is. Since the hydrophilic head group is often fixed (e.g., a sulfate or amine ionic group, or oligo-ethylene oxide nonionic block), the desire to increase the amphiphilicity has prompted an enlargement of the hydrophobic group such as the lengthening of an alkyl chain.

Due to the amphiphilic nature of a surfactant or lipid, the two parts of the same molecule interact very differently with either a polar solvent or surface or a nonpolar solvent or surface. There are two different ways to render favorable intermolecular contacts possible in surfactants while eliminating unfavorable ones: self-assembly in solution (see Figure 1) and adsorption at a surface or an interface (examples shown in Figure 2). The self-assemblies can involve only the amphiphilic molecules, or there can be a mixed aggregate formed together with a low molecular weight cosolute or a macromolecule [2]; in fact, self-assembly is quite sensitive to a range of cosolutes. During the recent years, we have learnt that amphiphile adsorption at different interfaces is also best considered as a self-assembly process promoted by the interface and that there is considerable organisation of amphiphilic molecules at interfaces.

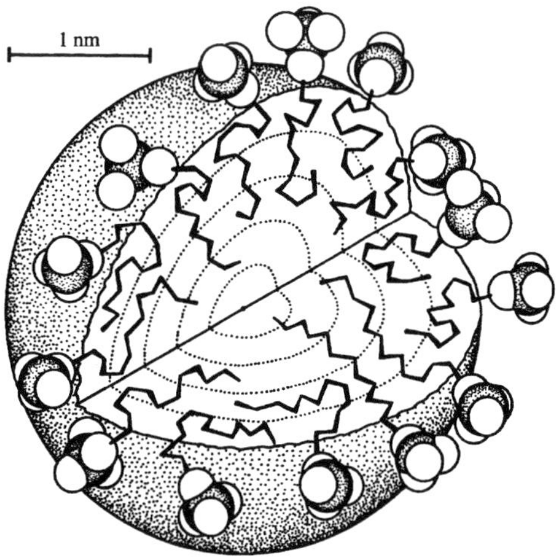

Figure 1. An illustration of a spherical micelle (for dodecyl sulphate) emphasizing the liquid like character with a disordered hydrocarbon core and a rough surface. (adapted from J. Israelachvili, Intermolecular and Surface Forces, Academic Press, London, 1985, p. 215)

Self-assembly involves the formation of domains of hydrophobic groups (in contact with a nonpolar solvent or surface if applicable) and domains of hydrophilic groups (in contact with water or another polar solvent or a hydrophilic surface). In general, in a system containing surfactant, water, and oil we have a segregation between water and oil domains and surfactant films (see Figure 3). The surfactant films can close on themselves and form discrete aggregates, or they can form "infinite" aggregates with connectivity over macroscopic distances, as shown schematically in Figure 4. A rich topological variation is possible as well as different intermicellar arrangements; examples of various self-assembled microstructures are given in Figure 5. This leads to a frequently complex phase behavior [5,6]. A general observation is that the phase behavior becomes richer as the size of the surfactant molecule increases.

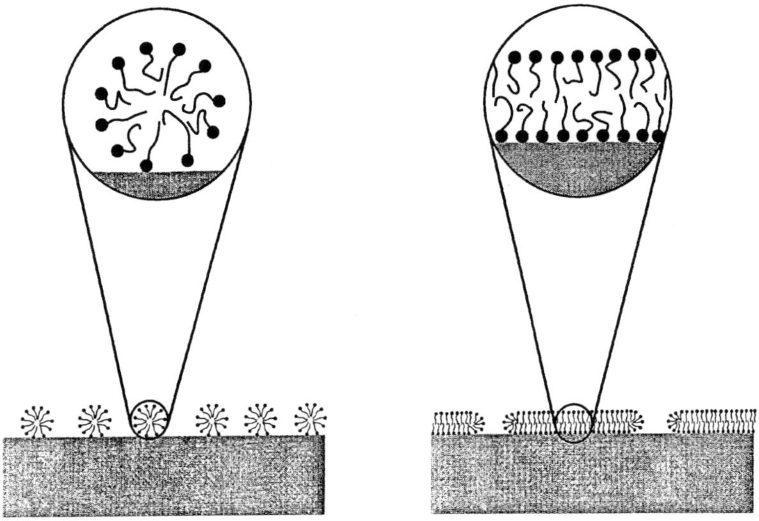

Figure 2. Surfactant molecules can self-assemble into discrete (spherical or cylindrical) micelles at a hydrophilic surface, in the absence of strong specific interactions between the surfactant head-group and the surface. (adapted from ref. [2], p. 60; © J. Wiley, 1998)

Both technical applications and biological functions may be enhanced by increasing the size of the amphiphilic molecule. Firstly, a stronger amphiphile has a low monomeric solubility, and thus has a stronger tendency to go to any interface. It is then more surface-active, and acts as a better detergent, a better stabiliser of dispersions or modifier of surfaces, and a better membrane-forming agent. Secondly, the richer self-assembly polymorphism gives a finer control over macroscopic and microscopic properties. Another point relates to the dynamic nature of the self-assemblies. Surfactant micelles and other aggregates are highly dynamic, with low residence times of surfactant molecules [1]. The residence times increase very strongly with increasing surfactant molecule size. An important incentive for increasing the molecular size is to decrease the lability of aggregates and making technical systems and biological structures more stable and less sensitive to perturbations.

3. OXYETHYLENE SURFACTANTS: SMALL BLOCK COPOLYMERS

Most surfactants have very small polar parts, a few Å, while the nonpolar part is an order of magnitude larger. For the important class of nonionic surfactants of the oxyethylene type the situation is different, the two parts being similar in size. These nonionic surfactants can be considered as short AB block copolymers and their behavior is a suitable starting point of our discussion; we adopt here the convention of denoting by A the (more) polar part. The nonionic surfactants show a complex phase behavior with several solution and liquid crystalline phases (see Figure 6 for a representative concentration-temperature phase diagram for a binary nonionic surfactant water system). Since the oxyethylene-water interaction is temperature dependent, and becomes less favorable at higher temperatures, we can control the self-assembly structure not only by the size of polar and nonpolar groups but

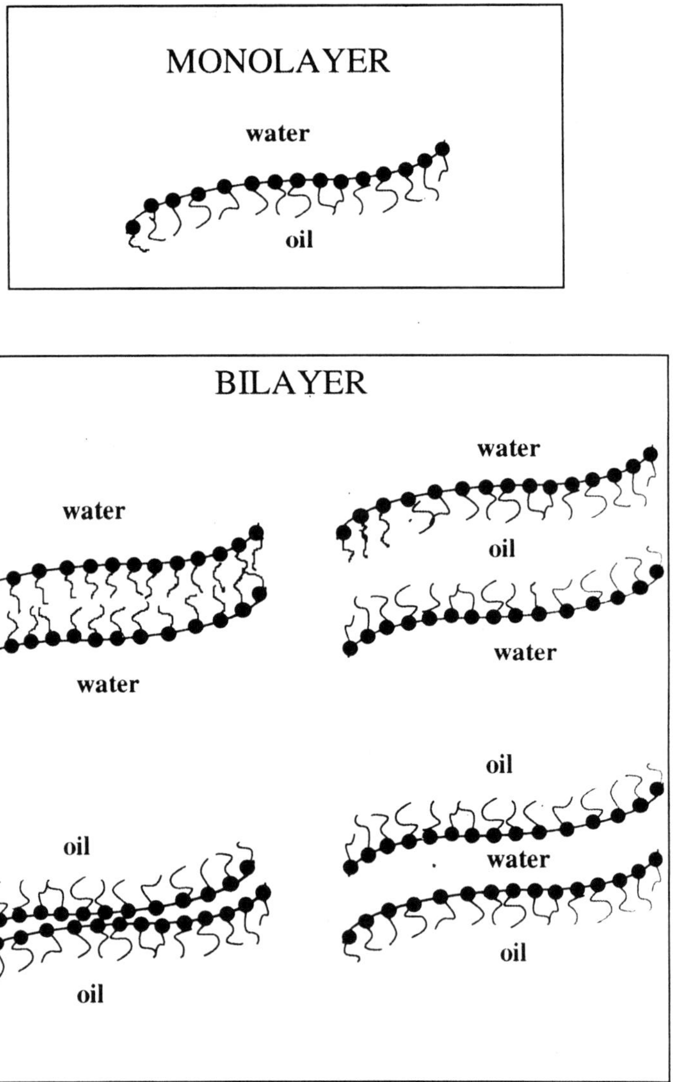

Figure 3. Surfactant monolayer (top panel) and bilayers (bottom panel). In any surfactant system we have a segregation into water-rich and oil-rich domains, as well as surfactant films. The latter can be pairwise correlated (into bilayers) or uncorrelated. (adapted from D.O. Shah, Ed., Micelles, Microemulsions, and Monolayers, Marcel Dekker, 1998, p. 105)

also by temperature. For binary surfactant-water systems, decreasing the polar head group size and increasing temperature give qualitatively the same change. For ternary systems with oil, we note that microemulsion structure is strongly dependent on temperature [4].

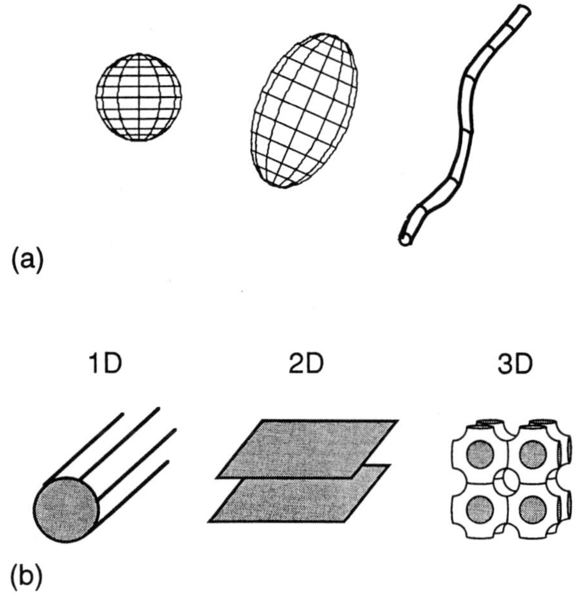

Figure 4. Examples of (a) discrete and (b) continuous surfactant self-assemblies. The latter can extend on 1 (cylinders), 2 (lamellae) or 3 (bicontinuous) dimensions. (adapted from D. O. Shah, Ed., Micelles, Microemulsions, and Monolayers, Marcel Dekker, 1998, p. 104)

The type of aggregate formed by a surfactant under specified conditions depends on the balance between hydrophilic and hydrophobic parts. This balance can be described by different parameters but the ones used more often are the surfactant number or critical packing parameter and the spontaneous curvature of the surfactant film (taken as positive if it is curved towards water) [1]. The surfactant number increases as the hydrophobic part gains in importance, while the spontaneous curvature decreases. For nonionics the surfactant number increases and the spontaneous curvature decreases as temperature increases (the temperature dependence of the spontaneous curvature is shown pictorially in Figure 7). The underlying less favorable oxyethylene-water interactions at higher temperatures have been discussed from different starting points. However, currently we understand these systems in terms of temperature induced conformational changes so that more polar conformers dominate at low temperatures and the less polar at high temperatures.

The simple diblock copolymer is the most direct analogue to conventional surfactants, but not the only one. Double-chain structures are found in many surfactants, like those used as textile softeners and developed for enhanced oil recovery, and many lipids are analogous to BAB block copolymers, while α,ω-surfactants are analogous to ABA copolymers.

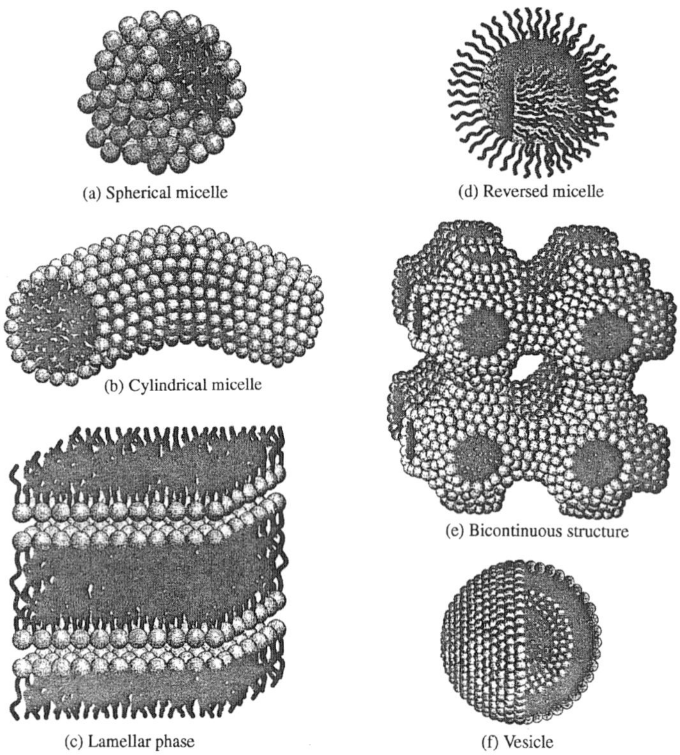

Figure 5. Surfactant self-assembly leads to a range of different structures of which a few are shown: (a) Spherical micelle with an interior composed of the hydrocarbon chains and a surface of the polar head-groups (depicted as spheres) facing water. The hydrocarbon core has a radius close to the length of the extended alkyl chain. (b) Cylindrical micelle with an interior composed of the hydrocarbon chains and a surface of the polar head-groups facing water. The cross section of the hydrocarbon core is similar to that of spherical micelles. The micellar length is highly variable so these micelles are polydisperse. (c) Lamellar phase consisting of surfactant bilayers. (d) Reverse micelle with a water core surrounded by the surfactant polar head-groups. The alkyl chains together with a non-polar solvent make up the continuous medium. (e) Bicontinuous structure with the surfactant molecules assembled into connected films characterized of two curvatures of opposite sign. (f) Vesicle built from bilayers similar to those of the lamellar phase, and characterized by two distinct aqueous domains, one forming the core and one the external medium. (adapted from ref. [2], p. 34; © J. Wiley, 1998)

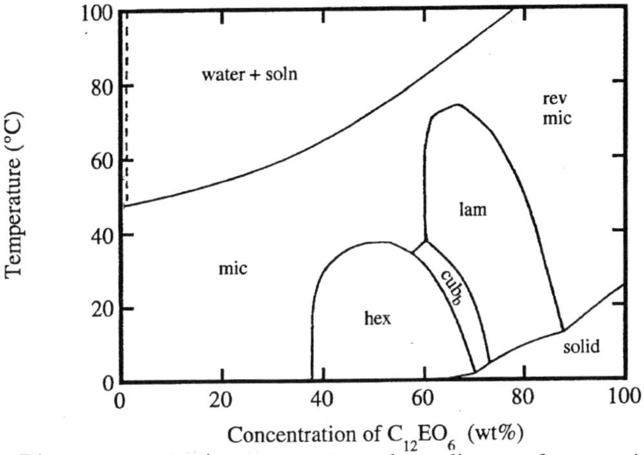

Figure 6. Binary concentration-temperature phase diagram for a nonionic surfactant with C12 hydrophobic chain and 6 oxyethylenes in the polar head group. Mic and rev mic denote micellar and reverse micellar solutions, respectively. Hex, cub, and lam, denote regions with hexagonal, cubic, and lamellar lyotropic liquid crystalline microstructure, respectively. (adapted from ref. [2], p. 99; © J. Wiley, 1998)

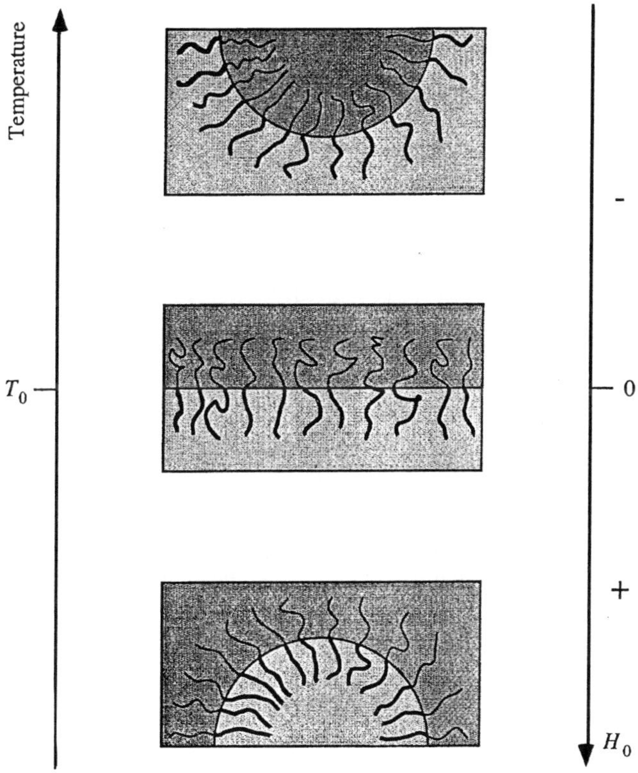

Figure 7. Spontaneous curvature, H, vs temperature for nonionic surfactants. H changes from positive at low temperature to negative at high temperatures. At some intermediate temperature, T_o, the spontaneous curvature is zero and the surfactant is termed balanced. (adapted from ref. [2], p. 101; © J. Wiley, 1998)

4. LARGE AMPHIPHILIC MOLECULES

The connection to surfactants, described above to some length both in order to introduce various self-assembly structures and to indicate the ubiquitous special temperature effects displayed for systems containing poly(oxyethylene) chains, is but one basis for the rapid development of the field of water-soluble amphiphilic polymers and for their scientific understanding. The relation to neat block copolymers and their rich phase behavior, and to block copolymers in the presence of selective solvents is another such basis. For neat block copolymers there is an increasing tendency for segregation between the blocks as the interblock interaction becomes less favorable (the interaction being described by a Flory interaction parameter, χ) as block molecular weight (~N) increases and as temperature decreases (generally, $\chi \sim 1/T$). Composition (that of the two blocks in the block copolymer molecule) - temperature phase diagrams can be generated in terms of the interaction parameter and molecular weight (Figure 8).

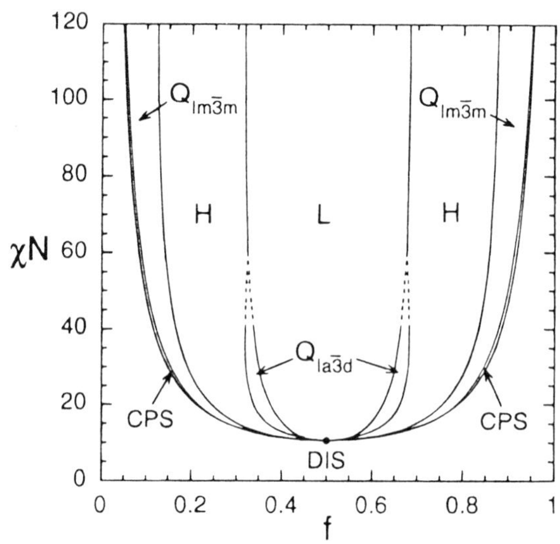

Figure 8. Mean-field phase diagram (χN vs *f*) for conformationally symmetric diblock copolymers. The notation for the different microstructures is: L, lamellar; H, cylinders in a hexagonal array; Q_{Ia3d}, bicontinuous cubic; Q_{Im3m}, micellar (discrete) cubic; CPS, close-packed spheres; and DIS, disordered melt. (adapted from M.W. Matsen and F.S. Bates, Macromolecules 29 (1996) 1091; © American Chemical Society, 1996)

For small amphiphilic molecules, the chemical structures are basically limited to the AB, ABA and BAB types, while enlarging the molecules gives much richer possibilities. It is also a major advantage in moving to higher molecular weight amphiphiles that the molecular architecture may be much more diverse: block, graft, and star copolymers are possible.

Amphiphilic polymers, which have already received major scientific attention and very extensive industrial use, are block copolymers of the AB, ABA or BAB types, and different graft copolymers (see Figure 9). Particular attention among the latter is directed towards hydrophobically-modified graft copolymers, where a low number (on the order of 1%) of hydrophobic groups have been grafted onto a linear water-soluble polymer, for example a polysaccharide or poly(acrylate). These polymers, used inter alia for rheology control in modern paints and coatings, may instructively be termed polymer-modified surfactants; thus it is very useful to consider them as conventional surfactants coupled together by the polymer back-bone (the converse structure also exists, where hydrophilic groups are grafted onto a nonpolar backbone). We should note the relation of these HM-polymers or polymer-modified surfactants with polysurfactants, like the since long known polysoaps. In fact a related principle of enlarging surfactant molecules of considerable current interest has been by polymerization of simple surfactants to create polymerized surfactants. Most studies have so far examined moderately sized molecules, like dimeric, or gemini, surfactants, trimeric surfactants etc.

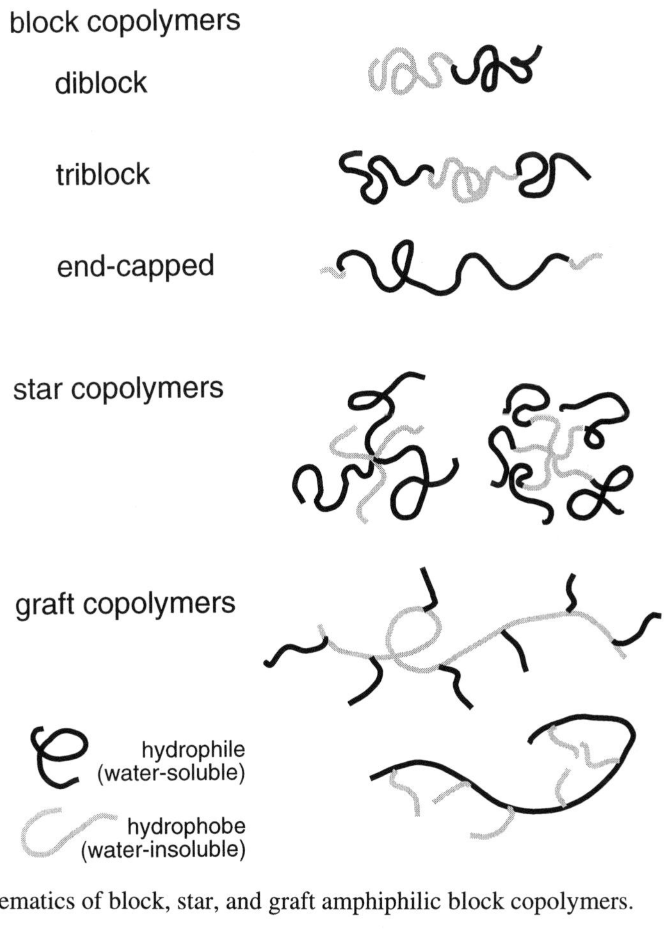

Figure 9. Schematics of block, star, and graft amphiphilic block copolymers.

We should also note that Nature makes frequent use of amphiphilic polymers. Some glycolipids can be considered as short diblock copolymers. Lipopolysaccharides are examples of strongly amphiphilic graft copolymers. One example is Emulsan, the trade name of a bacterially produced polyanionic lipopolysaccharide which is exceptionally efficient as emulsifier (Figure 10). In the polysaccharide field there are also many examples of copolymers which are composed of different hydrophilic parts, like proteoglucans in connective tissue built up of dermatan sulfate and chondrotin sulfate. Proteins have hydrophobic and hydrophilic domains and their character can be more or less block-like. One recent illustration of their amphiphilic character is that they cross-link to form amphiphilic graft copolymers (hydrophobically modified water-soluble polymers) very much in the same way as surfactant micelles. Sometimes Nature introduces special hydrophilic groups, like in glycoproteins, to enhance the amphiphilic character and we modify biopolymers to make them more amphiphilic.

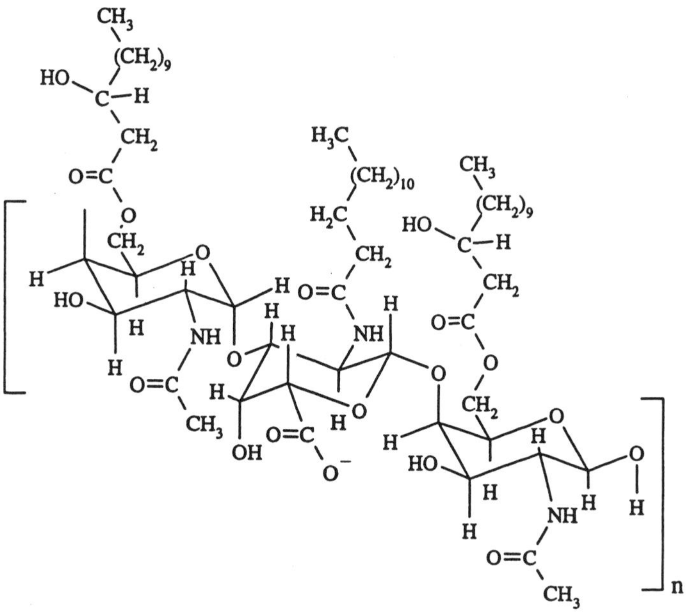

Figure 10. Chemical structure of Emulsan, a bacterially produced polyanionic lipopolysaccharide.

5. SELF-ASSEMBLY OF AMPHIPHILIC BLOCK COPOLYMERS

Large amphiphilic molecules such as diblock and triblock copolymers, as well as graft copolymers with sufficiently long grafts and flexible backbones, have been known to self-assemble in the form of micelles when dissolved in selective solvents [7,8] and also to adsorb on surfaces. The earlier studies were concerned primarily with block copolymers in organic solvents [7], but more recent studies have examined the formation of block copolymer micelles in aqueous solutions [8]. The energetics of micelle formation in organic solvents are generally different than these in aqueous solvents: block copolymer micelle formation in organic solvents is an exothermic (enthalpy-driven) process and micelles tend to "dissolve" at higher temperatures; in aqueous solvents micelle formation is often endothermic (entropy driven), a phenomenon more pronounced in oxyethylene-containing amphiphilic block copolymers where micelles form at elevated temperatures [8]. The ability of temperature to trigger the formation or dissolution of block copolymer micelles finds a number of technological applications. Block copolymer micelles differ from micelles formed by small amphiphiles in terms of size (polymeric micelles being larger) and degree of segregation between the blocks that form the micelle core and these that form the shell or corona (the surfactant tail and head group are strongly segregated), but most functional properties are common in micelles formed by both large and small amphiphilic molecules.

More recently, the self-assembly studies on block copolymers have been extended to cover a wide composition space in mixtures of block copolymers with selective solvents [9]. Such studies have identified various ordered block copolymer structures related to those found in block copolymers in the absence of solvents (see, e.g., Figure 8) as well as to the lyotropic liquid crystals formed in mixtures of small amphiphiles (surfactants and lipids) with water and (possibly) oil or cosurfactant (Figures 5 and 6). Scaling theories and mean-field lattice models developed for polymer blends have been extended to cover block copolymer solutions with encouraging results [9]. Still, the self-assembly of amphiphilic block copolymers remains an uncharted territory, and considerable effort is being extended on studies of the self-assembly of different classes of block copolymers (e.g., polyethers, polyelectrolytes) in various environments (e.g., in water, in non-aqueous solvents, or in the absence of solvents) [9]. A variety of experimental tools, ranging from static (e.g., small angle neutron scattering) to dynamic (e.g., rheology), are proving beneficial to this end. Concurrent with (and sometimes preceding) the fundamental efforts, applications of amphiphilic block copolymers are being explored in areas such as formulations, pharmaceutics, and separations [9].

6. OUTLOOK

Systems containing amphiphilic molecules, small (surfactants) or large (block copolymers), are notable in that they self-assemble (under appropriate conditions) to form morphologies of very well-defined structural order and nanometer scale dimensions. The challenge and the opportunity is in controlling and tuning the morphology, i.e., in determining the appropriate components and conditions leading to the desired microstructure. The importance of self-assembly studies becomes thus apparent. In addition to the numerous studies on small amphiphilic molecules already published (and expected to be published in the future), studies on large amphiphiles will become more prominent in the years to come and will significantly enhance our fundamental understanding and ability to manipulate self-assembly for the common good.

Large amphiphiles such as block copolymers are very flexible in attaining a great variety of micro-structures in the absence or in the presence of solvents and/or other additives. Furthermore, their macromolecular nature allows access to a wide range of length scales and time scales. Amphiphilic block copolymers thus have a potential of great contributions to diverse applications such as synthesis of functional nanoporous materials or formulations of coatings, paints, pharmaceuticals, and personal care products.

Acknowledgement. Financial support from the Swedish Natural Science Research Council (NFR) is gratefully acknowledged.

REFERENCES

1. D.F. Evans and H. Wennerström, The Colloidal Domain: Where Physics, Chemistry, Biology, and Technology Meet, VCH Publishers, New York, 1994.
2. B. Jönsson, B. Lindman, K. Holmberg and B. Kronberg, Surfactants and Polymers in Aqueous Solutions, J. Wiley & Sons, Chichester, 1998.
3. B. Lindman and H. Wennerström, Top. Curr. Chem., 87 (1980) 1.
4. J. Sjöblom, R. Lindberg and S.E. Friberg, Adv. Colloid Interface Sci., 95 (1996) 125.
5. R.G. Laughlin, The Aqueous Phase Behavior of Surfactants, Academic Press, London, 1994.
6. K. Larsson, Lipids - Molecular Organization, Physical Functions and Technical Applications, The Oily Press Ltd., Scotland 1994.
7. Z. Tuzar and P. Kratochvil, Surface and Colloid Science, E. Matjevic (ed.), Plenum Press, New York, 1993; Vol. 15.
8. P. Alexandridis and T.A. Hatton, The Polymeric Materials Encyclopedia, J.C. Salamone (ed.), CRC Press, Boca Raton, FL, 1996; p. 743.
9. P. Alexandridis and B. Lindman (eds.), Amphiphilic Block Copolymers: Self-Assembly and Applications, Elsevier Science, Amsterdam, 2000.

Modelling of the self-assembly of block copolymers in selective solvent

P. Linse

Physical Chemistry 1, Center for Chemistry and Chemical Engineering, Lund University, P.O. Box 124, S-221 00 Lund, Sweden

A number of theories and models with different degrees of sophistication and computational demand are today used to describe the self-assembly of block copolymers in solution. The different types of approach are reviewed and numerical examples are given which demonstrate the applicability of each one. The focus is on approaches useful for long chain molecules in a selective solvent, but illustrations of direct simulations of short-chain molecules are also included.

1. INTRODUCTION

Block copolymers in the melt and in solution tend to self-assemble and form microphases. In the melt and in a nonselective solvent, the repulsive interaction between the monomers on different blocks constitutes the driving force for the process. In a selective solvent, the bad solvency condition for one type of monomer and the good one for the other type is also of importance for the self-assembly. The structure and symmetry of the aggregates formed depend on the relative strength of the interactions as well as on the composition and the architecture of the block copolymer. The aim of the chapter is to present an overview of different theoretical treatments of the self-assembly of block copolymers in a selective solvent. We will cover the range from scaling relations obtained from simplified models to computer intensive simulation approaches. After the introduction where the scope of the overview is given, each type of theory is presented and numerical examples illustrating characteristic results are given. The chapter ends with a summary where the main differences of the types of theory are presented.

The driving force of the self-assembly depends on the particular system. Briefly, amphiphilic block copolymers can be divided into nonionic and ionic block copolymers. In both groups there exist polymers that are soluble in water and in organic solvents. For the nonionic block copolymers, the driving force behind the self-assembly depends on the solvent. In aqueous solution it is normally of entropic origin (e.g. PEO-PPO-PEO), whereas in organic solvents (e.g. PS-PEO) the enthalpic contributions dominate. Obviously, for block polyelectrolytes in aqueous solution and, in particular, for block ionomers in organic solvents, the electrostatic interaction between the free charges plays a dominant role for the self-assembly.

1.1. Scope of the review

A very important area related to the self-assembly of block copolymers in solution is the self-organization of block copolymers in melts. The types of structure appearing, as well as the theoretical approaches, are often similar in the two areas. Despite the recent advances in the understanding of block copolymers in melt [1], space does not allow to treat these systems. Similarly, the self-assembly of block copolymers in a nonselective solvent displays the same main features as in melts and is hence also treated here superficially. For the same reason, theories developed for describing polymeric gels formed by multifunctional block copolymers are also excluded.

Surfactants (short for surface active agents) constitute an important and related class of molecules. They are short chain molecules with amphiphilic character, and in solution they self-assemble in a manner similar to block copolymers. The driving forces for the self-assembly are the same, but some aspects are different due to the different chain lengths. Hence, the theoretical descriptions of the self-assembly of surfactants and of block copolymers in solution display large similarities, although a number of differences exist. In the following, the scope is on models and theories which have been applied to block copolymer systems. Some of them are the same as those used for describing surfactant systems, but there exist important theories which are only applied to surfactant systems and they are not considered here. However, direct simulation of the self-assembly constitutes the main exception. Due to the shorter chain length, computer simulations of the self-assembly of surfactant systems are simpler (but still not simple), and results pertaining to shorter chains more relevant for surfactant systems will be presented to illustrate what can be modelled for longer chains in the near future. Finally, although kinetic aspects are in many cases of large significance, we will here restrict ourselves to equilibrium systems only.

1.2. Structures and phases

The conventional or classical structures occurring in solutions of block copolymers as well as in solutions of surfactants are spherical, extended cylindrical, and planar aggregates. At low concentration, the spherical aggregates appear as isolated aggregates with no or low spatial correlation. At higher concentration liquid crystalline phases, such as closed packed cubic phases, hexagonal phases, and lamellar phases, may occur. The two former normally exist in two morphologies: a normal one where the solvent constitute the continuous phase and a reverse where the solvent is localized in the interior of the aggregates. Besides the classical structures, recent research has shown that intermediate structures as the double-diamond and the gyroid bicontinuous structures, as well as different perforated lamellae could possibly exist in narrow concentration and temperature ranges.

A large fraction of the theoretical effort has been devoted to describe the onset of forming the spherical aggregates, normally referred to as micelles. In particular, the lowest copolymer concentration at which micelles exist (the critical micellar concentration, cmc) is of large interest, but also the sizes of the micellar core and corona (the latter is also referred to as shell or brush height), the concentration of nonassemblied copolymers above the cmc, and the ordered phases occurring at high concentrations are of great interest to describe.

1.3. Types of theory

The different theories and models are presented in order of increasing computational demand. The theories are divided in (i) scaling approaches, (ii) semi-analytic mean-field models, (iii) numerical self-consistent mean-field models, and (iv) simulation approaches. The above division is of course somewhat arbitrary, and the borders between the subdivisions are not sharp.

The scaling approach provides us with simple but useful relations of how, e.g., the size of the micellar core and corona depends on the number of segments of the different blocks. The word segment is used to denote a part of a chain in the model and it has to be mapped on a monomer, repetitive unit, or similar, when the theory is applied. In the semi-analytic mean-field models, one normally assumes some block profile and from an expression for the free energy contributions, the aggregation number, the cmc, and the phase diagram can be extracted. The starting point of the numerical self-consistent mean-field models is normally random walks which represent the configurations of the chain molecules. These walks are performed in a potential field which, as such, depends on the configurations of the molecules. In addition to the results of the semi-analytic mean-field models, volume fraction profiles, distribution of segments, etc., are obtained. Finally, through direct simulations it is possible to go beyond the mean-field approximation, but the computational demand is high for polymeric systems.

2. SCALING APPROACHES

The idea behind the scaling "anzats" is to obtain relations which predict how quantities of interest depend on other variables without a full knowledge of the dependencies on all variables. On the basis of some simple model of the system, the leading free energy contributions are formulated. After minimization of the free energy with respect to one or a few variables, the leading dependencies of these quantities on the other variables are often readily extracted.

2.1. Micellization of AB-diblock copolymers in a selective solvent

We will consider first a solution of a monodisperse AB-diblock copolymer in a selective solvent of low molecular weight. The number of segments of the copolymer blocks are represented by N_A and N_B, respectively. We will derive the scaling relations of the size and the aggregation number of the micelles formed. Two limiting cases will be considered: the large core case [2] valid when $N_A >> N_B$ ("crew cut micelles") and the small core case [3,4] for $N_A << N_B$ ("hairy micelles"). In the former limit the analogy with theories for polymer brushes [2] is used, whereas in the latter case the theory is developed from star polymer theory [5,6].

The micelles are assumed to be monodisperse and their concentration so low that micelle-micelle interaction are negligible. A micelle is modelled as two concentrical spherical regions, where the molten lyophobic A-blocks reside in the central core, whereas the B-blocks, swollen by the solvent, are located in the shell. The incompatibility between the two types of blocks is assumed to be large so the core-shell interface, at which the AB junctions are localized, is sharp. In the case of the large core, the volume fraction of B segments is treated as constant in the shell (model I) (not always a realistic assumption, but nevertheless made), whereas in the case of the small core, the volume fraction of B segments is allowed to decay with increasing distance (model II), (see Figure 1).

Assuming that the core is incompressible and consists only of A segments, i.e, $\phi_A^{core} = 1$, we obtain the scaling relation

$$R_A^3 \approx pN_A a^3 \tag{2.1}$$

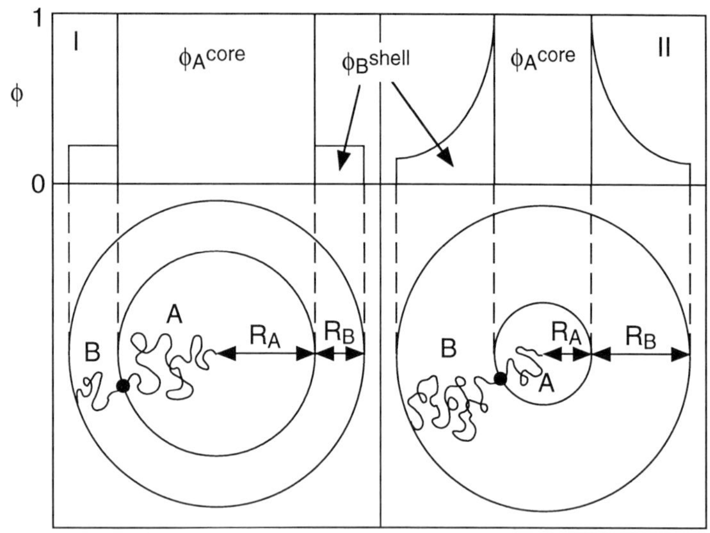

Figure 1. Illustration of model I (left) and II (right) of the AB-diblock copolymer micelle in a selective solvent (lower panel) and the volume fraction profiles of the polymer blocks (upper panel) applied for the large core case ($N_A >> N_B$) and the small core case ($N_A << N_B$), respectively.

where p denotes the aggregation number and a the length of a segment. (The symbol $\approx$ denotes equal within a numerical factor of order one.)

For model II, the scaling relation for R_B is obtained by analysing the conformations of the B-blocks in terms of blobs [7]. With the following assumptions: (i) the shell region consists of concentrical shells of blobs, (ii) there are p blobs in each shell, (iii) the blobs size ξ depends only on r, and (iv) the B-segments are in a good solvent, the volume fraction of the B segments can explicitly be expressed as $\phi_B^{shell} \approx p^{2/3}(r/a)^{-4/3}$. By integration of ϕ_B^{shell}, the scaling relation

$$R_B \approx p^{1/5} N_B^{3/5} a \qquad \text{(model II)} \qquad (2.2')$$

is obtained after applying $R_A << R_B$.

The two dominant contributions to the free energy are assumed to be the interfacial tension (enthalpic) and the chain stretching (entropic). Thus, the total free energy of a micelle is approximated according to

$$F = F_{interface} + F_{def}^{core} + F_{def}^{shell} \qquad (2.3)$$

where $F_{interface}$ describes the interfacial free energy, F_{def}^{core} the stretching free energy of the A-blocks, and F_{def}^{shell} the corresponding quantity for the B-blocks. The interfacial free energy is assumed to be proportional to the area of the interface, thus

$$F_{interface} \approx \gamma R_A^2 \qquad (2.4)$$

where γ is the surface tension. In this simple approach, the A blocks are assumed to be uniformly stretched with an end-to-end distance equal to the core radius, R_A, (obviously not a fully realistic picture) and the accompanied cost is expressed as

$$\beta F_{def}^{core} \approx p\left(\frac{R_A}{N_A^{1/2}a}\right)^2 \qquad (2.5)$$

where $\beta = 1/(kT)$ with k being Boltzmann's constant and T the temperature. In model I, a constant volume fraction of the B block is assumed and the uniformly stretching assumption gives similarly

$$\beta F_{def}^{shell} \approx p\left(\frac{R_B}{N_B^{1/2}a}\right)^2 \qquad \text{(model I)} \qquad (2.6)$$

Regarding model II, the blob picture is used again. To the leading order there are $\approx p^{1/2}$ blobs per chain and with a free energy penalty of kT per blob due to the confinement of the B-blocks at the core surface, we get

$$\beta F_{def}^{shell} \approx p^{3/2} \qquad \text{(model II)} \qquad (2.6')$$

Substitution of eqs (2.4)-(2.6) in eq (2.3) and elimination of R_A and R_B by employing eqs (2.1)-(2.2) leads to a free energy expression of a single micelle where the aggregation number p is the only unknown parameter according to

$$\beta F \approx \gamma N_A^{2/3} p^{2/3} a^2 + N_A^{-1/3} p^{5/3}(1 + N_A^{-1} N_B \rho^{-2}) \qquad \text{(model I)} \qquad (2.7)$$

$$\beta F \approx \gamma N_A^{2/3} p^{2/3} a^2 + N_A^{-1/3} p^{5/3} + p^{3/2} \qquad \text{(model II)} \qquad (2.7')$$

The aggregation number can now determined by minimization of the free energy per polymer chain, F/p. For the case of large cores (model I), the following scaling relations are obtained:

$$p \approx \gamma N_A \qquad (2.8)$$

$$R_A \approx \gamma^{1/3} N_A^{2/3} a \qquad (2.9)$$

where $N_A >> N_B$ has been applied corresponding to $F_{def}^{core} >> F_{def}^{shell}$. In the opposite limit with small cores (model II), we obtain:

$$p \approx \gamma^{6/5} N_A^{4/5} \quad (2.8')$$

$$R_A \approx \gamma^{2/5} N_A^{3/5} a \quad (2.9')$$

$$R_B \approx \gamma^{6/25} N_A^{4/25} N_B^{3/5} a \quad (2.10')$$

where $(N_A^{-2}p)^{1/6} << 1$ has been used corresponding to $F_{def}^{core} << F_{def}^{shell}$.

Thus, in the large core limit, the aggregation number is determined from the balance of the deformation free energy in the core and the interfacial free energy; the former tending to reduce p and the latter tending to increase p. Hence, the assumption of constant volume fraction in the shell does not affect the scaling relations, at least to the leading order. In the small core limit, the limitation of the growth of the micelle is, however, governed by the increased free energy of the blocks in the shell instead of in the core.

2.2. Solubilization in block copolymer micelles

The self-assembly of the block copolymers creates an environment in the core which is different from that in the solution. Solutes not soluble in the solvent can be solubilized (dissolved) in the core, and hence the solubility of a solute in a system can significantly be enhanced by the presence of block copolymers.

The blob picture of star polymers [5,6] has been employed by Nagarajan and Ganesh for deriving scaling relations where small and lyophobic molecules are solubilized in the micellar core [8]. Their extension involves the cases where the the lyophobic segments are nonswollen or swollen by the solubilizate leading to a radial volume fraction profile in the core according to $\phi_A^{core} = r^{-1}$ or $r^{-4/3}$, respectively. The corresponding incompressibility assumption of the core gives

$$R_A^3 \approx pN_A a^3 <\phi_A^{core}> \quad (2.11)$$

where $<\phi_A^{core}>$ denotes the average volume fraction of A in the core. With the assumption of unaffected surface tension due to the solubilization (for simplicity), the following scaling relations are obtained:

$$p \approx \gamma N_A \quad (2.12)$$

$$R_A \approx \gamma^{1/4} N_A^{3/4} a \quad (2.13)$$

$$R_B \approx \gamma^{1/5} N_A^{1/5} N_B^{3/5} a \quad (2.14)$$

for $\phi_A^{core} = r^{-1}$ and

$$p \approx \gamma^{10/11} N_A^{12/11} \quad (2.15)$$

$$R_A \approx \gamma^{2/11} N_A^{9/11} a \quad (2.16)$$

$$R_B \approx \gamma^{2/11} N_A^{12/55} N_B^{3/5} a \quad (2.17)$$

for $\phi_A^{core} = r^{-4/3}$, both in the limit of $R_A >> R_B$.

Thus, for "hairy" micelles the scaling approach predicts that the solubilization causes the aggregation number as well as the core and shell dimensions to depend stronger on N_A and weaker on γ than without solubilization.

2.3. Micellization of polyelectrolytes

An extension of the scaling relations for the small core limit to the case where the B-block is charged has been given by Dan and Tirrell [9]. With the assumption that the salt concentration is sufficiently high such that the persistence length of the B block is smaller than the blob size, the previous blob picture remains. In this regime, the leading terms of p and R_A are found to be independent of the salt concentration ϕ_{salt}, whereas the shell thickness scales as

$$R_B \approx \gamma^{6/25} N_A^{4/25} N_B^{3/5} \phi_{salt}^{-2/5} a \tag{2.18}$$

At low salt concentrations, where the persistence length exceeds the blob size, it is anticipated that both p and R_A display salt dependence.

3. SEMI-ANALYTIC MEAN-FIELD MODELS

The basic idea behind the core-shell models is the assumed physical separation of the two different types of blocks of the copolymer into two domains, the core and the shell (corona) of the micelle. On the basis of this separation, a free energy expression is constructed which represents the free energy of the micelle with respect to some reference state. The final step involves a minimization of the free energy with respect to a few parameters characterizing the micellar aggregate (as the domain sizes and the micellar volume fraction). In the case of only a few free energy terms, the simple scaling relations of the size of the domains are recovered.

3.1. Micellization of AB-diblock copolymers

The main variation within this class of models lies in the selection of free energy terms which describe different aspects of the self-assembly. In an early contribution, de Gennes [2] considered three terms: (i) the interfacial energy between the core and the corona, (ii) the deformation free energy of the A-block, and (iii) the deformation free energy of the B-block. The subsequent extension of Leibler et al. [10] included terms describing the mixing entropy of the solvent and the lyophobic block, the mixing entropy of micelles with solvent and non-assembled block copolymers, and the free energy of the homogeneous reference solution. Later Munch and Gast [11] extended the same theory to copolymer with different block sizes, and they also investigated the phase equilibrium between micelles and lamellar phases. Independently of Leibler et al., Noolandi and coworkers [12,13] developed very similar theories utilizing slightly different free energy expressions and obtained essentially the same results. In particular, they used a more elaborate model for obtaining the interfacial free energy, and they also included a term representing the free energy penalty of confining the A-B junctions.

In order to illustrate the principles of the core-shell model in more detail, the approach by Leibler et al. [10] and later used by Much and Gast [11] will be presented here, slightly reformulated.

The micellar solution is divided into spherical cells, each containing one spherical micelle with its accompanying solution. The volume of the cell is the inverse of the micellar number density. The lyophobic A-block constitutes the micellar core, whereas the shell contains the lyophilic B-block swollen with solvent as shown in Figure 2. The third region contains non-assembled (referred to as free) block copolymers and solvent. The notations used for the volumes of the three regions, the radius of the core, the thickness of the shell, and the volume fractions are also shown in Figure 2. The number of segments of the copolymer blocks are represented by N_A and N_B, respectively, and that of the solvent by N_S. All segments have the same length, a, and the same volume, a^3. The various interactions are described by Flory-Huggins χ-parameters [14]. For simplicity, the B segments and the solvent molecules are identical, i.e, $\chi_{BS}= 0$ and $\chi_{AB} = \chi_{AS}$, and hence the systems is characterized by only one interaction parameter, viz. χ.

The free energy of a single cell is expressed as

$$F = (F_{interface} + F_{def} - TS_{mix}^{shell}) + F_{mix}^{free} - TS_{mix}^{micelle} \tag{3.1}$$

where the terms in parentheses represent the free energy of a single micelle fixed in space. The first term in eq (3.1) denotes the interfacial free energy of the interface between the core composed by A-segments and the shell composed by B-segments and solvent which is given by

$$F_{interface} = 4\pi R_A^2\gamma \tag{3.2}$$

where $\gamma = kT\,(\chi/6)^{1/2}$ [15]. The second contribution arises from the elastic deformation of the copolymers in the micelle. It is assumed that all A-blocks are uniformly stretched/compressed from their Gaussian behaviour to an end-to-end distance equal to R_A and similarly for the B-blocks. Leibler et al. [10] employed

$$\beta F_{def} = \frac{3}{2}p\left[\left(\frac{R_A}{N_A^{1/2}a}\right)^2 + \left(\frac{R_A}{N_A^{1/2}a}\right)^{-2} + \left(\frac{R_B}{N_B^{1/2}a}\right)^2 + \left(\frac{R_B}{N_B^{1/2}a}\right)^{-2} - 4\right] \tag{3.3}$$

where p is the number of copolymers in the micelle. The following term in eq (3.1) represents the mixing free energy of solvent molecules with the B-segments in the shell

$$-S_{mix}^{shell}/k = \frac{V^{shell}}{a^3}\left[\frac{1}{N_S}\phi_S^{shell}\ \ln(\phi_S^{shell})\right] \tag{3.4}$$

Finally, the remaining two terms in eq (3.1) arise from the free energy of the homogeneous solution outside the micelles and the free energy of mixing the micellar aggregates in the micellar solution. Using Flory-Huggins mean-field model of homogeneous polymer solutions

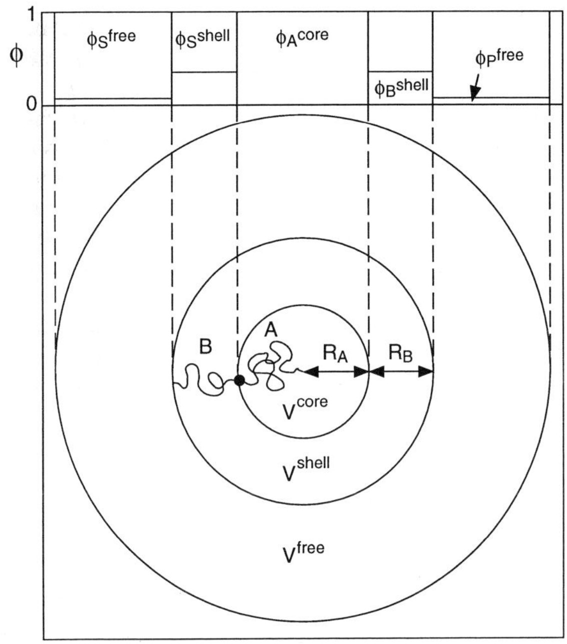

Figure 2. Illustration of the model showing the core and shell regions of a spherical micelle composed by an AB-diblock copolymer and the surrounding homogeneous polymer solution (lower panel). The volume fractions of the components in the three uniform regions are also displayed (upper panel).

[14], the former of the two is expressed as

$$\beta F_{mix}^{free} = \frac{V^{free}}{a^3}\left[\frac{1}{N_A+N_B}\phi_P^{free}\ln(\phi_P^{free}) + \frac{1}{N_S}\phi_S^{free}\ln(\phi_S^{free}) + \chi\phi_A^{free}(\phi_B^{free}+\phi_S^{free})\right] \tag{3.5}$$

where $\phi_A^{free} = N_A/(N_A+N_B)\phi_P^{free}$ and similarly for ϕ_B^{free}. The latter term is represented by

$$-S_{mix}^{micelle}/k = \phi_M^{-1}[\phi_M \ln(\phi_M) + (1-\phi_M)\ln(1-\phi_M)] \tag{3.6}$$

where $\phi_M = V^M/V$ denotes the micellar volume fraction with $V^M = V^{core} + V^{shell}$ and $V = V^M + V^{free}$. By employing the incompressibillity relations $V^{core} = (4\pi/3)R_A{}^3 = pN_A a^3$, $V^{shell} = (4\pi/3)[(R_A + R_B)^3] - R_A{}^3 = pN_B a^3/\phi_B^{shell}$, and $\phi_P^{free}V^{free} = \phi_P V - p(N_A+N_B)a^3$ as well as the fact that that volume fractions should sum up to one, F can be expressed in only three independent variables. For numerical computation, p (the aggregation number), ϕ_B^{shell} (the volume fraction of B segments in the shell), and ϕ_P^{free} (the volume fraction of free polymers) form a convenient set.

After minimization of F with respect of these variables, the equilibrium properties of the micellar solution of this model are obtained. In the original reports [10,11], the three coupled nonlinear equations obtained after partial differentiation were solved; however direct minimization is more straightforward and works fine.

3.2. Triblock copolymers

Extensions of the core-shell theory to cover cases with ABA-triblock copolymers have been proposed [16-18]. The main additional contribution is the cost for the middle block to form a loop. Ten Brinke and Hadziioannou [16] used the term $(3/2)\beta_l kT\ln(N_{loop})$ per polymer chain, where $\beta_l \approx 1$ and N_{loop} is the length of looping middle block. Later, Balsara et al. [17] suggested a smaller free energy cost, $(1/2)kT\ln(\pi\chi N_{loop})$, as the leading term. The general result is that the triblock copolymer displays an increased cmc as compared to a diblock copolymer with same A/B ratio and with half molecular mass. Balsara et al. [17] showed that the two cases lead to considerably differing increases in the cmc as compared to no loop formation. Recently, ordered structures occurring in ABC-triblock copolymers have received increasing attention [19,20]. A large number of possible morphologies have been suggested in melt and the these ideas should also hold for ABC-triblock copolymers in solution.

3.3. Polydisperse and mixed micelles

In the description so far, only monodisperse micelles have been considered. By utilizing theories developed for the self-assembly of short chain surfactants, Nagarajan and Ganesh [21] proposed a theory which contains micelles of all possible sizes beside the solvent molecules and singly dispersed copolymer molecules (unimers). The minimization of the free energy of this multicomponent system leads to an expression for the equilibrium size distribution of the micelles. Similarly, Gao and Eisenberg adapted a slightly different theory of describing the micellization of surfactants to the micellization of block copolymers [22].

The case of a mixture of two different diblock copolymers has been examined by Marques and coworkers [23,24]. For many conditions only mixed micelles are formed, but they also found situations where mixed micelles coexist with micelles formed by only one of the copolymers.

3.4. Solubilization

Nagarajan and Ganesh have also considered the case where lyophobic molecules are solubilized in the core of the micelles [25]. They found that the width of the distribution of the aggregation number is normally of minor consequence, and similarly the variation of the number of solutes in the micelles is small. The solubilization increases the aggregation number of the micelle and strongly increases the amount of solutes able to be dissolved in the phase. The effect of chain architecture on the solubilization has also bee examined [26].

3.5. Polyelectrolytes

Marko and Robin have derived expressions for the cmc as well as for the micellar size and aggregation number at the cmc for an AB block copolymer where A is insoluble and B is charged [27]. They used a similar approach as in section 3.1, but employed different free energy contributions. In particular, the driving force of the micellization is the bad solvency of the A block, which is opposed by the electrostatic repulsion within and among the charged blocks.

3.6. Selected results

The type of predictions obtained from the core-shell model, as described in section 3.1, will be illustrated for a symmetric diblock copolymer with $N_A = N_B = 50$ in a solvent with $N_S = 1$. The corresponding results of the more extended self-consistent lattice mean-field lattice theory (to be presented in section 4 below) are also included.

The volume fraction of free polymers (in the micellar free solution) and the volume fraction of micelles as a function of the total polymer volume fraction at $\chi = 1.45$ is given in Figure 3. At sufficiently low polymer volume fractions, the global free energy, as expressed by eqs (3.1)-(3.6), is obtained for $\phi_M \equiv 0$, i.e., there is only a homogeneous polymer solution without any micelles. However, at increasing ϕ_P, the global minimum occurs now for $\phi_M > 0$, hence micelles are present. The volume fraction where this transition occurs is referred to as the critical micellar concentration, and $\phi_P^{cmc} \approx 10^{-5}$ for the present parameter values. Above ϕ_P^{cmc}, the volume fraction of free polymers is almost constant; ϕ_P^{free} increases only 10% when ϕ_P increases 1000-fold.

Figure 4 shows that also the predicted aggregation number p is essentially independent of the polymer volume fraction and equals to ca. 55. The other parameter characterizing the micelle, ϕ_B^{shell}, is even less sensitive to ϕ_P. The weak influence of ϕ_P on p and ϕ_B^{shell} makes it possible to simplify the determination of these quantities by only considering the first three terms in eq (3.1) [10]. However, in order to determine ϕ_P^{cmc} and ϕ_P^{free}, the full expression is needed. Above the cmc, there is a partitioning of the polymers between the micelles and the intervening solution. Figure 4 also shows that the fraction of self-assembled block copolymers increases after the cmc to reach 90% at $\phi_P \approx 10\phi_P^{cmc}$.

The extension of the micellar core, the thickness of the shell, and the volume fraction of solvent in the shell are also of large interest. These quantities are shown in Figure 5 in terms of radial volume fraction profiles. For the present system, the core extends nearly 9 length units and the shell is 10.4 length units wide. Thus, both blocks are extended beyond their random coil configurations. The shell is highly swollen; ϕ_B^{shell} is only ca. 11%.

The degree of incompatibility between the A-segments on the one hand and the B-segments and solvent molecules on the other should affect the cmc. Figure 6 shows that the cmc is indeed very sensitive to χ. An increase of χ from 1.3 to 1.5 reduces ϕ_P^{cmc} from 10^{-2} to 10^{-6}. The rather high value of $N_A\chi$ ($\approx$65) required to obtain micelles in a solution of small molecules is related to the large loss in mixing entropy of the solution upon micellization. Already at $N_S = 5$, it is sufficient with $\chi = 0.435$ ($N_A\chi = 22$) to obtain $\phi_P^{cmc} = 10^{-2}$. As N_S is increased, the penetration of the solvent into the shell region is reduced as demonstrated by an increased ϕ_B^{shell} and a reduced shell thickness, which leads to less stretched B-blocks.

4. NUMERICAL SELF-CONSISTENT MEAN-FIELD MODELS

Mean-field descriptions of inhomogeneous systems containing chain molecules have been developed along two main directions. A lattice version was developed by Scheutjens and Fleer [28] from the earlier work by Flory [14]. It has been widely used to model adsorption at interfaces [29] and self-assembly of chain molecules (surfactants and polymers). Independently

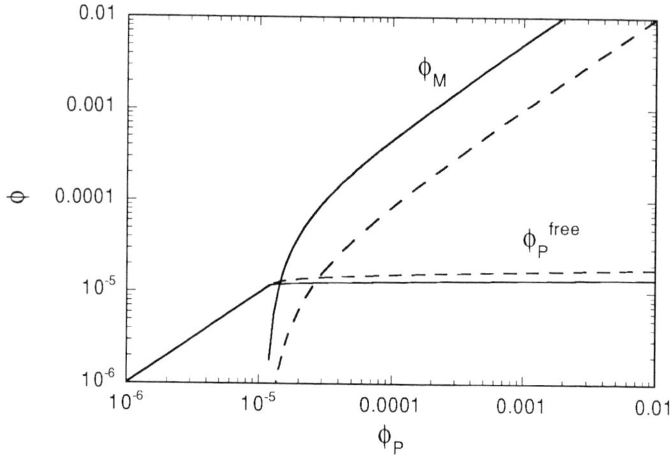

Figure 3. Volume fraction of free polymers ϕ_P^{free} and of micelles ϕ_M as a function of the total polymer volume fraction ϕ_P for a diblock copolymer with $N_A = N_B = 50$ and $N_S = 1$ from the core-shell model with $\chi = 1.45$ (solid curves) and the lattice mean-field theory with $\chi = 1.24$ and a hexagonal closed-packed lattice (dashed curves). Different values of χ were selected to give the same cmc. (In the core-shell model, ϕ_P^{free} refers to the volume fraction in the micellar free subvolume.)

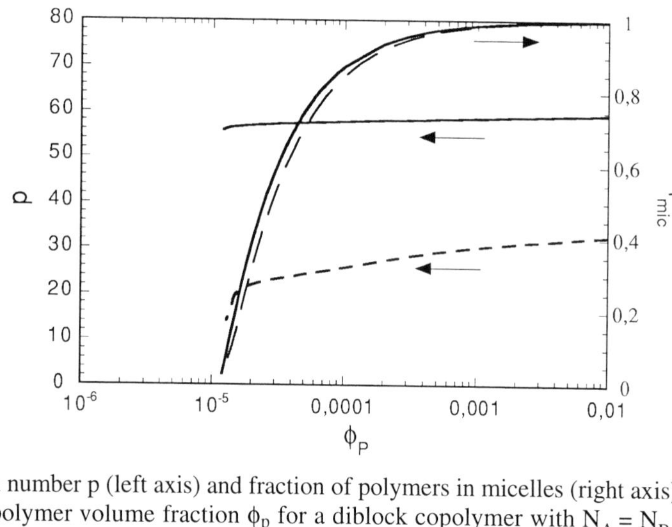

Figure 4. Aggregation number p (left axis) and fraction of polymers in micelles (right axis) as a function of the total polymer volume fraction ϕ_P for a diblock copolymer with $N_A = N_B = 50$ and $N_S = 1$ from the core-shell model with $\chi = 1.45$ (solid curves) and the lattice mean-field theory with $\chi = 1.24$ and a hexagonal closed-packed lattice (dashed curves).

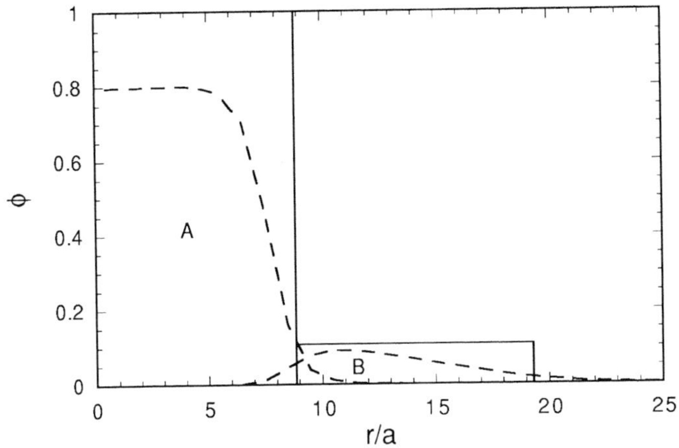

Figure 5. Volume fraction profiles of the A and B segments for a diblock copolymer with $N_A = N_B = 50$ and $N_S = 1$ at $\phi_P = 10^{-3}$ from the core-shell model with $\chi = 1.45$ (solid curves) and the lattice mean-field theory with $\chi = 1.24$ and a hexagonal closed-packed lattice (dashed curves). The radial distance is given in reduced units (segment lengths).

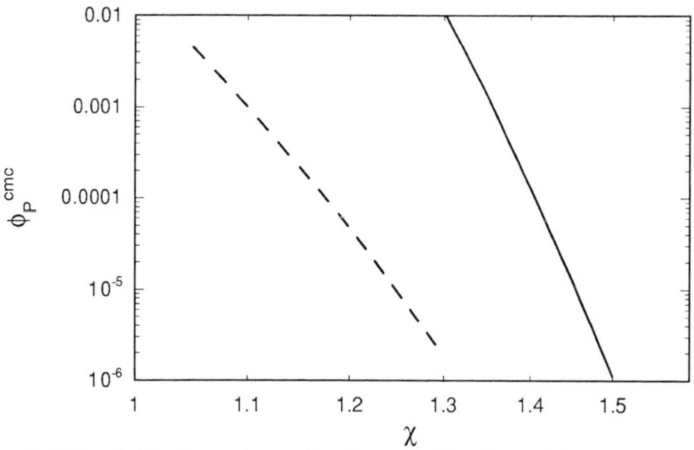

Figure 6. Critical micellar volume fraction as a function of the interaction parameter χ for a diblock copolymer with $N_A = N_B = 50$ and $N_S = 1$ from the core-shell model (solid curves) and the lattice mean-field theory with a hexagonal closed-packed lattice (dashed curves).

and in parallel, the corresponding continuum version, as derived from the functional integral formalism [30], has been described by Hong and Noolandi [31] and utilized in a number of investigations, in particular of block copolymers at fluid interfaces. Recently, both approaches were applied to the same model system and were compared directly [32]. Both approaches are presented below by using, as far as possible, equivalent expressions to highlight their similarities. For simplicity, only the case of the homopolymers is considered, but both approaches are readily generalized to copolymers [33].

4.1. Lattice approach

The division of the space into lattice sites, which are occupied by a solvent molecules or a segment of a chain molecule is fundamental in the lattice approach. Normally, spatial inhomogeneities are permitted in one direction and the random-mixing approximation is applied to all sites in each layer separately. The number of nearest neighbour sites, z, is determined from the lattice topology. A chain molecule is described as a connective walk on the lattice. The relative degeneration of a component x in conformation c is given by

$$\omega_{xc} = L_{s=1} \prod_{s=2}^{r_x} \lambda^{(s)} \tag{4.1}$$

where $L_{s=1}$ denotes the number of equivalent starting positions, $\lambda^{(s)}$ is a simplified notation for the transition probability of a random walk from the site where segment s-1 is located to the site where segment s is located, and is r_x the number of segments of component x.

The free energy expression of the system is given by

$$\beta(A-A^*) = \sum_x \sum_c n_{xc} \ln \frac{n_{xc} r_x}{\omega_{xc}} + \frac{1}{2} \sum_i L_i \sum_x \sum_{x'} \phi_{xi} \chi_{xx'} \langle \phi_{x'i} \rangle \tag{4.2}$$

where the first factor represents the entropy and the latter the interaction among the segments of the different components using Flory-Huggins interaction parameter χ. In eq (4.2), n_{xc} denotes the number of molecules of type x in conformation c, L_i the number of sites in layer i, and <...> an average over layer i-1, i, and i+1. From the maximization of the partition function (minimization of the free energy) with respect to the set of all conformations $\{n_{xc}\}$ under the constraint of filled lattice sites, the distribution of segment s of component x in layer i can be expressed by

$$n_{xsi} = \left\{ \frac{n_x}{\mathbf{s}^T.(\mathbf{W})^{r_x-1}.\mathbf{p}(x,1)} \right\} \{\mathbf{s}^T.(\mathbf{W})^{r_x-1}.\Delta_i\} \{\Delta_i^T.(\mathbf{W})^{s-1}.\mathbf{p}(x,1)\} \tag{4.3}$$

where the first factor is a normalization factor assuring that n_x molecules of type x is present. The last factor denotes the number of chains of type x with the start distribution $\mathbf{p}(x,1)$ among the layers which has its s:th segment in layer i and the second factor gives the appropriate weight of the remaining part of the chain to be somewhere on the lattice. In eq (4.3), W is a tridiagonal matrix comprising elements which contains factors describing the lattice topology and factors containing the mean-field potential for segments of component x, $\Delta_i^T = (0, ..., 0, 1, 0, ..., 0)$, and $\mathbf{s}^T = (1, 1, ...,1)$. The number of segments of component x in layer i is given by $n_{xi} = \sum_{s=1}^{r_x} n_{xsi}$. Finally, the volume fraction in a layer, ϕ, is obtained by diving n_{xsi} and n_{xi} by the number of lattice sites in that layer.

4.2. Continuum approach

In the continuum description, the partition function for a multicomponent system can be expressed as a functional integral over all space curves representing all chain molecules. Following Hong and Noolandi [31], the probability functional for a given space curve $P[\mathbf{r}_x(\bullet)]$ of a flexible chain molecules of type x (component x) is assumed to be of the Wiener form, i.e.,

$$P[\mathbf{r}_x(\bullet)] \propto \exp\left[-\frac{3}{2b^2}\int_0^{r_x} dt\, \dot{\mathbf{r}}_x^2(t)\right] \tag{4.4}$$

b being the Kuhn segment length of a chain and r_x the number of statistical segments. In an (external) field the distribution function of component x with one end at $\mathbf{r}_0$ and the other at $\mathbf{r}$ is given by the Green function $Q_x(\mathbf{r}, r_x \mid \mathbf{r}_0)$, which satisfies the modified diffusion equation

$$\frac{\partial Q_x}{\partial t} = \frac{b^2}{6}\nabla^2 Q_x - \omega_x Q_x \tag{4.5}$$

with $Q_x(\mathbf{r}, 0 \mid \mathbf{r}_0) = \delta(\mathbf{r} - \mathbf{r}_0)$, where $\omega_x[\mathbf{r}_x(t)]$ is the external potential.

From a free energy functional involving entropic terms (of the type $\rho_x(\mathbf{r}) \ln \rho_x(\mathbf{r})$, where $\rho_x(\mathbf{r})$ denotes the segment number density of component x at $\mathbf{r}$) and enthalpic terms (of the type $\rho_x(\mathbf{r})\rho_{x'}(\mathbf{r})\chi_{xx'}$, where $\chi_{xx'}$ is the Flory-Huggins interaction parameter between segments in components x and x'), minimization of the partition function with the constraint of constant local density gives the mean-field potential above denoted as an external potential.

From the Green functions, all relevant properties can be calculated. The density of segments of rank t of component x, $\rho_x(\mathbf{r},t)$, is, e.g., given by

$$\rho_x(\mathbf{r},t) = \left\{\frac{n_x}{\int d\mathbf{r}\int d\mathbf{r}_0 Q_x(\mathbf{r},r_x|\mathbf{r}_0)}\right\}\left\{\int d\mathbf{r}_0 Q_x(\mathbf{r},r_x-t\,|\mathbf{r}_0)\right\}\left\{\int d\mathbf{r}_0 Q_x(\mathbf{r},t\,|\mathbf{r}_0)\right\} \tag{4.6}$$

where the three factors have exactly the corresponding meaning as in eq (4.3). By integration over t, the polymer density $\rho_x(\mathbf{r})$ (equivalent to n_{xi}) is readily obtained from eq (4.6).

Thus, from the specification of the chain molecules in terms of their sequence of different types of segments and the interaction among all pairs of segment types, the distribution of the different segments in the inhomogeneous system is available. The calculations can be performed either with a specified number of components $\{n_x\}$ as indicated above, or the inhomogeneous system can be in equilibrium with a bulk of specified composition.

The main difference between the lattice and continuum approaches is the description of the chain conformations. In the lattice theory, these are relatively few due to the restriction to a lattice, whereas in the continuum theory they are infinite. However, the discretization of the equations makes the number of conformations finite, and in the limit of a mesh size of the discretization equal to the lattice size the approaches are essentially the same. Still, for more complex geometries, the continuum approach seems to be more tractable [19].

4.3. Different extensions

Over the years, the lattice approach has been extended to include, e.g., (i) charged components, (ii) branched components, (iii) bond angle restrictions, and (iv) the exclusion of direct backfolding [29]. Moreover, both the lattice and the continuum approaches have been applied to curved geometries and to the case where the segments have internal degrees of freedom. The latter extension has been very fruitful for modeling the reduced solubility of, e.g., poly(ethylene oxide) (PEO) in polar solvent [32,34-40], and has been used for modeling titrating polyelectrolytes [41].

4.4. Selected results

Some illustrative predictions of the self-assembly of $A_{50}B_{50}$ block copolymers in monomeric and selective solvent from the lattice theory were given in Figures 3-5. The qualitative predictions from the core-shell and the numerical self-consistent mean-field lattice approaches are the same. However, some major differences are discernable: (i) the cmc depends differently on χ due to different descriptions (free energy terms) of the system, (ii) compared for the same cmc (by selecting different χ-values), different aggregation numbers are predicted, and (iii) the lattice approach *predicts* an A-B separation where the interfacial width is substantial (for this model system).

Radial volume fraction profiles for a more "hairy" polymeric micelle formed by $A_{50}B_{500}$ block copolymers in solution are shown in Figure 7. The central lyophobic core of A segments is well represented by a block profile, whereas the lyophilic B segments form a highly solvated corona. Further analyses of Leermakers et al. [42] suggested a division of the profile of the B segments into four parts: (i) the "proximal" one where B segments are adsorbed onto the core, (ii) the "central" one where a power law regime exists, (iii) a "parabolic-like" regime further out and finally (iv) a "distal" regime where the profile approaches the bulk value approximately exponentially (cf. inset of Figure 7). Thus, such model calculations can provide guidance as to how to select profiles in more simple theories/models and in the evaluation of scattering experiments.

Since the free energy of the system and the chemical potentials of the components are available, we are also able to predict the stability of different ordered phases relatively to each other and to the disordered (homogeneous) one. By calculating the stability of phases with different structures, we can predict a phase diagram that contains different ordered phases. Figure 8 shows the prediction of such a phase diagram for the $(EO)_{21}(PO)_{47}(EO)_{21}$ triblock copolymer (Pluronic P94) in aqueous solution, made by Noolandi et al. [32]. At low polymer volume fraction we have the disordered L_1 phase. At increasing polymer concentration there are successive transitions to a cubic lattice of packed spherical aggregates (I_1), hexagonally packed cylindrical aggregates (H_1), and lamellar aggregates (L_α). At even higher concentrations the reverse phases appear where the water is localized in the center of the cylindrical aggregates

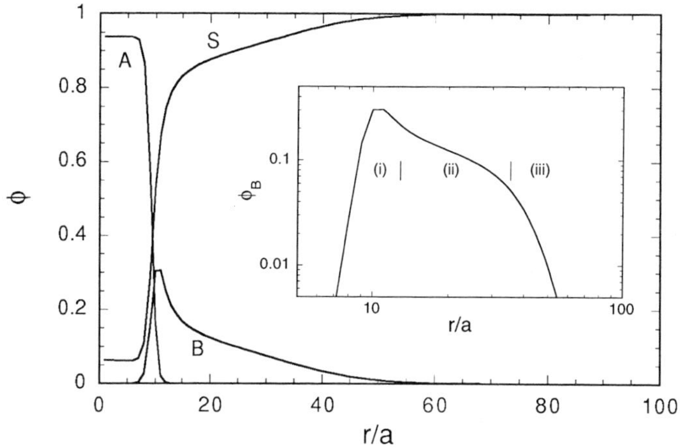

Figure 7. Radial volume fraction profiles of A and B segments and solvent through a spherical micelle of $A_{50}B_{500}$ molecules for $\chi_{AB} = \chi_{BS} = 0.5$, $\chi_{AS} = 2.0$, and $N_{agg} = 54$ using a face centered cubic lattice. The inset shows the profile of the B segments in a log-log representation and the extension of the different regions are indicated. The profile for the "distal" region is not visible.

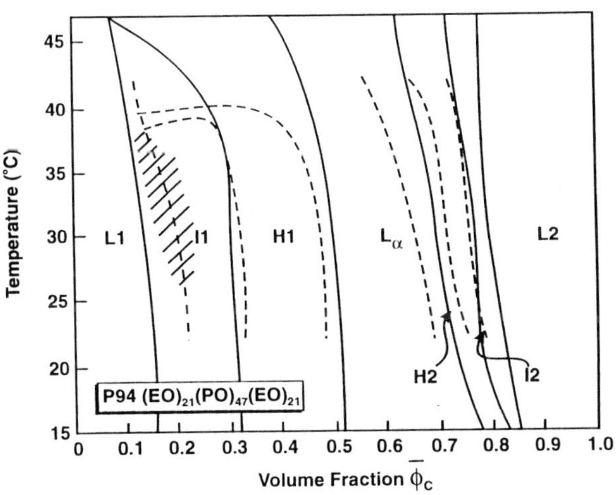

Figure 8. Calculated phase diagram for $(EO)_{21}(PO)_{47}(EO)_{21}$ in aqueous solution using the continuum (solid) and the lattice (dashed) descriptions. The hatched area shows the region with disordered solution of micelles. (Reprinted with permission from ref [32]. Copyright 1996 American Chemical Society.)

(H_2) and in the center of the reversed packed micelles (I_2). Finally, at very high polymer concentrations a new disordered phase appears (L_2). The two-phase regions between the one-phase regions were not considered. At increasing temperature the solubility of the polymer is reduced, promoting a change in the stability of the phases towards less curved surfaces (defined with water on the outside). Although the lattice and continuum approaches give slightly different phase boundaries, the general appearance is the same. The phase boundaries predicted by the two numerical methods seem to approach each other as the polymer chains becomes longer [32]. By using the theory of thermodynamics for small systems [43], a prediction of the cmc in the disordered L_1 phase can be made. The hatched area in Figure 8 shows the extension of the disordered micellar solution.

The volume fraction profiles of EO, PO, and H_2O for the H_1 phase are shown in Figure 9. It is clear that the interior of the rods is rich in the hydrophobic PO, whereas the outer parts consist of hydrated EO segments. At increasing temperature, the domain size increases and the volume fraction of the EO segments in the water region is reduced. The latter is a consequence of the reduced solubility as modeled by using internal degrees of freedom for EO. Since EO is not fully compatible with PO, the EO segments tends to form a new shell between the PO core and the continuous water region.

In the process of the determination of the most stable phase, the domain size of each phase is optimized. Figure 10 shows the predicted domain sizes for the equilibrium phases at two temperatures. Generally the domain sizes increase with increasing temperature. This is related to the smaller projected surface area for the PEO block due to the reduced solubility. Moreover, the domain size increases at transitions where the curvature is increased, but decreases within a phase as the polymer concentration is increased.

Most synthetic polymers are polydisperse. In the case of block copolymers, even a small polydispersity may profoundly affect the behaviour of the properties of the solution.

The lattice approach has been used to examine the effect of a polydisperse PEO-PPO-PEO triblock copolymer on the location of the cmc [39]. The polydisperse sample were represented by several components selected from the Schulz-Zimm distribution. Both the case of only a mass polydispersity and the case of mass-and-composition polydispersity were considered. Figure 11 shows how the location of the cmc changes as a function of the polydispersity ratio M_m/M_n for $(EO)_{37}(PO)_{56}(EO)_{37}$ triblock copolymer (Pluronic P105) in aqueous solution. Here M_m and M_n are the mass and number weighted mass averages, respectively. Depending on the criterion of the cmc (here the volume fraction of micelles in the solution is used) the cmc may change by several orders of magnitude already at the moderate polydispersity ratio of M_m/M_n= 1.3. Further investigations showed that the radial extension of the micelle *decreases* at increasing total polymer volume fraction due to the replacement of longer and less soluble polymer fractions with shorter and more soluble fractions [39]. Such observations have been made experimentally by Tuzar et al. [44] in a different system but the origin of the trends were never clarified.

It is straightforward to study the solubilization of lyophobic components in the center of the micelles [38,40,45]. One merely introduces an additional component with appropriate interaction parameters beside the block copolymer and the solvent. Figure 12 shows the radial volume fraction profiles for $(EO)_{30}(PO)_{61}(EO)_{30}$ triblock copolymer (Pluronic P104) in aqueous solution without and with a hydrophobic solute [40]. The model predicts that the solute is solubilized in the core of the micelle but tends to avoid the outer part of the core. Moreover,

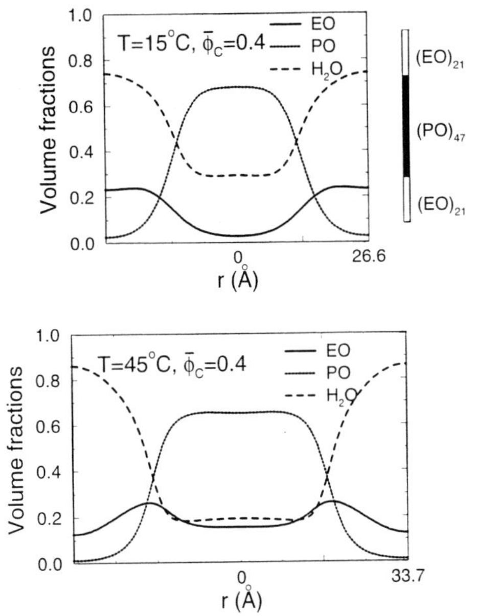

Figure 9. Radial volume fraction profiles of EO and PO segments and H_2O for the hexagonal phase of $(EO)_{21}(PO)_{47}(EO)_{21}$ in aqueous solution at two different temperatures at a total polymer volume fraction of 0.4. (Reprinted with permission from ref [32]. Copyright 1996 American Chemical Society.)

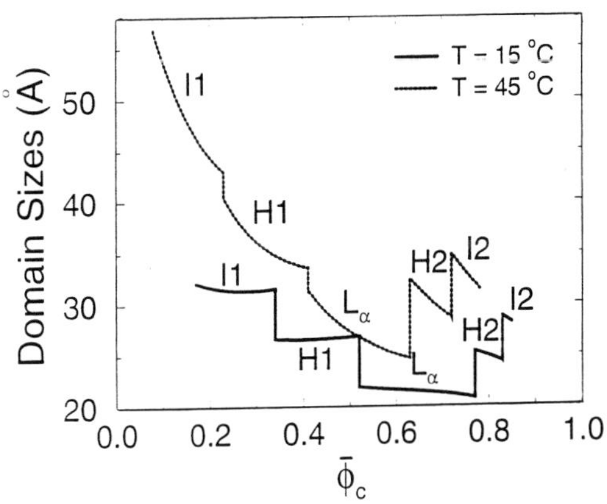

Figure 10. Domain sizes of the different equilibrium ordered phases for $(EO)_{21}(PO)_{47}(EO)_{21}$ in aqueous solution as a function of the polymer volume fraction at two different temperatures. (Reprinted with permission from ref [32]. Copyright 1996 American Chemical Society.)

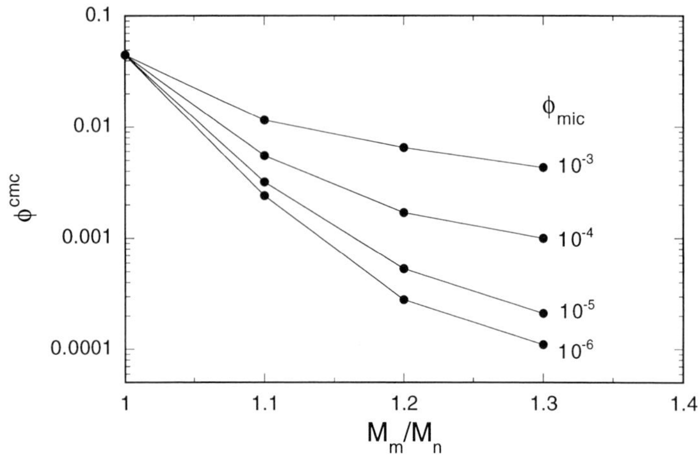

Figure 11. Critical micellar volume fraction as a function of the polydispersity ratio at different micellar volume fractions (criterion of the cmc) for $(EO)_{37}(PO)_{56}(EO)_{37}$ in aqueous solution at 320 K. The calculations were performed for a mass-polydisperse sample represented by 5 components. The symbols denote the calculated points, whereas the lines are given for increased readability. (Data from Linse [39].)

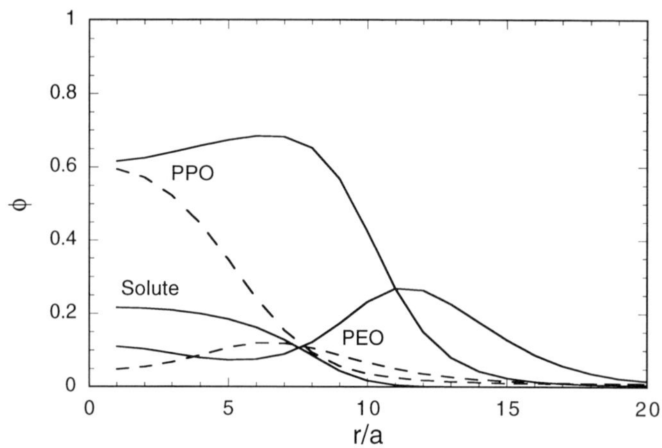

Figure 12. Radial volume fraction profiles of EO, PO, and solute for $(EO)_{30}(PO)_{61}(EO)_{30}$ in aqueous solution at 300 K without (dashed curves) and with (solid curves) solubilized hydrophobic solute at a polymer volume fraction $\phi_P = 0.05$. The volume fraction of the solute $\phi_{solute} = 2.9\text{x}10^{-5}$ is close to its saturation point. (A full account of the other conditions are given by Hurter et al. [40]. Note, $kT\chi_{POp\text{-}POn}$ is erroneously given in Table III in ref [40]; it should be 1.4 kJ mol^{-1}.)

the micellar aggregation number increases from ca. 10 to ca. 55 and the increase of the micellar size is obvious from the profiles. Thus, the scaling, semi-analytic, and numerical mean-field approaches all predict a micellar growth upon solubilization.

5. SIMULATION APPROACHES

We will restrict the use of the word simulation to techniques such as Monte Carlo simulation and molecular dynamics. These methods provide us with a wealth of information of model systems by making full statistical mechanical averages over the relevant degrees of freedom. The methods are computationally intensive and have grown in parallel with the computer development. Simulation methods have been applied to polymeric systems as well to other molecular systems. A recent book edited by K. Binder serves as an excellent introduction to simulation techniques and how these are applied to different subfields of polymeric systems [46].

The simulation of polymers is inherently computationally demanding due to the slow rearrangement of the molecules and this is in particular accentuated when investigating non-local properties. In order to facilitate the enumeration of the conformations, lattice systems are frequently used. It is today still not possible to routinely make direct simulations of the self-assembly of long block copolymers. However, substantial effort has been devoted to the self-association into micelles and liquid crystalline phases of short chain molecules. These simulations have been performed either on a lattice or off-lattice with simple spring-bead models. Moreover, the structure and dynamics of single micelles as well as mono- and bilayers have been investigated with (nearly) fully atomistic models [47-49], but in these studies the simulations were started from already assembled systems.

5.1. Micellar structure and the cmc

Pratt and coworkers performed early Monte Carlo simulations of single micelles formed by a fixed number of A_3B and A_5B chains on a diamond lattice [50,51]. Rodrigues and Mattice made further investigations of the structure of micelles composed of $A_{10}B_{10}$ [52] and $A_5B_{10}A_5$ chains [53]. These studies showed that (i) the interfacial region between the two types of beads is significant and (ii) that the shape of the aggregate is not perfectly spherical. Thus, the simulation methods, like the numerical self-consistent mean-field methods, provide information on the thickness of the interfacial region between the two blocks, whereas the other approaches often neglect the interfacial thickness. Moreover, the strength of the simulation methods is that the shape of the aggregate is a result of the system parameters and that all relevant fluctuations are included.

Subsequent lattice [54-61] and off-lattice [62,63] simulations involved a larger number of chain molecules making it possible to establish an equilibrium among free unimers and several micelles of different size. Beside structural information, these more extended studies can provide information of the distribution function of the aggregation number and the cmc.

The self-assembly of the diblock "copolymer" $A_{10}B_{10}$ on a primitive cubic lattice with six interacting neighbours has been extensively investigated by Mattice and coworkers [52,56-58] and by Wijmans and Linse [59-61]. These studies gives a detailed picture of the micellization. Figure 13 clearly illustrates the plateau of the chemical potential of the chains as a function of

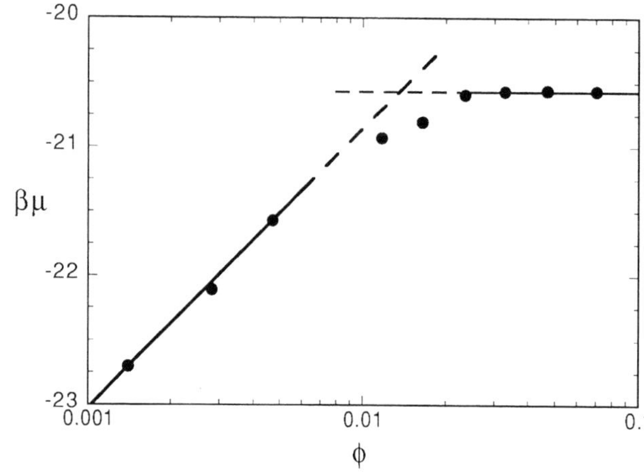

Figure 13. Chemical potential of $A_{10}B_{10}$ chains in solution as function of the volume fraction. Interaction energies: $w_{AB}/kT = w_{AS}/kT = 0.45$ and $w_{BS}/kT = w_{ii}/kT = 0$ (i = A, B, or S). (Data from Wijmans and Linse [61].)

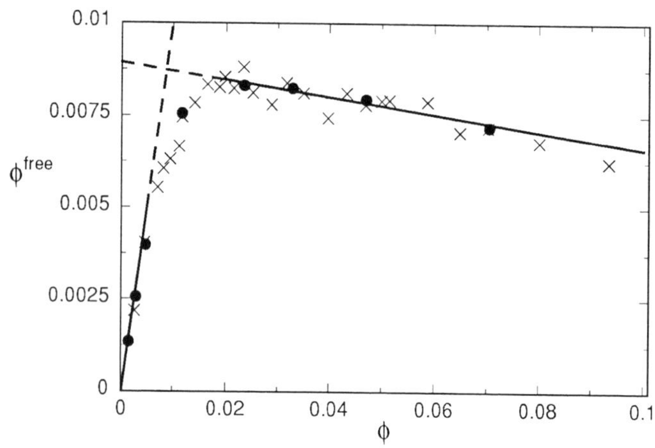

Figure 14. Volume fraction of free $A_{10}B_{10}$ chains in solution as a function of the volume fraction. Interaction parameters as in Figure 13. (Data from Wang et al. [56] (crosses) and Wijmans and Linse [61] (filled circles).)

the total volume fraction. The levelling off of the chemical potential is of course associated with the formation of micelles and the intersection of the extrapolated lines can be taken as the cmc. The volume fraction of free (non-associated) chains, ϕ^{free}, as a function of the total volume fraction is displayed in Figure 14. As for previous theories (cf. Figure 3), ϕ^{free} levels off when the micelles start to form. In fact, the simulation studies predict a reduction of ϕ^{free} at high micellar volume fractions. An extrapolation gives an estimate of the cmc in good agreement with that obtained from the chemical potentials. Figure 15 illustrates the bimodal distribution of aggregate sizes above the cmc as obtained from simulation. The solution is essentially composed by free chains, chains in small clusters composed by a few chains, and closed aggregates with ca. 20-40 chains. Additional analysis shows that micelles with an aggregation number of 30 are only weakly distorted toward a prolate shape from a spherical symmetry [59]. Similar characteristic micellar size distribution functions were also obtained from lattice simulations of short chain molecules by Desplat and Care [55], by Larson [64], and from extensive off-lattice simulations of eight bead long chain molecules with short range interactions by Smit et al.[63].

5.2. Ordered phases

A number of different lattice simulations have been performed to examine the ordered phases occurring in block copolymers solutions [64-73]. These studies cover chain lengths from 8 to almost 200 beads and deal with dilute as well as with more concentrated solutions with either a selective solvent, a nonselective solvent, or a binary solvent.

So far, close packed spheres of different cubic symmetries [65,66,69,71,72], hexagonally packed cylinders [64-66,69,71,72], the gyroidal structure [69], perforated lamellae [64-67,69,72], and lamellae [64-73], have been identified. Figure 16 shows a ternary phase diagram for the A_4B_4/A/B system obtained from lattice Monte Carlo simulations. Besides the extended classical lyotropic liquid crystalline phases, a narrow gyroidal phase was observed between the hexagonal and lamellar region. So far, it is not fully clear whether the gyroidal phase is a stable or only metastable phase [66]. In the disordered phases the simulation predicted an elongation of the micelles close to the to liquid crystalline phases, and the lamellar phase displayed perforations near the phase boundary toward the hexagonal phases. In such investigations, the size of the simulation box imposes nonphysical constrains on the domain sizes. This problem becomes less severe as the box length increases relative to the domain size. This influence has been systematically investigated for smaller chains as A_3B_3, A_4B_4, and A_3B_{12} [66].

The self-assembly and the concomitant microphase separation occurring in melts (as well as in nonselective solvents) are only driven by the interactions among the A and B segments themselves. However, lattice simulation of melts are technically difficult to perform, but by introducing vacancies, or equivalently, a nonselective solvent, it is possible to deal with melts and the effect of the the vacancies is taken account by a rescaling of the interaction parameter. Extensive simulations has been performed to investigate the critical χN and the stretching of the blocks of symmetric block copolymers. In particular, simulations [68,73], show that the order-disorder transition of block copolymer occurs at $\chi N > 11.5$ as predicted by mean-field theory in the limit of infinite chain lengths [74].

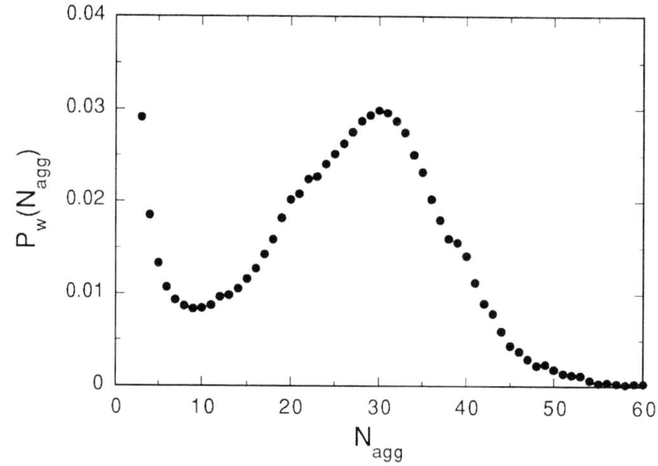

Figure 15. Normalized weight distribution of the aggregation numbers of micelles formed by $A_{10}B_{10}$ chains in solution at $\phi = 0.047$. The probability of free unimers is 0.17. Interaction parameters as in Figure 13. (Data from Wijmans and Linse [59].)

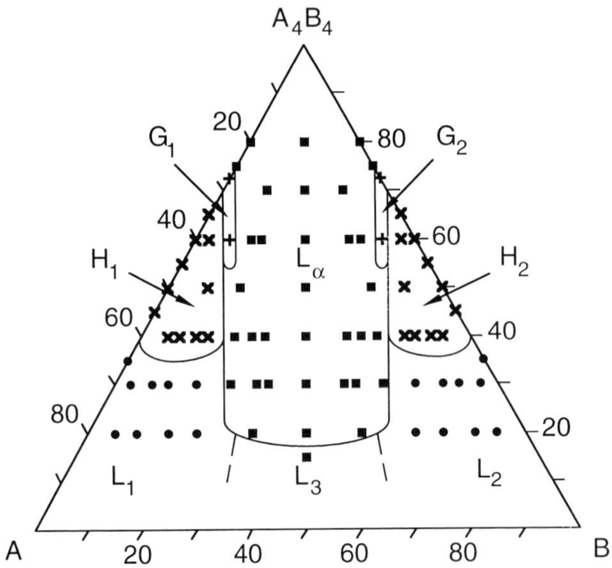

Figure 16. Ternary phase diagram for the A_4B_4/A/B system on a simple cubic lattice with 26 interacting neighbours with a contact energy $w_{AB}/kT = 0.1538$. L_1 and L_2 are disordered micellar phases, L_3 a disordered bicontinuous phase, H_1 and H_2 hexagonal phases, G_1 and G_2 gyroidal cubic phases, and L_α a lamellar phase. Only one half of the phase diagram was simulated, the rest is obtained by symmetry. (Data from Larson [69].)

6. CONCLUSIONS

The area of modelling block copolymers in solution is rapidly expanding and develops in close connection with the experimental progress. Some of the major approaches used to examine various aspects of the self-assembly of block copolymers in solution have been given. As we have seen, the development has been benefitted from the field of block copolymer melts and from the field of self-assembly of surfactants.

In brief, the scaling approach relies on a number of basic approximations which successively are relaxed in the more elaborated approaches. Typical for the scaling approach are the simple expressions relating a few variables describing the system which are obtained after a minimization of the free energy. In the semi-analytic mean-field models, the full functional dependencies among the variables are obtained, again, after a minimization of the free energy. Moreover, since a reference state is included, the theories are also able to predict the cmc, the micellar volume fractions etc, besides the aggregation number and the size of the micellar core and shell. The remaining two approaches are more fundamental in the sense that they explicitly contain chain molecules and that they are based on configurational averages. In the numerical self-consistent mean-field models, properties of the system are calculated for a given morphology. Volume fraction profiles appear as a results and are hence not a part of the assumption. In addition to the properties given above, predictions of the interfacial width and interfacial tension are made, and information on the distribution of individual segments is provided. Finally, in the simulation approaches, the mean-field approach is lifted and the morphology of the equilibrium structures is obtained directly. Furthermore, since fluctuations in all 3 dimensions are included, the results are improved over those from the numerical self-consistent mean-field models. However, (i) the simulation results are subjected to statistical uncertainty, (ii) the influence of boundary conditions, system size etc has to be assessed, and (iii) considerations of whether the system is in equilibrium or not has to be addressed.

Hence, the quality of the predictions are improved in the order that the types of approaches have been presented. Still, the scaling approach provides us with simple and very useful pictures of the system which is not the case for the more numerically intensive methods. Also, the computational effort increases in the same direction. Whereas the computer time is negligible for the minimization of the free energy in the semi-analytic mean-filed models (< 1 CPU second on a workstation), it is of the order of seconds or minutes for solving the set of non-linear equations obtained from the numerical self-consistent mean-field models. Finally, the computational effort for simulation of chain systems is of the order of days and upwards.

Nevertheless, the more computationally demanding approaches are expected to grow in importance. Fewer approximations are involved and a more detailed picture is provided. So far, most simulations have dealt with generic chains, but it is feasible to bring more chemistry into the models. An example illustrated here is the notion of internal degrees of freedom which makes it possible to model temperature dependent solvency from basic assumptions. This and similar approaches could be directly transferred into the models used in the simulation investigations. Another area of expected development is more complex models of the self-assembly of polyelectrolytes and ionomers. The increased number of system parameters and the computer intensive evaluation of the electrostatic interactions in direct simulations have, so far, hampered the progress here.

Finally, in practical all approaches, the short-range interaction are described by χ-

parameters or nearest neighbour interactions. In order to mimic some real system, suitable assignments have to be made. So far, the values of these parameters have been (i) estimated from more elaborate theories, (ii) extracted from simulation of small systems described on an atomic level, or (iii) obtained by fitting to experimental data. Due to the great simplification of the models, these parameters should be viewed as effective parameters with, at most, some physical relevance, making procedure (i) and (ii) less useful. This is, e.g., illustrated by Figure 6 where two different theories gives strongly different cmc for the same χ-value. The reason is of course (i) that these terms in which the χ-parameter enter are different, and (ii) these terms are balanced by other free energy terms which depend on the type of theory. The unrealistically low cmc $\phi \approx 10^{-34}$ for PEO-PPO diblock copolymer with 70% PEO as obtained by Nagarajan and Ganesh [21] by using χ-parameter from other sources constitutes a second example. Thus, it is clear that χ-parameters are not generally transferable between the different types of theories and there is also no guaranty of transferablilty between different applications of the same theory/model. However, our own experience is that for EO- and PO-containing polymers in aqueous solution, we have successfully been able to gradually build up values of χ-parameters from simpler system and employ those in more complex ones in a fruitful manner. In this scheme, phase diagrams of the binary PEO/water system were used to fit the values of the internal state parameters of EO and the $\chi_{EO,water}$-parameters (there are several parameters due to the presence of internal states) [34] and similarly for PPO in water [35]. Thereafter, interaction parameters between the EO and PO segments were fitted by using phase diagrams of the ternary PEO/PPO/water system and the previous obtained values from the binary systems [75]. Finally, the full set of parameters was used to predict a number of different properties of PEO-PPO-PEO triblock copolymers at different conditions with satisfactorily predictive power [32,36-39].

Acknowledgements

It is a pleasure to thank J. Noolandi, A.-C. Shi, and C. Wijmans for pleasant and fruitful collaboration and their permission to shown some of our recent research results. It is also a great pleasure to thank P. Alexandridis and L. Piculell for their kind and useful comments on the manuscript.

REFERENCES

1. M.W. Matsen and F.S. Bates, Macromolecules, 29 (1996) 1091.
2. P.G. de Gennes, in Solid State Physics, Ed L. Liebert, Academic, New York, 1978, suppl. 14.
3. A. Halperin, Macromolecules, 20 (1987) 2943.
4. C. Marques, J.-F. Joanny, and L. Leibler, Macromolecules, 21 (1988) 1051.
5. M. Daoud and J.P. Cotton, J. Phys (Paris) 43 (1982) 531.
6. T.M. Birshtein and E.B. Zhulina, Polymer, 25 (1984) 1453.
7. P.G. de Gennes, Scaling Concepts in Polymer Physics, Cornell University Press, Ithaca, N.Y., 1979.
8. R. Nagarajan and K. Ganesh, J. Chem. Phys., 98 (1993) 7440.
9. N. Dan and M. Tirrell, Macromolecules, 26 (1993) 4310.

10. L. Leibler, H. Orland, and J.C. Wheeler, J. Chem. Phys. 79 (1983) 3550.
11. M.R. Munch and A.P. Gast, Macromolecules, 21 (1988) 1360.
12. J. Noolandi and K.M. Hong, Macromolecules, 16 (1983) 1443.
13. M.G. Whitmore and J. Noolandi, Macromolecules, 18 (1985) 657.
14. P.J. Flory, Principles of Polymer Chemisty, Cornell University Press, Ithaca, N.Y., 1953.
15. E. Helfand and Y. Tagami, J. Polym. Sci. Part B, 9 (1971) 741.
16. G. ten Brinke and G. Hadziioannou, Macromolecules, 20 (1987) 486.
17. N.P. Balsara, M. Tirrell, and T.P. Lodge, Macromolecules, 24 (1991) 1975.
18. O. Procházka, Z. Tuzar, and P. Kratochvíl, Polymer, 32 (1991) 3038.
19. W. Zheng, Z.-G. Wang, Macromolecules, 28 (1995) 7215.
20. V. Abetz, R. Stadler, L. Liebler, Polymer Bulletin, 37 (1996) 135.
21. R. Nagarajan and K. Ganesh, J. Chem. Phys., 90 (1989) 5843.
22. A. Gao and A. Eisenberg, Macromolecules, 26 (1993) 7353.
23. D.F. Shim, C.M. Marques, and M. E. Cates, Macromolecules, 24 (1991) 5309.
24. P. Sens, C. M. Marques, and J.-F. Joanny, Macromolecules, 26 (1996) 4880.
25. R. Nagarajan and K. Ganesh, Macromolecules, 22 (1989) 4312.
26. R. Nagarajan and K. Ganesh, J. Colloid Inteface Sci., 184 (1996) 489.
27. J.F. Marko and Y. Rabin, Macromolecules, 25 (1992) 1503.
28. J.M.H.M. Scheutjens and G.J. Fleer, J. Phys. Chem., 83 (1979) 1619.
29. G.J. Fleer, M.A. Cohen Stuart, J.M.H.M. Scheutjens, T. Cosgrove, and B. Vincent, Polymer at Interfaces, Champan & Hall, London, 1993.
30. M. Doi and S.F. Edwards, The Theory of Polymer Dynamics, Clarendon Press, Oxford, 1986.
31. K.M. Hong and J. Noolandi, Macromolecules, 14 (1981) 727.
32. J. Noolandi, A.-C. Shi, and P. Linse, Macromolecules, 29 (1996) 5907.
33. O.A. Evers, J.M.H.M. Scheutjens, and G.J. Fleer, Macromolecules, 23 (1990) 5221.
34. G. Karlström, J. Phys. Chem., 89 (1985) 4962.
35. P. Linse and M. Björling, Macromolecules, 24 (1991) 6700.
36. P. Linse, Macromolecules, 26 (1993) 4437.
37. P. Linse, J. Phys. Chem., 97, (1993) 13896.
38. P. Linse, Macromolecules, 27 (1995) 2685.
39. P. Linse, Macromolecules, 27 (1995) 6404.
40. P.N. Hurter, J.M.H.M. Scheutjens, and T.A. Hatton, Macromolecules, 26 (1993) 5030.
41. R. Israëls, F.A.M. Leermakers, and G.J. Fleer, Macromolecules, 27 (1994) 3087.
42. F.A.M. Leermakers, C.M. Wijmans, G.J. Fleer, Macromolecules, 28 (1995) 3434.
43. D. G. Hall and B.A. Pethica. In Nonionic Surfactants, Ed. M.J. Schick, Marcel Dekker, New York, 1976.
44. Z. Tuzar, P. Stepánek, C. Konák, and P. Kratochvíl, J. Colloid Interface Sci., 105 (1985) 372.
45. P.N. Hurter, J.M.H.M. Scheutjens, and T.A. Hatton, Macromolecules, 26 (1993) 5592.
46. K. Binder, Monte Carlo and Molecular Dynamics Simulations in Polymer Science, Oxford University Press, New York, 1995.
47. B. Jönsson, O. Edholm and O. Teleman, J. Chem. Phys, 85 (1986) 2259.
48. K. Watanabe, M. Ferrario and M.L. Klein, J. Phys. Chem., 92 (1988) 819.
49. E. Egberts, H.J.C. Berendsen, J. Chem. Phys., 89 (1988) 3718.

50. S.W. Haan and L.R. Pratt, Chem. Phys. Lett., 79 (1981) 436.
51. B. Owenson and L.R. Pratt, J. Phys. Chem., 88 (1984) 2905.
52. K. Rodrigues and W.L. Mattice, J. Chem. Phys., 95 (1991) 5341.
53. K. Rodrigues and W.L. Mattice, Langmuir, 8 (1992) 456.
54. D. Brindle and C. M. Care, J. Chem. Soc. Faraday Trans., 88 (1992) 2163.
55. J.-C. Desplat and C. M. Care, Mol. Phys., 87 (1996) 441.
56. Y. Wang, W.L. Mattice and D.H. Napper, Langmuir, 9 (1993) 66.
57. Y. Zhan and W.L. Mattice, Macromolecules, 27, (1994) 677.
58. P. Adriani, Y. Wang, and W.L. Mattice, J. Chem. Phys. 100 (1994) 7718.
59. C.W. Wijmans and P. Linse, Langmuir, 11 (1995) 3748.
60. C.W. Wijmans and P. Linse, J. Phys. Chem., 100 (1996) 12583.
61. C.W. Wijmans and P. Linse, J. Chem. Phys., 106 (1997) 328.
62. B. Smith, P.A.J. Hilbers, K. Esselink, L.A.M. Rupert, M.N. van Os, and A.G. Schlijper, J. Phys. Chem., 95 (1991) 6361.
63. B. Smith, K. Esselink, P.A.J. Hilbers, M.N. van Os, L.A.M. Rupert, and I. Szleifer, Langmuir, 9 (1993) 9.
64. R.G. Larson, J. Chem. Phys., 96 (1992) 7904.
65. R.G. Larson, J. Chem. Phys., 91 (1989) 2479.
66. R.G. Larson, Chem. Eng. Sci., 49 (1994) 2833.
67. R.G. Larson, Mol. Sim., 13 (1994) 321.
68. R.G. Larson, Macromolecules, 27 (1994) 4198.
69. R.G. Larson, J. Phys. II (Paris) 6 (1996) 1441.
70. L.A. Molina, A. López Radríges and J.J. Freire, Macromolecules, 27 (1994) 1160.
71. L.A. Molina and J.J. Freire, Macromolecules, 28 (1995) 2705.
72. M.B. Ko and W.L. Mattice, Macromolecules, 28 (1995) 6871.
73. H. Fried and K. Binder, J. Chem. Phys., 94 (1991) 8349.
74. L. Leibler, Macromoleucles, 13 (1980) 1602.
75. M. Malmsten, P. Linse, K.-W. Zhang, Macromolecules, 26 (1993) 2905.

On the origin of the solution behaviour of ethyleneoxide containing polymers.

Gunnar Karlström

Department of Theoretical Chemistry, University of Lund
P.O.B 124, S-221 00 Lund, Sweden

The origin of the decreased solubility (clouding) of several ethyleneoxide containing polymers at elevated temperatures are discussed using available data both from theoretical and experimental investigations. It is found that the clouding process is closely related to the conformational equilibrium of the ethyleneoxide chains, and that this equilibrium depends on the properties of the free ethyleneoxide chain, the solvation of the chain, but also to a large extent on the effect that the ethyleneoxide chain has on the effective water - water interaction.

1.INTRODUCTION

Normally when two substances are brought into contact. At a sufficiently low temperature they do not mix, but when the temperature is increased the resistance to mixing is decreased. The phase behaviour is determined by the thermodynamic property DG_{mix}, which is the difference in Gibbs free energy of the solution and the two pure substances. This quantity can be divided into two parts, one energetic and one entropic according to

$$\Delta G_{mix} = \Delta H_{mix} - T\Delta S_{mix} \qquad (1)$$

The first term on the right hand side of this equation (ΔH_{mix})is normally positive and favours a situation where the two substances are kept apart, whereas the second term, the entropic one most often favours a solution. This manifests that ΔS_{mix} normally is positive, corresponding to that the solution is a less ordered system than the two pure substances. T is the absolute temperature.There are several models for calculation of ΔS_{mix} for ordinary non flexible molecules, and the most well known ones are probably the regular solution [1] model and the Carnahan- Starling model [2]. A common feature in both these models is that the maximum mixing entropy is of the order of R, where R is the gas constant. In the regular solution model it takes the value of Rln2 and in the Carnahan - Starling model it depends on the relative size of the mixed molecules. The smallest maximum value occurs for molecules of equal size and then it corresponds to that of the regular solution model. The regular solution model and the Charnhan - Starling model are derived for spherical molecules with nondirectional intermolecular forces and takeaccount of the increased volume accessible for the molecules due to the mixing process. This is often called the ideal entropy of mixing. For more complex molecules the situation is more complicated but, normally, the mixing entropy is dominated by the ideal term, and the solubility increases with temperature. A common generalisation of the regular solution model for polymers was derived independently by Flory and Huggins and is today generally known as the Flory-Huggins theory [3]. A characteristic feature of the Flory-Huggins model is that the ideal mixing entropy per polymer segment decreases with the molecular weight of the polymer. This means that the driving force for forming a solution gets smaller for higher molecular weight, and that the solubility of a polymer in solvent should decrease with the molecular weight of the polymer.

There are, however, several cases when a deviation from the behaviour is observed and here we will shortly discuss two. The first one occurs for aqueous solutions of hydrophobic material, such as benzene or cyclohexane, and the second is for solutions of polymers of intermediate polarity in polar solvents. The explanation to these phenomena on a molecular level is an interesting subject. In the first case it is generally agreed that the hydrophobic molecule induces a structuring of the water in its vicinity. This structuring is associated with an energetic gain and an entropic cost [4]. Without the structuring process one would have expected that the solution of a hydrophobic substance in water was accompanied by an entropic gain and an energetic cost. If the effect of the ordinary solution process and the structuring is added, then one obtains that for the entire solvation process at room temperature there is an entropic cost but that the process occurs with only very small changes in energy. At lower temperatures the system gains energy (ΔH_{mix} is negative) and at higher temperatures the solution process is associated with an energetic cost (ΔH_{mix} is positive). At higher temperatures both the entropic end energetic contributions act against the solution process. If Gibbs - Helmholtz´ equation

$$\left(\frac{\partial}{\partial T}\left(\frac{\Delta G}{T}\right)\right)_p = -\frac{\Delta H}{T^2} \tag{2}$$

is applied to the solution process then we see that the solubility increases with temperature for temperatures where ΔH_{mix} is positive and decreases with temperature in the opposite case. Thus for this type of system the solubility decreases with temperature at low temperature and then it increases at higher temperatures.

The thermodynamic reasoning above applies to any solvation process, with a solubility minimum at some temperature. Below this temperature ΔH_{mix} is negative and above it is positive. The origin, however, on a molecular level may be different. It must however be true that there must be an energy gain that can be associated with some degrees of freedom, which can be associated with the mixing process. For the hydrocarbon substances discussed above this degree of freedom was the relative orientation of neighbouring water molecules. Addition of hydrophobic material induced an increased structuring in these degrees of freedom.

Another class of molecules that show a similar solution behaviour is polymers that contains ethylene oxide groups. Typical examples are polyethyleneoxide (PEO) often called poly(ethyleneglycol) (PEG) [5], copolymers of polyethyleneoxide and polypropyleneoxide such as UCON [6] and PLURONICS [6], cellulose modified with ethyleneoxide and ethylchloride as ethyl-hydroxyethyl-cellulose (EHEC) [7]. Similar phase behaviour is also observed for aqueous solutions of nonionic surfactants of ethyleneoxide type [8,9]. It is the purpose of this chapter to look into the origin on a molecular level for this behaviour.

The solubility properties can also be illustrated with a figure showing the

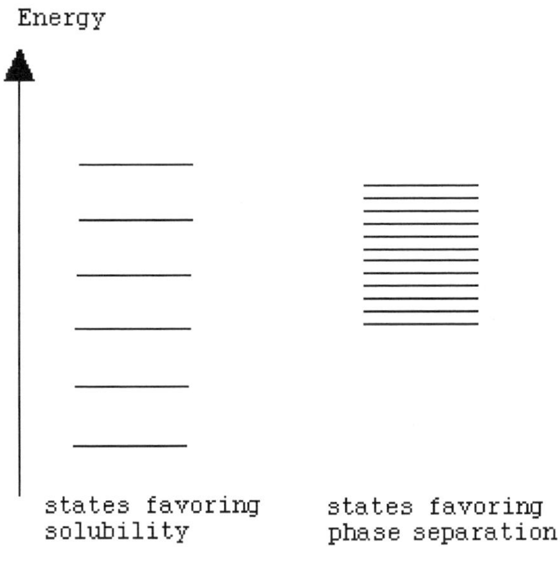

Figure 1. Schematic picture of the density of states for a clouding system. There are fewer states with low energy that favour solubility, and more states with higher energy that favour phase separation. The increased number of states can be due to chain conformations, absence of hydrogen bonding or absence of structured water

accessible number of states associated with the mixing process. In Figure 1 we present such a diagram. At low temperatures the low energy states favouring solubility are populated, but at an increased temperature entropy will populate the states favouring phase separation since they are more numerous. The observed thermodynamics obviously depend on the density of states in the two classes of structures. More complex behaviour can naturally be observed if more classes of states exists.

The outline of the rest of the chapter will be as follows. In the first fairly short section (2), the solubility behaviour of ethyleneoxide (EO) containing polymers and similar systems in polar solvents will be discussed. The focus in this section will be placed on the gross features of the solubility and not on the clouding process. Then, in the next section (3) we will discuss the experimental data available for solutions of ethyleneoxide containing molecules in polar solvents. In the following section (4) we will focus on the results from theoretical modelling of similar systems, and finally (5) we will summarize by giving a short description on a molecular level of the clouding process consistent with both experimental and theoretical data.

2. SOLUBILITY BEHAVIOR OF EO CONTAINING COMPOUNDS IN WATER AND OTHER POLAR SOLVENTS.

The solubility of PEO and related substances in water may seem remarkable in several ways. From the introduction we learned that the ideal part of the mixing entropy was smaller per segment for a long polymer chain, than for a short. Despite this fact, it is experimentally well known that the shortest PEO molecules, dimethylether and 1,2-dimethoxyethane (DME), only have a limited solubility in water [10], and that full solubility in water only occurs for longer chains. For even longer chains the normal, as predicted by the Flory-Huggins theory, behaviour is obtained. One difference is that the number of hydrogen atoms per carbon atom is larger in dimethylether and DME and that this makes the hydrophobic character of these substances larger. It is however difficult to believe that this fairly small increase in hydrophobicity would not be compensated by the larger ideal mixing entropy for these systems. Solubility is not solely determined by the properties of the solution, but it also depends on the properties of the pure substances. It may be that the behaviour discussed above depends on an increased stability of the pure dimethylether and DME. The extra stability of these systems may originate from better packing possibilities due to less restrictions imposed by the covalent structure of the molecules. Another anomaly in the solution behaviour of related compounds is that polymethyleneoxide (PMO), a polymer consisting of equally many oxygens as carbon atoms, is almost insoluble in water [11], whereas PEO which contains two carbon atoms per oxygen is soluble. For polymers that contain more hydrophobic material, as polypropyleneoxide (PPO) with 3 carbon atoms per oxygen, the solubility decreases again with increasing number of carbon atoms per oxygen [12]. The remarkable feature here is not the decreased solubility occurring for PPO and longer chains, but the higher solubility for PEO than for PMO.

Several explanations are possible, First we note that the packing conditions in the pure substances may be more favourable in PMO than in PEO and this is also reflected in the higher melting point temperature for PMO. Formulated in other words, the polymer - polymer interaction is so favourable in PMO that the solvent, water, can not dissolve the crystal. Further one may note that there are two CH_2 groups for each oxygen in PEO, but only one in PMO. This may mean that the oxygens in PEO are more polar than the oxygens in PMO, since each oxygen can receive electrons from two CH_2 - groups instead of one. An alternative explanation can be based on that the distance between the neighbouring oxygens in PMO is short (2.5Å) and that it may be difficult for water to solvate independently two neighbouring oxygens. The most likely explanation is certainly the first one given above.

We will conclude this section by shortly discussing the solubility properties of EO containing compounds in formamide, an other polar solvent. A proper starting point for such a discussion is to note that the an effect similar to the so called hydrophobic effect observed for hydrocarbon substances in water can not be observed for formamide for hydrocarbons. Instead formamide behaves as an ordinary polar solvent and the solubility of hydrocarbons in formamide increases with temperature [13,14]. PEO is completely soluble in formamide at relevant temperatures. However for modified systems such as micelles formed from nonionic surfactants of EO type and cellulose, that have been made soluble by covalently binding of EO and ethyl groups, a clouding is observed in formamide [15]. A similar behaviour is also observed for copolymers of EO and propyleneoxide (PO) in formamide [16].

From this it seems reasonable to conclude that the clouding process in this type of system has not necessarily the same origin as the hydrophobic effect.

3. EXPERIMENTAL INVESTIGATIONS ON THE CONFORMATION OF EO CONTAINING COMPOUNDS.

The purpose of this chapter is to illuminate the clouding process occurring for EO containing compounds. This is neither an easy nor a straight forward process to investigate by experimental means. We have, from the previous section, seen that a clouding process must be due to that one or in principle more degrees of freedom, on a molecular level, are associated a larger flexibility in the clouded system (the two separate phases) than in the isotropic solution. In principle there are only three types of interactions in a solution of a polymer in a solvent, and these are the solvent - solvent, the solvent - polymer and the polymer - polymer interactions. The last two terms may need some further further explanation. Naturally there may be a change in these two types of effective interactions with concentration and/or temperature just as all other interactions may change, but there may also be an other type of change, in that the average conformation of the polymer may change and this may change the effective polymer - solvent interaction. One may however also argue in the opposite way that if the effective solvent - polymer interaction changes this may change the average polymer conformation, and in principle if the solvent - solvent interaction is different for different conformations then changes in the solvent - solvent interaction may induce changes in the polymer conformation. After this fairly lengthy introduction to this section we will discuss the experiments performed to unravel the origin behind the clouding behaviour. There is however a weakness in the use of experimental techniques in that only the polymer conformation can be studied. The concept of water structure is not uniquely defined and certainly there are hydrogen bonds both between the water molecules and between the polymer chain and the water molecules, but these are difficult to study in a quantitative manner.

Large amounts of work have been devoted to characterize and understand the thermodynamic behaviour of the PEO - water system [11], but as discussed above all mechanisms that can model the clouding behaviour must have similar thermodynamics. The literature on this subject is in fact very large and in the recent work by Tasaki [17] several references are given. In that work it is also possible to find references to many other aspects of the PEO - water system not discussed in this chapter. A problem connected with this is that in the very large amount of experimental data available for the PEO - water system it is possible to find support for very diverging opinions. Two examples will be given. The first concerns the number of water molecules solvating an ether oxygen in the EO chain. The reported numbers vary form 1 [18], via 2 [19], 3 [20],4-5 [21] to 6 [21]. The other example concerns the average structure of the polymer chain in water. Several investigators claim that the chain is fairly rigid, with a helix as a dominating structure element [17,20b, 22], whereas others claim that it forms a random coil, and that there is a large conformational flexibility of the polymer chain [23]. The first group of investigators means that practically all C - C bonds prefer a gauche conformation and that a anti conformation is completely dominating around the C - O bonds. Fortunately on this last issue there seems to be a possibility to straighten things out. The experimental support for the idea of an EO helix in aqueous solutions is old [17,20b] and modern investigations of similar type have been interpreted in the opposite way [23b,23c]. The recent theoretical arguments for the helix model [17] is due to the use of erroneous potential functions for the intermolecular interactions, as will be discussed below.

Consequently we are left with a situation where only one of the proposed mechanisms can be tested using experimental techniques. This mechanism has however been extensively studied using several techniques, and these results will shortly be discussed below.

In the mid 70´s Podo and coworkers [24] studied the conformational equilibrium of DME in different solvents an as a function of temperature. The main conclusion from these investigations was that in polar solvents and at low temperature DME preferred polar conformations and in less polar solvents and high temperatures, the opposite was true. Similar results have later been obtained by Björling and coworkers [25] for PEO - chains and by Ahlnäs et al [26] for the head group of nonionic surfactants of EO - type in micelles. In the latter study it was found that for the EO - segments closest to the hydrocarbon-core of the micelle, the EO - segments favoured nonpolar conformations, whereas further away from this region polar conformations were more abundant. In recent years these observations have been verified by Matsuura and coworkers, who have studied the conformational equilibrium for short EO - chains with both IR and Raman technique [23b,23c]. Using high quality measurements they have been able to study the conformation around the different dihedral angles and found a strong dependence of the preferred conformation around the C - C bonds on temperature and polarity of the solvent as suggested above. They have also found that the conformation around the C - O bonds is much less dependent on the external conditions. Apart from verifying these old results, Matsuura and coworkers also have found that in water but not in formamide, at very low polymer concentrations, where the studied molecules can be expected not to interact with other polymer molecules, for very short PEO chains (3 - 5 monomeric units), that the probability for polar conformations reaches a maximum and decreases when the polymer concentration is further decreased. This effect is largest for chains with 3 monomeric units and decreases with increasing chain length to be very small for a pentamere. The explanation to this last observation is most likely found in that at these low concentrations the polymer changes its conformation in order to hide hydrophobic regions from the water. At higher polymer concentrations the same result can be obtained through interaction with other PEO chains.

From these observations it seems obvious to conclude that there is a link between the conformational properties and the solubility of EO containing compounds, and that an increased temperature will favour less polar conformations, leading to a reduced solubility. Whether this mechanism is powerful enough alone to be the origin of the clouding process can not be judged only from experimental observations.

4. THEORETICAL MODELLING OF EO CONTAINING COMPOUNDS.

Theoretical modelling can be made on different levels, and the insight that can be obtained from the modelling naturally depends on the applied level and the assumptions put into the modelling. In particular for the EO - water system three different approaches have been used to theoretically describe it.

4.1. Lattice models

The oldest, and crudest approach is based on Flory - Huggins theory [3]. The equations used within this model are derived on a lattice, and the interested reader can find a derivation of theequations in Hill´s book Statistical Mechanics[3c]. Within this approximation or model one writes

$$\Delta S_{mix} = -Rn_{tot}\left(\phi_1 \ln\phi_1 + \frac{\phi_2}{N}\ln\phi_2\right) \quad (3a)$$

and

$$\Delta H_{mix} = n_{tot}\ \phi_1\ \phi_2\ w_{12} \quad (3b)$$

In these equations R is the gas constant, f_1 and f_2 are the volume fraction of solvent and polymer respectively, N the degree of polymerization for the polymer, n_{tot} the total number of moles of solvent molecules and polymer segments, and w_{12} an effective interaction parameter between the polymer and the solvent. It is easy to show that clouding can never be obtained from a Flory - Huggins model assuming a fixed value of w_{12}. In fact in order to obtain clouding w_{12} must be positive and grow faster than linear with T. In order to obtain such an effective interaction two different models have been tested.

In the somewhat older approach Goldstein [27] suggested that one assumes that the interaction between water and an EO segment may either be of hydrogen bonded nature or of a non hydrogen bonded nature and further assumes that the first alternative is energetically favoured and entropically disfavoured compared to the second alternative. This means that when the temperature is increased the effective water polymer interaction becomes more repulsive, then this would lead to a phase separation. In his work Goldstein showed that a phase behaviour, similar to the experimental one determined by Sakei et al [28], could be obtained. The physics built into the model are that the probability for a hydrogen bond between the solvent and the solute at infinite temperature is much smaller than the probability for an nonbonded interaction, and that the hydrogen bonded regions between the solvent and the polymer are strongly attractive and that the nonbonded interactions between the solvent and the polymer are repulsive. A similar model was originally suggested by Hirschfelder, Stevenson and Eyring [29] as an explanation to the general clouding process.

An alternative approach, based on the empirical observation of changes in the preferred conformation of the EO - chain with solvent polarity and temperature but also on theoretical modelling of the EO - water system, was suggested by me [30]. The idea was that at low temperatures and polar solvents the EO containing molecules preferred a gauche conformation around the C - C bonds, due to the dipole moment of such conformations, whereas at higher temperatures and lower polarity of the solvent a less polar anti conformation was preferred. Using these assumptions it was also possible to model a phase behaviour, similar to the experimental one, if one assumed that the effective interaction between the polar conformations and the water was slightly repulsive and that the interaction between the nonpolar conformations and water was strongly repulsive. It was further necessary to assume that the nonpolar conformations were entropically favoured. Later investigations have however shown that most of the conformations are almost equally polar and only a few are less polar [34]. The advantage of this later model is that it apart from the phase behaviour also explains the conformational variations observed as a function of temperature, composition and solvent polarity.

The essence of these two models described above is that they suggest a molecular mechanisms that may give raise to a temperature dependence in the effective solvent - polymer interaction and that the processes described in the model will yield a phase separation at elevated temperatures. It is also interesting to note that the mathematical formulations of the two models are almost isomorphic, implying that it is not possible to distinguish between the two models using thermodynamic arguments. The actual origin to the observed phase behaviour can naturally only be found from all atomic molecular simulations with non empirical potentials, and below we will at some length describe the work of that type that has been done on relevant systems.

4.2. Continuum models of conformational equilibrium of DME.

A completely different approach to investigate the conformational equilibrium of the shortest possible PEO chain, has been used by several authors [31]. They have assumed that the solvent (water) is a medium, characterized by its dielectric permitivity and solved the quantum chemical equations describing the electronic structure coupled to the dielectric medium through Poissons equation. The outcomes of these studies are all in good internal agreement. They all suggests that polar structures with a gauche conformation around the C - C bond are stabilized relative to the anti form. The origin to this stabilization is obviously due to that polar structures give raise to a larger electric field that polarizes the dielectric medium and stabilizes the system.

4.3. Statistical mechanical modelling

The EO - water system is from many points of view an ideal system to study by theoretical simulation methods. There are however also many problems connected with this type of modelling. First we must realize that the simulation of actual two phase systems close to a critical point is quite out of reach today. Further we must realize that the equilibration of a long PEO chain in water is a slow process occurring on the millisecond time scale. Typical MD simulations study a system for 100 picoseconds or a few nanoseconds. On the other hand a chain contains many dihedral angles and even if a total equilibrium of the chain is not obtained in a simulation one could hope that a local equilibrium is obtained.

In practice there are several approaches to the investigation of EO containing systems in contact with water. Either a whole PEO chain is studied in an aqueous solution[17],or only the shortest PEO molecule (DME) is investigated [23f-g]. Other researchers have instead studied a hydrocarbon surface covered with a layer of EO groups [23e], in order to mimic a $C_{12}E_2$ (see below) lamella. A crucial point in these simulations is obviously the potential function used. Most investigators have worked with the so called OPLS model derived by Jorgensen [32], others have choosen to use a modified version of this model, where the polarity of the EO groups have been increased [17] compared to the OPLS model, with the motivation to obtain stronger hydrogen bonds between the polymer and water. In a series of works we have constructed inter and intra molecular potentials for the DME - water system [34,23f,g]. These potentials differ from the other potential used in that they are derived from quantum chemical calculations and that the atoms in the molecules are polarizable, and thus adjust their electrical properties to the surrounding and the conformation of the DME molecule. As we will see below, the result from the different simulations vary with the choosen potential in a way that in fact can be understood, and that it is even possible to obtain some insight from. It is thus fruitful so actually analyze the results from the different simulations together with information about the choosen potential function.

We will start this analysis by looking into the work by Tasaki [17]. He has used a modified OPLS model, where the dipole moment of an EO - segment has been increased to a value close to 2.5 D. This dipole moment is independent of the conformation of the EO - chain. The dihedral potential is taken from quantum chemical *ab initio* calculations and is similar to the one used by the other investigators and is probably associated with a small error compared to the errors we are discussing now. Accurate quantum chemical calculations on DME suggests that the dipole moment of an EO segment varies between 1.0 to 1.3 D, where the lowest value is obtained for a gauche conformation around the C - C bond and an anti conformation for the C - O bond and the highest value refers to an all anti conformation. From this we see that the polarity of the EO segments used in the study of Tasaki is far too high. For a more realistic system there would be associated a high energy cost with forming a gauche conformation around the C - C bond in vacuum. In his work Tasaki however associates the value from quantum chemical calculations with this cost, and consequently favours gauche conformations around C - C bonds. As a result Tasaki comes to the conclusion that PEO chains forms a helix in aqueous solutions, with gauche conformations around the C - C bonds and anti conformations around the C - O bonds. If Tasaki´s results are confronted with recent experimental data obtained by Matsuura [23b,c], it is evident that Tasakis simulations have described a system that is far too ordered. From Matsuura´s work it is quite clear that there is an equilibrium between different types of conformations and that polar solvent favours a gauche conformation around the C - C bonds, but that there also are a non negligible probability for anti conformations. It is further clear from Matsuuras investigations that there is a large probability for both gauche and trans conformations around the C - O bonds. Thus by making the EO segments more polar we may conclude that the gauche conformation is favoured, provided that the intramolecular dihedral potential is unchanged.

In another study by Kong et. al. a lamella created from $C_{12}E_2$ molecules is studied [23e]. $C_{12}E_2$ is a surfactant molecule with chemical formula $C_{12}H_{25}(OC_2C_2)_2OH$, that can form lamellar lyotropic liquid crystalline systems in contact with water. A large part of the behaviour of this system is dictated by a wish to hide the hydrocarbon region from the more polar water. The simulations are performed with unmodified OPLS potentials, but with united atoms, i.e. the nonpolar hydrogens are merged into the carbon atoms. In that work an equilibrium between the different conformations is observed in agreement with experiment both for the segments closest to the surface and the more distant ones. This gives strong support to the conclusions above regarding the work by Tasaki [17].

The probably most studied system with relevance for the behaviour of PEO in solution is the DME - water system, where most investigations performed so far have concentrated on the conformational equilibrium around the C - C bond. The first study of this type was performed by Liu et al, who studied a DME molecule[33], described using the MNDO model, a semiempirical quantum chemical method, in a solvent of explicit classical water molecules. They found that an aqueous solution stabilized a conformation which is anti around the C - O bonds and gauche around the C - C bond by 0.9 kcal/mole relative to a conformation which is anti around all bonds in gas phase. Similar results have later been obtained in a combined Monte Carlo/stochastic simulation of the same system, performed by Williams and Hall [23d]. The latter investigation also indicates that the influence from the solvent on the conformation around the C - O bonds is less effected by the solvent and that the influence of the solvent depends on the conformation around the C - C bond. Williams and Hall in the same work also studied the influence of the polarity of the solvent and reproduced the experimental observation the a more polar solvent favours the more polar conformations.

In a recent work we [23f] have studied the temperature dependence of the conformational equilibrium of the DME molecule using Monte Carlo simulations with a potential function based on *ab initio* quantum chemical calculations [34]. The potential function is constructed to handle the inter and intra molecular polarization effects on an equal level in order to ensure a correct treatment of the electrostatic. The simulations are performed at two temperatures 298 and 398 K. The main conclusions obtained from that study is that at the higher temperature the amount of the all anti configurations increases very rapidly with temperature, and that all other major configurations decrease their populations. In Table 1 we show the relative probability for the most probable conformations together with their temperature derivatives.

In Table 1 we have labeled the conformations with the dihedral angles

Table 1
The probabilities of different conformations and their temperature derivative at 398 K

	Conformation			
	aaa	aga	aag	agg´
Probability	0.291	0.226	0.483	0.257
Temperature derivative	0.010	-0.001	-0.006	-0.003

specified in a consecutive order starting from one end of the molecule. g´ indicates that this dihedral angle is in the opposite direction to the previous gauche angle in the chain.

To reveal the origin behind this behaviour we have analyzed the energetics in the solvation of the DME molecule and found that the increased stability at low temperature of the polar conformations mainly is due to favourable water - water interaction and to some extent due to the direct water - DME interaction. Naturally there is also an influence of the dihedral potential for the free DME molecule. In Table 2 the corresponding data are presented. An interesting observation is that the solvation of the polar DME conformation is stabilized by water - water interaction (water structuring) relativeto the non polar DME conformation. This is not in agreement with what could be expected from a comparison with the solvation of hydrocarbon substances, where the structuring is believed to be due to an increased water - water interaction around the non polar hydrocarbon molecule. A striking detail in Table 2 is that the direct water - DME interaction almost exactly cancels the gas phase dihedral potential.

Table 2
Partitioning of the solvation energy at 398 K for the different conformations of DME relative to the aaa conformation. All energies are given in kcal/mole.

	Conformation			
	aaa	aga	aag	agg´
Gas phase	0.0	0.8	2.1	0.4
DME - water	0.0	-1.0	-1.8	-0.5
water -water	0.0	-4.0	-5.1	-4.5
many body term	0.0	0.6	-0.2	0.4
total	0.0	-3.6	-5.0	-4.2

In order to investigate the origin of the abnormal conformational behaviour occurring in very dilute solutions of short chain PEO molecules an other approach has been tested. In a very recent work Engkvist and Karlström [23g] have simulated systems containing several short PEO molecules, containing 2 or 3 EO units, at different concentrations. Systems of this size are so large that simplified models are used. There are two assumptions built into the models. First it is assumed that the dihedral potential for the rotation around the C - C bond depends on the dielectric constant of the solution (i.e. on the EO concentration) in such a way that a gauche conformation around the C - C bond is favoured in more polar solutions. A united atom model is used. This means that each CH_2 and CH_3 group is modeled as one atom. It is further assumed that there is a repulsion between all atoms in the polymer chain provided that they are separated by more than 4 chain atoms or residing in different chains. Furthermore there is an attraction between different CH_x groups associated with a repulsion, as described above. The repulsive term depends on r as $1/r^{12}$ and the attractive one is more long ranged and depends on r as $1/r^6$. This interaction is supposed to model the hydrophobic interaction between the hydrophobic parts of the EO molecules. If this attraction is made slightly smaller than kT, then a variation of the preferred conformation with concentration as observed by Matsuura is observed. Thus one sees an increased probability for gauche conformation around the C - C bond with increasing water content. In the most dilute systems however there is an abrupt change and the gauche conformation suddenly becomes less probable for the molecules containing 3 EO units. This is obviously due to a wish to hide the hydrophobic parts of the molecule from water contact. This effect is not possible for the molecules containing 2 EO units (DME) due to its limited length. If the attraction is increased slightly then a phase separation occurs, meaning that the polymer no longer is soluble. If however the interaction is decreased to 0.5 kT then the decrease in gauche probability observed at low polymer content disappears. The interpretation of this is that there are two effects determining the optimal conformation of an EO chain in water. First a general dielectric effect, acting in such a way that a polar solvent favours a polar conformation, and second a wish to minimize the contact between the hydrophobic CH_2 and CH_3 groups with water. This last effect is most important for short chains that contain at least 3 EO units in high dilution of the polymer.

5. Conclusions.

It is naturally not possible to establish the exact origin behind the clouding behaviour from the simulations discussed above. This would require simulations of systems containing many fairly long PEO molecules, a study completely out of reach with todays computer resources. However the computer simulations have revealed an interesting interplay between conformational equilibrium, increased water - water interaction and the direct water - polymer interaction.

If the experimentally obtained information is analyzed together with the information from the computer simulations a fairly clear picture emerges which suggests that there is a link between the preferred conformation of the PEO chain and the solubility of the PEO molecules in water and other polar solvents, and that this conformational equilibrium is strongly influenced by the both the properties of the solvent and the temperature.

REFERENCES

1. See e.g. D.F. Evans and H. Wennerström, The Colloidal Domain Where Physics, Chemistry, Biology and Technology Meet, VCH Publishers, New York, 1994.
2. N.F. Carnahan and K.E. Starling, J.Chem.Phys.,51 (1969), 636.

3. a)M.L. Huggins, J. Phys. Chem., 46 (1942), 151.,b)P.J. Flory, J. Chem. Phys., 10 (1942), 51,c)See also T.L. Hill, An Introduction to Statistical Thermodynamics, Dover, NewYork, 1986.
4. a) K. Shinoda, J. Phys. Chem., 81 (1977), 1300.
b) C. Tanford, The Hydrophobic Effect. Wiley, New York, 2nd edn. 1980.
5. F.E. Bailey and J.V. Koleske, Poly(ethylene oxide), Academic Press, New York, 1976.
6. a) P.A. Harris, G. Karlström, F. Tjerneld, Biosepartion, 2 (1991) 237.
b) P. Alexandris and T.A. Hatton, Colloids Surf., 96 (1995)
7. A. Carlsson, Colloid Polym.Sci. 266 (1988), 1031.
8. J.C. Lang and R.D. Morgan, J. Chem. Phys. 73 (1980), 5849.
9. D.J. Mitchell,G.J. Tiddy,L. Warring,T. Bostock and M.P. McDonald, J. Chem. Soc. Faraday Trans. 79 (1983), 975.
10. Handbook of Chemistry and Physics, 63rd edition, edited by R.C. West, Chemical Rubber Company Cleveland,Ohio, 1982, page C-375 and C-296 .
11. R. Kjellander and E. Florin-Robertsson, J. Chem. Soc. Faraday Trans. 1, 77 (1981), 2053.
12 K. Zhang and M. Carlsson, J. Phys. Chem., 99 (1995), 5051.
13 G. Olofsson, J. Chem. Soc. Faraday Trans. 87 (1991), 3037.
14. D. Berling and G. Olofsson, Journal of Solution Chemistry, 23 (1994), 911.
15. a) T. Wärnheim, J. Bokström and Y. Williams, Colloid Polym. Sci., 266 (1988), 562. b) A.A. Samii, G. Karlström, B. Lindman, Langmuir, 7 (1991), 1067.
16. A.A. Samii, G. Karlström, B. Lindman,J. Phys. Chem., 95 (1991), 7887.
17. K. Tasaki, J. Amer. Chem. Soc., 118 (1996), 8459.
18. A. Macounachie, P. Vasudevan and G. Allen, Polymer, 19 (1978), 33.
19 H. Matsuura and K. Fukuhara, Bull. Chem. Soc. Jpn. ,59 (1986), 763.
20 a) M. Rösch, Kolloid Z. 147 (1956), 78, b) K.-J. Liu and J.L. Parson, Macromolecules 2 (1969), 529, c) J. Maxfield and I.W. Shephard, Polymer 16 (1975), 505, d) K.P. Antonsen and A.S. Hoffman in Poly(EthyleneGlycol) Chemistry, Edited by J.M. Harris, Plenum Press New York 1992.
21. P.-G. Nilsson, H. Wennerström and B. Lindman, J. Phys. Chem. 87 (1983), 1377.
22. a)K. Tasaki and A. Abe, Polym. J., 17 (1985), 641, b)H. Matsuura and K. Fukuhara, J. Mol. Struc. 126 (1985), 251.
23 a) W. Brown and P. Stilbs, Polymer, 23 (1982), 1780, b) S. Masatoki, M. Takamura, M .Matsuura, K. Kamogawa and T. Kitagaw, Chem. Lett. , 1995 ,991. c) H. Matsuura and T. Sagawa, J. Mol. Liq. 65/66 (1995), 313, d) D.J. Williams and K.B. Hall, J. Phys. Chem., 100 (1996), 8224, e)Y.C. Kong, D. Nicholson and N.G. Parsonage, Mol. Phys. 89 (1996) 835, f) O. Engkvist and G. Karlström, J. Chem. Phys., in press, g)O. Engkvist and G. Karlström, J. Phys. Chem., in press.
24 a) V. Viti, L. Indovina, F. Podo L. Radics and G. Nemety, Mol. Phys., 27 (1974), 541. b)F. Podo, G. Nemety, L. Indovina, L. Radics and V.Viti, Mol. Phys. 27 (1974), 521.
25 M. Björling, G. Karlström and P. Linse, J. Phys. Chem., 95 (1991), 6707.
26 T. Ahlnäs, G. Karlström and B. Lindman, J. Phys. Chem., 91 (1987), 4030.

27 R.E. Goldstein, J. Chem. Phys, 80 (1984), 5340.
28 S. Saeki, N. Kawahara, M. Nakata and M. Kaneko, Polymer, 17 (1976), 685.
29 J. Hirschfelder, D. Stevenson and H. Eyring, J. Chem. Phys., 5 (1937), 896.
30 G. Karlström, J. Phys. Chem., 89 (1985), 4962.
31 a) M. Andersson and G. Karlström, J. Phys. Chem., 89 (1985), 4957. b) F. Müller-Plathe and W.F. van Gunsteren, Macromolecules, 27 (1994), 6040, c) D.J. Williams and K.B. Hall, J. Phys. Chem., 100 (1996), 8224.
32 a)W.L. Jorgensen, J. Phys. Chem. 90 (1986), 1276, b) W.L. Jorgensen, J.D. Madura and C.J. Swenson, J. Amer. Chem. Soc., 106, (1984), 6638. c) W.L. Jorgensen and C.J. Swenson, J. Amer. Chem. Soc., 107 (1985), 1489, d)W.L. Jorgensen and M. Ibrahim, Chem. Phys. Letters, 103 (1981),3976.
33 H. Liu, F. Müller-Plathe and W.F. van Gunsteren, J. Chem. Phys.,102 (1995), 1723.
34 O. Engkvist, P.-O. Åstrand and G. Karlström, J. Phys. Chem., 100 (1996), 6950.

Block copolymers of ethylene oxide and 1,2-butylene oxide

C. Booth and G.-E. Yu
Manchester Polymer Centre, Department of Chemistry,
University of Manchester, Manchester M13 9PL, UK

V. M. Nace
The Dow Chemical Company, Texas Operations, Research and Development,
Freeport, Texas 77541, USA

Notation

E is used to denote an oxyethylene unit, OCH_2CH_2, and B to denote an oxybutylene unit, $OCH_2CH(C_2H_5)$. For example, a diblock copolymer is denoted E_mB_n, and a triblock copolymer $E_mB_nE_m$ or $B_nE_mB_n$. As a general notation for oxyethylene/oxybutylene copolymer systems we use E/B. In the same spirit, we use P to denote an oxypropylene unit, $OCH_2CH(CH_3)$, and C to denote a methylene (or methyl) unit.

1. Introduction

The preparation of block copolymers of ethylene oxide and 1,2-butylene oxide was first disclosed by the Wyandotte Chemicals Corporation in 1955 [1]. Later patents followed from the Dow Chemical Company [2]. This early work, which has been well reviewed [3,4], provides information on the preparation and performance properties of $E_mB_nE_m$ triblock copolymers. BASF-Wyandotte made $E_mB_nE_m$ copolymers (denoted Butronics) available for research [5-7], but they were not commercialised, and no systematic study of their self-assembly has been published.

Starting in the early 1980s, Szymanowski and co-workers at Poznán described the performance properties of E/B copolymers covering the range of simple linear architectures: i.e. diblock (B_nE_m) and triblock ($E_mB_nE_m$ and $B_nE_mB_n$). Preparative aspects are described in [8-10], and their many papers are reviewed in [11].

The preparation of E/B block copolymers and systematic study of their self-assembly in aqueous solution started in Manchester in the late 1980s, and continues to the present day [12-34]. At an early stage, cooperative work was undertaken with SUNY Stony Brook [16,23-26,29,31], and later with the Dow Chemical Company [26,27], the Surface Chemistry Institute, Stockholm [28], the Royal Institute of Technology, Stockholm [31], and the University of Bayreuth [32]. All the simple architectures (E_mB_n, B_nE_m, $E_mB_nE_m$, $B_nE_mB_n$, *cyclo*-B_nE_m) have been studied, and also tapered-block copolymers [28]. $B_nE_mB_n$ copolymers have been a particular interest at Stony Brook [26,29,35]. A brief summary of much of this work has appeared in a recent review of the physical chemistry of poly(oxyalkylene) surfactants [36].

Diblock and triblock E/B copolymers were introduced as commercial products by the Dow Chemical Company in 1993, with a first description appearing in the commercial literature in 1992 [37]. Several reports of the properties of these copolymers have appeared recently [27,38-41], including studies of self-assembly in water [27,40] and water/xylene mixtures [41], as well as relevant product brochures [37,42].

The aim in this contribution is to survey work on the self-assembly of E/B block copolymers, including the formation of lyotropic liquid-crystal gels. Particular emphasis is placed on the physical chemistry of the systems, but regard is also paid to the importance of synthesis and molecular characterisation in providing block copolymers of precise composition, chain length and architecture. Opportunity is taken to describe briefly the commercial range of E/B copolymers. The performance properties and applications of these interesting materials have been reviewed in some detail recently [11,43].

2. Commercial E/B copolymers

There are numerous structural possibilities for B/E copolymers. Degrees of freedom for their synthesis include initiator type, number of active hydrogen groups on the initiator, oxide feed order, oxide feed purity (random versus block), and block length. The commercial B/E copolymers are either $E_mB_nE_m$ triblock diols, resulting from the use of a dihydroxylic initiator, or diblock mono-ols, derived from a mono-alcohol initiator. Reverse triblock diols of type $B_nE_mB_n$ have been the subject of several studies [e.g. 26,35], but none are commercially available at this time.

Eight E/B block copolymers are presently offered by the Dow Chemical Company. Six are triblocks of type $E_mB_nE_m$ and two have diblock architectures. The system of nomenclature used in the product literature [37] has three variations: BXX-YYYY, BMXX-YYYY, and BLXX-YYYY. B indicates a poly(1,2-butylene oxide) hydrophobe, "XX" the approximate wt-% B, and "YYYY" the approximate number-average molar mass (M_n) of the copolymer. "M" indicates that the E block is methyl terminated and "L" stands for *n*-butyl termination of the B block. The triblocks are dihydroxyl-terminated, and the diblocks have one hydroxyl termination. All materials are on the United States TSCA inventory.

The commercial nomenclature relates to that used in this review as set out below.

Water-soluble triblocks	B40-1900 $E_{13}B_{10}E_{13}$	B20-3800 $E_{35}B_{10}E_{35}$	B40-2500 $E_{17}B_{14}E_{17}$	B20-5000 $E_{45}B_{14}E_{45}$
Water-soluble diblocks	BM45-1600 $E_{19}B_{10}$	BL50-1500 $C_4B_{10}E_{17}$		
Water-insoluble triblocks:	B70-4600 $E_{15}B_{46}E_{15}$	B50-6600 $E_{38}B_{49}E_{38}$		

The block lengths indicated are approximate averages of narrow distributions. The methyl termination of the E block of copolymer BM45-1600 is not indicated in present notation, but the more significant *n*-butyl termination is noted.

Other structural variations of B/E block copolymers may become commercial in the future, as new applications are identified.

2.1 Summary of properties

As wt-% E increases, the physical states of E/B copolymers change from clear liquid, to cloudy liquid, to paste, then to solid. The commercial copolymer surfactants are either pastes or solids at room temperature.

The high-M_n copolymers with low water solubilities can be useful as emulsifiers. The structures of the six water soluble copolymers were chosen for optimal performance as surfactants. The B_{10}-B_{14} range, which gives optimum wetting and surface/interfacial tension lowering, corresponds to a B-hydrophobe M_n of 700 to 1000 g mol^{-1} (Dalton). The optimal percentage of E is in the range 50-60 wt-%. Surface tension values range from 31 to 38 mN m^{-1} (dyne cm^{-1}) for 0.1 wt-% aqueous solutions. Interfacial tensions between aqueous and

n-dodecane phases range from 2 to 9 mN m^{-1} at 0.1 wt-%. Aqueous wetting performance, as measured by the Draves wetting time test, is good for the copolymers having 50-60 wt-% E, wetting times ranging from 3 to 10 seconds at 0.5 wt-%. The copolymers with 80 wt-% E have wetting times greater than 6 minutes.

B/E block copolymers are considered moderate-to-high-foaming surfactants. As wt-% E increases, initial and final Ross-Miles foam heights increase, resulting in more stable foams. In some applications, high foaming is desirable.

2.2 Comparison with commercial E/P block copolymers

E/P triblock copolymers ($E_mP_nE_m$) have been commercially available for some time, and have many applications. The P hydrophobe contains a higher proportion of oxygen than B and is more polar, as measured by inverse gas chromatography [39]. Consequently poly(B) is more soluble in oil phases than poly(P). This polarity difference partly explains why surface and interfacial tension lowering is more effective for E/B compared to E/P copolymers.

E/P analogues of four of the water-soluble commercial E/B copolymers have been studied in terms of property differences [40]. In general, the surface activity of a commercial E/B copolymer is higher than that of its E/P analogue. However, it must be pointed out that the surface activity of E/P copolymers is greatly enhanced when the P-block is of higher M_n (2000-3000 g mol^{-1}).

2.3 Biodegradability

Biodegradability is a very important sub-topic when discussing surfactant properties. Typically, E/B block copolymers are slowly degraded in standard closed-bottle oxygen uptake tests, but are not considered recalcitrant. A recent patent describes the use of ε-caprolactone as a comonomer for rendering E/B copolymers and other poly(oxyalkylene) materials more biodegradable [44].

3. Copolymer preparation

3.1 Anionic copolymerisation

E/B copolymers are prepared by sequential anionic polymerisation of the two monomers. For polymerisation of, e.g., a monofunctional poly(oxyethylene), the addition of ethylene oxide (EO) is initiated by a monofunctional alcohol (ROH), part of which is in the form of its alkali-metal salt (e.g., RO^-K^+). Initiation is instantaneous, and the subsequent reaction scheme is written, e.g.,

slow propagation $\quad MeE_mO^-K^+ + EO \rightarrow MeE_{m+1}O^-K^+$

fast equilibrium $\quad MeE_mO^-K^+ + MeE_nOH \leftrightarrows MeE_mOH + MeE_nO^-K^+$

where E_m represents a growing chain, the active centre is an ion pair, and rapid equilibration ensures that all chains are equally likely to grow. This ideal scheme would produce polymer with M_n = (mass of EO polymerised)/(moles of initiator used) and a Poisson distribution of chain lengths [45]: in practice the chain length distribution can be widened by slow initiation (see below). In the absence of deliberate or accidental termination, the system is living, and is ideal for preparation of block copolymers.

The same scheme applies to propylene oxide (PO) and 1,2-butylene oxide (BO), but with the addition of a hydrogen-abstraction reaction (transfer reaction), e.g.

transfer $\quad MeP_nO^-K^+ + PO \rightarrow MeP_nO^-K^+ + CH_2{=}CHCH_2O^-K^+$

This reaction reduces M_n and broadens the chain length distribution, since the unsaturated alcohol initiates new chains. Transfer is very important in the anionic polymerisation of PO [46], with consequences for the compositions of $E_mP_nE_m$ block copolymers, but is absent in BO polymerisation under laboratory conditions [12], although transfer is known in BO polymerisation under the more extreme conditions (e.g. high temperatures) used in industry [47].

Block copolymers of EO and BO are formed by sequential anionic polymerisation of one monomer followed by the other. As mentioned above, slow initiation widens the chain length distribution. Alcohols form oxyanions of different reactivity to glycols which can affect the first stage of copolymerisation [48]. However, initiators can be selected to have structures similar to the intended chain units (see Section 3.2). A large effect arises in the second stage of copolymerisation if EO is added to an active B block. The first addition of EO (to $B_nO^-K^+$) is slow compared to subsequent additions (to $B_nE_mO^-K^+$), giving a widened E-block length distribution, and possibly, if the E-block is short (e.g. < 40 units), uncapped B ends at the completion of the reaction [38]. In contrast, addition of BO to an active E-block is uncomplicated, since addition of the first BO to the primary E oxyanion is fast compared to subsequent additions. The same problem is found in sequential polymerisations involving PO, but is less severe. These aspects of the anionic copolymerisation of alkykene oxides have been reviewed [47].

3.2 Laboratory preparation: copolymers of different architecture

Block architecture in the sequential copolymerisation of EO and BO is controlled by the initiator used in the anionic polymerisation: examples are shown in Table 1

Table 1 Commonly used initiators for E/B block copolymers.

monofunctional	purpose	difunctional	purpose
$CH_3(OCH_2CH_2)_2OH$	E_mB_n	$H(OCH_2CH_2)_2OH$	$B_nE_mB_n$
$CH_3OCH_2CH(C_2H_5)OH$	B_nE_m	$HOCH_2CH(C_2H_5)OH$	$E_mB_nE_m$

Typically in the laboratory [e.g. 14,20,31] freshly-cut potassium is reacted with the initiator (mole ratio $[OH]/[K^+] \approx 10$). Monomers are dried over and distilled from powdered CaH_2. The first monomer is enclosed with an aliquot of initiator solution in an ampoule and polymerised to completion. The second monomer is then added and polymerised to form the copolymer. Polymerisation temperatures are in the range 40-80 °C. To avoid initiation by moisture, all manipulations are carried out either on a vacuum line or under dry N_2. Block copolymers prepared in this way have narrow chain length distributions: values of M_w/M_n from GPC are usually below 1.1, with the distributions for B_nE_m and $E_mB_nE_m$ copolymers being somewhat wider than those of E_mB_n and $B_nE_mB_n$: see Section 4.4.1 for examples.

3.3 Tapered block copolymers

Because the reactivity ratios of EO and BO in anionic copolymerisation are significantly different ($r_E = 4.1$, $r_B = 0.17$)[49], copolymerisation of a mixture of monomers to high conversion will result in marked composition drift, so much so that a mixture which is 80 mol-% EO in composition yields statistical copolymer of composition 5 mol-% B at low conversion and 100 mol-% B at high conversion. Because of the living polymerisation, this composition drift occurs along the chain, and a tapered block copolymer results. These copolymers are surface active in aqueous solution, and self-associate in much the same way as conventional block copolymers [28,50].

3.4 Cyclic block copolymers

Cyclisation of an $E_mB_nE_m$ diol by reaction with CH_2Cl_2 under Williamson conditions at high dilution was first reported in 1993 [51]. This work has continued [30,31] in conjunction with work on the preparation of large cyclic poly(oxyethylene)s [52], and large uniform crown ethers [53]. In the case of the copolymers, ring closure is *via* an acetal linkage, $-OCH_2O-$, and the formula of the cyclic copolymer prepared from $E_mB_nE_m$ (*cyclo*-B_nE_{2m}) can be written schematically as:

$$\left[\begin{array}{c} E_mB_nE_m \\ OCH_2O \end{array}\right]$$

The reactions involved are

(i) $—OCH_2CH_2O^-K^+ + CH_2Cl_2 \rightarrow —OCH_2CH_2OCH_2Cl + KCl$

(ii) $—OCH_2CH_2O^-K^+ + ClCH_2OCH_2CH_2O—$

$\rightarrow —OCH_2CH_2OCH_2OCH_2CH_2O— + KCl$

Conversions to cyclic products of 85% or more are usual. Ring closure by ether linkage is also possible [53,54], though this reaction is less efficient. The high efficiency of the acetal closure reflects the high reactivity of the chloroether formed initially. Chain-extended products formed as by-products can be largely removed by fractional precipitation. Details of preparative conditions and characterisation of products (GPC, NMR) can be found in [30,31].

4. Association properties in dilute aqueous solution

E/B copolymers readily form micelles in dilute aqueous solution. This self-assembly is initiated either at a given temperature by increasing the concentration beyond the critical micelle concentration (cmc) or at a given concentration by increasing the temperature beyond the critical micelle temperature (cmt). At high enough concentration or temperature micelles predominate in solution, and micelle molar masses and dimensions can be determined. A wide range of experimental techniques is available, as summarised recently by Zhou and Chu [36]. Those applied to dilute solutions of E/B copolymers are briefly described below. Details of the experimental methods and the analyses of data can be found in the papers referenced.

4.1 Experimental methods

4.1.1 Dynamic light scattering (DLS)

DLS is the most convenient way of demonstrating the presence of micelles. Because the intensity of scattered light is z-weighted ($z \propto cM_w$, c = mass concentration, M_w = mass-average molar mass), light scattering is sensitive to low levels of high-molar-mass solutes. Given a reliable method of data analysis (e.g. CONTIN [55]), DLS provides information on the intensity distribution of apparent translational diffusion coefficient (D_{app}) for the micelles and, through the Stokes-Einstein equation, the corresponding distribution of apparent hydrodynamic radius ($r_{h,app}$), i.e. the apparent radius of the hydrodynamically-equivalent hard sphere.

Examples of intensity fraction distributions of $\log r_{h,\,app}$ are shown in Figures 1 and 2. Unassociated copolymer molecules have apparent hydrodynamic radii in the range 1-2 nm. The single, narrow peaks at higher values of the radius seen for aqueous solutions of $E_{38}B_{12}$ and $E_{21}B_{11}E_{21}$ in Figure 1 indicate almost complete association into micelles. The two-peaked distributions seen for $B_5E_{39}B_5$ in Figure 2 indicate that those solutions contain both micelles and molecules, e.g. 40 wt-% micelles in the solution at 40 °C [24].

Averaging over the intensity distribution and extrapolating to zero micelle concentration gives true values of D and r_h. Alternatively, if the distribution is single-peaked and narrow, the cumulants method of analysis [56] can be used for this purpose.

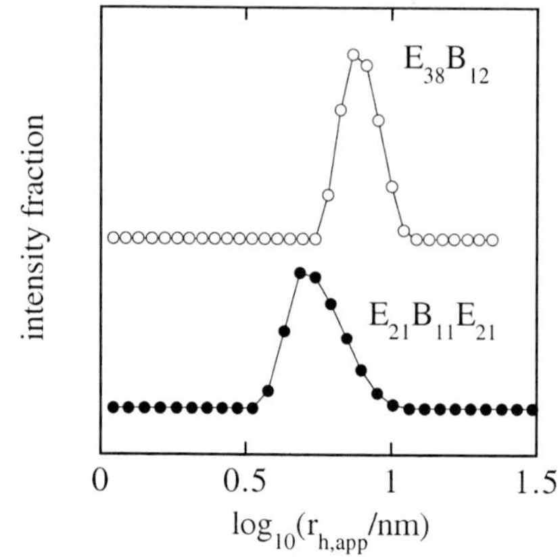

Figure 1. Intensity fraction *versus* logarithm of apparent hydrodynamic radius for solutions of $E_{38}B_{12}$ (20 g dm^{-3}, 40 °C) and $E_{21}B_{11}E_{21}$ (40 g dm^{-3}, 40 °C). The peaks correspond to micelles. Adapted from [23].

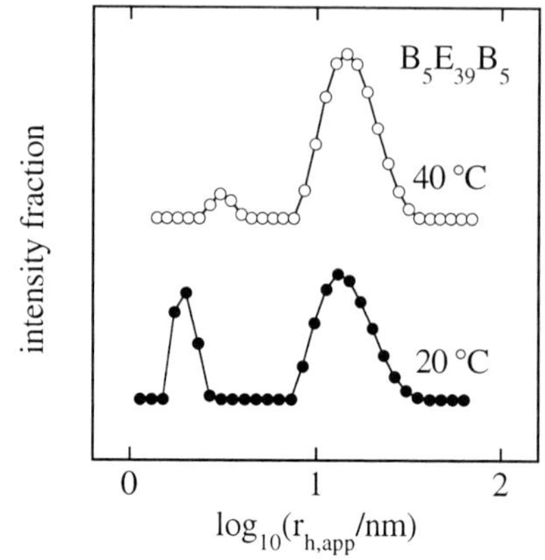

Figure 2. Intensity fraction *versus* logarithm of apparent hydrodynamic radius for solutions of $B_5E_{39}B_5$ (120 g dm^{-3}, 20 and 40 °C). The peaks correspond to molecules (or molecular clusters) at small values of $r_{h,app}$, and to micelles at large values of $r_{h,app}$. Adapted from [24].

4.1.2 Static light scattering (SLS)

Under favourable circumstances, the concentration dependence of excess light scattering intensity (excess over copolymer solution) can be used to determine the cmc. Plots of $I-I_S$ against concentration (c) for aqueous solutions of $E_{40}B_{15}E_{40}$ are shown in Figure 3. I is the scattering from solution relative to that from pure benzene, and I_S is the corresponding quantity for pure water.

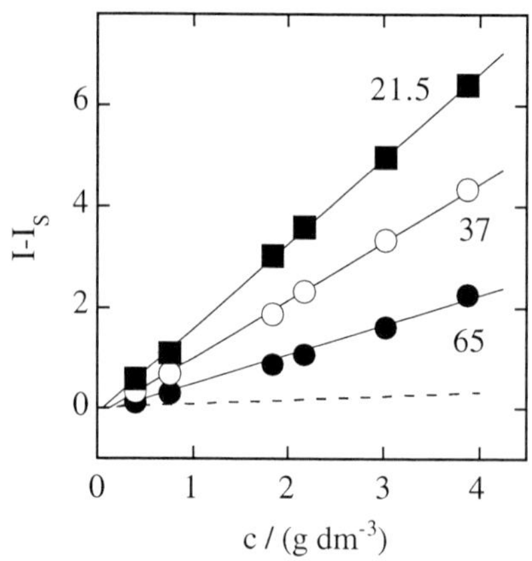

Figure 3. Excess scattering intensity ($I-I_S$) *versus* concentration (c) for aqueous solutions of $E_{40}B_{15}E_{40}$ at the temperatures indicated (°C). The dashed line indicates scattering from unassociated molecules. Adapted from [16].

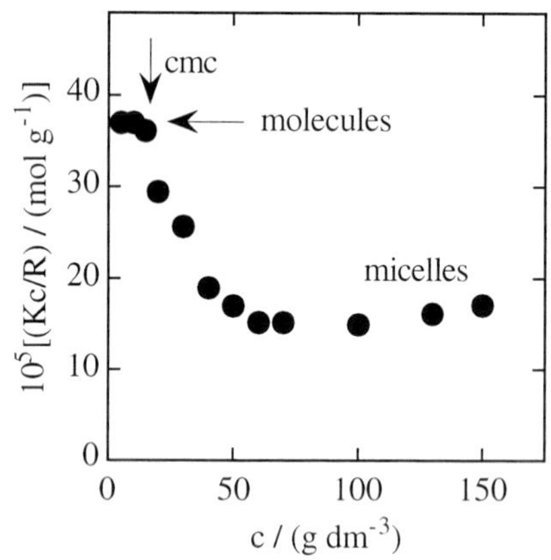

Figure 4. Debye function *versus* concentration for aqueous solutions of $B_4E_{40}B_4$ at 50 °C. The cmc is 16 g dm^{-3}. Adapted from [20].

An alternative plot of the SLS data (shown in Figure 4) is based on the Debye equation

$$\frac{Kc}{R} = \frac{1}{M_w} + 2A_2c + \ldots\ldots \qquad (1)$$

where K accounts for the optical properties of the system, and A_2 and higher terms account for intermicellar interaction (i.e. the structure factor). The cmc is located at the concentration where the Debye function departs from its value for unassociated molecules (unimers).

Measurement of the temperature dependence of excess scattering is a reliable method for determination of the cmt, which is detected as the point at which the scattering curve leaves the baseline established for unimers: see Figure 5. Anomalous scattering/temperature curves may be found: e.g., for copolymer $E_{21}B_8E_{21}$ [20], as shown in Figure 6. This effect was reported by Zhou and Chu [57] for copolymer $E_{13}P_{30}E_{13}$ (Pluronic L64), and prior to that in non-aqueous systems [58]. The intense scattering peaks are caused by phase separation (clouding) of less-soluble species in the poorer solvent at higher temperature, followed by solubilisation in the micelles once they are formed. This explanation places the cmt at the peak of the curve [20]. Impurities in the copolymers can also cause the effect, and these can often be removed by solvent extraction [59] or precipitation fractionation [60]. For purified copolymers, the effect is most likely to be seen for samples with short E blocks initiated by B (or P) oxyanions, where it is caused by the wide E-block-length distributions: see Section 3.

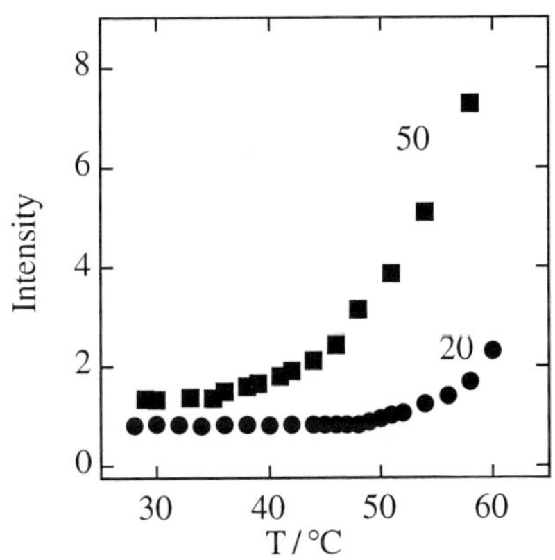

Figure 5. Intensity of light scattering *versus* temperature for aqueous solutions of $B_4E_{40}B_4$ with concentration 20 and 50 g dm^{-3} (as indicated). Adapted from [20].

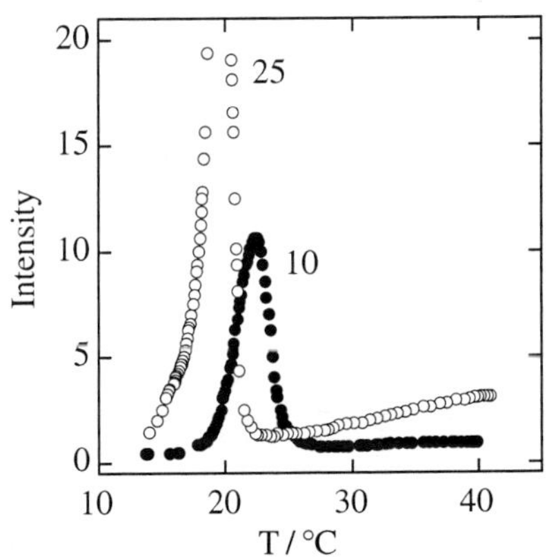

Figure 6. Intensity of light scattering *versus* temperature for aqueous solutions of $E_{21}B_8E_{21}$ with concentrations 10 and 25 g dm^{-3} (as indicated). Adapted from [20].

Because of the high hydrophobicity of a B unit, block lengths in E/B copolymers are usually short, the micelles are small, and extrapolation of light scattering data to zero angle is not usually required. Consequently, useful quantities are available from simple experiments provided that the scattering function can be extrapolated to zero micelle concentration. If that is possible, then the mass-average molar mass (M_w) of the micelles can be extracted and thereby (given M_w of the copolymer itself) the mass-average association number (N_w). If micellar dissociation is important in dilute solution the extrapolation may not be straightforward. Figures 7 and 8 show Debye plots for copolymers with different cmcs. Copolymer $E_{38}B_{12}$ has a very low cmc (0.03 g dm^{-3} in water at 30 °C [23]), and the customary linear extrapolation across the dilute solution range to zero micelle concentration can be made without significant error (Figure 7). Copolymer $E_{21}B_{11}E_{21}$ has a higher cmc (0.3 g dm^{-3} in water at 30 °C [23]) and micellar dissociation at low concentrations is evident from the turn-up in the Debye plot. Fitting data over a wide concentration range has proved to be a reliable method for treating scattering from such solutions. The curves drawn in Figure 8 are based on scattering theory for hard spheres [61]. The theoretically-based curves are seen to fit the data satisfactorily over a wide concentration range. The same curves are reproduced for low concentrations in Figure 7, where the unfilled points for copolymer $E_{21}B_{11}E_{21}$ are corrected for dissociation using c' = (c-cmc). In fact the data approximate the curve (effectively a straight line over the limited concentration range) if c' = (c-0.4) g dm^{-3} [23]. Using data obtained over a wide concentration range (as in Figure 8) gives useful guidance in the extrapolation to zero concentration.

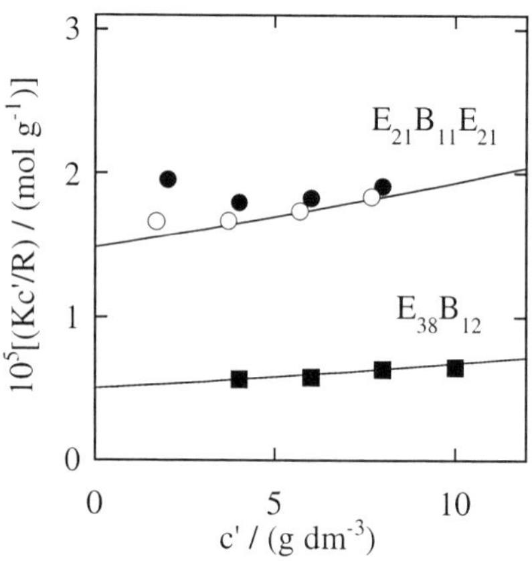

Figure 7. Debye plots for aqueous solutions of E/B copolymers at 30 °C. The curves are taken from Figure 8. The filled symbols represent data plotted using the copolymer concentration (i.e. c' = c). The unfilled symbols represent data plotted using c' = (c-cmc). Adapted from [23].

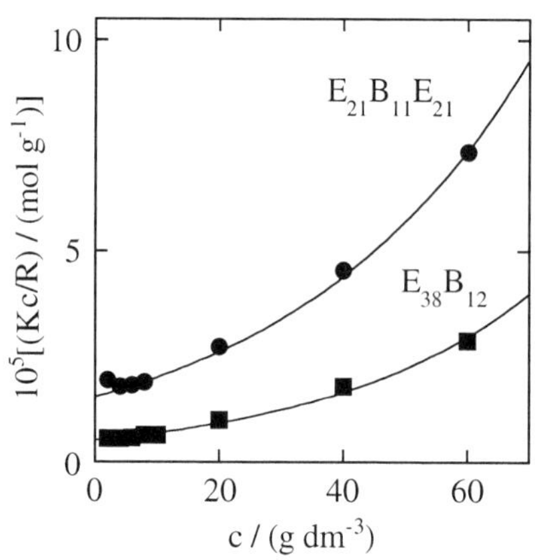

Figure 8. Debye plot for aqueous solutions of E/B copolymers at 30 °C. The curves are calculated assuming that the structure factor (S) for hard spheres applies, and using the Carnahan-Starling equation to calculate S. Adapted from [23].

Whatever concentration range is used for the Debye plot, the concentration dependence of excess scattering relates to the micelle-micelle interaction, i.e. to the average excluded volume of the micelles, and thereby to an effective hard-sphere volume equal to one-eighth of the excluded volume. The limiting slope of the curve at c = 0 gives the second virial coefficient, whence the excluded volume u ($A_2 = N_a u/2M^2$, N_a = Avogadro's constant) [62]. Since exclusion is a thermodynamic effect, this volume has been called the thermodynamic volume (v_t) to distinguish it from the hydrodynamic volume (v_h). The corresponding radius is the thermodynamic radius (r_t).

Strictly, the micellar parameters obtained from SLS are correct only if the refractive indices of the copolymer components are equal. In fact, the refractive indices of E and B blocks are very similar, i.e. within the experimental error of determination under normal circumstances, and the true molar mass is obtained [36,63]. Also, the analyses depend upon the micelle molar mass being independent of concentration at any given temperature. There is good evidence that this is so for $E_mP_nE_m$ copolymer micelles [64,65], and there is no reason to suppose different behaviour for similar E/B copolymer micelles.

4.1.3 Pulsed-gradient spin-echo NMR (PGSE-NMR)

DLS gives the translational mutual-diffusion coefficient of micelles in solution. The self-diffusion coefficient is available from PGSE-NMR. The concentration dependences of the two diffusion coefficients differ (see Figure 9), but they are identical in the limit of infinite dilution [66].

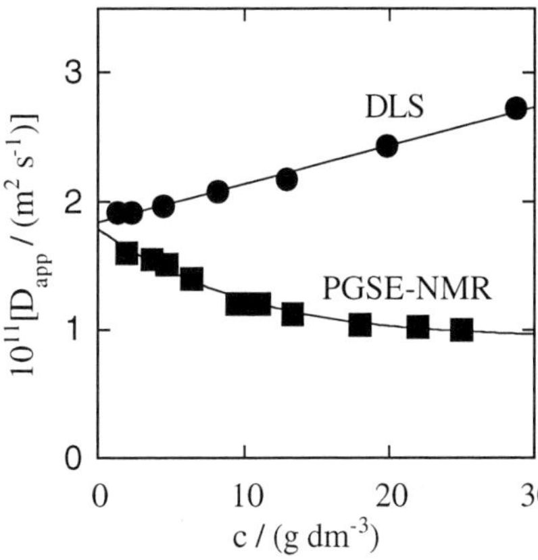

Figure 9. Apparent translational diffusion coefficient *versus* concentration for solutions of *cyclo*-$B_{27}E_{144}$. Results from DLS (25 °C) and PGSE-NMR (22 °C). Adapted from [31].

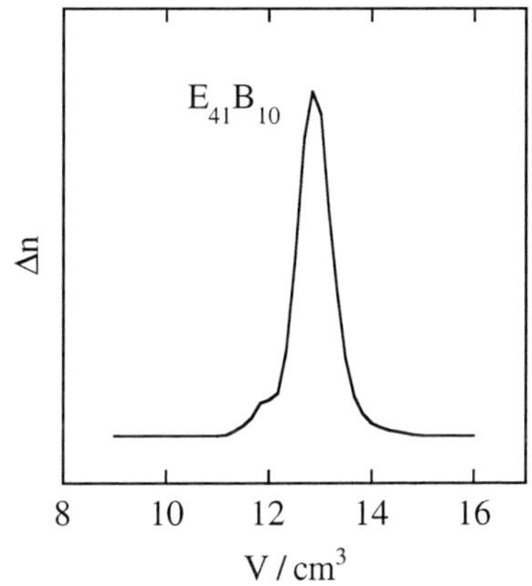

Figure 10. EGPC curve for a 0.5 g dm^{-3} aqueous solution of $E_{40}B_{10}$ at 37 °C. Through the universal calibration, the elution volume at the peak gives $r_{h,app}$ = 7 nm.

4.1.4 Gel permeation chromatography (GPC)

The apparent micellar hydrodynamic volume (hence $r_{h,app}$) can be determined by aqueous GPC. Calibration is through the 'universal calibration' [67]. It is necessary to avoid complete dissociation of the micelles as the solution is diluted in its passage through the GPC columns. One method, which is successful for copolymers with low cmcs, is to ensure that the concentration of copolymer in the injected solution is high enough that some part of the emerging solution remains above its cmc [13,15,17,68]. A second method, which is useful for any copolymer, is eluent GPC (EGPC) in which a micellar solution is used as eluent, and is probed by injecting a solution of the same copolymer at different concentration [69,70]. EGPC gives the distribution of hydrodynamic volume in the solute, similar to that available from DLS but mass-weighted rather than intensity-weighted. An EGPC curve of copolymer $E_{40}B_{10}$ in aqueous solution is shown in Figure 10: the hydrodynamic radius obtained (7 nm) is in good agreement with results from DLS [21]. EGPC has also been used to determine the cmt of $E_mP_nE_m$ copolymers [70,71].

4.1.5 Surface tension (ST)

A familiar method for determination of cmc is measurement of the surface tension (γ) of aqueous solutions over a wide range of concentration. The method detects completion of the Gibbs monolayer at the air/water interface, and is a secondary indicator of the onset of micellisation. Examples are shown in Figure 11. A fairly sharp break in γ *versus* c followed by a constant value of γ is typical of block copolymers with narrow block-length distributions. A minimum in γ indicates a surface active impurity. Fitting data by two straight lines, as is appropriate for a pure uniform solute, may be misleading.

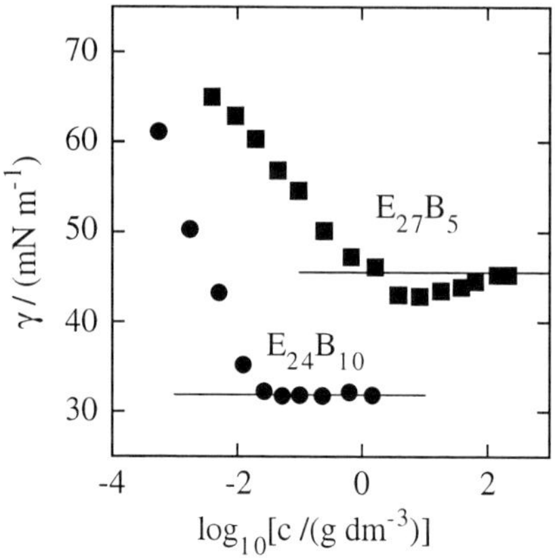

Figure 11. Surface tension (γ) *versus* log(c) for aqueous solutions of diblock copolymers. For clarity, the data for $E_{27}B_5$ are displaced upwards by 10 mN m^{-1}. The cmc is either at the point where γ reaches its steady value ($E_{24}B_{10}$) or is at the minimum in γ ($E_{27}B_5$). Adapted from [17].

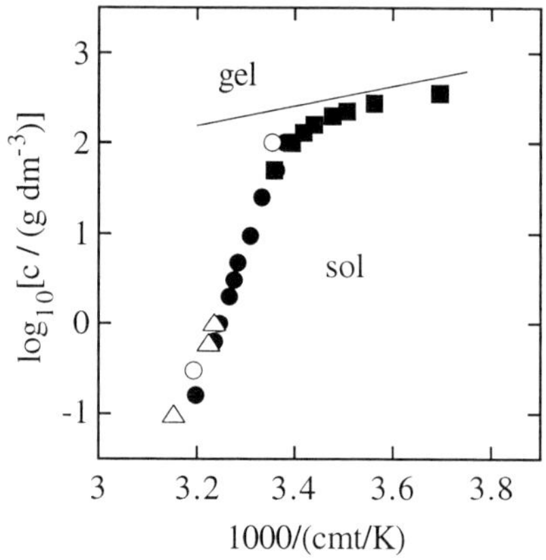

Figure 12. Logarithm of concentration *versus* reciprocal cmt for aqueous solutions of $E_mP_nE_m$ copolymer F127: (●) SLS; (■) DSC; (○) ST; (△) EGPC. Adapted from [60,70].

4.1.6 Differential scanning calorimetry (DSC)

Since micellisation is an endothermic event, DSC can be used to detect its onset (so defining the cmt) and to obtain a direct measure of the enthalpy of micellisation ($\Delta_{mic}H$). Note that $\Delta_{mic}H$ is not equivalent to the standard enthalpy of micellisation ($\Delta_{mic}H^o$) as obtained, e.g., from the temperature dependence of the cmc [see, e.g., 72]. Such experiments have not been reported for E/B copolymers. Results obtained by DSC and other methods for a fraction of copolymer F127 ($E_{93}P_{44}E_{93}$) are shown in Figure12 (plot of log(c) *versus* 1/cmt). The results emphasise the particular value of the DSC method for investigating micellisation in relatively concentrated solutions, including those approaching the critical gelation concentration (see Section 5) where other techniques break down, and where the standard enthalpy of micellisation can be much smaller than that at lower concentrations as a result of modification of the water structure [60].

4.1.7 Comparison of methods

The results plotted in Figure 12 serve to indicate the good agreement found between various methods of determination of cmc and cmt. Results from different laboratories may show wide variation [60], probably as a result of variation in samples of ostensibly the same composition. A more recent comparison, including results from two laboratories, illustrates this point [71].

4.2 Critical micelle concentrations and micelle characteristics

Critical micelle concentrations are listed in Table 2. Values are given for one temperature only. Other results, to be found in the papers cited, show the expected decrease in cmc with increase in temperature (see Section 4.3). Micellar association numbers and radii are similarly listed in Table 3. The effect of B-block length in determining the cmc and N_w is apparent. The effect of E-block length is not so well defined by the available data: a small increase in cmc and decrease in N_w with increase in E-block length would be consistent with the behaviour of E/P [36] and E/C [73] copolymers in aqueous solution.

Table 2
Critical micellisation concentrations (cmc) for aqueous solutions of E/B copolymers[a]

Copolymer	T / °C	cmc / g dm^{-3}	Method	Reference
$E_{50}B_4$	40	45	ST	17
$E_{28}B_5$	30	6	SLS-cmt	19
$E_{27}B_5$	40	8	ST	17
$E_{27}B_7$	30	6	SLS-cmt	19
$E_{41}B_8$	40	0.11	ST	33
$E_{49}B_8$	40	0.3	ST	17
$E_{18}B_9$	25	0.035	ST	40
$E_{24}B_{10}$	40	0.023	ST	17
$E_{50}B_{13}$	20	0.03	ST	17
B_8E_{41}[b]	40	0.21	ST	33
B_9E_{16}	25	0.018	ST	40
$B_{35}E_{135}$	26	0.07	ST	13
cyclo-B_8E_{42}	40	2.6	ST	30
cyclo-$B_{27}E_{144}$	25	0.005	SLS-cmc	31
$E_{21}B_8E_{21}$	40	2	SLS-cmt	20
$E_{33}B_{10}E_{33}$	25	5.9	SLS-cmt	27,40
$E_{13}B_{10}E_{13}$	40	0.08	SLS-cmt	27,40
$E_{21}B_{11}E_{21}$	32	0.2	SLS-cmt	23
$E_{43}B_{14}E_{43}$	25	0.43	SLS-cmt	27,40
$E_{40}B_{15}E_{40}$	21	0.20	SLS-cmc	16
$E_{58}B_{17}E_{58}$	30	0.30	ST	14
$E_{72}B_{27}E_{72}$	25	0.01	ST	31
$E_{71}B_{28}E_{71}$[c]	30	0.16	SLS-cmc	15
$E_{132}B_{53}E_{132}$[c]	30	<0.02	SLS-cmc	15
$B_4E_{40}B_4$	40	30	ST	20,24
$B_5E_{39}B_5$	40	4.3	SLS-cmt	24
$B_5E_{91}B_5$	35	4.4	SLS-cmc	35
$B_7E_{40}B_7$	20	1.0	SLS-cmc	24
$B_{12}E_{260}B_{12}$	20	0.040	SLS-cmc	29

(a) ST: surface tension; SLS-cmc: extrapolation of static light scattering data to zero scattering from micelles; SLS-cmt: detection of light scattering from micelles on heating a solution.
(b) B_8E_{41} was denoted $E_{41}B_8$ in reference 20.
(c) Impure sample: see [15].

Table 3
Mass-average association numbers (N_w), thermodynamic radii (r_t) and hydrodynamic radii (r_h) for E/B copolymers in aqueous solution[(a)]

Copolymer	T / °C	N_w	r_t / nm	r_h / nm	Method	Reference
$E_{28}B_5$	30	10	2.0	(5)	SLS/DLS	19
$E_{27}B_7$	30	16	2.5	–	SLS/DLS	19
$E_{30}B_8$	30	60	5.1	(7)	SLS/DLS	19
$E_{41}B_8$	35	31	4.3	7.2	SLS/DLS	33
$E_{49}B_8$	30	18	3.8	–	SLS/DLS	17
$E_{18}B_9$	25	70	4.4	6.3	SLS/DLS	27
$E_{40}B_{10}$	30	37	4.9	(7)	SLS/DLS	21
$E_{32}B_{10}$	30	60	5.2	(7)	SLS/DLS	19
$E_{24}B_{10}$	30	70	4.9	6.1, 8	SLS/DLS, GPC	17
$E_{38}B_{12}$	30	74	6.7	8.3	SLS/DLS	23
$E_{50}B_{13}$	30	57	6.6	9.4, 12	SLS/DLS, GPC	17
$E_{29}B_{14}$	30	140	7.2	–	SLS	19
$E_{27}B_{16}$	30	–	–	(9)	DLS	19
B_8E_{41}[(b)]	35	78	6.5	(10)	SLS/DLS	20,25,33
$B_{35}E_{135}$	26	300	–	23	SLS/DLS, GPC	13
cyclo-B_8E_{42}	45	16	2.8	4.4	SLS/DLS	30
cyclo-$B_{27}E_{144}$	25	60	10.8	13.3	SLS/DLS	31
	22	–	–	13.2	PGSE-NMR	31
$E_{21}B_8E_{21}$	50	5.8	2.2	(3.5)	SLS/DLS	20
$E_{33}B_{10}E_{33}$	40	8	3.2	5.1	SLS/DLS	27
$E_{13}B_{10}E_{13}$	40	23	3.0	5.2	SLS/DLS	27
$E_{21}B_{11}E_{21}$	40	33	5.1	6.2	SLS/DLS	23
$E_{43}B_{14}E_{43}$	40	16	4.7	6.6	SLS/DLS	27
$E_{40}B_{15}E_{40}$	37	18	4.7	6.7	SLS/DLS	16
$E_{58}B_{17}E_{58}$	30	13	5.6	6.9	SLS/DLS	14
$E_{72}B_{27}E_{72}$	25	48	10.3	11.6	SLS/DLS	31
	22	–	–	12.2	PGSE-NMR	31
$E_{71}B_{28}E_{71}$[(c)]	30	41	10.8	10, 11	SLS/DLS, GPC	15
$E_{132}B_{53}E_{132}$[(c)]	30	400	–	53	SLS/DLS, GPC	15
$B_4E_{40}B_4$	50	11	–	6.6	SLS/DLS	24
$B_5E_{39}B_5$	40	19	–	7.4	SLS/DLS	24
$B_5E_{91}B_5$	45	14	–	(8)	SLS/DLS	35
$B_6E_{46}B_6$	25	–	–	(6)	DLS	26
$B_7E_{40}B_7$	20	18	–	6.7	SLS/DLS	24
$B_{12}E_{260}B_{12}$	20	24	–	12.4	SLS/DLS	29

(a) SLS, static light scattering; DLS, dynamic light scattering; GPC, gel permeation chromatography. Values in parentheses are $r_{h,app}$ determined for dilute solution.
(b) B_8E_{41} was denoted $E_{41}B_8$ in reference 20.
(c) Impure sample: see [15].

The effect of temperature on micelle characteristics is illustrated for selected systems in Table 4. The results show the expected increase in N_w with increase in temperature, i.e. as the solvent becomes poorer. Hydrodynamic radii are not very temperature dependent, which has been explained as a compensation between an increase in N_w and a decrease in swelling of the E-block fringe [74]. Thermodynamic radii usually increase slightly with temperature, but may decrease, as shown by the results for $E_{40}B_{10}$ in aqueous K_2SO_4: the exact behaviour depends on the relative importance of the increase in N_w and the decrease in solvent quality.

Table 4
Effect of temperature on micellar parameters[(a)]

	T / °C	N_w	r_t / nm	r_h / nm
$E_{40}B_{10}$ in water	20	23	4.0	8.5
	30	37	4.9	7.3
	45	49	5.5	7.3
$E_{40}B_{10}$ in 0.4 mol dm^{-3} K_2SO_4	20	45	5.0	7.8
	30	55	4.9	7.5
	45	80	4.6	7.6
$E_{38}B_{12}$ in water	30	74	6.7	8.3
	40	82	6.9	8.7
	50	89	7.0	8.7
$E_{21}B_{11}E_{21}$ in water	30	24	4.5	5.8
	40	33	5.1	6.2
	50	43	5.6	6.4

(a) Data from [21,23].

4.3 Thermodynamics of micellisation

If the association number (N) is single-valued (or the average of a narrow distribution of values), then the association equilibrium can be written

$$A \leftrightarrows \frac{1}{N} A_N \qquad (2)$$

and, given mass action, the equilibrium constant is

$$K = [A_N]_{eq}^{1/N}/[A]_{eq} \qquad (3)$$

If the association number is large, then the equilibrium constant $K \approx 1/[A]_{eq}$ and, taking $[A]_{eq}$ to be the cmc in mol dm^{-3}, the standard Gibbs energy and enthalpy of micellisation are given by:

$$\Delta_{mic}G^o = -RT\ln K \approx RT\ln(cmc) \quad \text{and} \quad \Delta_{mic}H^o \approx R\,[d\ln(cmc)/d(1/T)] \qquad (4)$$

The process referred to is copolymer chains in their standard state of ideally dilute solution at unit concentration (1 mol dm^{-3}) going to copolymer chains in their micellar state. The equation for $\Delta_{mic}H^o$ requires that N is independent of temperature.

The requirement that N is large and independent of temperature is restrictive, and creates a problem in applying equations (2) to (4) to E/B and E/P copolymer systems. Hall [75] has indicated that N should be larger than 50. Results summarised in Table 3 for E/B copolymers, and elsewhere for E/P copolymers [36], show that this condition is not always met. Nor is N independent of T: see Table 4. Consequently, care must be taken in relating the standard thermodynamic quantities to equilibrium (2) and its temperature dependence.

Exact equations can be written down for K [see, e.g. 20,75]. If the advancement of the equilibrium (α) is small at the cmc (e.g. 0.01), a good approximation is

$$K \approx [\frac{\alpha}{N}]^{1/N} (\text{cmc})^{(-1+1/N)} \qquad (\text{cmc in mol dm}^{-3}) \qquad (5)$$

The usual equation ($N \rightarrow \infty$) is $K_{app} \approx (\text{cmc})^{-1}$. Numerical calculation can be used to illustrate the magnitude of the difference between K and K_{app} resulting from small values of N. If cmc = 10^{-4} mol dm^{-3} at 300 K and N = 10, then $K \approx 2\times10^3$ and $K_{app} = 10^4$ (80% error), with corresponding values of $\Delta_{mic}G^o$ = −19 and −23 kJ mol^{-1} (20% error). Similar estimations of the error in $\Delta_{mic}H^o$, which requires estimation of the change in cmc and N with T, suggest errors of comparable magnitude. Of course, the value of $\Delta_{mic}H^o$ obtained by the usual analysis correctly describes the temperature dependence of the cmc but, except in favourable circumstances, does not correctly describe the temperature dependence of K.

Figure 13 provides an example of a plot of log(c) against 1/cmt, (equivalent to log(cmc) against 1/T) for solutions of $E_{21}B_8E_{21}$. Over the temperature range investigated (10-40 °C), N < 6 and is temperature sensitive [20]. Clearly the thermodynamic parameters obtained for this copolymer must be treated with caution.

Values of the thermodynamic parameters of micellisation obtained *via* eqn. (1) for E/B copolymers are listed in Table 5. The results serve to emphasise that the micellisation of E/B copolymers in water is entropy-driven, in keeping with the hydrophobic effect [76].

Table 5
Standard Gibbs energies, enthalpies and entropies of micellisation for aqueous solutions of E/B copolymers.

Copolymer	T / °C	$\Delta_{mic}G^o$ kJ mol^{-1}	$\Delta_{mic}H^{o(a)}$ kJ mol^{-1}	$T\Delta_{mic}S^o$ kJ mol^{-1}	reference
$E_{28}B_5$	40	−17	80 (16)	97	19
$E_{27}B_7$	40	−17	80 (11)	97	19
$E_{41}B_8$	40	−26	78 (10)	104	33
$E_{24}B_{10}$	40	−29	30 (3)	59	17
B_8E_{41}	40	−23	70 (9)	93	33
cyclo-B_8E_{42}	40	−18	55 (7)	73	30
$E_{21}B_8E_{21}$	40	−17	95 (12)	112	20
$E_{13}B_{10}E_{13}$	25	−19	119 (12)	138	27
$E_{33}B_{10}E_{33}$	25	−16	87 (9)	103	27
$E_{43}B_{14}E_{43}$	25	−23	125 (9)	148	27
$E_{40}B_{15}E_{40}$	37	−27	24 (2)	51	16
$E_{58}B_{17}E_{58}$	40	−27	34 (2)	61	14
$B_4E_{40}B_4$	40	−11	67 (8)	78	20,23
$B_5E_{39}B_5$	40	−20	114 (11)	134	23
$B_5E_{91}B_5$	35	−18	81 (8)	99	35
$B_7E_{40}B_7$	20	−19	170 (12)	189	23
$B_{12}E_{260}B_{12}$	20	−31	25 (1)	56	29

(a) Values in parentheses are $\Delta_{mic}H^o$ per B unit.

Of interest in Table 5 are the low values of $\Delta_{mic}H^o$ found for the longer B blocks of each architecture, i.e. 1-3 kJ (mol of B units)$^{-1}$ compared with 7-16 for the shorter blocks. This effect, which is outside possible error, has been attributed to the long blocks (especially those in the E_mB_n and $B_nE_mB_n$ architectures) being in a tightly-coiled conformation, so that interaction with water *via* hydrophobic bonding is reduced compared to that for the shorter blocks [17,19]. A copolymer with its hydrophobic block in such a tightly-coiled state has been called a 'monomolecular micelle'.

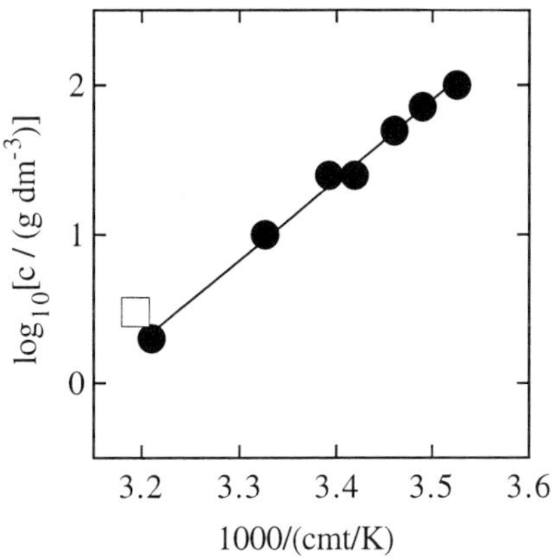

Figure 13. Logarithm of concentration *versus* reciprocal cmt for aqueous solutions of triblock copolymer $E_{21}B_8E_{21}$. The unfilled data point is log(cmc) from surface tension plotted against reciprocal temperature. Adapted from [20]

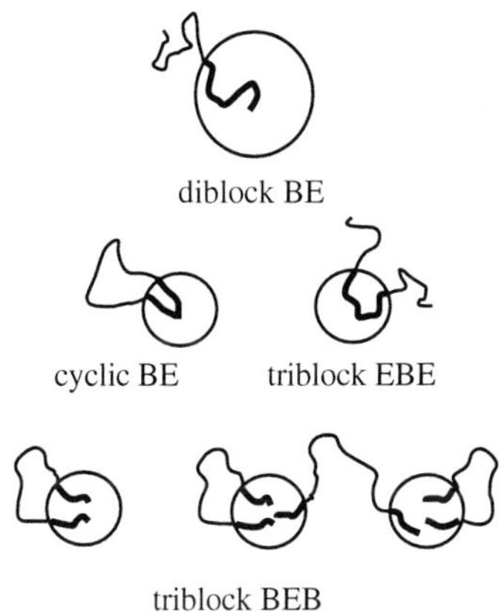

Figure 14. Schematic representation of chain conformation in micelles formed from linear diblock copolymer, cyclic diblock copolymer, and linear triblock copolymers (as indicated).

4.4 Effects of block composition and architecture

The results available for E/B copolymers allow a broader examination of the effect of chain architecture on self-assembly than is possible in other systems, e.g. E/P copolymers [36]. There are some complications in making comparisons (discussed below), and there are refinements in synthesis to take in hand, but much of the picture is in place. A general framework for discussion is presented in Figure 14. The scheme assumes copolymers containing the same numbers of E and B units. Note that only the $B_nE_mB_n$ copolymers will form linked micelles, and eventually micelle clusters.

Compared to diblock copolymers, the triblock and the cyclic copolymers are entropically disfavoured in the micellar state, the restriction being that two block junctions per molecule must be located in the core/fringe interface. Comparing at constant composition and chain length (as assumed in Figure 14) this means that the cmc of a linear diblock copolymer should be significantly lower than those of other architectures. Additionally, the maximum possible radius of spherical micelles formed from a linear diblock copolymer will be twice (or more) that of micelles formed from copolymers of the other architectures, i.e. the association number and radius of the linear diblock copolymer will be larger: N_w by $2^3 = 8$ and r_h (or r_t) by 2. A cyclic copolymer in a micelle suffers the greatest restriction of all, since both of its blocks must loop, whereas only one block of the linear triblock copolymers must loop, either in the core ($E_mB_nE_m$) or the fringe ($B_{n/2}E_mB_{n/2}$). However, because the cyclic copolymer is restricted in both unassociated (solution) and micellar states, the additional contribution to the Gibbs energy of micellisation from this effect is likely to be small. As described below, these simple ideas prove to be broadly correct. A more complete description is available from Linse's theoretical work [77].

4.4.1 Effect of copolymerisation sequence

Figures 15 and 16 show results for copolymers $E_{41}B_8$ and B_8E_{41} prepared by copolymerisation in the sequence indicated. As described in Section 3, the effect of polymerising the E block last (B_8E_{41}) is to broaden the E-block-length distribution, otherwise the block-length distributions are narrow. This is clearly seen in the GPC curves shown in Figure 15. An example of the effect on surface properties, taken from [33], is shown in Figure 16, where plots of surface tension versus concentration show a sharp break defining the cmc for copolymer $E_{41}B_8$, but a gradual approach to constant γ for copolymer B_8E_{41}. The lower surface tensions of B_8E_{41} also result from its wide E-block-length distribution. $E_mB_nE_m$ copolymers are prepared starting with the difunctional B block, and are affected in the same way [24,27]. Similar, but less severe, effects would be expected in the properties of $E_mP_nE_m$ copolymers.

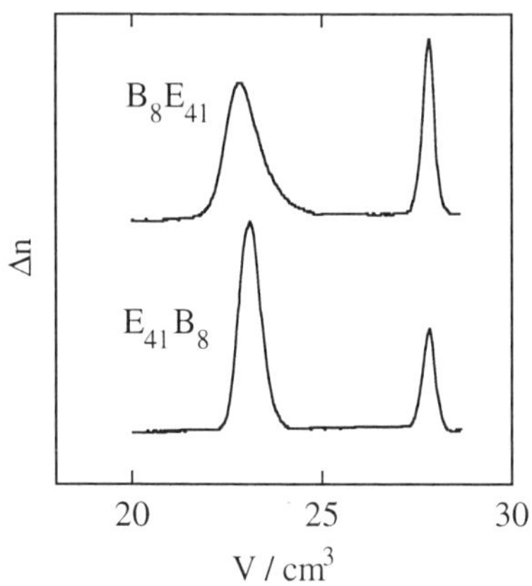

Figure 15. GPC curves for $E_{41}B_8$ (M_w/M_n = 1.04) and B_8E_{41} (M_w/M_n = 1.14) in THF eluent. Adapted from [33].

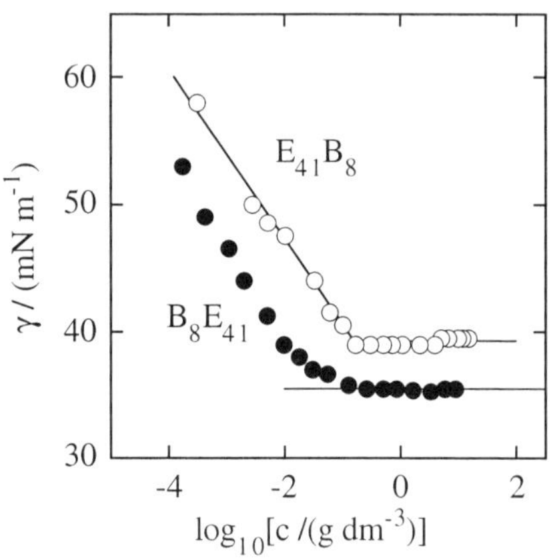

Figure 16. Surface tension *versus* concentration for aqueous solutions of $E_{41}B_8$ and B_8E_{41} at 30 °C. Adapted from [33].

4.4.2 Effect of block architecture

Log(cmc) is plotted against the number of B units in the copolymer (n) in Figure 17, and association number (N_w) is similarly plotted in Figure 18. Values are for solutions at 30 °C, and are restricted to copolymers with compositions in the range 60-90 wt-% E. It is assumed that the effect of E-block length is small compared with that of B-block length.

Critical micelle concentration. The data for E_mB_n and $B_{n/2}E_mB_{n/2}$ copolymers can be fitted by straight lines. The slopes for the two data sets are similar: –0.44 (E_mB_n) and –0.48 ($B_{n/2}E_mB_{n/2}$), corresponding to an increment in $\Delta_{mic}G^o$ of approximately 2.7 kJ (mol B units)$^{-1}$, i.e. similar to that found per methylene group for E_mC_n block copolymers [73] (see also Section 4.6). The displacement between the two straight lines of approximately two decades at given n reflects the entropy penalty described above. The plot for $E_{m/2}B_nE_{m/2}$ copolymers is complicated by differences in breadth of the E-block-length distribution.

Points at B ≤ 10 represent copolymers with relatively small E-block lengths (average m ≤ 40), and so have a proportion of chains with very short E blocks (diblock character), hence a lower cmc than expected. This effect makes it difficult to place $E_{m/2}B_nE_{m/2}$ copolymers precisely in a ranking order: the results available so far suggest

$$E_mB_n << B_{n/2}E_mB_{n/2} \leq E_{m/2}B_nE_{m/2}$$

It should be useful to synthesise $E_mB_nE_m$ copolymers by a route which produces narrow E-block-length distributions.

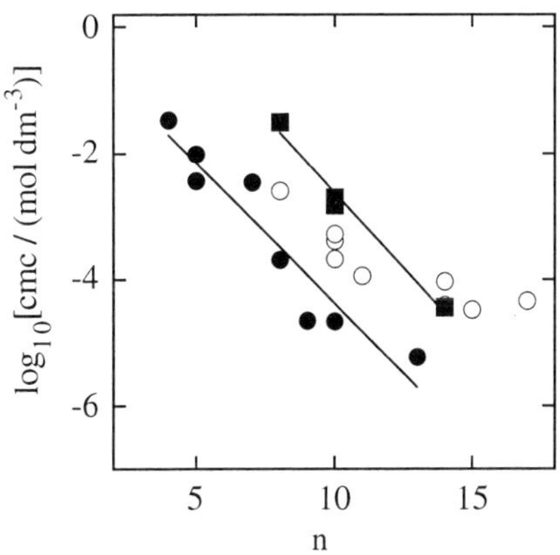

Figure 17. Log(cmc) versus number of B units in the copolymer for aqueous solutions of E/B block copolymers at 30 °C: (●) E_mB_n; (■) $B_{n/2}E_mB_{n/2}$; (○) $E_{m/2}B_nE_{m/2}$. Adapted from [23,24,27,35].

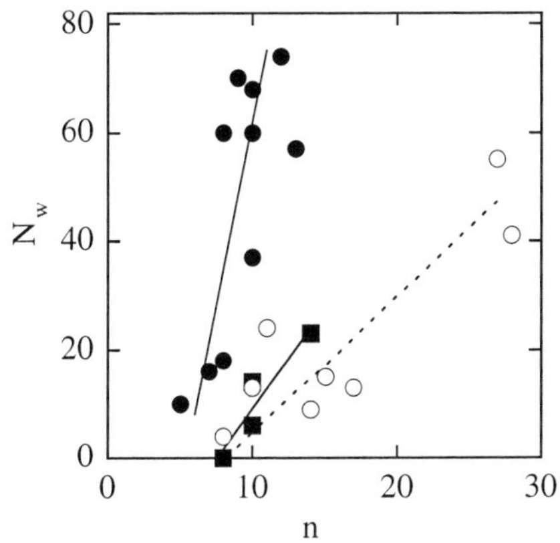

Figure 18. Association number (N_w) versus number of B units in the copolymer for aqueous solutions of E/B block copolymers at 30 °C: (●) E_mB_n; (■) $B_{n/2}E_mB_{n/2}$; (○) $E_{m/2}B_nE_{m/2}$. Adapted from [23,24,27,35].

Association number. As explained above, for well-developed micelles the major determinant of size is the geometrical restriction on the size of the core, i.e. the core radius must be proportional to n for a diblock copolymer, and n/2 for either type of triblock copolymer. Since N_w directly reflects core volume then, for a given value of n, values of N_w would be expected to be in ratio 8:1 for diblock *versus* triblock. In Figure 18 the lines drawn to smooth the results show that at n = 10, $N_w \approx 60$ for the diblocks, and N_w = 6-9 for the triblocks, i.e. in reasonable agreement with expectation. The association numbers found for $E_{m/2}B_nE_{m/2}$ copolymers with n ≤ 10 are higher than might be expected, which reflects their wide E-block distributions.

Hydrodynamic radius. The core in E/B block copolymer micelles is very small, hence hydrodynamic radii depend largely on the size of the fringe, i.e. on E-block length. The increase in r_h with increase in B-block length seen in Table 3 is attributed to the fact that longer E-blocks accompany longer B blocks in the present range of samples. This was necessary in order to maintain solubility over a reasonable temperature range.

4.4.3 Effect of cyclisation

The self-association behaviours of two cyclic E/B copolymers have been investigated: *cyclo*-B_8E_{42} (prepared from $E_{21}B_8E_{21}$) and *cyclo*-$B_{27}E_{144}$ (prepared from $E_{72}B_{27}E_{72}$). The linking reaction to form the cyclic copolymer couples the E blocks, and so narrows the E-block-length distribution somewhat [78]. Results taken from Tables 2–4 are set out in Table 6 for ease of comparison. The values of $\Delta_{mic}H^o$ for the B_8 copolymers cannot be closely compared, since the association number of $E_{21}B_8E_{21}$ is very low. Taking all results into account, it is concluded that micellisation of a cyclic diblock copolymer is favoured over that of its corresponding linear triblock, the equality of cmcs for the B_8 copolymers being explicable as an effect of E-block-length distribution [30,31].

Table 6
Comparison of cyclic and triblock E/B copolymers in aqueous solution[(a)]

Property	T / °C	*c*-B_8E_{42}	$E_{21}B_8E_{21}$	T / °C	*c*-$B_{27}E_{144}$	$E_{72}B_{27}E_{144}$
cmc / g dm^{-3}	40	3	3	25	0.005	0.01
$\Delta_{mic}H^o$ / kJ mol^{-1}		55	95		—	—
N_w	50	16	6	40	62	49
r_t / nm	50	2.9	2.2	40	10.4	10.0
r_h / nm	50	4.0	4.4	40	12.8	11.3

(a) Data from [20,30,31]

4.4.4 Micellar linking

As discussed above, micellar linking is possible with $B_nE_mB_n$ copolymers. This is illustrated in Figure 19, which shows intensity-fraction distributions of log($r_{h,app}$) from DLS for aqueous solutions of diblock copolymer B_8E_{41} alone and mixed (50 wt-%) with triblock copolymer $B_{12}E_{76}B_{12}$. The second peak is attributable to linked clusters of (on average) 20-30 micelles. The clusters are transient in nature, since the whole solute is in dynamic equilibrium between

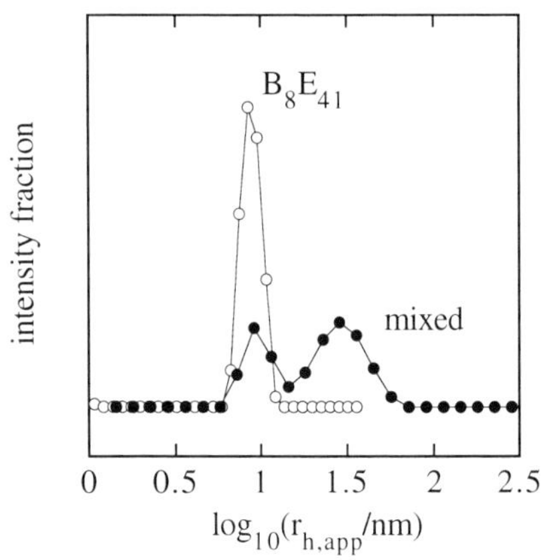

Figure 19. Intensity fraction *versus* logarithm of $r_{h,app}$ for (●) B_8E_{41} (50 g dm^{-3}) and (○) 50:50 wt-% mixture of $B_8E_{41}$and $B_{12}E_{76}B_{12}$ (38 g dm^{-3}) in aqueous solution at 25 °C. Adapted from [25].

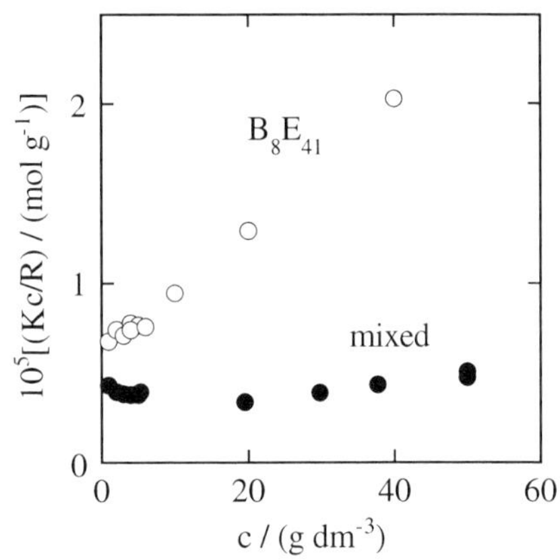

Figure 20. Debye plot for (○) B_8E_{41} (50 g dm^{-3}) and (●) 50:50 wt-% mixture of B_8E_{41} and $B_{12}E_{76}B_{12}$ (38 g dm^{-3}) in aqueous solution at 25 °C. Adapted from [25].

molecules, micelles and clusters. The effect of micellar linking on the Debye plot is shown in Figure 20: micelles of B_8E_{41} alone behave as hard spheres with a large positive value of A_2 (positive initial slope), whereas the mixed micelles behave as 'sticky' spheres with a negative value of A_2 (negative initial slope). The attractive force arises from the reduction in Gibbs energy on linking.

4.5 Effect of salt

Effects of adding inert salt (K_2SO_4) to aqueous solutions of copolymer $E_{40}B_{10}$ are illustrated in Figure 21 (SLS) and 22 (DLS). The most marked effect of adding salt is to reduce the excluded volume of the micelles. The concentration dependence of Kc/R depends directly on the excluded volume [62], while that of D depends on excluded volume plus hydrodynamic parameters [66]. The concentration dependences shown in Figures 21 and 22 are consistent with the E-block fringe approaching the theta condition (zero contribution to excluded volume) as the concentration of salt is increased.

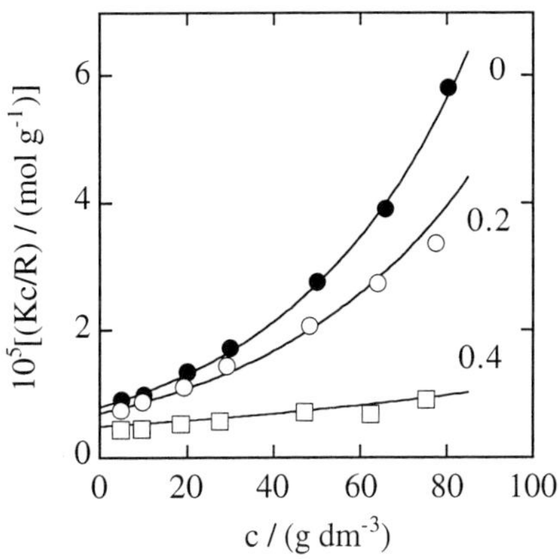

Figure 21. Effect of adding K_2SO_4 at the concentration (mol dm^{-3}) indicated on SLS from aqueous solutions of $E_{40}B_{10}$. Debye function *versus* concentration. T = 45 °C. Adapted from [21].

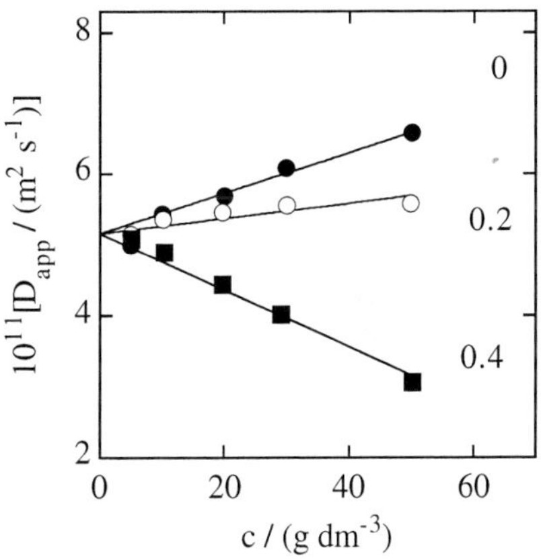

Figure 22. Effect of adding K_2SO_4 at the concentration (mol dm^{-3}) indicated on DLS from aqueous solutions of $E_{40}B_{10}$. Apparent diffusion constant *versus* concentration. T = 45 °C. Adapted from [21].

Selected values of M_w, r_t and r_h are listed in Table 7, together with the thermodynamic expansion factor $\delta_t = v_t/v_a$, where v_a is the 'dry' volume of the micelles. Extrapolation of δ_t (45 °C) to v_c/v_a (where v_c is the core volume of the micelles) indicates that theta conditions would be reached in 0.5 mol dm^{-3} aqueous K_2SO_4 at 45 °C [21]. A similar extrapolation against temperature indicates theta conditions for 0.4 mol dm^{-3} aqueous K_2SO_4 at 60 °C. It is known that 0.45 mol dm^{-3} aqueous K_2SO_4 at 35 °C is a theta solvent for poly(oxyethylene) itself [79,80], and that higher temperatures compensate for lower salt concentrations.

4.6 Comparison with other systems

Many of the effects described in the preceding sections are seen in other systems: e.g. micellar linking in $C_nE_mC_n$ copolymer solutions [81-84] and the effect of salt on micellisation of $E_mP_nE_m$ copolymers [summarised in 36]. The comparisons made in this section are restricted to critical micelle concentration and association number, and to copolymers with hydrophobic blocks of B, P, C and L units [L = DL-lactide, OCOCH(CH_3)]. Data for C_nE_m (i.e. alcohol ethoxylates) and E_mC_n copolymers are included without distinction, since differences in behaviour are known to be small in E/C systems provided that the E-block length is long [73,85].

Table 7
Effect of inert salt (K_2SO_4) on the mass-average association number (N_w), thermodynamic radius (r_t) and hydrodynamic radius (r_h) of copolymer $E_{40}B_{10}$ in aqueous solution [21].

K_2SO_4 conc. mol dm^{-3}	T / °C	r_h / nm	N_w	r_t / nm	δ_t
0	45	7.3	49	5.5	3.5
0.2	45	7.3	55	5.3	2.8
0.32	45	7.4	63	5.0	2.0
0.4	45	7.6	80	4.6	1.3

E_mL_n copolymers are of interest because their L blocks undergo hydrolysis, i.e. are biodegradable. As described below, judged by cmc and ignoring minor refinements, the hydrophobicities for the four units are in approximate ratio

B	C	L	P
4	3	3	1

4.6.1 E_mB_n, C_nE_m and E_mL_n

In Figures 23 and 24, log(cmc) and N_w are plotted against number of hydrophobic units for diblock copolymers E_mB_n, C_nE_m and E_mL_n. In Figure 23, the slopes of the straight lines are similar, indicating that B, C and L groups make similar incremental contributions to the Gibbs energy of micellisation. The results for C_nE_m copolymers can be refined to take account of the fact that the alkyl blocks are uniform, and that the junction CH_2 group of the alkyl block behaves as if it were part of the E block [23].

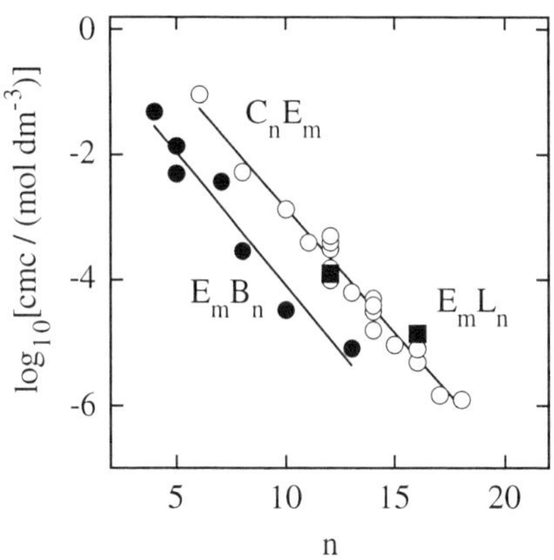

Figure 23. Log(cmc) *versus* hydrophobic-block length for aqueous solutions at 25 °C of (●) E_mB_n, (○) E_mB_n and (■) E_mL_n copolymers. Adapted from [23,85,86].

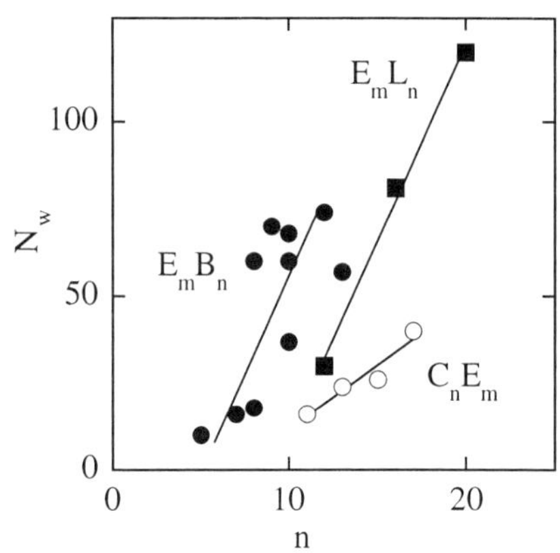

Figure 24. Association number (N_w) *versus* hydrophobic-block length for aqueous solutions of diblock copolymers at 30 °C. Adapted from [23,85,86].

Considering N_w (Figure 24), the major determinant is the length of the hydrophobic block, giving similar slopes for B and L blocks where each unit contributes three backbone atoms, and a lower slope for the E_mC_n copolymers where each unit contributes one backbone atom.

4.6.2 $E_mB_nE_m$ and $E_mP_nE_m$

In Figures 25 and 26, log(cmc) and N_w are plotted against number of hydrophobic units for triblock copolymers $E_mB_nE_m$ and $E_mP_nE_m$. The data for the $E_mP_nE_m$ copolymers are taken from [60,87,88]. The data sets for log(cmc) can be brought into coincidence by plotting those for $E_mB_nE_m$ against 4n [23], i.e. one B unit is equivalent to four P units. The same adjustment also brings the data for N_w into approximate agreement.

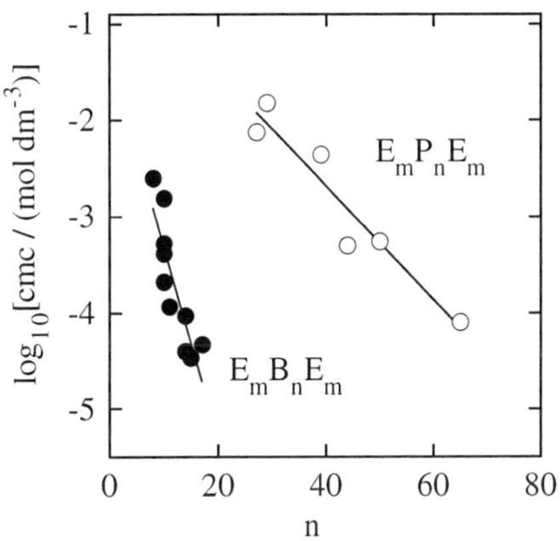

Figure 25. Log(cmc) *versus* hydrophobic-block length for aqueous solutions of (●) $E_mB_nE_m$ and (○) $E_mP_nE_m$ copolymers at 25 °C. Adapted from [23].

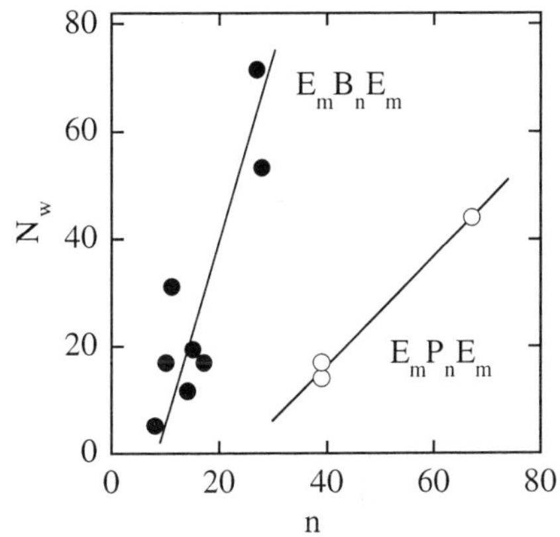

Figure 26. Association number (N_w) *versus* hydrophobic-block length for aqueous solutions of (●) $E_mB_nE_m$ and (○) $E_mP_nE_m$ copolymers at 25 °C. Adapted from [23].

5. Phase behaviour

The phase behaviour established for concentrated aqueous solutions of $E_mP_nE_m$ copolymers [see recent reviews 36, 89] has its counterpart in E/B copolymers. The first phase diagram reported [14] for aqueous solutions of an E/B copolymer ($E_{58}B_{17}E_{58}$) is reproduced in Figure 27. The rigid-gel phase is isotropic under the polarising microscope, and can now be characterised as a cubic liquid-crystal phase formed from spherical micelles of much the same size as those in dilute micellar solution [21]. More recent work on aqueous solutions of E/B copolymers has confirmed hexagonal and lamellar phases [32,34,90]. Considering the cubic phases of E/B and E/P copolymer systems, there is general agreement that the low-temperature boundary results from packing micelles which are formed from unassociated molecules in increasing proportion as the copolymer solution is heated. As the temperature is increased further the solvent becomes progressively poorer and a number of processes occur.

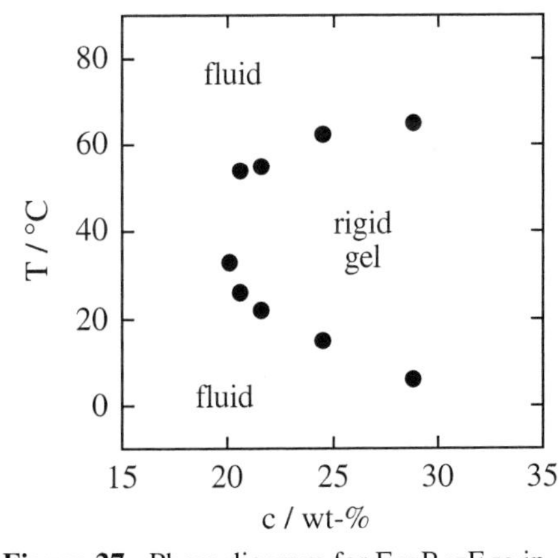

Figure 27. Phase diagram for $E_{58}B_{17}E_{58}$ in water. Adapted from [14].

(i) The molecule-micelle equilibrium shifts further towards the micellar state.
(ii) The association number (N_w) increases.
(iii) The excluded volume of the micelle (i.e. the effective size for packing) eventually decreases as the theta temperature of poly(oxyethylene) is approached.

This last process provides a mechanism for the release of packing constraints and the formation of the fluid at high temperature.

Studies of the micellisation and gelation of copolymer P85 ($E_{27}P_{39}E_{27}$) [64,91] provide a second mechanism for formation of the fluid at the high-temperature boundary, i.e. a transition from spherical to cylindrical micelles, with consequent release of packing constraints.

5.1 Experimental methods

Tube-inversion. The mobile/immobile (sol/gel) transition temperature is readily determined by confining a solution in a tube and inverting the tube from time to time as the temperature is changed. The results shown in Figure 27 were obtained in this way. A more sophisticated version of this test which uses a falling-ball device has been described [92]. The method is sensitive to the yield stress of the gel.

Differential scanning calorimetry. The formation of a rigid-gel phase is a thermal event detectable by DSC, preferably using a high precision instrument [87,93]. A DSC curve obtained for 30 wt-% aqueous solution of $E_{41}B_8$ is shown in Figure 28: the upper and lower sol/gel transitions (cubic phase) are clearly seen.

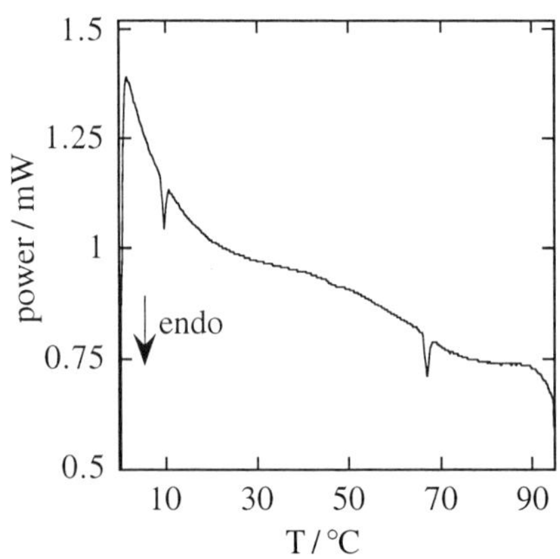

Figure 28. DSC curve for a 30 wt-% aqueous solution of $E_{41}B_8$. The sharp minima correspond to upper and lower gel boundaries (cubic phase) by tube inversion. Adapted from [32].

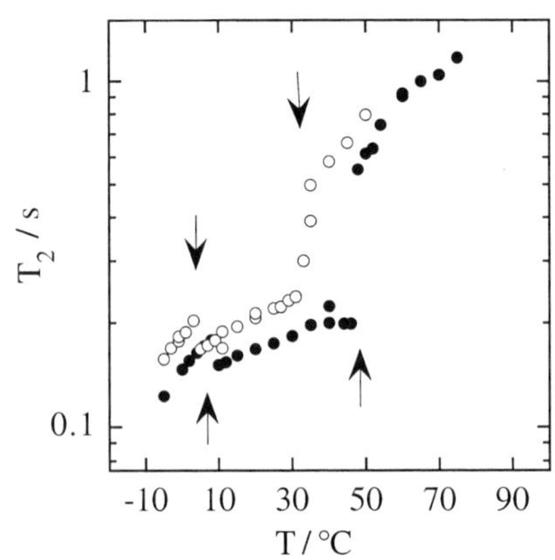

Figure 29. Transverse relaxation time of the E protons (log scale) *versus* temperature for aqueous solutions of (●) $E_{21}B_{11}E_{21}$ (29 wt %) and (○) $E_{32}B_{10}$ (19 wt-%). The arrows show the upper and lower gel boundaries (cubic phase). Adapted from [22].

NMR spectroscopy. The variation with temperature of the ^{1}H NMR transverse relaxation time (T_2) of the E-block has been used to detect sol/gel transitions [22]. Examples are shown in Figure 29. T_2 relates to the slow relaxation of large-scale molecular motions: the NMR measure of local motions (T_1) is essentially unaffected by the phase change.

Hot-stage polarised-light microscopy. Textures observed under polarised light are characteristic of anisotropic phases, allowing distinction to be drawn between hexagonal and lamellar phases. Examples of such micrographs for E/B and E/P copolymers can be seen in, e.g., [32,87].

Rheometry. Because of its high sensitivity to large-scale relaxation processes, rheometry provides a powerful method of investigating phase behaviour. In particular, examination of a solution under oscillatory shear provides a means of detecting 'soft gels' [89,92], to which other methods are largely insensitive. As an example, the variation with temperature of the storage (G') modulus of aqueous solutions of $E_{41}B_8$ is shown in Figure 30. The 44 wt-% solution shows only hard gel (G' > 10^3 Pa), the 15 wt-% solution only soft gel (G' < 10^3 Pa), while the 27 wt-% solution shows both hard and soft gel regions. The notation hard and soft gel is that used by Hvidt and coworkers [89,92].

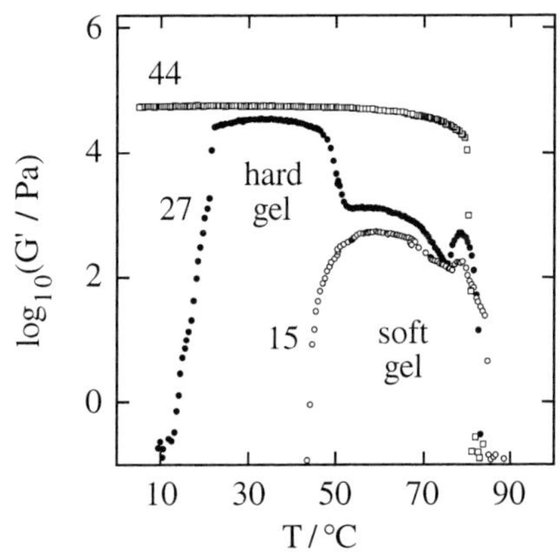

Figure 30 Storage modulus (G') *versus* temperature for solutions of $E_{41}B_8$ with the concentrations (wt-%) indicated. Adapted from [32].

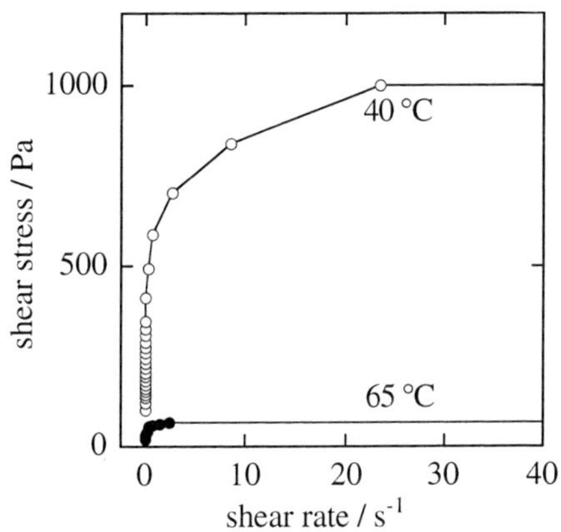

Figure 31. Shear stress *versus* shear rate for a 27 wt-% solution of $E_{41}B_8$ in the hard-gel (40 °C) and soft-gel (65 °C) regions. Adapted from [32].

Using the rheometer in continuous shear mode allows measurement of yield stress, as seen for a 27 wt-% solution of $E_{41}B_8$ at selected temperatures in Figure 31. Outside these gel regions the solutions show Newtonian flow.

5.2 Phase diagrams

5.2.1 Diblock copolymers

A phase diagram obtained [32] for copolymer $E_{41}B_8$ is shown in Figure 32. The hard-gel boundary is defined by results from tube-inversion, DSC and rheometry. Agreement between the various methods is excellent. The transition from the isotropic (cubic) to a birefringent hexagonal phase, detected by polarised-light microscopy, is through an ill-defined biphasic region. It has been suggested that the extensive soft gel region for $E_{41}B_8$ results from formation of a fractal network of spherical micelles which is highly swollen by the water phase [32]. An alternative reason for soft-gel formation, appropriate for high temperatures, is the formation of cylindrical micelles [92].

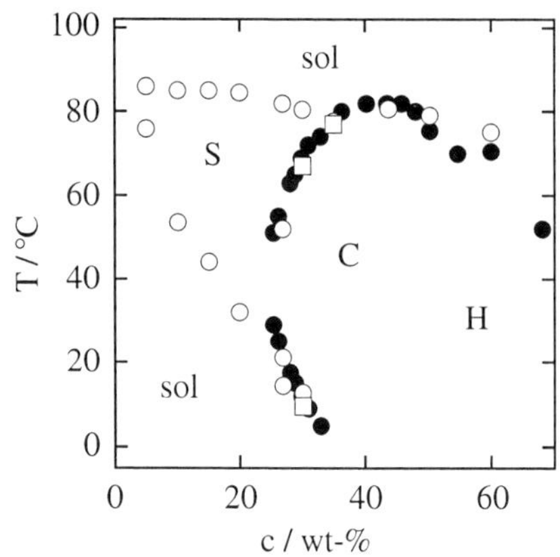

Figure 32. Phase diagram for aqueous solutions of $E_{41}B_8$. (●) Tube-inversion, (□) DSC, (○) rheometry. Adapted from [32].

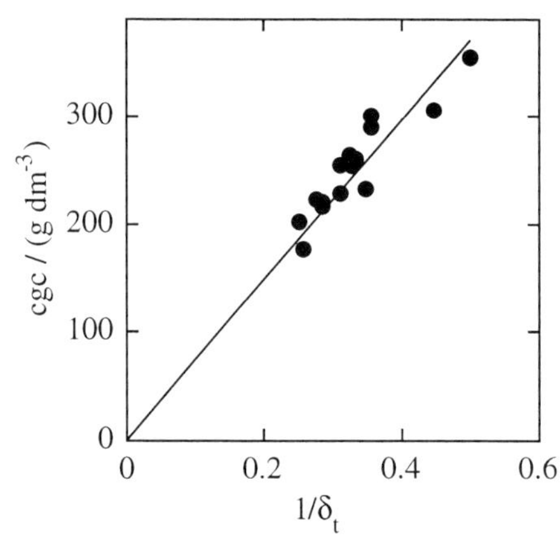

Figure 33. Critical gel concentration *versus* $1/\delta_t$ for aqueous solutions of diblock E/B copolymers. Results from [17-21,32,90].

Phase diagrams for other diblock copolymers have largely been investigated by the tube-inversion and polarised-light microscopy methods [18-21,90]. All have been found to have an isotropic (cubic) hard-gel phase.

The transition to a hard gel occurs when the volume fraction of spheres in the system (ϕ) reaches a critical value. This would be $\phi_c \approx 0.68$-0.74 for close-packed cubic structures, or $\phi_c \approx 0.52$ for a primitive cubic structure, the latter corresponding approximately to the equilibrium condition of Hoover and Ree [94]. With an expansion factor, δ, defined as $\delta = v_s/v_a$, where v_s is the effective hard-sphere volume of a micelle in solution and v_a its 'dry' volume, then the critical copolymer concentration for gelation, cgc in g dm^{-3}, is given by

$$\mathrm{cgc} = \frac{10^3 \phi \rho_a}{\delta} \tag{6}$$

where ρ_a is the density of 'dry' copolymer. If the hard-sphere volume v_s is equated with the thermodynamic volume v_t then $\delta = \delta_t$ (see, e.g., Section 4.5). Accordingly, for a given structure (given ϕ_c), the determining factor for gelation is δ_t, which in turn is largely determined by the exclusion behaviour of the E-block fringe. This depends sensitively on the structure of the micelle (including the association number) and the quality of the solvent. For large spherical micelles, δ_t becomes smaller as the solvent is made poorer, e.g. by increasing the temperature or by adding salt. These effects have been demonstrated [21]: examples are given in Table 7. As a consequence, phase boundaries relate only indirectly (through δ_t) to copolymer composition, chain length and architecture. Characterisation of the micelles (N_w, r_t) is required if the hard-gel boundary is to be predicted.

The advantage of this approach is seen in Figure 33, where critical gelation concentrations found for aqueous solutions of diblock copolymers are plotted against $1/\delta_t$, with values of δ_t for the micelles in dilute solution obtained *via* static light scattering. The data refer to the upper gel-sol transition, for which the micelle concentration can be equated unequivocally with the total copolymer concentration. The line shown in Figure 33 corresponds to eqn. (6) with $\phi_c = 0.70$, i.e. the data are consistent with cubic close packing.

5.2.2 $E_mB_nE_m$ copolymers

Published phase diagrams for aqueous solutions of $E_mB_nE_m$ copolymers [14-16] cover only a moderate concentration range. Figure 27 is an example. Recent work [34,90] has covered a wider concentration range, up to 80 wt-%. Hard-gel boundaries for three triblock copolymers with the same central block length (B_{10}) but various E block lengths, $E_{16}B_{10}E_{16}$, $E_{21}B_{10}E_{21}$ and $E_{35}B_{10}E_{35}$, are shown in Figure 34. Considering the effect of concentration, copolymers $E_{35}B_{10}E_{35}$ and $E_{21}B_{10}E_{21}$ form an isotropic cubic phase first (consistent with spherical micelles in the sol) then, as concentration is increased, a birefringent hexagonal phase. Copolymer $E_{16}B_{10}E_{16}$ forms a hexagonal phase first (consistent with cylindrical micelles in the sol), and a lamellar phase at higher concentration. As for diblock copolymer $E_{41}B_8$ (see above), the phase behaviours of these copolymers have been investigated by a range of techniques [34].

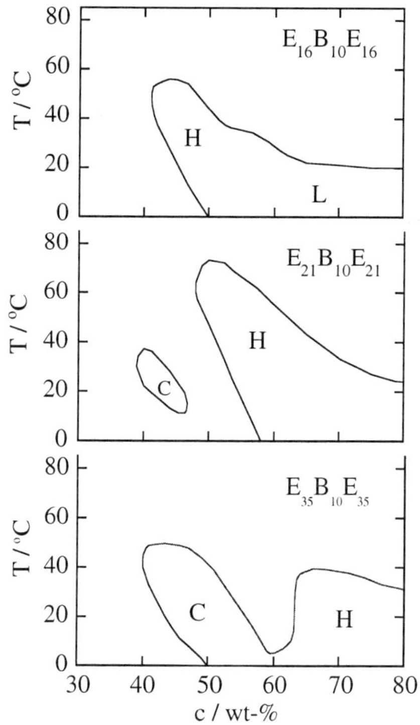

Figure 34. Phase diagrams (tube-inversion method) for triblock $E_mB_{10}E_m$ copolymers, as indicated. Adapted from [34].

5.2.3 $B_nE_mB_n$ copolymers

A phase diagram obtained [90] for triblock copolymer $B_7E_{40}B_7$ is shown in Figure 35. Tube-inversion and polarised-light microscopy were used to identify the hard-gel boundary and to locate the cubic, hexagonal and lamellar phases within it. The phase diagram is complicated by the cloud point curve (solid line), which passes smoothly through micellar sol and gel phases alike. Clouding of aqueous solutions of $B_nE_mB_n$ copolymers is consistent with an attractive micellar interaction through linking (see Section 4.4.4 and [24,25,29]). The poorly-defined gel region

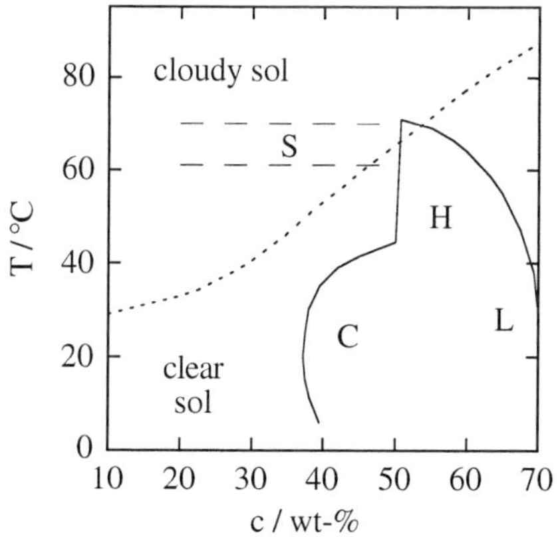

Figure 35. Phase diagram for aqueous solutions of $B_7E_{40}B_7$. The dashed line separates clear solutions and gels from cloudy ones. The boundary enclosing the cubic (C), hexagonal (H) and lamellar (L) phases is indicated. A region of gel extending to low concentrations (denoted S) was difficult to define by the tube-inversion technique. Adapted from [90].

extending to low concentrations is reminiscent of that found [92,95] for aqueous solutions of copolymer P94 ($E_{21}P_{47}E_{21}$).

5.3 Thermodynamics of gelation

The temperature dependences of the processes serve to show that gel formation at the lower boundary (cold gelation [96]) is endothermic, while that at the upper boundary (hot gelation [96]) is exothermic. DSC has been used to show that the enthalpy changes associated with cubic-gel formation are small: e.g. $\Delta_{gel}H \approx -140$ J (mol copolymer)$^{-1}$ for gelation on cooling (high-T boundary) of $E_{41}B_8$ (30 wt-%, T = 67 °C, see Figure 32 for DSC curve). An alternative way of exploring the thermicity of gelation is from the temperature dependence of the cgc, i.e. in much the same way as the standard enthalpy of micellisation is derived from the temperature dependence of the cmc. For gelation, the standard states are non-interacting micelles in 1 mol dm^{-3} solution and micelles in the packed gel, so the standard enthalpy of gelation ($\Delta_{gel}H^o$), which relates to the equilibrium process, differs from the DSC value. A plot of log(cgc) against 1/T for the hot gelation of $E_{41}B_8$ is shown in Figure 36. The slope of the tangent at T = 67 °C leads to $\Delta_{gel}H^o \approx -200$ J (mol copolymer)$^{-1}$. Values of similar magnitude, but positive, are found for gelation on heating (low-T boundary). All are very much smaller in magnitude than values of $\Delta_{mic}H^o$, which are in the range 30-80 kJ (mol copolymer)$^{-1}$ for E/B copolymers (see Table 5). Compared with micellisation, gelation is almost athermal.

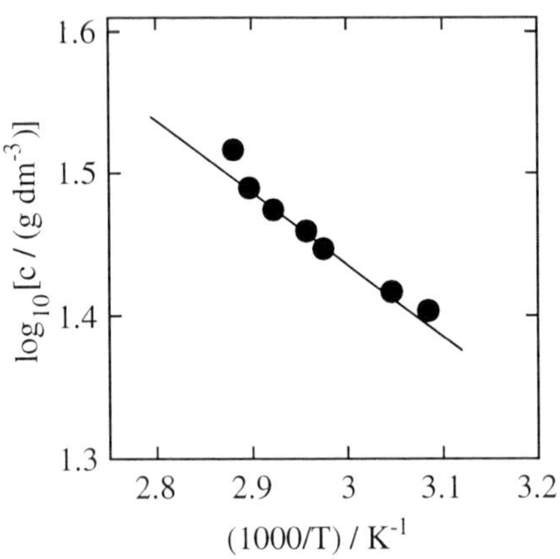

Figure 36. Logarithm of critical gel concentration versus inverse temperature for aqueous solutions of $E_{41}B_8$. Adapted from [32].

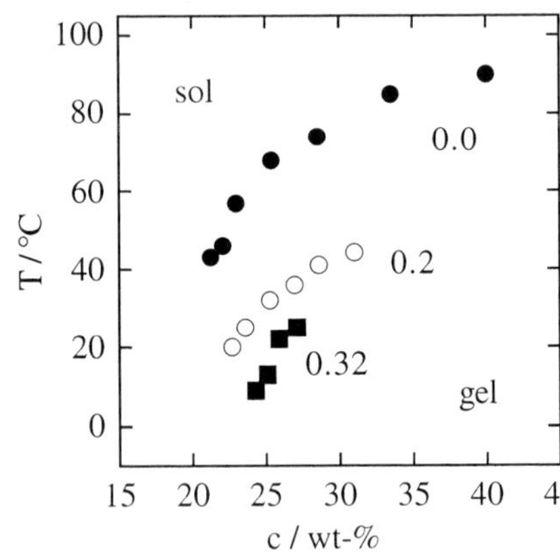

Figure 37. Effect of salt on the upper sol-gel boundary for aqueous solutions of $E_{40}B_{10}$. Concentrations of K_2SO_4 (mol dm^{-3}) are indicated. Adapted from [21].

5.4 Effect of salt

Figure 37 shows the effect of salt on the phase diagram (cubic gel region) for diblock copolymer $E_{40}B_{10}$. For clarity only the top boundary is shown. In fact the lower boundary is much less sensitive to addition of salt. This is understandable, since the effect of salt in reducing the micellar expansion factor (δ_t) is compensated at the lower boundary by more favourable conditions for formation of micelles in the poorer solvent (i.e. a lower cmc), whereas no such compensation is possible at the upper boundary. The depression of the upper boundary can be large: e.g. from 65 °C (water) to 10 °C (0.32 mol dm^{-3} K_2SO_4) for 25 wt-% solutions of $E_{40}B_{10}$.

5.5 Comparison with other diblock copolymers

In Figure 38 comparison is made of the hard-gel boundaries (cubic gel region) of aqueous solutions of diblock copolymer $E_{40}B_{10}$, an alcohol ethoxylate $C_{17}E_{40}$ [85], and an ethylene oxide/DL-lactide copolymer ($E_{77}L_{26}$) [97]. These copolymers were chosen because they have similar sol-gel boundaries at high temperature. The difference in their phase diagrams at low temperature is striking: that for the E/B copolymer has a lower boundary, whereas those of the other two copolymers do not. Similar phase diagrams have been reported for other C_nE_m [85,98] and E_mL_n [97] copolymers, as well as a triblock oxyethylene/ caprolactone ($CL_6E_{90}CL_6$) copolymer [99]. The low-T boundary in E/B copolymers (and E/P copolymers) is attributed to the micelle-molecule equilibrium, i.e. to micellar dissociation on reducing the temperature. It seems that this mechanism does not operate for micelles with alkyl or polyester blocks.

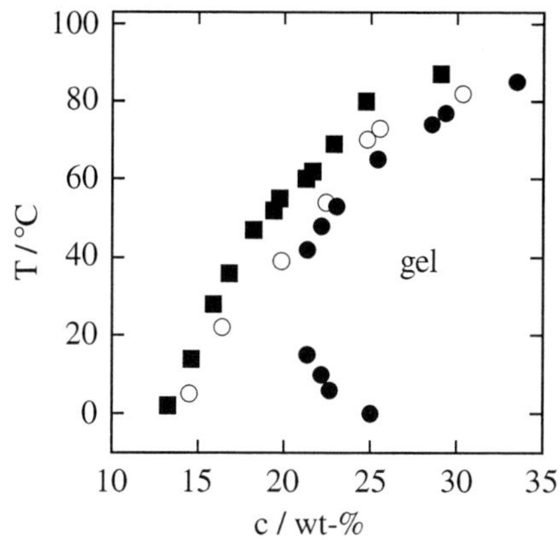

Figure 38. Phase diagrams (cubic region) for aqueous solutions of (●) $E_{40}B_{10}$, (○) $C_{17}E_{40}$ and (■) $E_{77}L_{26}$. Adapted from [21,85,97].

Crystallisation and liquid crystallisation of alkyl chains in the micelle core of E_mC_n and $C_nE_mC_n$ copolymers is known to occur [82,100], and this provides a stabilising mechanism as temperature is reduced. A similar effect is possible for a CL block. Crystallisation is not possible for a stereo-irregular poly(DL-lactide) block, but the glass transition temperature of poly(lactide) is fairly high ($T_g \approx 60$ °C), with the possibility of kinetic stabilisation of the micelle core below this temperature.

Several applications of aqueous block copolymer gels make use of the low-T boundary: e.g. the subcutaneous injection of drug-loaded copolymer solutions which gel at body temperature and provide a reservoir for slow release [16,101]. The importance of E/B and E/P copolymers in this application is obvious. An advantage of E/B over E/P copolymers is their much smaller size for similar gelation effect: e.g. $E_{40}B_{10}$, $M_n \approx 2500$ g mol^{-1} and $E_{98}P_{67}E_{98}$ (F127), $M_n \approx 12500$ g mol^{-1}. This allows rapid elimination of spent E/B copolymers from the host.

Acknowledgements

We thank Dr. D. Attwood and Dr. F. Heatley for their helpful advice, and our colleagues in other laboratories who made their papers available before publication.

References

1. L.G. Lundsted (to the Wyandotte Chemicals Corp.), British 722,746 (1955).
2. J.S. Spriggs (to the Dow Chemical Co.), U.S. 2,828,345 (1958); A.S. Teot (to the Dow Chemical Co.), Canadian 698,568 (1964).
3. I.R. Schmolka in Nonionic Surfactants, Surfactant Science Series, Vol. 1, M.J. Schick (ed.), Marcel Dekker, New York, 1967, Ch. 10.
4. L.G. Lundsted and I.R. Schmolka in Block and Graft Copolymers, Vol. 2, R.J. Ceresa (ed.), John Wiley and Sons, New York, 1976, Ch. 1 and 2.
5. J.H. Lee, J. Kopecek and J. Andrade, Polymeric Materials Science and Engineering, 57 (1987) 613.
6. M. Gangoda, B.M. Fung and E.A. O'Rear, J. Colloid Interface Sci., 116 (1987) 230.
7. I.R. Schmolka, J. Am. Oil Chem. Soc., 68 (1990) 206.
8. J. Myszkowski, J. Szymanowski, W. Goc and K. Alejski, Tenside Deterg., 19 (1982) 7.
9. J. Szymanowski, J. Myszkowski, K. Prochaska and K. Szafraniak, Tenside Deterg., 19 (1982) 11.
10. J. Szymanowski, J. Myszkowski, K. Szafraniak and J. Nowicki, Tenside Deterg., 19 (1982) 14.
11. V.M. Nace in Nonionic Surfactants, Polyoxyalkylene Block Copolymers, Surfactant Science Series Vol. 60, V.M. Nace (ed.), Marcel Dekker, New York, 1996, Ch. 4.
12. F. Heatley, G.-E. Yu, W.-B. Sun. E.J. Pywell, R.H. Mobbs and C. Booth, Eur. Polym. J., 26 (1990) 583.
13. W.-B. Sun, J.-F. Ding, R.H. Mobbs, D. Attwood and C. Booth, Colloids and Surfaces, 54 (1991) 103.
14. Y.-Z. Luo, C.V. Nicholas, D. Attwood, J.H. Collett, C. Price and C. Booth, Colloid Polym. Sci., 270 (1992) 1094.
15. C.V. Nicholas, Y.-Z. Luo, N.-J. Deng, D. Attwood, J.H. Collett, C. Price and C. Booth, Polymer, 34 (1993) 138.
16. Y.-Z. Luo, C.V. Nicholas, F. Heatley, D. Attwood, J.H. Collett, C. Price, C. Booth, B. Chu and Z.-K. Zhou, J. Chem. Soc., Faraday Trans., 89 (1993) 539.
17. A.D. Bedells, R.M. Arafeh, Z. Yang, D. Attwood, F. Heatley, J.C. Padget, C. Price and C. Booth, J. Chem. Soc., Faraday Trans., 89 (1993) 1235.
18. A.D. Bedells, R.M. Arafeh, Z. Yang, D. Attwood, J.C. Padget, C. Price and C. Booth, J. Chem. Soc., Faraday Trans., 89 (1993) 1243.
19. S. Tanodekaew, N.-J. Deng, S. Smith, Y.-W. Yang, D. Attwood and C. Booth, J. Phys. Chem., 97 (1993) 11847.
20. Z. Yang, S. Pickard, N.-J. Deng, R.J. Barlow, D. Attwood and C. Booth, Macromolecules, 27 (1994) 2371.
21. N.-J. Deng, Y.-Z. Luo, S. Tanodekaew, N. Bingham, D. Attwood and C. Booth, J. Polym. Sci., Part B, Polym. Phys., 33 (1995) 1085.
22. J. Godward, F. Heatley, S. Smith, S. Tanodekaew, Y.-W. Yang and C. Booth, J. Chem. Soc., Faraday Trans., 91 (1995) 3461.
23. Y.-W. Yang, N.-J. Deng, G.-E. Yu, Z.-K. Zhou, D. Attwood and C. Booth, Langmuir, 11 (1995) 4703.
24. Y.-W. Yang, Z. Yang, Z.-K. Zhou, D. Attwood and C. Booth, Macromolecules, 29 (1996) 670.
25. Z. Yang, Y.-W. Yang, Z.-K. Zhou, D. Attwood and C. Booth, J. Chem. Soc., Faraday Trans., 92 (1996) 257.
26. Z.-K. Zhou, B. Chu, V.M. Nace, Y.-W. Yang and C. Booth, Macromolecules, 29 (1996) 3663.
27. G.-E. Yu, Y.-W. Yang, Z. Yang, D. Attwood, C. Booth and V.M. Nace, Langmuir, 12 (1996) 3404.
28. Y.-W. Yang, G. Brine, G.-E. Yu, F. Heatley, D. Attwood, C. Booth and M. Malmsten, Polymer, 38 (1997) 1659.

29. Z.-K. Zhou, Y.-W. Yang, C. Booth and B. Chu, Macromolecules, 29 (1996) 8357.
30. G.-E. Yu, Z. Yang, D. Attwood, C. Price and C. Booth, Macromolecules, 29 (1996) 8479.
31. G.-E. Yu, Z.-K. Zhou, D. Attwood, C. Price, C. Booth, P.C. Griffiths and P. Stilbs, J. Chem. Soc., Faraday Trans., 92 (1996) 5021.
32. H. Li, G.-E. Yu, C. Price, C. Booth, E. Hecht and H. Hoffmann, Macromolecules, 30 (1997) 1347.
33. G.-E. Yu, M. Ameri, Z. Yang, D. Attwood, C. Price and C. Booth, J. Phys. Chem., Part B, 101 (1997) 4394.
34. G.-E. Yu, H. Li, C. Price and C. Booth, to be published.
35. Z.-K. Zhou, B. Chu and V.M. Nace, Langmuir, 12 (1996) 5016.
36. Z.-K. Zhou and B. Chu in Nonionic Surfactants, Polyoxyalkylene Block Copolymers, Surfactant Science Series Vol. 60, V.M. Nace (ed.), Marcel Dekker, New York, 1996, Ch. 3.
37. Product Brochure, B-Series Polyglycols. Butylene Oxide/Ethylene Oxide Block Copolymers, The Dow Chemical Co., Freeport, Texas (1992).
38. V.M. Nace, R.H. Whitmarsh, M.W. Edens, J. Am. Oil Chem. Soc., 71 (1994) 777.
39. V.M. Nace and J.C. Knoell, J. Am. Oil Chem. Soc., 72 (1995) 89.
40. V.M. Nace, J. Am. Oil Chem. Soc. 73 (1996) 1.
41. P. Alexandridis, U. Olsson and B. Lindman, Langmuir, submitted.
42. Product Brochure, Property Differences in BO/EO and PO/EO Block Copolymers, The Dow Chemical Co., Freeport, Texas (1994).
43. M.W. Edens in Nonionic Surfactants, Polyoxyalkylene Block Copolymers, Surfactant Science Series Vol. 60, V.M. Nace (ed.), Marcel Dekker, New York, 1996, Ch. 6.
44. V.M. Nace (to the Dow Chemical Company), Biodegradable Alkylene Oxide-Lactone Copolymers, US 5 525 702 (1996).
45. P.J. Flory, Principles of Polymer Chemistry, Cornell U.P., Ithaca, 1953, p. 336.
46. G.-E. Yu, A.J. Masters, F. Heatley, C. Booth and T. G. Blease, Macromol. Chem. Phys., 195 (1994) 1517.
47. R.H. Whitmarsh in Nonionic Surfactants, Polyoxyalkylene Block Copolymers, Surfactant Science Series Vol. 60, V.M. Nace (ed.), Marcel Dekker, New York, 1996, Ch. 1.
48. B. Weibull and B. Nycander, Acta Chem. Scand. 8 (1954) 847.
49. S. Dickson, G.-E. Yu, F. Heatley and C. Booth, *Eur. Polym. J.*, 29 (1993) 281.
50. G.-E. Yu, D. Mistry, S. Ludhera, F. Heatley, D. Attwood and C. Booth, *J. Chem. Soc., Faraday Trans.*, in proof.
51. Z.-G. Yan, Z. Yang, C. Price and C. Booth, Makromol. Chem., Rapid Commun., 14 (1993) 725.
52. G.-E. Yu, P. Sinnathamby, C. Price and C. Booth, Chem. Commun., (1996) 31.
53. Z. Yang, G.-E. Yu, J. Cooke, K. Viras, H. Matsuura, A.J. Ryan and C. Booth. J. Chem. Soc., Faraday Trans., 92 (1996) 3173.
54. T. Sun, G.-E. Yu, C. Price, C. Booth, J. Cooke and A.J. Ryan, Polymer Commun., 36 (1995) 3775.
55. S.W. Provencher, Makromol. Chem., 180 (1979) 201.
56. D.E. Koppel, J. Chem. Phys., 57 (1972) 4814.
57. Z.-K. Zhou and B. Chu, Macromolecules, 26 (1987) 3089.
58. T.P. Lally and C. Price, Polymer, 15 (1974) 325.
59. N.K. Reddy, P.J. Fordham, D. Attwood and C. Booth, J. Chem. Soc., Faraday Trans., 86 (1990) 1569.
60. G.-E. Yu, Y.-L. Deng, S. Dalton, Q.-G. Wang, D. Attwood, C. Price and C .Booth, J. Chem. Soc., Faraday Trans., 88 (1992) 2537.
61. N.F. Carnahan, K.E. Starling, J. Chem. Phys. 51 (1969) 635; A.J. Vrij, Chem. Phys., 69 (1978) 1742.
62. P.J. Flory, Principles of Polymer Chemistry, Cornell U.P., Ithaca, 1953, p. 532.
63 P. Kratochvil, Classical Light Scattering from Polymer Solutions, Elsevier Science

Publishers, Amsterdam, 1987.
64. K. Mortensen and J.S. Pedersen, Macromolecules, 26 (1993) 805.
65. K. Mortensen, Prog. Colloid Polym. Sci., 91 (1993) 69.
66. H. Vink, J. Chem. Soc., Faraday Trans. 1, 81 (1985) 1725.
67. Z. Grubisic, P. Rempp and H. Benoit, J. Polym. Sci., Polym. Phys. Edn., 5 (1967) 753.
68. H.H. Teo, M.G. Styring, S.G. Yeates, C. Price and C. Booth, J. Colloid Interface Sci., 114 (1986) 416.
69. Q.-G. Wang, C. Price and C. Booth, J. Chem. Soc., Faraday Trans., 88 (1992) 1437.
70 Q.-G. Wang, G.-E. Yu, Y.-L. Deng, C. Price and C. Booth, Eur. Polym. J., 29 (1993) 665.
71. G.-E. Yu, H. Altinok, S.K. Nixon, C. Booth, P. Alexandridis and T.A. Hatton Eur. Polym. J., 33 (1997) 673.
72. Y.-L. Deng, G.-E. Yu, C. Price and C. Booth, J. Chem. Soc., Faraday Trans., 88 (1992) 1441.
73. N.K. Reddy, A. Foster, M.G. Styring and C. Booth, J. Colloid Interface Sci., 136 (1990) 588.
74. D. Attwood, J.H. Collett, C.J. Tait, Int. J. Pharm., 26 (1985) 25.
75. D.G. Hall in Nonionic Surfactants, Physical Chemistry, Vol. 23; M.J. Schick (ed.), Marcel Dekker, New York, 1987, p. 247.
76. C. Tanford, The Hydrophobic Effect, John Wiley and Sons, New York, 1980.
77. P. Linse, Macromolecules, 26 (1993) 4437.
78. P.J. Flory, Principles of Polymer Chemistry, Cornell U.P., Ithaca, 1953, p. 333.
79. F.E. Bailey and R.W. Callard, J. Appl. Polymer Sci., 1 (1959) 56, 373.
80. D.R. Beech and C. Booth, J. Polym. Sci., Part A2, 7 (1969) 575.
81. P.R. Knowles, R.B. Stubbersfield and C. Price, Makromol. Chem., Macromol. Symp., 40 (1990) 203.
82. P.R. Knowles, R.J. Barlow, F. Heatley, C. Booth and C. Price, Macromol. Chem. Phys., 195 (1994) 2547.
83. K. Persson, G. Wang and G. Olofsson, J. Chem. Soc., Faraday Trans., 90 (1994) 3555.
84. M. Gradzielski, A. Rauscher and H. Hoffmann, J. Physique IV, 3 (1993) 65.
85. M. Ameri, J.H. Collett, D. Attwood and C. Booth, *J. Chem. Soc., Faraday Trans.*, 93, (1997) 2545.
86. S. Tanodekaew, R. Pannu, F. Heatley, D. Attwood and C. Booth, Macromol. Chem. Phys., 198 (1997) 927.
87. G. Wanka, H. Hoffmann and W. Ulbricht, Macromolecules, 27 (1994) 4145.
88. P. Alexandridis, J.F. Holzwarth and T.A. Hatton, Macromolecules, 27 (1994) 2414.
89. M. Almgren, W. Brown and S. Hvidt, Colloid Polym. Sci., 273 (1995) 2.
90. Y.-W. Yang, PhD Thesis, University of Manchester, 1996.
91. O. Glatter, G. Scherf, K. Schillen and W. Brown, Macromolecules, 27 (1994) 6046.
92. S. Hvidt, E.B. Jorgensen, W. Brown and K. Schillen, J. Phys. Chem., 98 (1994) 12320.
93. G. Wanka, H. Hoffmann and W. Ulbricht, Colloid Polym Sci., 268 (1990) 569.
94. W.G. Hoover and F.H. Ree, J. Chem. Phys., 49 (1968) 3609.
95. P. Bahadur and K. Pandya, Langmuir, 8 (1992) 2666.
96. I.R. Schmolka, J. Biomed. Mater. Res., 6 (1972) 571.
97. S. Tanodekaew, J. Godward, F. Heatley and C. Booth, Macromol. Chem. Phys., accepted.
98. G.T. Dimitrova, Th.F. Tadros, P.F. Luckham and M.R. Kipps, Langmuir, 12 (1996) 315.
99. L. Martini, D. Attwood, J.H. Collett, C.V. Nicholas, S. Tanodekaew, N.-J. Deng, F. Heatley and C. Booth, J. Chem. Soc., Faraday Trans., 90 (1994) 1961.
100. F. Heatley, H.H. Teo and C. Booth, J. Chem. Soc., Faraday Trans. 1, 80 (1984) 981.
101. I.R. Schmolka, in Polymers for Controlled Drug Delivery, P. Tarcha (ed.), CRC, Boca Raton, FL, 1991, p. 189.

Self-Assembly of Block Polyelectrolytes†

Lifeng Zhang, Karine Khougaz,+ Matthew Moffitt and Adi Eisenberg*

Department of Chemistry, McGill University, 801 Sherbrooke Street West, Montréal, Québec, Canada H3A 2K6

1. INTRODUCTION

The self-assembly of low-molecular weight amphiphiles in water is well established, and has been found to dramatically influence the physico-chemical properties of surfactant solutions through the formation of cooperative structures. When the resulting structures consist of a spherical insoluble core solubilized by hydrophilic moieties, they are known as micelles, and these have become the focus of both academic and industrial interest [1]. The tendency of surfactants to associate above a certain concentration (the critical micelle concentration, or cmc) can be attributed to the nature of the molecules, which consist of one or more hydrophobic chains linked to a polar or ionic head group. The primary driving force for self-assembly is generally ascribed to the hydrophobic effect; the entropically unfavorable ordering of water molecules in the vicinity of hydrocarbon chains is alleviated by the association and microphase separation of the insoluble moieties, forming hydrophobic domains of colloidal dimensions which are solubilized by surrounding hydrophilic head groups. The final size of the micelles is an expression of an optimum thermodynamic state, in which the sum of such factors as chain stretching, interfacial tension, and repulsive interactions between head groups, are minimized. Although micelles of spherical geometry are most common, other morphologies, such as rods and lamellae, have also been observed at different surfactant concentrations above the cmc.

Analogous to the micellization of low-molecular weight surfactants is the self-assembly of diblock copolymers in a selective solvent. Above a critical concentration of the copolymer, the insoluble blocks undergo microphase separation, forming swollen or compact cores surrounded by an extended shell, or corona, of the soluble blocks. These aggregates are structurally and phenomenologically similar to spherical associates of surfactants, and are therefore commonly termed block copolymer micelles. It should be noted, however, that these structures are generally less labile than their surfactant counterparts, with slower kinetics of exchange between associated and non-associated chains. Aggregates of block copolymers in selective solvents and in the solid state have been

* To whom correspondence should be addressed. email: eisenber@omc.lan.mcgill.ca

+ Present address: Centre for Therapeutic Research, Merck Frosst Canada Inc., P. O. Box 1005, Pointe Claire, Quebec, H9R 4P8.

† Funding from The Natural Sciences and Engineering Research Council of Canada (NSERC) is gratefully acknowledged.

characterized extensively, using a wide range of techniques such as size-exclusion chromatography (SEC), static and dynamic light scattering (SLS and DLS), small-angle x-ray scattering (SAXS), small-angle neutron scattering (SANS), and transmission electron microscopy (TEM). Comprehensive reviews of experimental and theoretical studies of block copolymer micellization have been written by Price [2], and Tuzar and Kratochvil [3,4].

More recently, several studies have been concerned with aqueous solutions of block copolymer micelles (i.e., systems in which the selective solvent is water). Since they are water-soluble, such micelles have a number of potential applications to fields such as pharmacology, agriculture, and ecology. The commercial availability of polypropylene oxide (PPO)-polyethylene oxide (PEO) block copolymers has made them extremely attractive materials for numerous theoretical and experimental investigations [5-15]. In aqueous solutions, these polymers form aggregates in which the more hydrophobic PPO blocks constitute the micelle core, protected by the solubilized PEO blocks. Aqueous micelles of polystyrene-*b*-poly(ethylene oxide) (PS-*b*-PEO) have also been studied in some detail [16].

An interesting class of water-soluble block copolymers are the block polyelectrolytes, in which the solubilized coronal chains contain ionic moieties. Aggregates of this type combine the microphase separation of colloidal block copolymers with the complicated behavior of polyelectrolytes in aqueous solution. Ionic block copolymers become block polyelectrolytes when dissolved in aqueous media, and may contain one of a number of different ionic blocks, including poly(4-vinylpyridinium alkyl halides), poly(metal acrylates), poly(metal methacrylates), and sulfonated polystyrene. The growing field of block polyelectrolytes was initiated by the early studies of Selb and Gallot, and their reviews on ionic block copolymers serves as a comprehensive summary of their results [17]. We will therefore give only an abbreviated summary of their work in the present chapter.

The colloidal behavior of block polyelectrolytes is largely governed by the existence of repulsive electrostatic interactions between charged polymer units. Ionic repulsive forces within the corona have a strong influence on micelle growth, allowing these systems to be modified through such controllable factors as pH and the addition of low-molecular weight salts. The ionic nature of block polyelectrolytes therefore makes them extremely versatile materials, although they are quite complicated from an academic perspective. For instance, electrostatic interactions between micelles in solution may result in ordering, liquid crystalline behavior, or gel formation, at sufficiently high concentrations. Such order formation is problematic for studies of colloidal behavior, and one is usually required to work at high degrees of dilution to minimize intermicellar effects. The influence of electrostatic forces on block polyelectrolytes, and ionomer colloids in general, has also proved interesting from a theoretical perspective, and these systems have been modeled by different groups [18-22].

Along with micelles containing ionic units, the present review encompasses the behavior of block copolymers with ionizable acid blocks, as these are classified as block polyelectrolytes under conditions of sufficiently high pH. Aqueous solutions of block copolymers such as polystyrene-*b*-poly(acrylic acid) (PS-*b*-PAA) and polystyrene-*b*-poly(methacrylic acid) (PS-*b*-PMAA) are therefore discussed in some detail. As well, we touch upon the effect of added ions to micelles of PS-*b*-PEO; although nonionic in nature, the PEO block is found to mimic some aspects of polyelectrolytes, insofar as its colloidal behavior is influenced by the addition of low-molecular weight salts.

As mentioned earlier, the kinetics of exchange between single chains and aggregates are quite slow in the case of block copolymer micelles, when compared with the relatively labile associates of surfactants in water. Unlike their surfactant counterparts, micelles of block copolymers can be passed through an SEC column intact, without significant adsorption of single chains [2]. These experiments indicate that, for block copolymers in selective solvents, micelle-single chain exchange occurs over a time scale comparable to, or longer than, the passage of the sample through the column. This is due to relatively high viscosities in the cores of block copolymer micelles, and the consequentially long residence times of single chains within the aggregates.

In aqueous solutions of PS-*b*-PAA and polystyrene-*b*-poly(sodium acrylate) (PS-*b*-PANa), which are the primary focus of this review, the PS cores at room temperature are well below their glass transition temperature (ca. 100° C). As well, large differences in solvent-polymer interaction parameters (χ) of hydrophilic and hydrophobic blocks result in high interfacial tension between micelle cores and the aqueous medium, which minimizes the swelling of the cores with solvent. At normal temperatures and in pure water, therefore, the PS cores are glassy, such that exchange between micelles and single chains is essentially frozen. Without a dynamic equilibrium with the unimers, the self-assembled structures are not true associates (or even "micelles" in the strictest sense) and are best described as block copolymer aggregates. The term "block copolymer micelle", however, has become generally accepted, and is therefore employed in the present discussion.

In some cases, block copolymer micelles in aqueous media are formed by dissolving the single chains in a good solvent for both blocks, followed by the gradual decrease in the solvent quality by the slow addition of water. A recent example is the micellization of PS-*b*-PAA, to be discussed later in the present work, by the addition of water to a solution of single chains in DMF [23]. At the point of micellization, the cores are significantly swollen with the non-aqueous co-solvent, such that the micelles form under conditions of dynamic equilibrium. Further addition of water forces remaining co-solvent out of the core, such that the dynamics of exchange gradually decrease until the micelles are essentially "frozen" or "dead". In fact, the terms "frozen micelle" or "dead micelle" may be considered appropriate descriptions for non-equilibrium aggregates which were formed under conditions of dynamic equilibrium.

It is important to note that the conditions for dynamic exchange between micelles and single chains become considerably different at extremely low concentrations (i.e., in the region of the cmc). This is evinced by the observation that micelles of PS-*b*-PANa in pure water, which are "dead" at normal concentrations, dissociate into single chains upon dilution to concentrations below the cmc [24,25]. The observation of cmc's in block polyelectrolytes indicates the existence of dynamic micelle-single chain equilibria in the limit of extreme dilution, and provides further justification for the use of the term "micelle" in reference to these colloidal systems.

We distinguish between two types of block polyelectrolytes, stars and crew-cuts, based on the relative lengths of the soluble and insoluble blocks (Chart I); it should be noted that the demarcation between these two types of aggregates is far from clear. In the first part of this paper, we discuss block polyelectrolytes in which the insoluble blocks are relatively short; these form star-like spherical micelles with small, compact cores and large, extended coronae. Several aspects of star-like micelles are described, including variations in critical micelle concentrations, micelle sizes, and the solubilization of organic molecules as functions of block copolymer and solution properties. In the second part of the paper, we discuss the formation of crew-cut micelles from block polyelectrolytes with relatively long insoluble blocks; these aggregates have large cores surrounded by a relatively thin shell of hydrophilic coronal chains. In our overview of crew-cut systems, we first address detailed studies of spherical micelles which form in certain ranges of block copolymer compositions. Finally, we address the wide range of non-spherical morphologies that have been observed in aqueous solutions of crew-cut aggregates. Although the number of papers in the literature on crew-cut micelles is quite small, it may be pointed out that this variety of block polyelectrolytes occupies a relatively large portion of the present review; we believe that the degree of recent interest in such systems justifies the extent of our coverage.

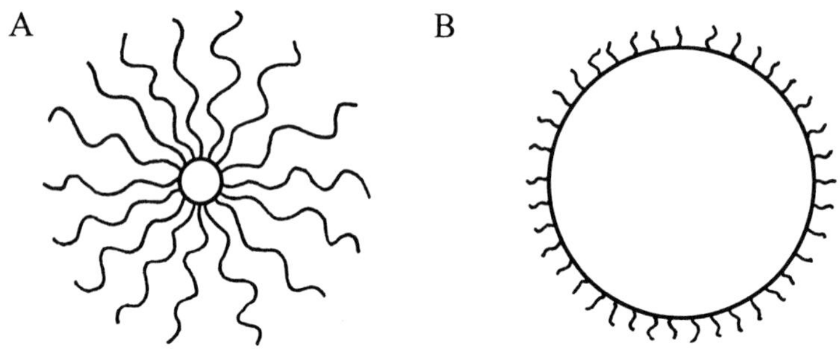

Chart I. Schematic illustration of (A) a star and (B) a crew-cut micelle

2. STAR-LIKE BLOCK POLYELECTROLYTES

2.1. Critical micelle concentrations

The critical micelle concentrations of PS-*b*-P4VP·EtBr water/methanol/LiBr system have been investigated through changes in the ratio of polymer concentration to the intensity of scattered light (c/I) as a function of polymer concentration [17]. A curvature in such plots at low polymer concentrations or an increase in the c/I ratio with decreasing concentration suggests the presence of lower molecular weight particles and yields information on the micelle-single chain equilibrium. The curvature in these plots was not observed for copolymers having > 30 PS units. It was suggested that this result may be due to very low cmc values which could not be detected by the light scattering instrument, or to the formation of frozen micelle structures that do not dissociate. An increase in the temperature was found to shift the single chain-micelle equilibrium in the direction of single chains. The temperature effects were dependent on the salt content and PS molecular weight. The equilibrium shift with temperature was found to be smaller as the salt content and PS block length increased. As the PS block length increased, the influence of temperature on the micelle-single chain equilibrium became less significant.

Critical micelle concentrations of polystyrene-*b*-poly(sodium acrylate) (PS-*b*-PANa) block polyelectrolytes have been determined using steady-state fluorescence spectroscopy of a pyrene probe [24,25]. This method monitors changes in the excitation and emission spectra of pyrene, as the probe is solubilized into the hydrophobic environment of micelle cores at the onset of aggregation [26]. In a typical experiment, spectral features of the probe are plotted as a function of polymer concentration, and the cmc is determined from sudden changes in the slope.

The dependence of the cmc on insoluble and soluble block lengths has been investigated for a wide range of block polyelectrolytes in water [24]. The PS block length was found to have a significant influence on the cmc values. For instance, at a constant PANa block length of 1000 units, an increase in the PS block length from 6 to 110 units resulted in a decrease in the cmc from 1.6 x 10^{-5} to 5 x 10^{-8} M (a factor of 320).

The influence of the soluble polyelectrolyte block on the cmc was also investigated [25]. Figure 1 illustrates cmc values for two block copolymer series having constant PS block lengths of 11 and 23 units, and PANa block lengths ranging from 15 to 1000 and from 29 to 1400 units, respectively. Clearly, the cmc as a function of soluble block length (N_{PANa}) exhibits a parabolic profile, in which the cmc increases with the PANa block length for relatively short soluble blocks, attains a maximum value, then decreases as the PANa block length increases further.

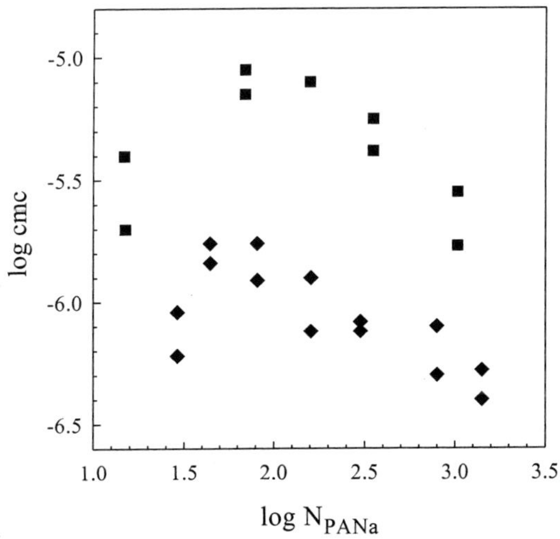

Figure 1. Effect of ionic block length on the cmc values for the PS-b-PANa series having PS block lengths of 11 (■) and 23 (◆) units.

The parabolic nature of the plots shown in Figure 1 is explained in terms of the influence of the PANa block on the solubility of single chains for different ranges of ionic block length [25]. For convenience, the left and right sides of the parabolas are referred to as the short and long ionic block regions (SIBR and LIBR, respectively). Considering the SIBR first, the observed increase in the cmc with the length of the soluble block reflects the concurrent increase in the solubility of single chains for relatively short ionic blocks. The cmc behavior in the LIBR is related to the polyelectrolyte nature of the soluble block; due to its ionic character, an increase in the length of the ionic block effects a concurrent increase in the ionic strength of the solution. As the ionic strength increases, the solubility of the diblock polyelectrolyte will be lowered by ionic screening effects, resulting in a greater tendency to form micelles and a lowering of the cmc. Based on the cmc behavior in the LIBR, these screening effects seem to dominate for longer polyelectrolyte blocks. The observed maxima in the plots indicate a balance in the forces governing cmc behavior in the two regions of ionic block lengths.

The effect of the PANa block length on the cmc was found to be influenced by the length of the PS block in the short ionic block region (SIBR) [25]. The PANa block length-dependence on the cmc, described by the slope d(log cmc)/d(N_{PANa}), was found to decrease with increasing PS block length, approaching zero for relatively long PS blocks (ca. 100 units). This result has been found in other block copolymer systems [17,26] and suggests that, for short insoluble block lengths, the ionic block length influences the micellization while, for long insoluble blocks, micellization is dominated by thermodynamic factors related to the core-forming block.

The influence of a low molecular weight salt (NaCl, C_s = 0.10 to 2.5 M) on the cmc values was investigated for three series of block polyelectrolytes [25]. In this study, the PS block lengths were 6, 11 and 23 units, while the ionic block length varied from 15 to 1400 units. The cmc values were found to decrease linearly with the square root of the salt concentration ($C_s^{1/2}$) for most of these samples.

A plot of the slope d(log cmc)/d($C_s^{1/2}$) versus d(N_{PANa}) was found to exhibit sigmoidal behavior; initially small values of the slope were found to rise at a certain ionic block length, then level off at a relatively high value. For instance, for a series with 11 PS units, the slope increased from 0.21 to 0.25 for 15 to 160 PANa units, from 0.25 to 0.39 for 160 to 300 PANa units and from 0.39 to 0.47 for 300 to 1000 PANa units. These results can be discussed in terms of the chain conformations of the PANa block. For relatively short PANa chain lengths, the polyelectrolyte blocks are expected to resemble rigid rods. In this highly extended conformation, all of the ionic segments are located in the vicinity of the core, where the high concentration of ionic units and counterions produces an environment of high ionic strength which is not significantly perturbed by the addition of salt. As the ionic chain length increases past a certain value, the PANa block no longer resembles a rigid rod; the increased number of available conformations make the cmc appreciably sensitive to the addition of salt. At very long ionic block lengths, the effect of added salt then levels off as the PANa block approaches a random coil conformation. These observations are in qualitative agreement with a theory proposed by Ronis [18].

A recent micellization model describes the effect of chain length polydispersity on the cmc values [27]. This model assumes that at c < cmc, the insoluble block adopts a collapsed spherical conformation. The association energy of these spheres is described by the Hamaker equation, such that sphere-sphere interactions are proportional to the radius of the sphere, or $N_B^{1/3}$, where N_B is the length of the insoluble block. The single chain fraction (S) as a function of total block copolymer composition has been determined for different polydispersity indexes (P.I.); such plots show that cmc values for a distribution of chain lengths decrease with increasing P.I., and are generally less distinct than the sharp transitions found in monodisperse samples.

For monodisperse polymers, the cmc is described by log cmc = $a\ N_B^{1/3} + b$, where a and b are constants. The dependence of the cmc on the polydispersity has been evaluated from the magnitude of the a constant. For the block polyelectrolyte system, PS-*b*-PANa/water, the value of a is relatively low (-0.68), and the effect of polydispersity on the cmc is not significant. In comparison, the cmc values for the weakly segregated system polystyrene-*b*-polyisoprene/n-hexadecane exhibits a larger dependence on the polydispersity (a = -1.64). The magnitude of a values for various block copolymers was found to be proportional to interaction parameters between the solvent and the core-forming block ($\chi_{sol\text{-}core}$) [28].

2.2. Aggregation numbers and micelle sizes

The aggregation of PS-*b*-P4VPEtBr has been studied as a function of the water/methanol solvent composition [17]. Three distinct regions were distinguished depending on the water content. At low water contents only single chains were found to exist. As the water content increased, micelle formation commenced due to the decreased solubility of PS in the solvent mixture. The degree of association passed through a maximum, which was attributed to a balance of two forces affecting the copolymer solubility, i.e., the increased solubility of the P4VPEtBr and the decreased solubility of the PS blocks. This maximum corresponded to a minimum of single chain solubility. In the third region, Z decreased due to the greater affinity of the solvent for the P4VPEtBr blocks, which increased the solubility of the copolymer. The range of the three regions depended on the lengths of the blocks, in particular that of PS.

The molecular weight and architecture of the block polyelectrolyte were also found to influence aggregation behavior. The aggregation numbers increased with increasing PS block length and decreased with increasing P4VP·EtBr block length; the influence of the insoluble PS block length was found to be stronger than that of the soluble block. The aggregation of diblock copolymers were found to be higher than the triblock copolymers having terminal P4VPEtBr blocks, due to the higher solubility of the latter system. The effect of different quarternizing agents for PS-*b*-P2VP copolymers in water has also been investigated by Schwab and Heilweil [29]. Copolymers having longer alkyl chain lengths had lower viscosities than those with shorter chains, which was attributed to the increased hydrophobicity of the longer alkyl blocks.

The effect of low molecular weight salt on the micellization of PS-b-P4VPEtBr has also been studied. When the PS block lengths were below a certain value (ca. 30 PS units), the addition of salt resulted in an increase in the aggregation number and a decrease in cmc value; these effects were found to level off above a certain salt concentration.

Aqueous solutions of block polyelectrolyte micelles formed from PS-*b*-PANa have been investigated by static light scattering [30], in order to determine the effect of added salt on the micelle aggregation numbers. The aggregation numbers (Z) were found to increase with salt content at low C_s, and remain constant above ca. 0.10 M NaCl. This same effect has also been observed by Selb and Gallot in the PS-*b*-P4VPEtBr system [17], and has been theoretically predicted by a the model of Dan and Tirrell [19], which describes the screening of electrostatic repulsions in the micelle corona by added salt. At low salt content, electrostatic correlations dominate the chain configurations and restrict micelle growth. As the salt content increases, electrostatic repulsions are screened, and the aggregation number increases. At moderate salt content, micellar properties are dominated by the core-forming block, and Z does not vary significantly with C_s.

The dependence of the radius of gyration (R_g) and the solvent quality on salt content was also evaluated for PS-*b*-PANa/water block polyelectrolytes [30]. An increase in the extension of the PANa corona with decreasing salt content was observed for PS(6)-*b*-PANa(180), in which R_g at 0.025 M was twice the value at 2.5 M NaCl. Information on the solvent quality for the micelles was obtained from the second virial coefficient (A_2) as a function of added salt. The A_2 values for the block polyelectrolytes were similar in magnitude and agreed with literature values for poly(sodium acrylate) in the appropriate range of molecular weights. The A_2 values decreased with salt content and a linear dependence with C_s^{-1} was observed. The coronal chains were found to be under θ conditions at ca. 0.3 M NaCl.

The effects of the soluble and insoluble block lengths on the aggregation numbers of PS-*b*-PANa block polyelectrolytes were investigated in 2.5 M NaCl [30]. The PS and PANa block lengths ranged from 6 to 71 units, and 40 to 780 units, respectively. The PS block length was found to significantly influence the aggregation numbers; for instance, when the PS block length was increased from 11 to 50 units, Z increased from 54 to 270 (ca. a factor of 5). It was found that for short PS blocks, the aggregation numbers decreased with increasing PANa block length; for instance, as the length of the ionic block increased from 89 to 400 units for a constant PS block of 6 units, Z decreased from 59 to 18. However, for micelles having longer PS blocks, the aggregation numbers were less dependent on the length of the soluble block; all samples having PS blocks of 23 units showed aggregation numbers of 148 ± 6, irrespective of the soluble block lengths. These results lend further credence to the claim that, for sufficiently long PS blocks, micellization is dominated by the self-assembly of core-forming segments, as previously noted in our discussion of cmc values.

Comparisons were made between the core radii of block polyelectrolytes in 2.5M NaCl and in the solid state [30]. The core radii for the micelles in solution were calculated from aggregation numbers, assuming an unswollen, spherical core:

$$V_c=(4/3)\pi R_c^{\,3} = ZN_BM/(\rho N_{av}) \qquad (1)$$

where N_B is the number of repeat units, M is the repeat unit molecular weight, ρ is the density of the PS core, and N_{av} is Avogadro's number. The R_c values from similar samples in the solid state were obtained from small-angle x-ray scattering (SAXS) [31]. Good agreement was obtained between these two R_c values, supporting the assumption of a compact, solvent-free core.

Aggregation numbers, core radii, and radii of gyration for the block polyelectrolytes in 2.5 M NaCl have been compared with theoretical scaling relations [32-37]. Figure 2 shows a plot of R_c as a function of $N_{PS}^{3/5}$, a relation predicted from the star-like micelle models [32,33]; the linearity of the fit shows that the data agrees well with these models. Block polyelectrolytes measured in the solid state by SAXS are also shown in the figure. It should be noted that the comparison of thermodynamic models to these kinetically-frozen aggregates can be justified, insofar as the micelles were formed under equilibrium conditions. As well, the existence of a cmc in these systems indicates that a dynamic exchange between micelles and single chains is operative under conditions of extreme dilution.

The aggregation behavior of PS-*b*-PMAA and PMAA-*b*-PS-*b*-PMAA block copolymers has been investigated in 80/20 vol% dioxane/water mixtures [38]. Several block copolymers having various molecular weights and compositions were characterized. Micellar molecular weights was evaluated by three techniques, SLS, DLS coupled with viscosity and DLS coupled with sedimentation velocity, and results obtained by these different methods were found to agree reasonably well. The aggregation numbers of diblock copolymers were found to be much larger than those of triblock copolymers with comparable copolymer mass and composition. Sedimentation velocity and combined measurements of R_g and R_h were used to measure the polydispersity of the micelles. The polydispersity was low for small micelles, and higher for larger micelles which had a tendency to form clusters. It should also be noted that in this solvent mixture, the cores of the micelles were found to be swollen with the dioxane component.

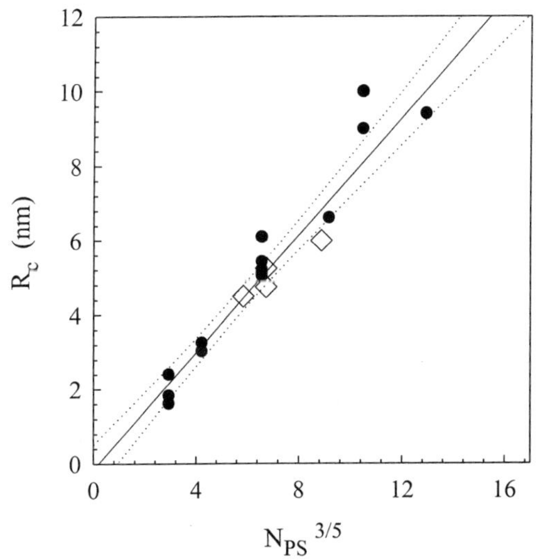

Figure 2. Radius of the micelle core plotted as predicted by the star models for PS-b-PANa in water (●) and PS-*b*-PMACs and PS-*b*-P4VP·MeI (◊) in the solid state.

The Z and R_h values have been determined as a function of PS and PMAA block lengths. The Z values increased significantly with increasing PS block length and decreased with the length of the PMAA block length [38]. The effect of the PMAA length on the R_h was moderate, and could be attributed to contributions from two factors, i.e. .a decrease in the aggregation number and a concurrent increase in the shell depth as the length of the PMAA block increased. Scaling relations for the aggregation numbers and R_h values as a function of soluble and insoluble block lengths were obtained. For the diblock copolymers these relations are:

$$Z = 0.40 N_{PS}^{1.89} N_{PMAA}^{-0.86}$$

$$R_h = 0.347 N_{PS}^{0.71} N_{PMAA}^{0.09} \text{ nm} \qquad (2)$$

and for the triblock copolymers,

$$Z = 0.041 N_{PS}^{1.62} N_{PMAA}^{-0.41}$$

$$R_h = 0.162 N_{PS}^{0.56} N_{PMAA}^{0.30} \text{ nm} \qquad (3)$$

Recently, the micellization of poly(*tert*-butyl acrylate)-b-P2VP (PBA-b-P2VP) was characterized in acidic aqueous solutions [41]. For pH < 4.8, this copolymer was found to form micelles having compact PBA cores and protonated P2VP coronae. For higher pH values, the P2VP corona collapsed resulting in precipitation of the copolymer. The stretching of the P2VP shell was found to decrease with increasing pH and ionic strength of the solution. For instance, the R_h was reduced from 63 nm at low pH and low salt content to 32 nm at the highest pH value investigated (pH 4.75, C_s = 0.1M). The solubilization of benzene in the micellar cores and the partitioning of pyrene between the micelles and aqueous phase was also investigated for this system by fluorescence spectroscopy. It is interesting to note that a soluble onion-type micelle was formed by the addition of P2VP-*b*-PEO to the PBA-*b*-P2VP copolymer at pH > 4.8 [40]. This structure consists of a PBA core, an outer core formed by the two P2VP blocks which coprecipitated and a soluble PEO corona.

2.3. Phase separation at extreme dilution

The behavior of PS-*b*-PANa block polyelectrolyte micelles at extreme dilution was investigated in aqueous NaCl solutions by static light scattering, and it was observed that micellar solutions at constant temperature and at NaCl concentrations of ≥ 2.0 M became turbid on dilution [41]. This turbidity was monitored via scattered light intensity as a as a function of concentration. For low salt content (C_s < 2.0 M), the scattered light intensity decreased linearly with decreasing concentration. However, for higher salt concentrations, the turbidity was found to increase significantly upon dilution. This anomalous turbidity was not observed in the PANa homopolymer. The concentration of turbidity onset (C_{onset}) was greater than the cmc, as determined by fluorescence measurements, and was found to increase with salt concentration.

The phenomenon of turbidity upon dilution of block polyelectrolytes was attributed to the separation of the micellar phase from solution. It was proposed that phase separation occurred due of the reduced hydration of the polyion, the decreased electrostatic repulsion between micelles, and an increase in the amount of ion binding in highly dilute salt solutions.

2.4. Degree of neutralization of the hydrophilic shell

Closely related to the system PS-*b*-PANa/water are block copolymer micelles in which the soluble coronae are made up of ionizable blocks such as poly(acrylic acid) (PAA) and poly(methacrylic acid) (PMAA). These micelles obtain the characteristics of block polyelectrolytes under conditions of high pH, as the acidic blocks become neutralized. To investigate the effect of ionization on the hydrophilic shells of such micelles, the hydrodynamic radius of PS-b-PMAA in water was probed by dynamic light scattering (DLS) as a function of the degree of neutralization (α) of the acid blocks [42]. For low polymer concentrations and low ionic strength, the apparent hydrodynamic radius (R_h) as measured by DLS increased monotonically with α. This was attributed to unscreened electrostatic repulsions between ionized moieties, which result in stretching of coronal chains. At higher polymer concentrations, the plot of R_h versus α was more complicated, and a maximum in R_H was observed at $\alpha = 0.25 - 0.35$.

Analysis of the autocorrelation function indicated the presence of two diffusion modes, with diffusion coefficients which differed by a factor of ca. 30. The fast diffusion mode was attributed to the collective diffusion which have also been observed in polyelectrolyte solutions [43] and charged latex particles [44]. The slow mode may correspond to the diffusion of clusters of structurally organized micelles.

2.5. Micelle hybridization

In order to gain insight into the dynamics and thermodynamics of the unimer-micelle equilibrium for the PS-*b*-PMAA system, the hybridization of two micelle populations was investigated by sedimentation velocity [45]. The hybridization process was studied as a function of the architecture and composition of the copolymer as well as composition of the dioxane/water mixture. When two micellar populations were mixed in the same solvent mixture, the rate of hybridization decreased as either of the block sizes increased. This result was explained by a decrease in the rate of escape of unimers from the micelles, which decreased by increasing the size of the PS block extricating itself from core and by increasing the width of the PMAA shell, complicating passage of unimers from the micelle.

The effect of solvent composition on hybridization was also studied [45]. Hybridization was found to occur only for a narrow range of solvent composition with low water content. As the water content of the solvent increased, the decreased swelling of the core by the dioxane component decreased the mobility of chains within the micelles, and decreased the rate of unimer-micelle exchange. In the extreme case, when the micelles are in pure water, the core becomes glassy and the dynamics of exchange are frozen. For two dynamic micellar populations, it was found that the reentry of large molecules into small micelles was more feasible than the reentry of small unimers into large micelles. In some cases, the hybridization of copolymers with very different insoluble block lengths resulted in mixtures of two different types of hybrid micelles.

2.6. Solubilization of organic molecules

The potential of block polyelectrolyte micelles as water-soluble microcapsules for drug delivery has sparked several studies on the solubilization of organic molecules in the micelle cores [42,46,47]. The quenching of fluorescent probes by fumeric acid in the hydrophobic domains of poly(9-vinylphenanthrene)-*b*-poly(methacrylic acid) (PVPh-*b*-PMAA) in water was investigated as a function of the nature of the PMAA chains [46]. At a pH of 7 and 9, the anionic quencher was unable to reach the core due to electrostatic repulsions in the neutralized PMAA chains, and no quenching was observed. At lower pH, the PMAA chains were only partly neutralized, allowing fumeric acid to penetrate the shell and quench the fluorescent probes. The solubilization of toluene in block copolymers of poly(*t*-butylstyrene)-*b*-poly(styrene sulfonate) (PTBS-*b*-PSS) in water has also been investigated [47]. Solubilization resulted in an increase in the viscosity of the micelle solution, the magnitude of which was a function of the toluene to polymer ratio.

The solubilization of benzene in PS-*b*-PMAA micelles was examined by DLS and fluorescence spectroscopy as a function of pH, ionic strength and copolymer concentration [42]. The state of ionization of the PMAA shell was found to have a significant effect on colloidal stability and thus the solubilization capacity of the micelles. At low pH, the solubilization capacity was found to be relatively low, and copolymer precipitation was observed under some conditions of benzene loading. At high pH, the solutions were more stable, and large amounts of organic liquid could be solubilized without precipitation of the copolymer. The location of solubilized benzene was also found to pH dependent. At low pH (< 9), solubilization occurred simultaneously in the PS cores and the PMAA shells, which collapsed partially with the induced formation of hydrophobic domains. At high pH of 9-11 the repulsion of ionized carboxylic groups inhibited the formation of hydrophobic domains in the shell, and benzene solubilized predominately in micellar cores. The mobility of fluorescent probes, as investigated by eximer formation and time-dependent fluorescence anisotropy measurements, was found to be sensitive to the amount of solubilized benzene.

The solubilization of a wide range of organic molecules in PS-*b*-PMAA block copolymer micellar solutions at pH 7 has been studied [48]. The R_h of the micelles was not found to vary appreciably after solubilization of organic molecules; however, the intensity of scattered light changed considerably. A method was developed to determine the extent of solubilization by the micelles (δ), by measuring the intensity of scattered light before (R_{mic}) and after solubilization (R_{sol}). The amount of solubilization was described as

$$\delta = \frac{m_{sol}}{m_{mic}} = \frac{(dn/dc)_{mic}}{(dn/dc)_{sol}}(\sqrt{R_{sol}/R_{mic}} - 1) \tag{4}$$

where m_{sol} and m_{mic} are the masses of a single micelle with and without solubilizate and $(dn/dc)_{mic}$ and $(dn/dc)_{sol}$ refer to the refractive increment of the organic molecule and the block copolymer micelle, respectively. The amount of solubilization was found to depend on the interactions between the organic molecule and the micellar cores. This interaction was evaluated from the Flory-Huggins interaction parameter between PS and the organic molecules (χ) as determined by inverse gas chromatography [49]. It was found that as the χ values decreased, the extent of solubilization increased. This trend suggests that molecules which have a strong interaction with PS (i.e. low χ), such as chloroform, are solubilized substantially by the block copolymer micelles. On the other hand, aliphatic hydrocarbons which are poor or non-solvents for PS are solubilized to a much smaller extent.

A thermodynamic model describing the extent of solute solubilization in the block copolymer micelles has been developed [48]. The extent of solubilization was essentially governed by a balance of two factors, i.e. the interactions between the core and the solubilizate, which favor solubilization, and the surface tension at the core boundary, which opposes solubilization. The model predicts that larger cores will swell more than smaller cores, due to the interfacial tension-induced pressure inside the cores, which decreases with increasing core radius. This increase in core swelling was observed in the experimental data. In general, the theoretical calculations were found to agree reasonably well with experimental solubilization results obtained for the block copolymer micelles investigated.

3. CREW-CUT BLOCK POLYELECTROLYTES

3.1 Formation of crew-cut micelles in aqueous solutions

3.1.1. Preparation of crew-cut micelles. Although their existence had been predicted by de Gennes some time ago, it is only quite recently that crew-cut micelles have received attention from experimentalists in this laboratory [23,50-58] and in others [59,60] In the present review, we cover mainly our recent work on crew-cut micelles of polystyrene-*b*-poly(acrylic acid) (PS-*b*-PAA) [23,51-57]. While star micelles can be prepared by the direct dissolution of a highly asymmetric block copolymer, the large weight fraction of the insoluble block makes this method impracticable for the preparation of crew-cut micelles. Crew-cut micelles have therefore been prepared by the initial dissolution of the diblock copolymers in N,N-dimethylformamide (DMF), which is a common solvent for both PS and PAA. Subsequently, water is added slowly to the polymer/DMF solution, such that the quality of the mixed solvent for the PS block gradually decreases. At a certain water content, the aggregation of PS blocks results in micelle formation, as indicated by solution turbidity. The addition of water is continued until well after micelle formation is complete. The resulting colloidal solutions are then dialyzed against distilled water to remove DMF from the mixed solvent.

3.1.2. Phase separation behavior and micelle formation. Because crew-cut micelles are prepared by adding water to a homogeneous polymer/DMF solution, a micro-phase separation process is involved in the course of micellization, resulting in the formation of a micellar pseudo-phase. Static light scattering was used to explore the micro-phase separation behavior as a function of added water content [56]. It was found that the PS-*b*-PAA diblock copolymer solutions undergo microphase separation when the added water content reaches some critical value. The critical water content for phase separation (or cwc) depends on both the polymer concentration and the molecular weight. Figure 3 shows a plot of cwc against the logarithm of the polymer concentration for three PS-*b*-PAA copolymers of different molecular weights. One finds that the cwc decreases linearly with the log of polymer concentration, and also decreases with the molecular weight of the copolymer. Moreover, the slope of the lines, d[cwc]/d[log c_0], decreases with the molecular weight. For a specific polymer concentration, the cwc decreases with an increase in the PS block length, implying that micellization occurs at lower water content for copolymers with longer PS blocks. These effects can be related to the solvent-PS (Flory-Huggins) interaction parameter, χ, which increases with the water content of the DMF/water mixture; from a thermodynamic point of view, the micelles form at χ values which decrease with increasing the PS block length and/or polymer concentration.

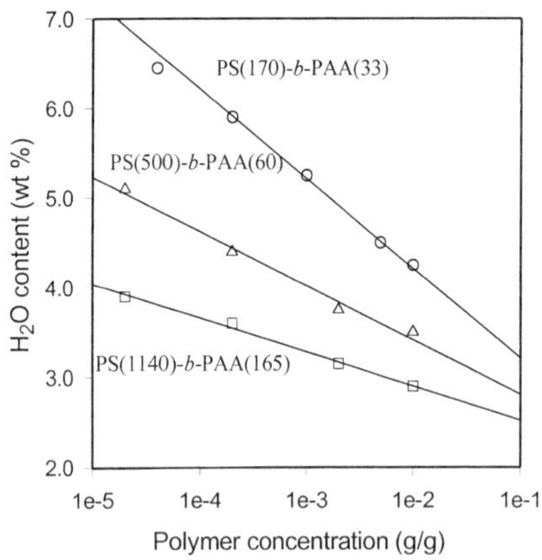

Figure 3. A plot of the cwc against the logarithm of the polymer concentration for three PS-*b*-PAA copolymers of different molecular weights.

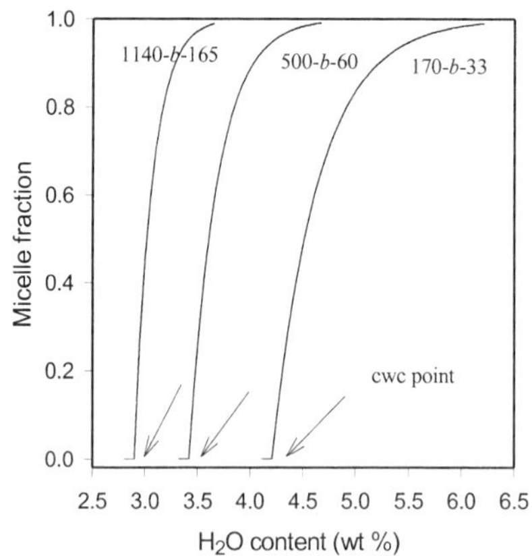

Figure 4. A plot of the micelle fraction against the water content for 1 wt% concentration solution of different block copolymers in DMF.

According to the relationship between the cwc and the initial copolymer concentration c_0 in Figure 3, a general equation can be written as

$$\text{cwc} = -A \log c_0 + B \qquad (5)$$

where A and B are constants for a specific copolymer. At the critical water content, the concentration of the copolymer represents its critical micellization concentration (cmc). As the water content in the solution increases, the cmc decreases, and more and more polymer chains form micelles. The micelle fraction can be defined as a ratio of associated polymer to the total polymer concentration, $(c_0 - \text{cmc})/c_0$. Therefore, for a solution of a certain copolymer concentration, the change in the micelle fraction as a function of the added water content can be calculated according to the following equation [5,6]:

$$(c_0 - \text{cmc})/c_0 = 1\text{-exp}\,[2.303(\text{cwc-}H_2O\%)/A] \qquad (6)$$

When $H_2O\%$ is lower than the cwc, the ratio of $(c_0 - \text{cmc})/c_0$ equals zero, because all of the copolymer chains are unassociated. Figure 4 shows a plot of the micelle fraction against water content for 1 wt% solutions of different block copolymers in DMF. It is clear that copolymers of different molecular weights form micelles in different regions of water content. For copolymers with relatively long PS chains, micellization is complete within a narrow range of water content; in contrast, copolymers with fairly short PS blocks undergo micellization over a broad range of added water. This suggests that, even for a single copolymer solution, micelles may be formed at a value of the χ parameter which is slightly different from micelle to micelle. However, as will be seen in the next section, in the early stages of micellization the structure of the micelles is still labile, and a dynamic equilibrium between micelles and unimers exists. The structural parameters of the micelles can therefore still change in response to the solvent quality.

The micellization from solutions of mixed polymers has also been explored [56]. The polymer mixtures were made from either two different block copolymers or one copolymer and one PS homopolymer. It was found that solutions of mixed polymers show a cwc that is largely determined by the component with the lower cwc.

3.2. Spherical crew-cut micelles

3.2.1. Scaling relations and structural characteristics. At most of the compositions investigated in this study, PS-*b*-PAA block copolymers formed spherical micelles [23,52]. A TEM picture of micelles formed from the copolymer PS(500)-*b*-PAA(60) is presented in Figure 5. The size distribution of the micelles is narrow (average diameter: 30.3 nm, standard deviation: 2.4 nm). Because the PAA content in the copolymers is low, the coronal layer on the surface of the core is very thin, once the micelles have dried on the TEM grid (ca 0.30 nm for the micelles in Figure 5). Thus, the particle sizes measured directly from the prints represent the actual core dimensions to a very good approximation. It was found that the number-average core radii (R_{core}) depend on the chain lengths of both the insoluble PS block (N_{PS}) and the soluble PAA block (N_{PAA}), and can be fitted empirically to the relation

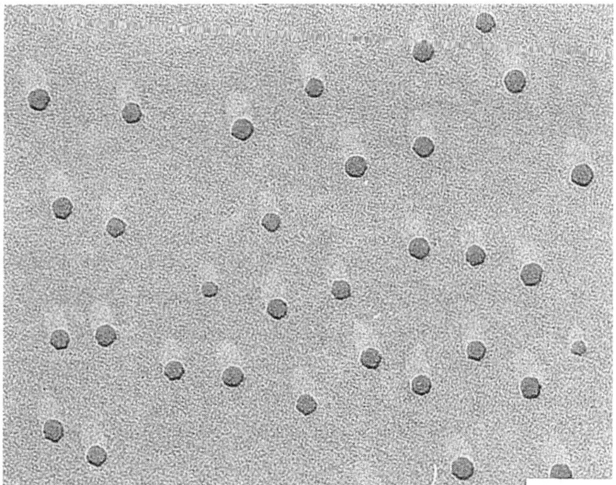

Figure 5. Spherical micelles from the block copolymer PS(500)-*b*-PAA(60) (average diameter: 30.3 nm and standard deviation: 2.4 nm).

$$R_{core} \sim N_{PS}^{0.4} N_{PAA}^{-0.15} \quad (7)$$

where N_{PS} and N_{PAA} are the degrees of polymerization of the PS and the PAA blocks of the copolymers.

One of the important aspects of this work is that the number average micelle core radius, R_{core}, appears to depend not only on the length of the insoluble PS block, but also on the length of the soluble PAA block. Also, the core radius shows a weaker dependence on the insoluble block length than those predicted by scaling theories [61], which may be attributed to two aspects of the present system. The first is the relatively low interfacial energy between the core and the solvent during micellization; as predicted by de Gennes, Leibler et al [34,35], lower interfacial tension generally results in weaker scaling of core radii with the length of the core-forming block. The second consideration is that the present scaling relation applies to isolated aggregates, which are kinetically-frozen, while theoretical predictions are based on thermodynamic equilibrium conditions.

Assuming that the micelle cores are compact, and that the density of polystyrene in the cores is homogeneous, the scaling relation of equation 7 allows one to calculate the surface area per coronal chain at the core/shell interface (A_c), along with the degree of PS chain stretching (S_c) in the cores [52]. S_c is defined as the ratio of the micelle core radius to the PS chain end-to-end distance in the unperturbed state (proportional to $N_{PS}^{1/2}$). Clearly, at least some of the associated PS chains must extend from the surface to the center of the micelle core. Both A_c and S_c depend on the composition of the block copolymers, and the following relationships have been obtained

$$A_c \sim N_{PS}^{0.6} N_{PAA}^{0.15} \quad (8)$$

$$S_c \sim N_{PS}^{-0.1} N_{PAA}^{-0.15} \quad (9)$$

Once again, the above equations apply to isolated spherical micelles in aqueous solution. Since the cores are expected to be swollen with DMF in the early stages of micellization, S_c values calculated from the radius of compact cores are expected to be lower than those corresponding to the initial equilibrium structures.

The large dependence of A_c on the PS block length is related to the low degrees of stretching of PS chains in the cores. Only when the core radius is increased to the contour length of the PS block will A_c become independent of the insoluble block length [52]. The dependence of A_c on the PAA block length can be ascribed to the strength of repulsive interactions between coronal chains during micellization. Compared to the star-like systems, crew-cut micelles show a low density of coronal chains on the core surface, and a consequently large surface area per coronal chain (A_c). This implies that both repulsive interactions between coronal chains and the deformation energy of the PS blocks in the cores are important factors during the micelle formation.

3.2.2. Coronal dimensions. The hydrodynamic radii (R_h) of spherical crew-cut micelles have been determined by dynamic light scattering (DLS). Based on the values of

R_{core} and R_h, the thickness of the coronal shell in solution can be obtained, and this quantity has been used to calculate the degree of extension of the PAA chains [23]. Since the PAA segments are ionizable, their conformations are expected to be affected by changes in the degree of neutralization, related to the pH of the solution, and by the presence of added electrolyte. When micelles were prepared by dialysis against distilled water, the resulting solutions were found to have a pH of ca. 7, indicating that the PAA segments were ionic. DLS of these samples revealed that the coronal chains were highly extended (60 ~ 97 % of the contour length of the PAA block). When similar micelles were prepared by dialysis against deionized water, the resulting solutions showed a pH of ca. 4.5. For PAA homopolymer at this pH, it is known that the degree of neutralization of the polymer chains is very low. Under these circumstances, the degree of extension of coronal chains was only 25 %. It was also found that in the presence of 1 M NaCl, the ionized coronal chains are more highly stretched than those at pH 4.5, but much less stretched than the unscreened ionized chains.

We consider the high degree of extension of coronal chains observed at pH ca. 7 to be due to the charged nature of PAA, the low degree of curvature of the surface to which the chains are attached, the relatively low macroscopic concentration of micelles in the DLS experiments, and the polydispersity of the PAA blocks in the copolymers. When inert electrolytes are added, or when the pH of the solution is low, the chains adopt a more random conformation, because the range of electrostatic repulsion between coronal chains is reduced.

3.2.3. Study of the thermodynamics of micellization in DMF/water mixtures. The thermodynamics of crew-cut micelle formation from PS-*b*-PAA copolymers in DMF/H_2O mixtures have been studied by static light scattering [57]. From the measurements of scattered light intensity as a function of temperature, the temperature-dependence of the critical micelle concentration (cmc) was determined at different copolymer concentrations. The results were used to calculate the standard Gibbs free energies ($\Delta G°$), the standard enthalpies ($\Delta H°$) and the standard entropies ($\Delta S°$) of micellization [62].

All $\Delta G°$, $\Delta H°$ and $\Delta S°$ values were found to be negative, and to be strongly dependent on the composition of the solvent mixtures, the lengths of both blocks, and the concentration of added NaCl. In particular, as the water content increases, $\Delta G°$ decreases because of the decreasing of the cmc. Moreover, as the water content increases, both $\Delta H°$ and $\Delta S°$ increase. By extrapolating the data to higher water content, it was found that both $\Delta H°$ and $\Delta S°$ become positive at 15 and 7.3 wt% added water, respectively. Therefore, it appears that in the region of lower water content, $\Delta H°$ is mainly responsible for micellization; at higher water contents, the major driving force for micelle formation becomes $\Delta S°$. Both PS and PAA block lengths have an effect on the thermodynamics of micellization. As expected, the effects of the core-forming block and of the corona-forming block are opposite. The addition of NaCl has a similar effect on $\Delta G°$ of micellization to that of a decrease in the PAA block length. This is consistent with the results of the morphogenic effect of added ions on the aggregates, as will be discussed in section 3.4.1. General equations for $\Delta G°$, $\Delta H°$ and $\Delta S°$ as a function of all studied parameters can be written as

$$\Delta G°(kJ/mol) = 20.4 - 0.00859N_{PS} + 0.00938N_{PAA} - 10.2w_{H2O} - 3.65M_{NaCl}$$

$$\Delta H°(kJ/mol) = -151 + 0.0906N_{PS} - 0.0692N_{PAA} + 7.38w_{H2O} - 22.9M_{NaCl} \quad (10)$$

$$\Delta S° (J/mol) = -573 + 0.333N_{PS} - 0.264N_{PAA} + 59.0w_{H2O} - 64.5M_{NaCl}$$

Where w_{H2O} is the water content in wt%, and M_{NaCl} is the added NaCl concentration in millimoles. It should be noted that these equations are only valid within the range of experimental conditions that were investigated in ref. 57.

3.2.4. Dynamics of chain-exchange between micelles. In the course of water addition to the copolymer/DMF solution, the solvent quality for the PS blocks decreases gradually. It is expected that, in the early stages of copolymer micellization, the micelle cores are highly swollen with DMF. A dynamic equilibrium between micelles and the copolymer unimers is therefore operative. This allows the exchange of polymer chains between micelles, and changes in micelle structure in response to changing solvent quality. However, as the water content increases, DMF is gradually removed from the cores and the mobility of the PS chains decreases. The exchange of chains thus becomes kinetically controlled. At a certain water content, the micelle structure eventually becomes kinetically-frozen on the time scale of the experiment.

The influence of the water content on the rate of chain exchange has been investigated, by mixing two different micelle populations at various water contents and studying the change of the core size distribution by TEM [57]. The two copolymers were PS(1140)-*b*-PAA(165) and PS(170)-*b*-PAA(33); the micellization of these copolymers as a function of added water content is shown in Figure 6. After a predetermined quantity of water was added to each solution, aliquots of each of the two solutions were withdrawn and mixed. Then, the mixed solutions were stirred for one day to allow polymer chain exchange and partition over that time scale. If the mobility of polymer chains is significant at a particular water content over the time scale of the experiment, then chain exchange between different micelles will homogenize the micelle populations. On the other hand, if the chain exchange process is very slow, the micelle populations will remain bimodal. The addition of water to the mixed solutions was continued until the water content was 25 wt%. The aggregates were isolated into water by dialyzing the resulting solutions.

Figures 6-A and B show the size distributions of spherical micelles from separate solutions of PS(1140)-*b*-PAA(165) and PS(170)-*b*-PAA(33). The size distributions of each micelle sample are seen to be monomodal, with average diameters of 37 nm and 25 nm, respectively. Due to their narrow size distributions, the micelle populations are separated reasonably well. The micelle populations made by mixing two different copolymer micelle solutions at different water contents are shown in Figure 6-C to E. When the micelle solutions are mixed at 5 wt% water content and allowed to equilibrate for one day (Figure 6-C), the micelle population is monomodal, with an average core diameter of 41 nm. The average diameter is larger than those made from either PS(1140)-*b*-PAA(165) or PS(170)-*b*-PAA(33) alone, due to an increased polydispersity of PS blocks in the mixture of the two copolymers [61]. When the water contents at mixing are 7 and 9 wt% (Figure 6-D and E), the micelle population is bimodal, indicating that the rate of polymer chain exchange among the micelles of different sizes has slowed down significantly. When the two micelle solutions are mixed at an added water content of 11 wt% (Figure 6-E), the micelles show two separate populations, each of which is similar to the distributions in Figure 6-A and B. It was concluded that although the two micelle solutions are mixed with each other and allowed to interact for one day, chain exchange over this time scale is negligible. This suggests that micelle structures are kinetically locked when the added water content is ca. 11 wt%.

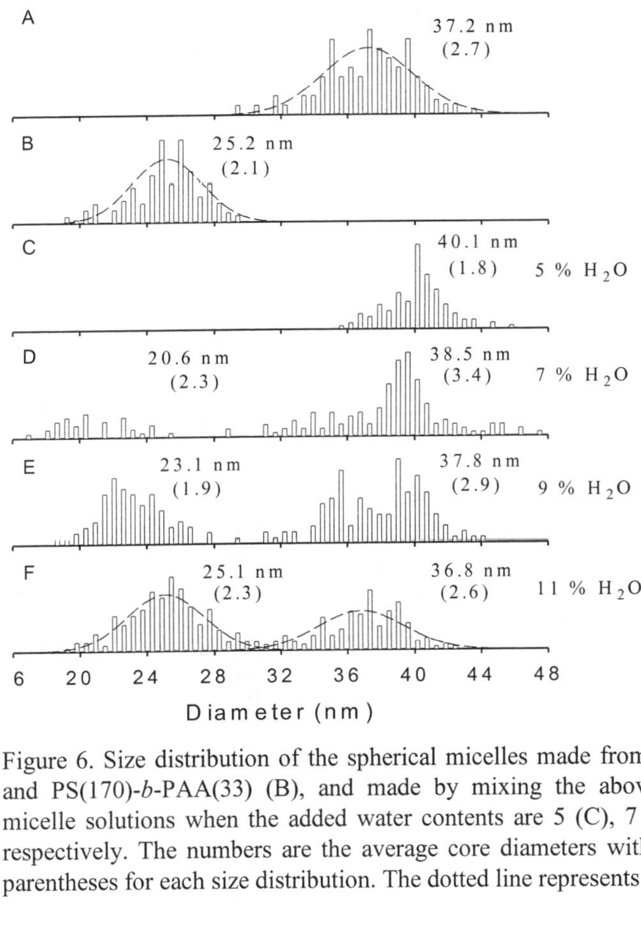

Figure 6. Size distribution of the spherical micelles made from PS(1140)-*b*-PAA(165) (A) and PS(170)-*b*-PAA(33) (B), and made by mixing the above two different copolymer micelle solutions when the added water contents are 5 (C), 7 (D) 9 (E) and 11 wt% (F), respectively. The numbers are the average core diameters with the standard deviations in parentheses for each size distribution. The dotted line represents a Gaussian fit.

3.3. Morphological transitions as a function of the copolymer composition

3.3.1. Various morphologies and their characteristics. While the copolymers with relatively high PAA block contents form spherical micelles, the morphology of the aggregates is found to change as the PAA content decreases [51,52]. For example, spherical micelles are formed as water is added to 2% solutions of PS(180)-*b*-PAA(28) in DMF. Rod-like micelles, as shown in Figure 7A, are obtained from 2% solutions of the block copolymer PS(180)-*b*-PAA(15). The rod-like micelles have a relatively narrow distribution of diameters, but a widely variable length. As the PAA block content drops further at the same polymer concentration, vesicular aggregates are formed. An example of vesicles from PS(410)-*b*-PAA(16) is shown in Figure 7B. The PS wall thicknesses of the vesicles are very uniform and independent on the overall sizes of the vesicles. Lamellar aggregates, another kind of bilayer structure, were also observed occasionally in the samples which form rod-like micelles or vesicles. When the PAA block length becomes very short, e.g. PS(200)-*b*-PAA(4), the copolymers form highly polydisperse spheres, with diameters exceeding one micron, as shown in Figure 7C. A study of the internal structure of these spheres by TEM has shown that the aggregates consist of a large number of reverse micelles, in which PS blocks form the corona and the PAA blocks form the polar micelle core (Figure 7D). However, the surface of these aggregates must be hydrophilic because of the presence of PAA chains. These aggregates have been termed large compound micelles (LCMs).

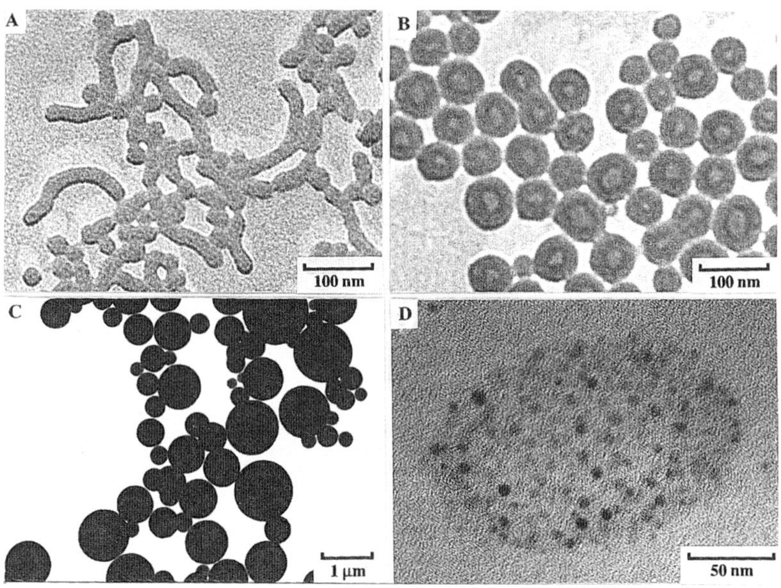

Figure 7. (A) rod-like micelles from PS(180)-*b*-PAA(15), (B) vesicles from PS(410)-*b*-PAA(16), (C) large compound micelles from PS(200)-*b*-PAA(4), and (D) the internal morphology of the micelles in (C).

While the surface area per coronal chain at the core/corona interface of spherical micelles, A_c, depends on the composition of the copolymers (*cf.* equation 8), it also decreases as the morphology changes from spherical to cylindrical, and to vesicular [52], as does the degree of stretching of PS segments, S_c. For example, the block copolymers 200-*b*-21, 200-*b*-15 and

200-*b*-8 have the same PS block, and form, respectively, spheres, rods and vesicles. The degree of stretching of the PS chains in these different morphologies are 1.41, 1.26 and 0.99, respectively. If the last two copolymers were to form spherical micelles, the stretching would be expected to increase as the PAA block length decreases. According to equation 9, the values of S_c would then be 1.47 and 1.62, respectively (rather than 1.26 and 0.99).

3.3.2. Thermodynamics of morphological transitions. Although various morphologies have been observed in many self-assembled systems, such as small molecule amphiphiles in solution, block copolymers in bulk phase, block copolymer/homopolymer blends, amphiphilic block copolymers on water surfaces, and polystyrene amphiphilic dendrimers in solution, the present system provides the first instance in which a number of the aggregate morphologies has been observed directly for block copolymers in a low molecular weight solvent.

Considering only the effects of interfacial tension, a large micellar aggregation number is favored in order to minimize this contribution to the total free energy. However, the aggregation number of spherical micelles is limited by the increase of the deformation (stretching) energy of the core-forming blocks as the radius increases. Therefore, the morphological transitions are believed to be mainly controlled by the balance of three forces: 1. the stretching of the PS chains in the core, 2. the surface tension between the PS core and the solvent and 3. repulsive interactions among the corona chains. The strength of interactions between coronal chains is related to the PAA block lengths (or coil dimensions). As the length of PAA block decreases, the inter-corona chain repulsion decreases, and the core size increases according to the scaling relation in equation 1. However, the increase in the core size cannot continue indefinitely, due to the entropic penalty of chain stretching in the core. At some point, the system becomes unstable, and the morphology changes, first from spherical to rod-like, and then to lamellar or vesicular, presumably to reduce the thermodynamic penalty of chain stretching. Therefore, the degree of stretching of PS blocks decreases with each morphological transition, as shown in the previous section. Nagarajan has recently pointed out that the amount of solvent in the core will also effect the morphology [64]. This is very relevant to the present system at the point of aggregate "freezing".

3.4. Other factors controlling the aggregate morphology

3.4.1. Ion-induced morphological changes. The morphology of block copolymer aggregates formed from a single block copolymer sample can be controlled via the addition of micromolar $CaCl_2$, $Ca(Ac)_2$, NaOH or HCl, or milli-molar NaCl [53,55]. In addition to the various aggregates that have been previously described, a new morphology, termed large compound vesicles (LCVs), have been prepared using added salt.

The salt effect can be demonstrated by considering a 1 wt% PS(410)-*b*-PAA(13) solution, which forms spherical micelles with an average core diameter of 31 nm (Figure 8A). When the added $CaCl_2$ concentration is 71 μM (R=0.025; R is the molar ratio of added $CaCl_2$ molecules to acrylic acid repeat units), the micelles are still spherical, but their average core diameter has increased to 39 nm (Figure 8B). The copolymers form vesicles when the $CaCl_2$ concentration is 90 μM (R=0.032) (Figure 8C). Finally, spherical large compound vesicles (LCVs) form when the $CaCl_2$ concentration is 130 μM (R = 0.045)

(Figure 8D). The structure of the LCVs has some resemblance to that of aggregated soap bubbles. Their formation may be a result of kinetic control involving the collisions of individual vesicles, and a subsequent PS wall-fusion process. A number of novel features of LCVs may make them especially useful as vehicles for drug-delivery. Furthermore, the spontaneous formation of LCVs may provide a practical method for making stable artificial tissue-like composites and soft biomaterials with tunable properties.

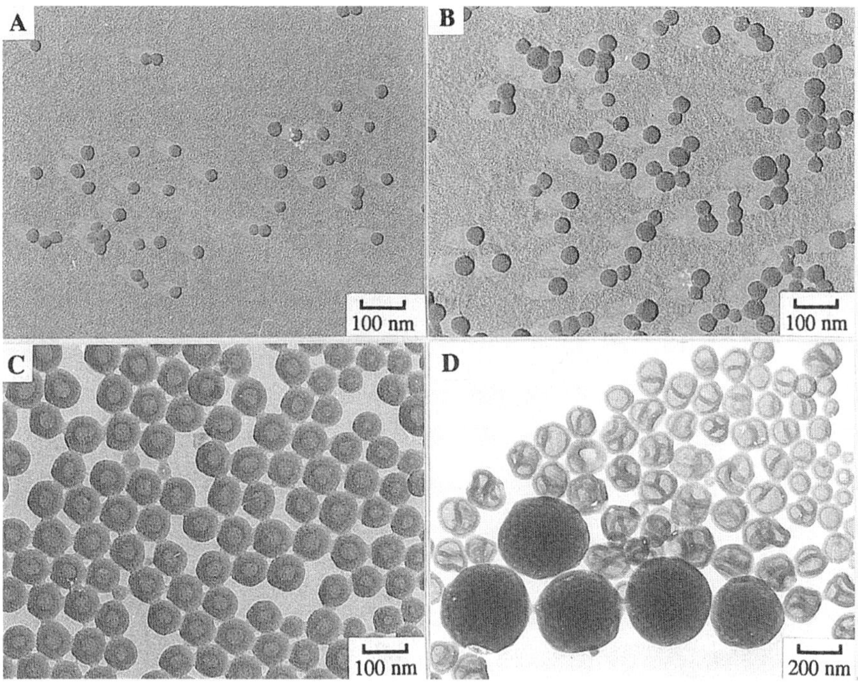

Figure 8. Aggregates from PS(410)-*b*-PAA(13) without any additive (A), and with added $CaCl_2$ to different final concentrations: (B) 71 μM (R = 0.025), (C) 90 μM (R = 0.032), and (D) 130 μM (R = 0.045).

The effect of added ions on the aggregate morphology is parallel to a change in the soluble block length. In order to understand the mechanism of the morphogenic effect of added ions, proton NMR has been used to study the change in the micro-environment of the PAA segments during micelle formation [55]. It was found that added water accumulates preferentially near the PAA blocks to some extent. Thus, although the micellization usually starts at a water content of around 4 wt%, the local water concentration around the PAA blocks can be much higher than the overall water concentration in the solution. The PAA blocks probably behaves in a manner similar to that in aqueous solutions, i.e. they are partially ionized during the formation of the micelles. Because the PAA segments are partially charged, the repulsive interactions (both steric and electrostatic) between coronal chains are operative. Therefore, the morphogenic effect of added ions can be ascribed to changes in repulsive forces between coronal chains in the presence of added ions.

In the case of added HCl, the ion effect is attributed to the protonation of the PAA blocks. The addition of strong acid HCl shifts the equilibrium to a lower degree of PAA neutralization, and greatly reduces the electrostatic or steric repulsion among the PAA chains. Therefore, the decreased repulsion between coronal chains has an effect on the aggregate morphology similar to a decrease in the length of the PAA block. However, added acetic acid shows almost no effect on the aggregate morphology; since it is a weak acid, it has little effect on the degree of neutralization of the PAA blocks. On the other hand, the addition of NaOH has an effect on the aggregate morphology similar to an increase in the PAA block length; the added NaOH neutralizes the PAA segments, and the degree of neutralization and the repulsion among the corona chains are therefore increased. The strong morphogenic effect of added $CaCl_2$ originates from strong Ca^{2+} binding to the carboxylic acid groups of the PAA segments. A simultaneous morphogenic effect of HCl may also be involved, arising from the formation of calcium acrylate and HCl due to Ca^{2+} binding. Finally, the weaker morphogenic effect of added NaCl can be ascribed to the screening of the partially-charged PAA segments by the low-molecular weight salt.

3.4.2. Effect of PS homopolymer addition. The effect of added homopolystyrene on the aggregate morphology has also been studied [52]. In order to incorporate homopolystyrene into the micelle core, polymer/DMF solutions with a known weight ratio of PS-*b*-PAA diblock copolymer to homopolymer were prepared. Deionized water was then added to the solutions. At a critical water content, micellization of the block copolymer commences; at the same time, homopolystyrene is solubilized into the micelle core. According to the phase separation behavior of mixed polymer solutions, the added PS homopolymer should have a higher cwc for precipitation than that of the block copolymer for micellization. Otherwise, the homopolymer chains will precipitate first from the solution as the water content increases, and will not become incorporated into the PS core. A higher cwc can be achieved by decreasing the polymer concentration and/or the molecular weight. Evidence for the solubilization of added homopolystyrene into the micelle core was observed, as a clear micelle solution and a narrow distribution of micelle sizes were obtained, without precipitation of the homopolystyrene.

Upon addition of PS homopolymer, it was found that block copolymers which form spherical micelles without added homopolystyrene still yield spherical micelles. The homopolymer chains are generally distributed throughout the core when their content is relatively low. The spherical micelles containing homopolystyrene have a low polydispersity, with core radii that are slightly larger than those found in micelles without homo-PS. As the content of added homopolystyrene increases, the core sizes become more polydisperse. Homopolymer accumulation in micelle core center can be inferred. The addition of homopolystyrene changes the morphologies of the aggregates from bilayers or cylinders to spheres. For example, vesicles are formed from a 2 wt% PS(410)-*b*-PAA(16) solution, but spherical micelles are obtained upon addition of 10 wt% PS(180) homopolymer to the same copolymer. These changes in morphology are mainly ascribed to the reduction in the degree of stretching of the PS blocks, due to the accumulation of small amounts of homopolystyrene at the center of the core.

3.4.3. Effect of the initial copolymer concentration. The morphology of the aggregates has also been found to depend on the initial copolymer concentrations in DMF [54]. For example, for PS(410)-*b*-PAA(25) block copolymer, when the copolymer concentration is 2 wt%, spherical micelles are formed. Rod-like micelles are formed from a 2.6 wt% copolymer solution. When the concentration is 3.0 wt%, vesicles start to form with coexisting rod-like micelles. If a 4.0 wt% solution is used, most of the aggregates are vesicles with highly polydisperse sizes, with outer diameters ranging from 50 to 500 nm.

3.5. Multiple morphologies of PS-*b*-PEO aggregates

In this section, we review very briefly recent studies of the micellization of polystyrene-*b*-poly(ethylene oxide) (PS-*b*-PEO) block copolymers. Although PS-*b*-PEO is not strictly an ionic copolymer, the micellization of these materials shows some similarities to those of block polyelectrolytes. Extensive aqueous solution studies have been performed on spherical micelles made from PS-*b*-PEO copolymers in which the soluble PEO blocks are much longer than the insoluble PS blocks [16,65-67] It was shown that the apparent molecular weight (or the aggregation number) of the micelles increases with the copolymer molecular weight, and with decreasing PEO content [65]. Both TEM and light scattering experiments show that bimodal distributions of micelle sizes exists [66,67]. Recently, it has been shown that non-spherical aggregates can be prepared from PS-*b*-PEO copolymers which have PS blocks that are much longer than the PEO blocks [58]. In particular, the morphologies of the aggregates are a function of the soluble block length. As the PEO content decreases, one obtains, as the dominant morphology, spherical micelles, then rods, then lamellae, and finally vesicles. It was also found that the morphologies are strongly dependent on the concentration of added ions or inert electrolytes [53], allowing one to obtain a wide range of new morphologies, including tubules, irregular bilayer surfaces, aggregated rods, aggregated vesicles, and "pin cushions" (i.e., large compound micelles with protruding rods).

4. CONCLUDING REMARKS

This review has summarized some primary findings in the field of block polyelectrolytes, including discussions of both star-like and crew-cut micelles. Star micelles formed from PS-*b*-PANa have been shown to be an interesting and highly stable class of charged colloidal particles. These micelles were characterized in aqueous media by fluorescence spectroscopy and by static light scattering. Information on the cmc, aggregation number and radius of gyration has been obtained for a wide range of block copolymer systems, including PS-*b*-PANa in aqueous solutions, PS-*b*-PMAA and PMAA-*b*-PS-*b*-PMAA in dioxane/water mixtures, and PS-*b*-P4VPEtBr in water/methanol mixtures. For the PS-*b*-PANa micelles, the effects of the PS and PANa block lengths and concentration of added salt on the micellar properties were investigated. The solubilization of small organic molecules into the micelle cores has also been discussed. These studies are important steps toward the application of aqueous micelles as vehicles for drug delivery.

Crew-cut micelles formed from PS-*b*-PAA diblock copolymers have been shown to exhibit a number of interesting phenomena. In the region of spherical micelles, a scaling relation for the core radius has been empirically determined ($R_{core} \propto N_{PS}^{0.4} N_{PAA}^{-0.15}$). The morphology of the crew-cut aggregates changes progressively from spheres to cylinders, to vesicles or lamellae, to large compound vesicles and eventually to large complex micelles, as the PAA content of the PS-*b*-PAA copolymers decreases. It has also been shown that the morphology of these aggregates can be controlled by varying the polymer concentration, or through the addition of added homopolymer. As well, the morphologies observed in systems of PS-*b*-PAA and PS-*b*-PEO have been found to be sensitive to the addition of extremely small amounts of added ions, and this factor has opened the door to an even greater number of aggregate structures. The characteristics of crew-cut micelles provide unprecedented insight into the factors controlling the architecture of block polyelectrolyte colloidal aggregates, in a composition regime which exhibits much diversity.

REFERENCES

1. B. Lindman and H. Wennerstrom, in *Topics in Current Chemistry*; Springer-Verlag: Berlin 1980; Vol. 87, p 1; R. G. Laughlin, *The Aqueous Behavior of Surfactants*; Academic Press: London, 1994.
2. C. Price, in *Developments in Block Copolymers-1*; I. Goodman, Ed.; Elsevier App. Sc. Pub. U.K.: 1982, p 39.
3. Z. Tuzar and P. Kratochvil, *Advances in Colloid and Interface Science* **1976**, *6*, 201.
4. Z. Tuzar and P. Kratochvil, in *Surface and Colloid Science*; E. Matijevic, Ed.; Plenum Press. New York: 1993, p 1.
5. J. Rassing and D. Attwood, *Int. J. Pharm.* **1983**, *13*, 47.
6. G. Wanka, H. Hoffmann, W. Ulbricht, *Colloid Polym. Sci.* **1990**, *268*, 101.
7. K. Mortensen and J. S. Pedersen, *Macromolecules* **1993**, *26*, 805.
8. P. Linse, *Macromolecules* **1993**, *26*, 4437.
9. A. V. Kabanov, I. R. Nazarova, I. V. Astafieva, E. V. Batrakova, V. Y. Alakhov, A. A. Yaroslavov, V. A. Kabanov, *Macromolecules* **1995**, *28*, 2303.
10. M. Almgren, W. Brown, M. Almgren, W. Brown, S. Hvidt, *Colloid Polym. Sci.* **1995**, *273*, 2.
11. P. Alexandridis and T. A. Hatton, *Colloids Surf. A* **1995**, *96*, 1.
12. B. Chu and Z. Zhou, *Surfactant Sci. Ser.* **1996**, *60*, 67.
13. M. W. Edens, *Surfactant Sci. Ser.* **1996**, *60*, 185.
14. K. Mortensen, *Condens. Matter* **1996**, *8*, 103.
15. G. Wanka, H. Hoffmann, W. Ulbricht, *Macromolecules* **1994**, *27*, 4145.
16. R. Xu, M. A. Winnik, G. Riess, B. Chu, M. D. Croucher, *Macromolecules* **1992**, *25*, 644.
17. J. Selb and Y. Gallot, in *Developments in Block Copolymers*; I. Goodman, Ed.; Elsevier Applied Science London: 1985, Vol. 2, p 27; in *Polymeric Amines and Ammonium Salts,* E. J. Goethals, Ed., Pergamon Press: New York, 1980, p 205.
18. D. Ronis, *Physical Review A,* 1991, *44*, 3769; *Macromolecules* **1993**, *26*, 2016.
19. N. Dan and M. Tirrell, *Macromolecules* **1993**, *26*, 4310.
20. D. Ronis, *Physical Review E* **1994**, *49*, 5438.

21. I. A. Nyrkova, A. R. Khokhlov, M. Doi, *Macromolecules* **1993**, *26*, 3601.
22. J. F. Marko and Y. Rabin, *Macromolecules* **1992**, *25*, 1503.
23. L. Zhang, R. J. Barlow, A. Eisenberg, *Macromolecules* **1995**, *28*, 6055.
24. I. Astafieva, X.-F. Zhong, A. Eisenberg, *Macromolecules* **1993**, *26*, 7339.
25. I. Astafieva, K. Khougaz, A. Eisenberg, *Macromolecules* **1995**, *28*, 7127
26. M. Wilhelm, C. L. Zhao, Y. Wang, R. Xu, M. A. Winnik, J. L. Mura, G. Riess; M. D. Croucher, *Macromolecules* **1991**, *24*, 1033.
27. Z. Gao and A. Eisenberg, *Macromolecules* **1993**, *26*, 7353.
28. K. Khougaz, X.-F. Zhong, A. Eisenberg, *Macromolecules* **1996**, *29*, 3937.
29. F. C. Schwab and I. Heilweil, *Polym. Prep., A.C.S. div. Poly. Chem.* **1983**, *24*, 65
30. K. Khougaz, I. Astafieva, A. Eisenberg, *Macromolecules* **1995**, *28*, 7135.
31. D. Nguyen, S. K. Varshney, C. E. Williams, A. Eisenberg, *Macromolecules* **1994**, *27*, 5090.
32. E. B. Zhulina and T. M. Birshtein, *Vysokomol. Soedin.* **1985**, *27*, 511.
33. A. Halperin, *Macromolecules* **1987**, *20*, 2943.
34. P.-G. de Gennes, *Solid State Physics,* J. Liebert, Ed.; Academic Press, New York, 1977; Suppl. 14, p 1.
35. L. Leibler, H. Orland, J. C. Wheeler, *J. Chem. Phys.* **1983**, *79*, 3550.
36. M. D. Whitmore and J. Noolandi, *Macromolecules* **1985**, *18*, 657.
37. R. Nagarajan and K. Ganesh, *J. Chem. Phys.* **1989**, *90*, 5843.
38. A. Qin, C. Tian, C. Ramireddy, S. E. Webber, P. Munk, Z. Tuzar, *Macromolecules* **1994**, *27*, 120.
39. K. Procházka, T. J. Matin, P. Munk, S. E. Webber, *Macromolecules* **1996**, *29*, 6518.
40. K. Procházka, T. J. Matin, P. Munk, S. E. Webber, *Macromolecules* **1996**, *29*, 6526.
41. R. J. Barlow, S. Zimmerman, K. Khougaz, A. Eisenberg, *J. Poly. Sci., Polym. Phys. Ed.* **1996**, *34*, 1197.
42. D. Kiserow, K. Prochazka, C. Ramireddy, Z. Tuzar, P. Munk, S. E. Webber, *Macromolecules* **1992**, *25*, 461.
43. M. Sedlak, *Macromolecules* **1993**, *26*, 1158
44. P. Stepanek and C. Konak, *Adv. Colloid Interface Sci.* **1984**, *21*, 195.
45. M.. Tian, A. Qin, C. Ramireddy, S. E. Webber, P. Munk, Z. Tuzar, K. Prochazka, *Langmuir* **1993**, *9*, 1741.
46. Y. Morishima, Y. Itoh, T. Hashimoto, S.-I. Nozakura, *J. Polym. Sci., Polym. Chem. Ed.* **1982**, *20*, 2007.
47. P. L. Valint, Jr. and J. Bock, *Macromolecules*, **1988**, *21*, 175.
48. M. Tian, E. Arca, Z. Tuzar, S. E. Webber, P. Munk, *J. Polym. Sci., Polym. Phys. Ed.* **1995**, *33*, 1713.
49. M. Tian and P. Munk, *J. Solution Chem.* **1995**, *24*, 267; *J. Chem. Eng. Data* **1994**, *39*, 742.
50. Z. Gao, S. K. Varshney, S. Wong, A. Eisenberg, *Macromolecules* **1994**, *27*, 7923.
51. L. Zhang and A. Eisenberg, *Science* **1995**, *268*, 1728.
52. L. Zhang and A. Eisenberg, *J. Am.Chem. Soc.* **1996**, *118*, 3168.
53. L. Zhang, K. Yu, A. Eisenberg, *Science* **1996**, *272*, 1777.
54. L. Zhang and A. Eisenberg, *Macromolecular Chem. Phys., Macromol. Symp., in press.*
55. L. Zhang and A. Eisenberg, *Macromolecules* **1996**, *29*, in press.
56. L. Zhang, H. Shen, A. Eisenberg, *Macromolecules* **1997**, *30*, in press.
57. H. Shen, L. Zhang, A. Eisenberg, manuscript in preparation.

58. K. Yu and A. Eisenberg, *Macromolecules* **1996**, *29*, 6359.
59. C. Honda, K. Sakaki, T. Nose, *Polymer* **1994**, *35*, 5309.
60. A. Guo, G. Liu, J. Tao, *Macromolecules* **1996**, *29*, 2487.
61. See: Z. Tuzar, P. Kratochvil, in *Surface and Colloid Science*, Plenum Press, New York, 1993, Vol. 15, p 1, and the references therein.
62. J. R. Quintana, M. Villacampa, I. A. Katime, *Macromolecules* **1993**, *26*, 601.
63. D. Nguyen, X.-F. Zhong, C. E. Williams, A. Eisenberg, *Macromolecules* **1994**, *27*, 5173.
64. R. Nagarajan, in *Solvents and Self-Organization of Polymers, NATO ASI Series E*; S. E. Webber, P. Munk, Z. Tuzar, Eds.; Kluwer Academic Publishing: 1996; Vol. 327, p 121.
65. K. Nakamura, R. Endo, M. Takena, *J. Polym. Sci., Polym. Phys. Ed.* **1976**, *14*, 135; 1287.
66. G. Riess and D. Rogez, *Polym. Prepr. (Am. Chem. Soc., Div. Polym. Chem.)* **1982**, *23*, 19.
67. T. N. Khan, R. H. Mobbs, C. Price, J. R. Quintana, R. B. Stubbersfield, *Eur. Polym. J.* **1987**, *23*, 191.

Formation of Amphiphilic Block Copolymer Micelles in Nonaqueous Solution

T. Liu, L.-Z. Liu and B. Chu

Department of Chemistry, State University of New York at Stony Brook
Stony Brook, NY 11790-3400, USA

In contrast to aqueous solution, the nature of nonaqueous solvents for amphiphilic block copolymers can vary from organic solvents to supercritical fluids, and even homopolymers (e.g., one of the corresponding low-molecular-weight homopolymers as a solvent). The driving force of micellization of an amphiphilic block copolymer in both aqueous and nonaqueous solution is the solvent selectivity for different blocks. However, the micellization of a block copolymer in an organic solvent with hydrogen bonding properties is usually enthalpy-driven. Moreover, the variation in organic solvents leads to a diversity in the association behavior. In this chapter, the micellization behavior of amphiphilic block copolymers in selective organic solvents is reviewed.

Supercritical fluids, especially supercritical CO_2, are becoming increasingly more attractive as environmentally benign solvents. The investigation of supramolecular formation of block copolymers in supercritical CO_2 has potential because it is closely related to the development of new routes for polymer synthesis. In dilute solution, the micellization of amphiphilic block copolymers in supercritical CO_2 can be quite different when compared with regular solvents because of the unique properties of supercritical fluids. For example, the solvent quality for supercritical fluid is pressure dependent while it is much less so for aqueous or nonaqueous organic solvents.

A brief introduction on the micellization of amphiphilic block copolymers in dilute low-molecular weight homopolymer solutions is also included in this chapter in order to show the similarity of micelle formation of a block copolymer in polymer blends to the more conventional nonaqueous solutions.

Acknowledgements: BC gratefully acknowledge the support of their work by the Department of Energy (DEFG0286ER45237.012), and the National Center for Human Genome Research (1R01HG0138601). He also wishes to thank many of his colleagues and scientists who have graciously sent him their reprints and preprints, in particular to A. Eisenberg, I.A. Katime, A.P. Gast, T. Nose, C. Frank, S.P. Nunes and J.L. Fulton.

I. MICELLIZATION IN ORGANIC SOLVENTS

1. INTRODUCTION

The self-assembly behavior of amphiphilic block copolymers in organic solvents has continuously drawn attention in the last several decades, although much less work has been reported in organic solvents when compared with that in an aqueous solution. Amphiphilic diblock copolymers in aqueous solution can self-associate into micellar structures with the hydrophobic blocks forming the core and the hydrophilic blocks forming the corona. The formation of micelles with the so called "core-shell" structure often obeys a close-association mechanism and is entropy-driven. For triblock copolymers with a middle soluble block and two end blocks in a poor or nonsolvent, flower-like micelles which also obeys a close-association mechanism or even supramolecular formation with open structures which tend to obey an open-association mechanism can occur.

In view of the dominant role for the solvent in the aqueous case, it is important to determine whether similar considerations could be applied to systems based on organic solvents. The answer can be more complex because of the diverse nature of different organic solvents as well as the chemical composition and chain architecture of block copolymers. In some cases, the association mechanism and the micellar structure may be quite different from those in aqueous solution, as the solvent quality, instead of hydrophobic or hydrophilic nature of the copolymer blocks, plays a dominant role in the micellar (or supramolecular) formation of block copolymers in selective organic solvents.

The field concerning micellar formation of block copolymers in organic solvent media is less developed. However, over a hundred papers and several reviews [1-4] related to this topic have been published within the last ten years, mostly dealing with polystyrene-based or Pluronic type block copolymers. In part I, some basic properties of the micelle-forming phenomena of block copolymers in organic solvents, the similarities and differences between the aqueous-based systems and the organic-based systems, and pertinent physical methods employed in the study are reviewed. Some of the topics, such as the kinetics and dynamics of micellization [5-12] are being reviewed extensively by other authors in this book. Thus, we provide some references but omit more detailed discussions.

2. PHYSICAL METHODS USED FOR CHARACTERIZATION OF BLOCK COPOLYMER MICELLES

Numerous physical methods have been utilized in studying block copolymer micelles in organic solvents. Some have only marginal value and provide a limited amount of narrowly focused information. Usually a combination of several physical methods are needed to characterize block copolymer micelles. A brief description of the more pertinent physical methods is presented in this section.

2.1. Scattering Techniques

Scattering techniques have always been proven to be among the most powerful methods for investigation of micellar structures and properties. These involve the use of a probe radiation which may be (visible) laser light, X-rays or neutrons, as we are dealing with sizes in the nanometer scale to those approaching the wavelength of visible light. By detecting the interactions between the particles and the probing radiation, physical properties, such as micellar size, shape and internal structure, can be investigated. In order to be able to vary the magnitude of the scattering wave vector q over a large range, a probing radiation, e.g., light, X-rays or neutrons is chosen depending on the size range of interest. Here the pertinent parameter q is defined as:

$$q=(4\pi/\lambda)\sin(\theta/2) \qquad (1)$$

with λ and θ being respectively the wavelength of the probing radiation (e.g., 400-600 nm for laser light, 0.15 nm for X-rays and 0.5-1.2 nm for neutrons [1]) and the scattering angle (e.g., 15°-150° for laser light, units of degrees and angular minutes for neutrons and X-rays, respectively). Smaller structural details can be observed by choosing a higher q value, such that $q{\cdot}R>1$ with R being the characteristic micellar size.

Light has the longest wavelength and hence is limited to the low q range, implying that we can use it to measure the weight-average molecular weight (M_w) from extrapolated zero-angle time-averaged absolute scattered intensity at infinite dilution, the second virial coefficient (A_2) and the radius of gyration (R_g) in the range $q{\cdot}R_g<1$. Due to the difference in the refractive index increments of different copolymer blocks, determination of M_w and R_g depends on the micellar structure, except for polymers with similar blocks (e.g., polyethylene oxide (PEO) and polypropylene oxide (PPO) blocks have similar refractive indices).

Small angle X-ray scattering (SAXS) and small angle neutron scattering (SANS) which can cover larger q ranges are capable of detecting smaller structural details such as dimensions of the micellar core, or the thickness of the core/shell interface at high polymer solution concentrations. However, SAXS and SANS experiments are often difficult to achieve in the dilute solution regime because of low contrast. For SAXS, an appropriate solvent should be chosen in order to obtain a larger electron density difference; while for SANS, deuteration of solute or solvent is often used to achieve the needed contrast in scattering length.

In dynamic light scattering (DLS), which is also called quasi-elastic light scattering (QELS) or photon correlation spectroscopy (PCS), the intensity-intensity time correlation function $G^{(2)}(t)$ can yield information on the translational diffusion coefficient (D_t), and, in some cases for anisotropic particles, also the rotational diffusion coefficient (D_r). With the Stokes-Einstein relation:

$$R_h=kT/6\pi\eta_0 D_t \qquad (2)$$

we can calculate the hydrodynamic radius R_h. In Eq. (2), k is the Boltzmann constant, T is the absolute temperature and η_0 is the solvent viscosity. Laplace inversion of $G^{(2)}(t)$ yields information on the characteristic linewidth (and the resulting hydrodynamic size) distribution. It is noted that due to noise and upper lower limits of integration in the Laplace equation, the inversion is an ill-conditioned problem and should be treated with great care even by appropriate methods of regularization or maximum entropy. In a mixture of unimers (u) and micelles (m), the intensity contribution (I_u and I_m) coming from the bimodal size distribution, together with the association number, can provide us an estimate of weight fractions of unimer and micelle species. Some information on the compactness of the spherical micelles can be estimated by considering the micellar size ratio of R_g/R_h.

2.2. Viscometry

Viscosity is widely used partly because of its simplicity. It can be employed to measure hydrodynamic properties of micellar systems by determining the value of $[\eta]$, the intrinsic viscosity [13], which is a measure of the hydrodynamic volume of particles in solution. The Huggins equation has the form:

$$\eta_{sp}/c=[\eta]+k_H[\eta]^2c \qquad (3)$$

with η_{sp} being the specific viscosity, c being the polymer concentration and k_H being the Huggins coefficient. By studying the concentration dependence of viscosity we can determine the critical micelle concentration (CMC) as well as the changes in micellar size and mass. The more compact micelles often show lower intrinsic viscosity and higher Huggins coefficient.

2.3. Nuclear Magnetic Resonance (NMR)

NMR spectroscopy is useful to characterize the local environment of block copolymers in solution by detecting the chemical shifts of certain atoms, e.g. ^1H, ^{13}C. It is very useful to obtain sub-micellar properties, such as chain distribution and chain mobility within the micelles [14]. The pulsed field-gradient spin-echo nuclear magnetic resonance (PFGSE-NMR) has shown to be very effective to characterize the micellar size, shape, structure and intermicellar interactions [15-18]. Due to the importance of NMR spectroscopy in studying micellar solutions, a large amount of work has been reported, both in aqueous and organic solvents. A specific review has been published in 1987 [19].

2.4. Fluorescence

Fluorescence probes can be used to provide insight into the probe mobility and the local environment of block copolymer micelles in solution. Steady-state fluorescence, time-resolved fluorescence and fluorescence depolarization can be used to characterize the micelle formation and equilibrium with unimers, as well as the micellar association number and micropolarity [20,21]. These more recent techniques have proven to be valuable for the investigation of micellar properties at the molecular level [22]. Detailed discussions are presented in other chapters of this book.

Nonradiative direct energy transfer (NDET) experiment is an extension of fluorescence spectroscopy. Some fluorescence-sensitive groups, either an electron donor like phenanthrene or an electron acceptor like anthracene, are placed at the junctions of different blocks when the block copolymers are synthesized [23-25]. By detecting the resultant fluorescence signals, not only the formation of the micelles, but also the properties of the interface between the micellar core and the shell, can be investigated.

2.5. Size Exclusive Chromatography (SEC) (or Gel Permeation Chromatography (GPC)) and Sedimentation Velocity

Both methods give straightforward information on the unimer $\leftrightarrow$ micelle equilibrium. By detecting different sedimentation-velocities or the hydrodynamic volumes (from SEC), the slow kinetics of micellization and critical micelle temperature (CMT) values can be estimated.

However, at least until recently, no quantitative information on the kinetics of unimer $\leftrightarrow$ micelle equilibrium can be obtained from either one of these two methods [1]. In most cases, the sedimentation velocity method is more reliable because it can take into consideration the unimer-micelle equilibrium during transport while SEC measurements need a series of assumptions. Detailed discussions have been presented by Tuzar and Kratochvil [1].

2.6. Miscellaneous Methods

The classical techniques, including viscometry, surface tension, dye solubilization and differential scanning calorimetry (DSC), deal with measurements of macroscopic properties. These methods, though less glamorous, are by no means uninformative. While lacking information on supramolecular structures, the macroscopic properties are extremely useful and are often more reliable to characterize the phase behavior, including the CMC and the CMT and thermodynamic properties of micellization.

Vapor pressure and osmometry techniques are standard methods to determine the number-average molecular weight. Transient electric birefrigence (TEB) can be used to investigate the shape of non-spherical micelles or aggregates in solution [26-28]. Transmission electron microscopy (TEM) can be used to measure directly the size and the shape of micelles, and in some cases, the micellar core/shell structure. The stopped-flow method can measure the kinetics (and dynamics) of micellization. Positron annihilation has been used to determine the presence of small aggregates in non-aqueous media [29] and the structure changes in condensed phases [30-32] have been reported.

3. ASSOCIATION CHARACTERISTICS OF BLOCK COPOLYMERS IN ORGANIC SOLVENTS

3.1. Solvent effect

In comparison with aqueous solutions, the micellization of amphiphilic block copolymers in selective organic solvents shows more diverse behavior mainly due to the nature of different solvents. The driving force of micellization is to minimize the free energy in polymer solutions. If the solvent is selectively good for one kind of blocks but poor for another one, micelles will be the favorite structure in solution at high enough polymer concentrations. However, if the solvents are weakly selective or if the interfacial energy between the solvent and the poorly soluble polymer block can be reduced by other interactions, e.g., charge transfer, then the comparatively small energy gain in micellization may not necessarily compensate the entropy loss due to micelle formation. In this case, the existence and the nature of the supramolecular formation can become delicate.

Ravey *et. al.* suggested to divide the solvents into three groups [33]. In the first class, there exist only monomers or possibly a few dimers in the whole range of polymer concentration, e.g., $C_{12}(EO)_4$ in polar solvents such as MeOH and $CHCl_3$, where strong hydrogen bond exists. $C_{12}(EO)_n$ (n=4-6) in benzene is another case. The charge-transfer type interaction between benzene and EO blocks tends to keep the polar part of the amphiphile shielded by benzene so that no micelle is formed even by adding a small amount of water. The Pluronic L64 ($PEO_{13}PPO_{30}PEO_{13}$) also exists as unimers in o-xylene [34], However, micellization can be induced by adding a small amount of water.

For some other systems, e.g., $C_{12}(EO)_4$ in heptane and cyclohexane (with a small amount of water) at 20 °C [33], SANS measurements revealed that some small aggregates existed in solution; however, the scattered intensity I(q=0) as a function of polymer concentration did not exhibit any sharp transition, suggesting that the aggregation number of such aggregates increased with polymer concentration. Such an aggregation phenomenon is probably a step-wise process. The conformation of both parts of the amphiphile are in extended form, and in the aggregates, the polymer chains choose hank-like or lamellar packing (Figure 1).

In the last class of solvents the CMC or pseudo-CMC existed. The latter concept was introduced for some cases that dimers also existed before the CMC; above which the aggregation number would be much higher and showed little concentration dependence, which implied a closed association mechanism. Most of the micelles have core-shell structures, just like those in aqueous solution; other structures, such as lamella and hank-like, were also reported [33].

Prochazka *et. al.* [20] reported the micellization of PS-poly(hydrogenated isoprene) (PS-PHI) in the mixture of 1,4-dioxane and heptane. The former solvent was selective for PS blocks while the latter one was selective for PHI blocks. In the solvent containing 0-30% of dioxane, micelles could form with PS being cores; for the solvent mixture containing 30-50% of dioxane, no micelle could form; and if the dioxane content increased further, reversed micelles would form.

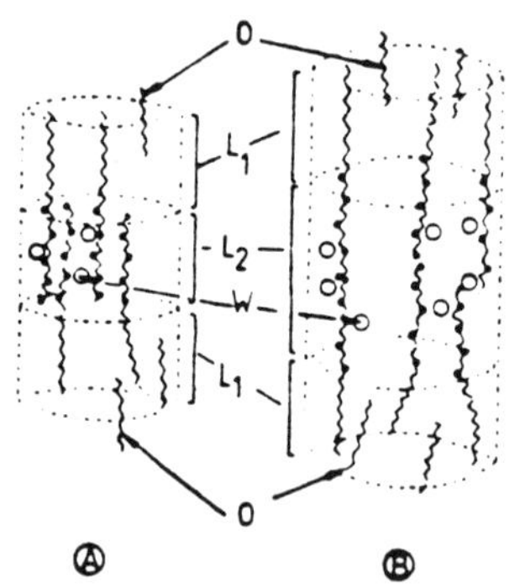

Figure 1. Schematic models of hank-like (A) and lamellar (B) aggregates in nonpolar media. L_1 and L_2 are hydrophobic and hydrophilic chain of the block copolymer, respectively. O=solvent molecule (oil); W=water molecules (isotope mixtures).
Reprinted from Reference 33, with permission.

For ion-containing block copolymers, the interaction between polymer ions and solvent molecules become stronger and more important. Desjardins *et. al.* [35] reported that for polystyrene-b-poly(methacrylic acid) (PS-PMAA) and PS-poly(metal methacrylate) block copolymers in solvents selectively good for PS, the neutralized materials form micelles in solvents such as THF, DMF, dioxane, CCl_4, toluene and cyclohexane, but the acid materials formed micelles only in solvents such as CCl_4, toluene and cyclohexane which did not interact with the carboxylic acid functionalities of the methacrylate blocks.

The solvent plays a key role in the formation and properties of micelles of block copolymers, a detailed discussion is presented as follows.

3.2. Critical conditions of micelle formation

The critical micelle concentration (CMC) and the critical micelle temperature (CMT) are basic parameters for characterizing the micellization with a closed-association mechanism. A large amount of data have been reported in both aqueous and non-aqueous solutions.

The CMC is defined as the polymer concentration, above which the formation of micelles occurs. Analogously, the CMT is the transition temperature, above or below (in organic solvents, usually below) which the formation of associated structures becomes appreciable. For a given system, the CMC and the CMT are interconnected. The temperature at which the CMC is determined can simply be considered as the CMT for that polymer concentration. To study the micelle formation, we can either measure the influence of temperature on a particular physical property at a given concentration or study the concentration dependence at a constant temperature.

The CMC of block copolymers in organic solvents usually displays a temperature dependence, with the magnitude depending on the block composition and the solvent quality [36]. The CMC generally increases with increasing temperature, opposite to that in an aqueous solution. In aqueous solution, the carbon-hydrogen chains become more hydrophobic at higher temperatures. Thus, the CMC decreases with increasing temperature. In organic solvents, higher temperatures can usually improve the solvent quality to both blocks in a block copolymer. Thus, the CMC value increases with increasing temperature. For some systems, like PS-(poly ethelene/propylene) (PS-PE/PP) (M_w=1.05x10^5 g/mol, 34 wt% PS) in a mixture of 1,4-dioxane and n-dodecane (10 vol% n-dodecane) [37], the temperature dependence could be quite strong. The copolymer has a CMC value of 3.5x10^{-16} g/cm^{-3} and 5.8x10^{-4} g/cm^{-3} at 25 °C and 85 °C, respectively.

In both aqueous and nonaqueous solutions, the micellization of block copolymers depends strongly on their composition, i.e., the block length ratio, the total molecular weight and the chain architecture [3,38]. The CMC decreases dramatically with increasing block length ratio which favors the insoluble block. On the contrary, by keeping the length of the insoluble block constant, the CMC value changes little and usually becomes a little higher with increasing length of the soluble block. For example, Hilfiker *et. al.* [39] studied PS-polyisoprene (PS-PI) in aniline, a selectively good solvent for PS. With M_w of the PI block remained at 19K g/mol, the CMC increased a little, from 2.5x10^{-5} g/mol to 3.3x10^{-5} g/mol when M_w of PS block increases from 28K g/mol to 41K g/mol. Quintana *et. al.* reported that the micellar size for PS-PE/PP in ketones would change by adding a certain amount of homopolymers (PS) which were the same as the insoluble PS block in PS-PE/PP. However, there was no effect on the CMC value of the block copolymers [40,41]. Adding enough amount of PS homopolymers to PS-PE/PP copolymer in toluene, which was a good solvent for both blocks, could induce micellization [42]. The same authors also concluded that the CMC would decrease with increasing total molecular weight of the copolymer while keeping the block length ratio constant [43].

Abundant work has been performed to study the solvent effects on the CMC value. The larger the difference in solubility between the polymer blocks, say for example in a diblock copolymer, i.e., the more selective the solvent becomes, the easier the micelles can form. Katime and co-workers systematically studied the CMC of PS-PE/PP diblock copolymers in different organic solvents [37,43-47]. In Table 1 [44], the CMC value of PS-PE/PP decreases with increasing carbon number of the solvents (n-alkanes, which are selectively good for PE/PP blocks) at 25 °C. However, at 85 °C, the CMC has a maximum value in n-octane. Increasing or decreasing the chain length of n-alkane will lead to a smaller CMC value. By successively adding dioxane, a selectively good solvent for PS, to the dodecane solution, the CMC value becomes higher until the two solvents have similar volumes in the solution mixture. Then, the block copolymer becomes soluble as unimers in the solution mixture (Table 2) [47]. Further addition of dioxane to the solution mixture induces a reverse micelle formation with PE/PP blocks being the core and PS blocks being the shell.

Table 1.
Values of CMCs (g/mL) for block copolymer PS-PE/PP (SEP1: 1.05×10^5 g/mol, 34±3 wt% PS) in alkanes.

Solvent	CMC (25 °C)	CMC (85 °C)
n-hexane	3.0×10^{-4}	1.0×10^{-2}
n-heptane	1.1×10^{-4}	6.6×10^{-2}
n-octane	6.3×10^{-6}	2.5×10^{-1}
n-decane	2.4×10^{-6}	1.1×10^{-1}
n-dodecane	6.4×10^{-7}	3.3×10^{-2}
n-hexadecane		$>2.7\times10^{-3}$

Reprinted from Reference 44, with permission.

Table 2.
Values of CMCs (g/mL) calculated for block copolymer (SEP1) in n-dodecane, 1,4-dioxane and mixtures of both solvents at 25 and 85 °C.

n-dodecane	CMC (25 °C)	CMC (85 °C)
0		$<4\times10^{-7}$
10	3.5×10^{-16}	5.8×10^{-4}
20	1.0×10^{-9}	1.4×10^{-1}
40	1.7×10^{-3}	2.6
50	$>4\times10^{-3}$	
60	$>4\times10^{-3}$	
70	1.7×10^{-3}	3.6×10^{-1}
80	5.7×10^{-5}	4.1×10^{-2}
100	6.4×10^{-10}	3.3×10^{-5}

Reprinted from Reference 47, with permission.

Block ionomers, which can behave like block polyelectrolytes in aqueous solution, often have very low CMCs in organic solvents. The low CMC values are difficult to determine [48,49]. However, the CMCs can be evaluated by fitting the micellar and the transition regions in a light scattering experiment according to an equation proposed by Debye [50]. For systems with significant polydispersity in the insoluble block, Gao and Eisenberg developed a new method based on the micellization model [51,52], with which one can compare the extremely low values of CMCs of block ionomer micelles over a wide range of CMCs observed for other self-assembled species.

The micellization of block ionomers depends strongly on the polarity [53] and the neutralization of ionic blocks [54]. PS-poly(acrylic acid) (PS-PAA) block copolymers initially in single chain form in both dioxane and benzene/MeOH (90/10 vol%) were neutralized to different degrees by adding CsOH dissolved in MeOH. Monitored by LLS experiments, the micelles began to form at 7% neutralization. Under this condition, the length of PAA block did not affect the value of the critical degree of neutralization.

Zhong and Eisenberg studied the effect of counter ions on the properties of monochelic PS [55], which could be considered as a special block ionomer with only one repeating unit for the ionic block. In this case, the size of the counter ions are comparable to that of the ionic block. Thus, the effect of counter ions becomes significant. The lowest CMC value was found in a polymer solution with K^+ being the counter ions. K^+ and COO^- are more comparable in size, when compared with those of Li^+, Na^+, Rb^+, Cs^+, etc.. At the same time, the micellar association number reached a maximum value. Zhong *et. al.* introduced a new concept "critical micelle length" (CML) for block ionomers in organic solvents [56]. The CML in a solvent is defined as the unimer repeating units of an ionic block in a block ionomer, below which the micelles will not form at any polymer concentration. For PS-poly(sodium acrylate) in THF, DMF, N-methylpyrrolidinone or toluene, the values of CML were found to vary between 2 and 3 units of the ionic blocks and to be insensitive to the length of PS block between 190 and 2300 units. The effects on counter ions and solvents were found to be generally negligible. The authors determined the CML value by computer modeling using values of relative amounts of micelles and single chains for block copolymers of different PS block length and different ionic block lengths.

3.3. Thermodynamics of micellization

Both the open-association mechanism and the closed-association mechanism were observed in the micellization of amphiphilic block copolymers in selective organic solvents. For the latter case, the CMC exists, the association number tend to be relatively large and the micelles have relatively narrow size distributions The standard enthalpy and the Gibbs free energy of micellization can be calculated by the equations:

$$\Delta G° = RT\ln(\text{CMC}) \quad (4)$$

$$\Delta H° = R[d\ln(\text{CMC})/d(1/T)] \quad (5)$$

where CMC is in mole fraction units. Assuming that $\Delta H°$ is independent of temperature within a certain temperature range, we can integrate Eq.(5) to get

$$\ln(\text{CMC}) = \Delta H°/RT + \text{constant} \quad (6)$$

Obviously, this is an approximate expression, because it has not considered the change in the association number, if any, with changing temperature. The inverse CMT values are also applicable to Eq.(6).

For block copolymers in selective organic solvents, one usually finds a positive slope of the CMC versus temperature, implying that $\Delta H°$ has a negative value. Among all the data available from diverse systems, negative $\Delta H°$ values are the norm and dominant the association process.

It is interesting to compare the thermodynamic parameters of similar block copolymers in aqueous solution, where ΔS° and ΔH° are often positive [3]. The positive ΔS° values are due to the destruction of the ordered hydrogen bond structure formed by water molecules around hydrocarbon chains. At higher temperatures, the polymer chains become more hydrophobic due to the disruption of hydrogen bonds, and the solvent quality for the block decreases. The solvent-polymer chain interactions are weaker in organic solvents, which means that the micelle formation would mainly be an enthalpy process. Evidence has shown that the higher the association number, the more negative ΔS° will be [57]. The driving force of micellization in organic solvents should be the negative ΔH°, i.e., enthalpy driven. The above conclusions were well tested by Price *et. al.* [58-64].

Katime and co-workers studied several PS-PE/PP block copolymers in 4-methyl-2-pentanone [46]. Several polymer samples, with different M_w (60.7K, 87.3K and 260K g/mol) but similar weight ratio of two blocks (~30 wt% for PS) were measured for comparison. The ΔG° became more negative at lower temperature. The effect of total molar mass on ΔH° and ΔG° seemed not so large with ΔG° being a bit higher with increasing polymer molar mass.

The same research group did a series of experiments on the solvent effect of thermodynamic properties of micellization [45-47,57,65-67]. Table 3 summarizes part of their results related to PS-PE/PP block copolymers. ΔH°, ΔG° and ΔS° decrease with increasing alkane chain length at 25°C. However, a sharp decrease was observed between n-heptane and n-octane for ΔH° and ΔS°. The authors explained that n-alkanes became poorer solvents for PS with increasing n-alkane carbon number. As the PS segment/n-alkane interactions became more repulsive, the micellar core became more compact. The PS/n-alkane interactions were then totally replaced by PS/PS and n-alkane/n-alkane interactions so that the enthalpy of the process became more negative and a sharp transition occurred. N-hexane and n-heptane showed remarkably less repulsive PS/solvent interactions than higher n-alkanes [68]. At 85 °C, the Gibbs free energy of micellization showed a minimum (-38.5 kJ/mol) for n-octane, in coincidence with the lowest CMC value of the same system.

Table 3.
Thermodynamic data for the micellization of PS-PE/PP (SEP1) block copolymer in n-alkanes at 25 and 85 °C.

solvent	ΔG°(25 °C) kJ/mol	ΔG°(85 °C) kJ/mol	ΔH° kJ/mol	ΔS° kJ/mol·K
n-hexane	-48.7	-48.1	-52.0	-0.011
n-heptane	-51.1	-42.5	-93.9	-0.144
n-octane	-58.3	-38.5	-156.7	-0.330
n-decane	-60.6	-40.8	-158.9	-0.330
n-dodecane	-63.9	-44.6	-160.1	-0.323
n-hexadecane		<-52.0		

Reprinted from Reference 65, with permission.

3.4. Association number, micellar size and micellar structure

One of the most important features of the closed-association process is that the association number (n_w) remains constant near the CMC. The n_w can be determined by measuring the average mass of micelles through static light scattering (SLS) experiments or fluorescence probing. Both DLS and PFGSE-NMR measurements can provide information on the micellar size in terms of the hydrodynamic radius (R_h). For common block copolymer micelles which are considered to have core-shell structures, the micellar core size can be obtained by SAXS in an appropriate solvent with a large enough electron density difference between the core and the surrounding medium or by SANS with deuterated block or solvent. Both methods yield a value of radius of gyration (R_g) through the Guinier plot [69]. Recently cryo-transmission electron microscopy has been employed in studying the micellar structure in organic solvents. It can directly visualize the core structure and the phase behavior [70]. By assuming a nonswollen, liquid-like micellar core with known polymer density in the core, the core size (r_c) can be calculated approximately from n_w. Furthermore, the relation between R_h and r_c or $R_{g,core}$ can provide a rough estimate on the thickness of the micellar corona. The assumption of a nonswellon micellar core is valid only when the solvent-phobic blocks of a block copolymer are totally insoluble in that solvent. These frozen-core micelles may not have the usual unimer-micelle equilibrium as the equilibration time can be very long. In a weak-selective solvent, the micellar core still contains a small amount of solvent and the core-shell interface is not so sharp as one often assumes.

The temperature effect on the n_w of a block copolymer is comparatively simple in organic solvents. A lower n_w is expected at higher temperatures because the solvent quality tends to increase faster with the more insoluble block. The association of PS-PI in a mixture of dioxane and MeOH showed a strong temperature dependence, with n_w being enhanced by about 9 times, from 6 to 52, when the temperature went down from 23.0 °C to 16.7 °C [71]. Eisenberg and co-workers used LLS and SEC to study the n_w of ionomers by changing the block length, the solvent and the counter ions. A collective summary is listed in Tables 4-6 [35,73]. The n_w increases with increasing length of the insoluble block and decreasing length of the soluble block, similar to those in aqueous solution. The effect of soluble block is more pronounced when the insoluble block is longer. When the insoluble blocks (PAA or PA-metal in Tables 4-6) become very short (6-7 units in Table 4), changing the length of the soluble PS block seems to have negligible effect on the n_w. However, when the length of PA-metal blocks becomes shorter (e.g., 2.6 units in Table 6), the n_w goes up again. If this result were universal for ionomers, the n_w as a function of the length of ionic blocks in organic solvents could have a minimum value. The counter ions also have noticeable effect on both the CMC and the n_w. The effect due to the soluble PS blocks can be drawn in Table 4 with the PS blocks having 170 units for each block copolymer. By changing the length of the $(PMA)_x$-Na block from 9 units to 25 units in CCl_4 solution, the n_w increased by a factor of 44 while for the $(PMA)_x$-H block under the same condition, the n_w increased by a factor of only about 5.
However, with a longer PS block of 440 units, the difference in n_w made by different counter ions became smaller: only about 16 times for Na^+, 8.5 times for Cs^+ and 4 times for H^+. With an even longer PS block (1,100 units), the rate of increase in n_w due to increasing PMA length between Na^+ and H^+ could be neglected [73].

Phoon *et. al.* [74] reported that ion-containing polymers often formed micelles with larger overall size despite their smaller n_w. This effect can be explained as arising from the repulsive effect of the ionic charges on the polymer sequences in the corona.

The selectivity of a solvent for block copolymers plays an important role on the value of n_w. Antonietti *et. al.* [75] reported the association behavior of PS-poly(4-vinylpyridine) block copolymers in selective organic solvents. The n_w decreased with increasing "solvent" quality for the insoluble 4-PVP block. However, the influence of the solvent quality for PS was hardly

Table 4.
Characteristics of block ionomer micelles in DMF and CCl_4 solutions.

sample	DMF solutions					CCl_4 solutions		
		SEC		DLS		DLS		
	[η],dL/g	$R_h \pm \sigma^a$, nm	n_w	R_h,nm	n_w	[η],dL/g	R_h,nm	n_w
PS(170)PMA(9)-Na	0.200	6.9±2.5	6	7	6	0.301	12	20
PS(170)PMA(25)-Na	0.180	16.9±4.2	85	27	343	0.273	42	883
PS(170)PMA(9)-H						0.255	11	19
PS(170)PMA(25)-H						0.205	18	91
PS(440)PMA(7)-Na	0.456	9.9±2.0	3	10	3	0.626	18	12
PS(440)PMA(18)-Na	0.322	19.0±4.6	28	18	36	0.412	24	45
PS(440)PMA(40)-Na	0.263	30.3±7.9	134	31	138	0.316	37	197
PS(440)PMA(7)-Cs	0.440	10.2±2.4	3	14	9	0.797	20	14
PS(440)PMA(18)-Cs	0.332	20.1±5.9	32	21	20	0.413	25	48
PS(440)PMA(40)-Cs	0.277	28.5±7.0	102	28	88	0.332	33	121
PS(440)PMA(7)-H						0.479	20	24
PS(440)PMA(18)-H						0.446	24	42
PS(440)PMA(40)-H						0.361	29	91
PS(1100)PMA(6)-Na	0.900	25.7±6.1	10	23	7	1.02	29	13
PS(1100)PMA(21)-Na	0.665	32.5±7.0	28	30	23	0.941	34	24
PS(1100)PMA(42)-Na	0.603	37.5±7.6	47	36	42	0.816	44	55
PS(1100)PMA(59)-Na	0.539	38.0±8.8	54	37	49	0.798	45	60
PS(1100)PMA(6)-H						0.921	35	26
PS(1100)PMA(21)-H						0.736	45	69
PS(1100)PMA(42)-H						0.700	51	104
PS(1100)PMA(59)-H						0.660	53	119

[a]: σ is the relative standard deviation of the distribution of radii.
Numbers in the brackets are the length of polymer blocks.
Reprinted from Reference 35, with permission.

Table 5.
Summary of SLS results for PS(x)-b-PANa(y) in THF.

PS(x)-b- PANa(y)	n_w	R_g (total) (nm)	$A_2 \times 10^{-4}$ (mL mol/g^2)
190-b-10	130	42	0.35
190-b-24	590	85	1.1
630-b-4.2	20	54	0.42
630-b-18	79	50	0.50
630-b-31	100	50	0.56
2300-b-4.6	23	58	
2300-b-31	41	65	0.40
2300-b-69	60	68	0.54

Data were taken from Reference 73, with permission.

seen. The association number n_w decreased slowly if a second solvent which was good for both kind of blocks was added to a block copolymer micellar solution. One of the typical experiments was done by Villacampa *et. al.* [67], who added 2-chlorobutane to the (PS-PE/PP)/4-methyl-2-pentanone system. The n_w decreased from 500 (no 2-chlorobutane) to 200 (40% 2-chlorobutane), and finally the block copolymer micelles disrupted to single molecules in a 1:1 mixed solvent. The same authors also showed [76] that for diblock copolymers in different selective solvents, the n_w would be affected by the location of the larger blocks. A higher n_w could be expected if the larger blocks were in the micellar cores.

The size of micelles is mainly determined by the size of micellar cores, especially in the case when the insoluble blocks are very short. Gouin *et. al.* [72] proved that for poly(4-vinylpyridine)-PS-poly(vinylpyridine) ($PV_{10}PS_{500}PV_{10}$ and $PV_{20}PS_{500}PV_{20}$) block copolymer micelles in THF, the average radius of gyration of a PS midblock containing 500 units was constant and close to the value of a PS chain of the same length in the melt.

The micellar core size is dependent on the solvent-phobic nature and polymer concentration. Nonswollen micellar cores can be found when the solventphobic blocks of a block copolymer are totally insoluble in that solvent [7]; however, in a weak-selective solvent, the micellar cores will become swollen, and often, the overall micellar size will increase [7,42,77]. In organic solvents, higher temperature will usually increase the solvent quality for both of the copolymer blocks. Therefore, more swollen micellar cores can be expected [78].

Khougaz *et. al.* [54] found that for PS-PAA in dioxane or benzene/MeOH (90/10 in volume), neutralizating the polymer with CsOH would change the micellar core size. With higher degree of neutralization, more solvent molecules would be expelled from the cores. Yeung *et. al.* [22,79] studied the concentration dependence of PS-PE/PP block copolymer micellar cores in heptane, by using intrinsic excimer fluorescence. The results showed that the volume fraction of PS in the micellar core increased from 0.40 to 0.92 when the polymer concentration increased from 0.001% to 0.50%, with the R_g values decreasing at the same time. However, Siqueira *et. al.* [80] concluded from SAXS and TEM measurements that PS-PBMA block copolymer micellar cores did not change in size or shape at different polymer concentrations in 2-propanol, a nonsolvent for PS blocks and θ solvent for PBMA blocks.

Wu *et. al.* studied the change in the n_w by adding water, which was selective for PEO blocks and was essentially immisable with o-xylene, to the Pluronic L64 ($PEO_{13}PPO_{30}PEO_{13}$)/o-xylene solution [34]. L64 can not form micelles in pure o-xylene, which is a good solvent for both blocks. However, by adding a small amount of water the micellization process could occur because water which resided with the hydrophilic PEO blocks reduced the solvent quality of o-xylene to the PEO/H_2O complex. The volume of the micellar cores can be calculated from the combined volume of PEO chains and water. Table 7 shows that the n_w increases with increasing amount of water (in term of Z which is defined as H_2O/EO molar ratio). For Z<1.4, the ratio of R_{HS}/R_h (~0.8) and of R_η/R_h (~0.9), are almost constants with R_{HS}, R_η being the equivalent hard sphere radius, the equivalent hydrodynamic radius from viscosity measurements, respectively. Assuming a spherical structure, the value of R_h-R_c which is treated as a measurement of the shell thickness should be a constant because of the constant PPO chain length, that is just the case for Z<1.4. For Z>1.4, the apparent shell thickness increases considerably with increasing Z, suggesting that the sphere model is no longer valid.

The properties of the interface between the micellar core and the corona has also attracted attention. Godward *et. al.* [14] used ^{1}H NMR longitudinal and transverse relaxation measurements to show that an intermediate layer might exist in the (PS-PE/PP)/ [$^2H_1$]octane system at high temperatures. Spin diffusion following selective inversion of protons from one component has been used to probe the extent of mixing of slightly swollen PS cores and the micellar fringe consisting of highly solvated PE/PP. At 80 °C, a small degree of spin diffusion was detected, indicating that a mixed phase existed. A simple analysis showed that 50% of PS blocks lay in the boundary while only 3% of PE/PP blocks were there.

Table 6.
Summary of results obtained for PS(660)-b-PANa(x) and PS(660)-b-PAA(x) by SLS and DLS in different solvents.

Composition	Toluene			THF			Chloroform		
PS(660)-b-x	n_w	$A_2 \times 10^4$ (mL·mol/g)	R_h(nm)	n_w	$A_2 \times 10^4$ (mL·mol/g)	R_h(nm)	n_w	$A_2 \times 10^4$ (mL·mol/g)	R_h(nm)
PANa(x)									
2.6	72	0.56	23	49	0.70	25	20	0.70	24
4.7	31	0.65	23	27	0.75	24	15	0.77	22
6.4	32	0.58	24	29	0.50	25	13	0.58	25
8.9	35	0.52	27	37	0.56	27	20	0.79	27
14	38	0.64	27	39	0.38	26	31	0.68	25
18	40	0.37	29	46	0.40	28	41	0.65	28
21	45	0.36	30	47	0.31	26			29
PAA(x)									
4.7	9.6	0.69							
6.4	11	0.75							
8.9	23	0.28							
14	39	0.39							

Data were taken from Reference 73, with permission.

Table 7.
Results of L64 in o-xylene in the presence of water.

Z	N	R_{HS}(nm)	R_h(nm)	R_{HS}/R_h	R_η(nm)	R_η/R_h	R_C(nm)	R_h-R_C(nm)
0.2	102	7.2			7.5		4.0	
0.4	111	6.7	9.3	0.72	7.7	0.83	4.2	5.1
0.6	125	6.9	9.0	0.77	8.0	0.89	4.4	4.6
0.8	134	7.1	8.9	0.80	8.2	0.92	4.7	4.2
1.0	142	7.3	9.1	0.80	8.4	0.92	4.8	4.3
1.2	149	7.5	9.2	0.80	8.5	0.92	5.0	4.2
1.4	160	7.8	9.9	0.79	8.7	0.88	5.2	4.7
1.6	171	8.1	11.5	0.71	8.9	0.77	5.4	6.1
1.8	184	8.4	12.9	0.65	9.1	0.71	5.7	7.2
2.0	199	8.8	14.7	0.60	9.4	0.64	5.9	8.8
2.2	215	9.0	17.1	0.53	9.6	0.56	6.2	10.9
2.4	231	9.3	20.5	0.45	9.8	0.48	6.4	14.1

Reprinted from Reference 34, with permission.

3.5. Micellar shape transition

Micelles could change shape by changing external conditions, such as temperature or solvent composition. As early as 1974, Utiyama *et. al.* [81] found that PS-PMMA (38% PS) block copolymer micelles could exist as prolates. Another example of prolate shaped micelles is PS-(4-PVP) block copolymers in organic solvents, as reported by Antonietti *et. al.* [75]. The micelles were spherical when the n_w was between 20 and 500. For micelles with higher n_w (500-800), stretching of the corona chains resulted in a destabilization of the spherical shape. Anisometric micelles were formed when a critical energy loss was exceeded and they were best described by prolate ellipsoids with an axial ratio of about 4.

Wu *et. al.* reported that for Pluronic L64 ($PEO_{13}PPO_{30}PEO_{13}$) in o-xylene, micellization could be induced by adding water [26]. The shape of the micelles would change from spherical to ellipsoidal for Z (H_2O/EO molar ratio)>1.3. The degree of asymmetry of the micelles was relatively small. For Z>2, and by using a combination of physical techniques (LLS, TEB and SAXS), an oblate ellipsoid could fit all the experimental parameters. By minimizing the area occupied per surfactant molecule at the micellar surface, it could be argued that an ellipsoidal (oblate) shape was thermodynamically favored. The prolate shape could be expected only when the repulsion between head groups was very strong.

3.6. Anomalous micellization

Anomalous micellization, first reported by Lally and Price [82] in their study of the temperature-induced micellization of triblock copolymer PS-PB(polybutylene)-PS in ethyl acetate, has been proven to be a common, but undesirable phenomenon for block copolymers in both aqueous and organic solvents. Most reports described the phenomenon in terms of some unusually large particles appearing in solution in the vicinity of CMC or CMT (i.e., just before the onset of micellization), causing an anomalous increasing in the scattered intensity and the dissymmetry of the scattered light, which was defined as the ratio of scattered intensity at the scattering angles of 45° and 135°).

Several explanation on the origin of the anomalous micellization have been reported. For examples, three types of large particles: worm-like micelles [63,83-85], ellipsoids [81] and sphere [86,87] have been reported in different solutions, however, without determining their structure. Price *et. al.* [63] found that the behavior of these worm-like micelles depended strongly on the thermodynamic history of the sample. After heating the solution of PS-PI in N,N-Dimethylacetamide to 70 °C then cooling down to 30 °C, the large particles disappeared.

Tuzar and Kratochvil summarized relevant experimental results and hypothesis in their review [1]. Several suggestions on the formation of anomalous micellization were presented. Tuzar *et. al.* in one of their early papers [87] investigated the solubilization of insoluble polybutadiene (PB) and PS-PB-PS copolymers by micelles of PS-PB-PS with PB cores. They noticed that the unusually large particles appeared just before the onset of micellization, if either the homopolymer PB or the copolymer PS-PB-PS with a higher butadiene content, and both insoluble at a particular temperature, was present in the solution of PS-PB-PS copolymers. When the normal micelles began to form, the insoluble component could be incorporated into the cores or could form mixed micelles. Zhou and Chu further indicated that the anomalous micellization might be due to the heterogeneity in chemical composition of the block copolymers [88,27]. It happened when the phase separation of a small fraction of copolymers with longer insoluble blocks occurred before the onset of micellization of the major component. They reported that for PEO-PPO-PEO triblock copolymers in aqueous solution, the large particles could be removed by filtration; no new large particles appeared again, and only a small part of the sample formed such aggregates, which had a significantly higher content of hydrophobic PPO units in comparison with the average composition of the sample through NMR analysis [88].

Price *et. al.* raised another hypothesis claiming that the abnormal increase in the scattered intensity might be attributed to the polymer concentration fluctuations [84], related to the micelle-free chain equilibrium, rather than to the actual presence of large particles. However, this hypothesis has not been substantiated. One of the authors, Quintana, has changed his opinion in later publications [65,44,57,89,40].

Some other conclusions have also been presented to interpret the large particles to be undissolved copolymers [90,91] or agglomerates of micelles [92].

In recent years, further study on the nature of anomalous micellization is still in process [44,57,89,40,93]. Katime and co-workers investigated the anomalous behavior in their study of PS-PE/PP block copolymers in organic solvents. For copolymer SEP2 (M_w=105K g/mol, 35 wt% PS) in 5-methyl-2-hexanone, LLS showed that a very small peak appeared by plotting asymmetry factor versus temperature [57]. For SEP3 (M_w=145K g/mol, 26 wt% PS), the asymmetry factor value showed a much larger sharp peak when the temperature was lowered, and this behavior was reproducible (Figure 2). Considering the larger weight percent of PE/PP in SEP3, the higher asymmetry factor value could be explained based on the model proposed by Zhou and Chu. The reproducible nature of the phenomenon differs greatly from the observations by Price *et. al.*. In another paper from the same research group [40], the solubility of insoluble homopolymers polyisobutene (PIB) in SEP2/5-mehtyl-2-hexanone, they found that the amount and the length of PIS strongly affected the anomalous behavior. There was no abnormal large particles in solution with 4 wt% PIB3 (M_w=20.7K g/mol); however, with 10 wt% PIB3 in solution, the anomalous micellization was quite significant. By decreasing the PIB molar mass to 14K g/mol, even 90 wt% of PIB in solution could only show a minor anomalous behavior. This experiment further confirmed that the insoluble PIB was the main component responsible for the observed anomalous behavior.

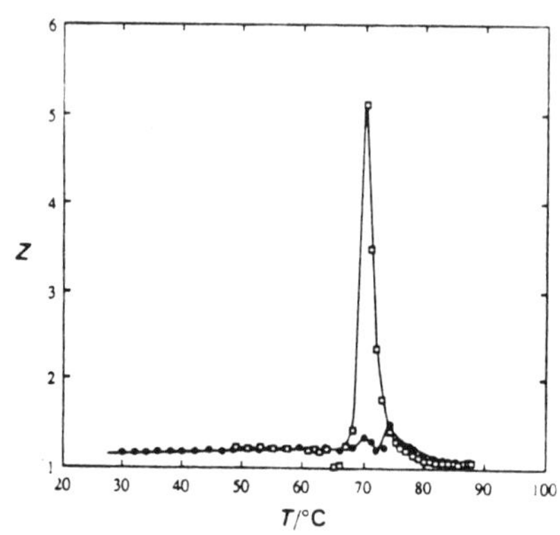

Figure 2. Variation of asymmetry factor (z) as a function of temperature for a solution of copolymer SEP3 in 5-methylhexan-2-one, at a concentration of 1.0×10^{-3} g/mL, on raising (filled circle) and lowering (open square) the temperature.
Reprinted from Reference 57, with permission.

Cogan and Gast reported that in the water-induced micellization of the PS-PEO/cyclopentane system [93], the amount of water was also related to the anomalous micellization, which was observed in solutions of $PS_{65}PEO_{80}$ and $PS_{170}PEO_{1730}$ block copolymers having water contents below 70 and 25 PPM, respectively. They believed that the driving force for forming these large aggregates could involve the crystallizability of PEO insoluble blocks.

3.7. Effect of chain architecture

The chain architecture of block copolymers is expected to have influence on their self-assembly behavior in both aqueous and organic solvents. For AB diblock copolymers in a selective solvent good either for A or B, or ABA triblock copolymers in a solvent selective for terminal blocks, the association into micelles are usually straightforward, obeying the closed association mechanism and yielding the star-like micelles [94] with a core-shell structure which is composed of a relatively compact micellar core and a swollen protective corona. In some special cases, other structures can also form. Hilfiker *et. al.* used SAXS to study the association structure of PS-polyisoprene (PS-PIP) in aniline [95,39], a selective solvent for PS. By properly fitting the scattering curves, they concluded that when the PS blocks were short enough (PS:PI=28:19 in weight), vesicle-like structures could form (Figure 3). When the PS chain length became longer (PS:PI≥41:19 in weight), normal micelles were found. As mentioned in the previous section, the critical micelle length (CML) exists in some block ionomer systems [56,73], a minimum length for the ionic block is needed to form micelles.

For the ABA block copolymer in a solvent selective toward the middle block, a variety of possible associated structures could be expected. Flower-like micelles could form if the two end blocks constitute a part of the same micellar core whereas the central block takes a loop conformation [96,27]. The extended soluble central block may function as a bridge to connect the small clusters composed of insoluble end blocks, thus leading to a branched structure at low concentrations or a gel-like network at higher concentrations. An intermediate pattern is also possible in that some of the coronal blocks show a looping geometry, but the other copolymer chains may have one of their end blocks located in the core and the other dangling in solution. A balance exists between two factors: an entropy loss due to the loop conformation of the micelle block would favor the formation of branched structure, and an interfacial energy penalty that accompanies the transfer of the poorly solvated blocks from the cores into solution would favor the flower-like micellar structure.

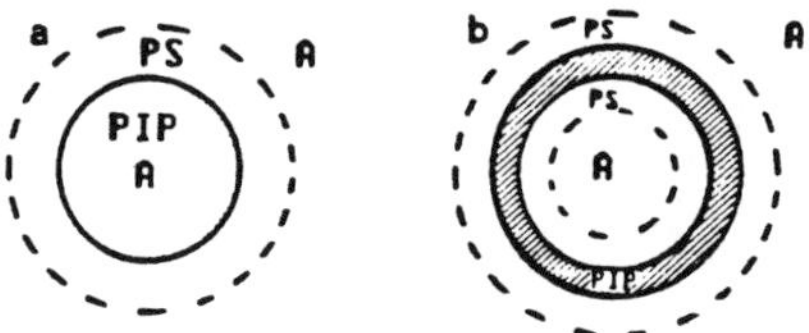

Figure 3. Schematic representation of a micelle (a) and a vesicle (b). Straight lines denote sharp boundaries and broken lines denote diffuse boundaries. A, aniline.
Reprinted from Reference 95, with permission.

Balsara *et. al.* first reported the existence of flower-like poly(vinylpyridine)-PS-poly(vinylpyridine) (PVP-PS-PVP) micelles in toluene, a solvent selectively good for the PS block [96]. The diffusional characteristics of these micelles were similar to those of micelles formed by PVP-PS diblock copolymers in toluene, implying a similarity in the overall architecture of the micelles. It was found that the molecular weight of the middle block (PS) played a crucial role in determining the CMC and the size of the micelles.

The flower-like micelles formed by triblock copolymers may crosslink to each other to form supramicellar structures at high enough concentrations. Some of the insoluble end blocks belong to one polymer molecule may reside in two difference micellar cores and the soluble middle blocks function as bridges. An example was given by Wu *et. al.* [97] in their studies of the micellization behavior of L64 ($PEO_{13}PPO_{23}PEO_{13}$) in o-xylene by introducing a small amount of water. The micellar crosslinking became stronger when a larger amount of water was introduced. The concentration dependence of the crosslinking was also obvious, suggesting that higher micellar concentration (i.e., smaller distance between adjacent micelles) favored the supramicellar structure (Figure 4).

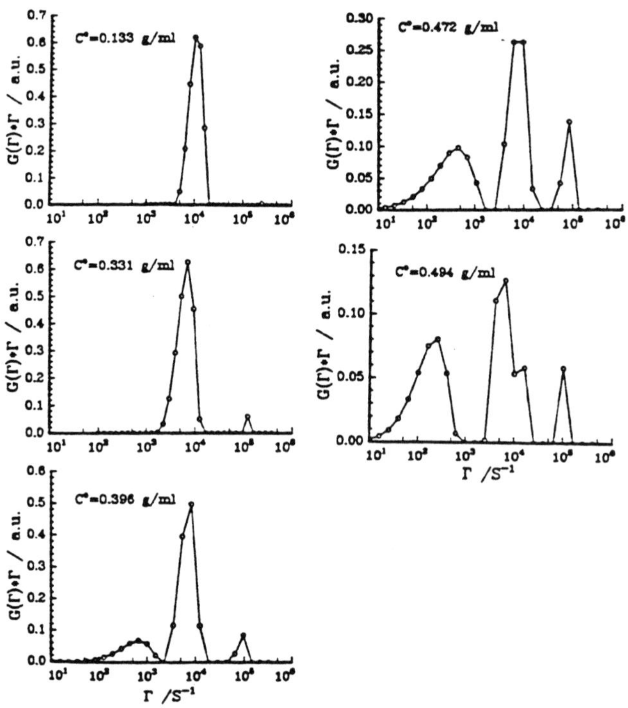

Figure 4. Relaxation rate distributions at indicated L64 concentrations obtained by the CONTIN analysis of DLS measurements. From right to left, the peaks represented single copolymer molecules, micelles and intermicellar crosslinking (if the third peak existed). Reprinted from Reference 97, with permission.

Adam *et. al.* reported that PS-PI-PS triblock copolymers in n-heptane, a good solvent for the PI block and non-solvent for the PS block, formed aggregates with an open structure [98]. Detailed discussion can be found in section 3.8.

Table 8.
Comparison in micellar parameters between diblock micelles and triblock micelles in DMA in dilute region.

	$PBS_{385}S_{770}PBS_{385}$	$PS_{700}PBS_{700}$
CMC (mg/mL)	0.58±0.05 (19.9 °C)	$(2.0\pm0.5)\times10^{-4}$ (22.4°C)[a]
	1.8±0.14 (25 °C)	1.0±0.2 (50.4 °C)
M_w (g/mol)	$(4.5\pm0.3)\times10^{6}$ (19.9 °C)	$(5.8\pm0.4)\times10^{6}$ (45.0 °C)
	$(3.4\pm0.3)\times10^{6}$ (25.0 °C)	$(1.0\pm0.1)\times10^{7}$ (25.0 °C)
average n_w	22±1.5 (19.9 °C)	31±2 (45.0 °C)
	17±1.5 (25.0 °C)	54±5 (25.0 °C)
A_2 (cm^3 mol/g^{-2})	$-(1.3\pm0.2)\times10^{-6}$ (19.9 °C)	$-(1.9\pm0.6)\times10^{-6}$ (45.0 °C)
	$-(1.4\pm0.5)\times10^{-5}$ (25.0°C)	$(2.0\pm0.4)\times10^{-5}$ (25.0 °C)
D_0 (cm^2/s)	$(5.8\pm0.2)\times10^{-5}$ (19.9 °C)	$(8.0\pm0.1)\times10^{-6}$ (45.0 °C)
	$(6.6\pm0.3)\times10^{-6}$ (25.0 °C)	$(4.7\pm0.14)\times10^{-6}$ (25 °C)
R_h (nm)	37±1 (19.9 °C)	40.5±0.5 (45.0 °C)
	36±1.6 (25 °C)	50.0±1.5 (25.0 °C)
k_D (cm^3/g)	-(48±10) (19.9 °C)	
	-(51±2) (25 °C)	
ΔH^0 (kJ/mol)	-(163±2)	-(270±10)

[a] Extrapolated value.
Reprinted from Reference 27, with permission.

For block copolymers having the same chemical composition, the association behavior is strongly influenced by their chain architecture. Zhou *et. al.* studied the micellar parameters of poly(tert-butylstyrene)$_{385}$-PS_{770}-poly(tert-butylstyrene)$_{385}$ (PBS_{385}-PS_{770}-PBS_{385}) and $PBS_{700}PS_{700}$ in N,N-dimethylacetamide, which was a selectively good solvent for PS [27]. A brief summary is listed in Table 8. The CMC value of the triblock copolymer was much higher (~5,000 times at 20 °C) and its association number was about 3 times smaller than those of $PBS_{700}PS_{700}$. Similar results were also reported by other authors [40,45,43,99,100]. Although the entropy penalty associated with the looping geometry of the middle block did not preclude the possibility of micelle formation, the tendency towards self-assembly was greatly decreased. Following ten Brinke and Hadziioannou [101], the entropy penalty due to the looping formation is given by:

$$G_{loop}/kT=(3/2)\beta\ln N_\beta \qquad (7)$$

where N_β is the number of the repeat units in the middle block. The coefficient β that depends on the length of the two constituent blocks has a limiting value of 0.3~0.4 [102]. The free energy term associated with the interface formation between the insoluble core and the solvent is approximately

$$G_{int}/kT=\pi a^2 N^{2/3}\sigma \quad (8)$$

where a is the size of the insoluble unit, N is the number of insoluble units in the terminal block and σ is the relevant interfacial tension (in kT unit). For some systems whose interfacial tension are still unknown, an estimation can be made by using Antonoff's relation:

$$\sigma_{1/2}=\sigma_{1(2)/air}-\sigma_{2(1)/air} \quad (9)$$

where 1(2) and 2(1) refer to two mutually saturated phases forming the interface.

3.8. Open association mechanism

The entropy-driven process is typical in micellization of block copolymers in aqueous solution. It favors the shielding of the hydrocarbon moiety of amphiphilic block copolymers from the solvent. The total efficient shielding requires a certain minimum number of polymer molecules per micelle and micelles exist only above a certain minimum total polymer concentration (CMC) which occurs when the adsorbed surface layer becomes saturated [4]. Micellization may also happen without a minimum concentration, then the association number could vary with changing polymer concentration. This mechanism is known as an open association mechanism. Usually, a step-wise association can be expected:

unimer↔dimer↔trimer↔......

An example was given by Ravey *et. al.* [33], when they used SANS to study the association behavior of $C_{12}(EO)_4$ block copolymers in heptane and cyclohexane with a small amount of water. In these systems, the scattered intensity I(q=0) as a function of polymer concentration did not exhibit any sharp transition, which was an indication of existence of the CMC, but still revealed the presence of small aggregates, whose association number increased with polymer concentration having the n_w varying from 5 to 8 for $C_{12}(EO)_4$/heptane solutions.

A transition of micellization mechanism has been reported by Saito *et. al.* [103]. A concentration-jump method was employed to study the poly(α-methylstyrene-b-methylmethacrylate) (PMeS-PMMA) in mixtures of benzene and cyclohaxane (volume ratio 4/6 and 2/8). The solubility of the polymer micelles could be detected by measuring the viscosity of polymer solutions as a function of time after rapid dilution. For micelles obeying the close association mechanism, the stabilizing time T of viscosity was very small, because the micelles could quickly reach a new equilibrium. On the contrary, a much longer T would be expected for micelles obeying the open association mechanism. In the 2/8 solvent, mixture PMeS-PMMA (M_n=22K g/mol, 49 wt% PMeS) followed an open association mechanism at concentrations lower than 4 wt%. However, in the 4/6 solvent, micellization of two samples (M_n=63K g/mol, 68 wt% PMeS and M_n=280K, 55 wt% PMeS) showed a transition from the closed association process at low concentrations to an open association process at higher concentrations. The above observation suggests that the micellization mechanism strongly depends on the solvent quality.

Raspaud *et. al.* also reported an open association mechanism for a triblock copolymer, PS-PI-PS in tetrahydrofuran, a good solvent for PI and nonsolvent for PS [98]. The entropy loss of forming flower-like micelles cannot be compensated by the change in enthalpy. Thus, loose-structured aggregates were formed instead of flower-like micelles. However, by using LLS and viscometry measurements, they found the existence of a critical aggregation concentration (CAC) of 1.6×10^{-3} g/mL, similar to the CMC in micelles obeying the closed association mechanism. The open association process does not have a critical concentration for micelle formation. So this result requires further study.

3.9. Micellization of heteroarm star-block copolymers

A series of new block copolymers with a novel architecture, known as "heteroarm star copolymers", have been synthesized recently [104-107]. The macromolecules are star-shaped polymers having a central dense cross-linked poly(divinylbenzene) core which bears two kinds of branches of equal number n (A_nB_n). They can be considered as a number of diblock copolymers which have been joined together at their A-B junction points.

Tsitsilianis and Kouki presented the results of heteroarm star copolymer PS-poly(tert-butylacrylate) (PS_n-$PtBA_n$) in three solvents [108]: THF, a common good solvent for all branches; acetone, a poor solvent for the PS blocks and good for the PtBA blocks; and MeOH, a marginal solvent for PtBA blocks and a precipitant for the PS blocks. The conformational transition and/or micellization phenomenon depended on the selectivity of the solvent. In a weakly selective solvent (acetone), a temperature-induced conformational transition was observed. The Huggins coefficient k_H showed a maximum value, similar to those of simple linear block copolymers. Micellization did not happen in such solvents. In a strongly selective solvent (MeOH), association phenomenon was observed. The authors proposed a spherical structure, comprising of three domains (Figure 5). Domain A (core) was constituted of PS chains and domain B (shell) was formed by the same number of PtBA chains. A third domain C existed between the two other domains and was formed by poly(divinylbenzene) star copolymeric cores. This domain had very high segment density. The association number was about 27 at 35 °C. Further systematic work is needed to characterize the association phenomenon of this type of copolymers. So far they exhibit analogous behaviors as those in linear block copolymers.

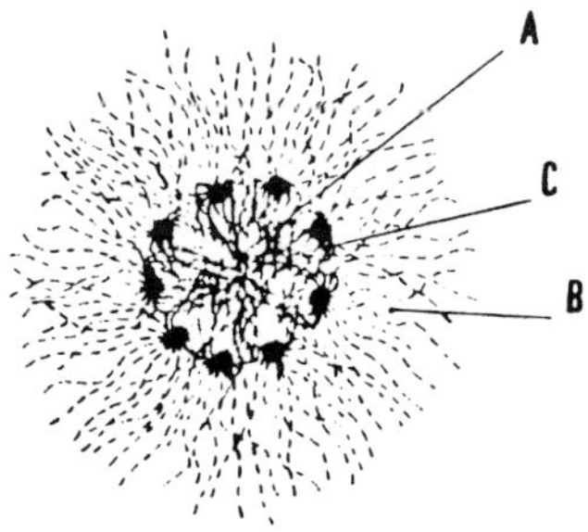

Figure 5. Two-dimensional schematic representation of micelle structure consisting of a number of A_nB_n heteroarm star copolymer molecules. A: core; B: shell; C: A, B interface. Reprinted from Reference 108, with permission.

4. PHASE BEHAVIOR

Low molecular weight nonionic surfactants associate into micelles above the CMC in water and form different liquid crystal phases at higher concentrations. In addition, the aqueous solution of amphiphilic block copolymers shows a lower cosolute temperature called the cloud point where the one-phase solution is separated into a two phase region. The phase behavior of nonionic block copolymers in aqueous solution has been investigated extensively [3]. However, there has been little attention paid to exploring the phase behavior of block copolymers in nonaqueous solvents. Such studies can aid in our understanding of the underlying mechanism responsible for the phase behavior.

Lindman and co-workers studied the phase behavior of PEO-PPO block copolymers in different polar organic solvents, and made comparison with those in aqueous solution [109,110]. The EO-containing block copolymers display a phase separation at elevated temperatures. A general explanation for the decrease in solubility of those copolymers at higher temperatures is a rapid increase in the effective attraction between different solute molecules [110]. Several models have been presented to explain the origin of this effect. Kjellander and Florin [111] claimed that the water formed an ordered structure around the EO chains at low temperatures. At higher temperatures, this ordered structure broke down, due to the unfavorable entropy contribution. Goldenstein proposed another model [112] originally based on Tirschfelder *et. al.* [113]. The formation of hydrogen bonds between water molecules and ether oxygen of EO groups and their decomposition at higher temperatures are responsible for the cloud point. Karlstrom explained the decrease in solubility in terms of a change in the conformational structure of the EO chains as a function of temperature [114]. Table 9 is a summary of the cloud points in water and in formamide solutions. The copolymers with higher EO weight percentage has a much higher cloud point [109]. In other organic solvents, with both hydrogen-bonding and non-hydrogen bonding, with both higher and lower dielectric permittivity than water and formamide, clouding does not exist by increasing the temperature. The authors found that the ratio of molecular dipole moment μ, to the molecular volume V, has larger values in PEO-PPO/water and PEO-PPO/formamide systems. They drew the conclusion that small and polar solvent molecules would induce clouding when the temperature was increased.

Table 9.
Some properties related to the phase behavior of PEO-PPO in water and in formamides.

Solvent Molecule	Molecular dipole moment μ(D)	Molecular volumes V (Å)	μ/V	μ^2/V	Dielectic permittivity ε	Hydrogen bonding
Water	1.85	29.89	0.0619	0.1145	78.5	Yes
Formamide	3.73	65.98	0.056	0.2109	109	Yes
N-methyl formamide	3.83	97.01	0.039	0.1512	182.4	Yes
N,N Dimethyl formamide	3.82	127.91	0.0292	0.1141	37	No
DMSO	3.96	117.78	0.034	0.1331	49	No
N-methylaceamide	3.73	126.81	0.0294	0.1097	175.7	Yes

Reprinted from Reference 109, with permission.

The effect of electrolytes on the cloud point of PEO-PPO block copolymers was also studied by the same authors [110]. Both NaCl and NaI affected the clouding temperature in water and formamide. The addition of NaCl lowered the clouding temperature in both solvents, whereas the addition of NaI raised the lower cosolute temperature. These effects are referred to by the concept of "salting in" and "salting out" cosolutes [115]. At low salt concentrations (e.g. about 0.3 wt% for NaCl), the depression of the lower cosolute temperature was more pronounced in formamide (15 °C) than in water (1 °C) (Figure 6). In formamide solutions, there was essentially no further variation in clouding temperature above 0.3% NaCl concentration. The qualitative similarities of the effects of electrolytes on cloud point in water and in formamide suggested that the underlying mechanism was the same. The added NaCl may probably interact with the solvent and make the solvent effectively more polar, thus increasing the difference in polarity between the solvent and the polymer, which favors phase separation. The partitioning of NaI, with a large negative ion I^-, will be slightly in favor of the polymer, which means that the difference in polarity between the solvent and the solute is diminished and thus leading to an increase in the cloud point.

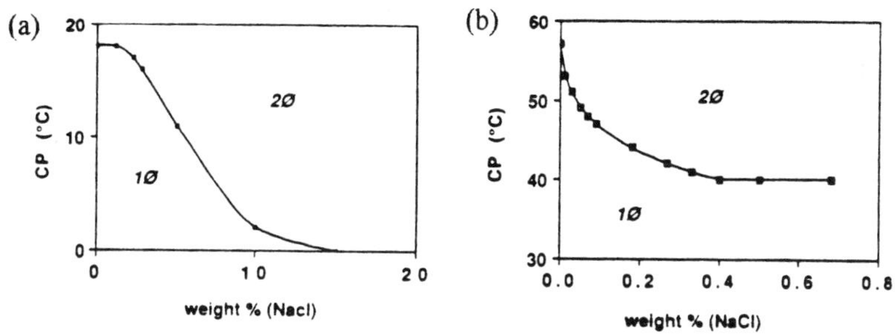

Figure 6. Partial phase diagram for 1% polymer A (PEO-PPO, M_w=3438 g/mol, EO/PO molar ratio=0.33) in water (a) and in formamide (b) as a function of weight percent of NaCl added (1Φ, one-phase region; 2Φ, two-phase region)
Reprinted from Reference 110, with permission.

Nunes and co-workers reported the cloud point of diblock copolymer PS_{110}poly(butyl methacrylate)$_{170}$ ($PS_{110}PBMA_{170}$) in 2-propanol [116], a θ solvent for PBMA blocks and a precipitant for PS blocks. The cloud point curve of $PS_{110}PBMA_{170}$ was quite similar to that of $PBMA_{2050}$ and $PBMA_{520}$ homopolymers in the same solvent, and can be superimposed to a high degree of accuracy if one compared the results of $PS_{110}PBMA_{170}$ with those of $PBMA_{170}$. This observation indicated that the PS domains were very effectively protected from any interactions with the solvent by the PBMA corona.

Gast and co-workers explored the nature of the disorder-order transition in micellar suspensions of PS-PI in decane by means of SAXS [117]. The phase diagram was considered by using a plot of PS core volume fraction (ϕ_{PS}) versus $\langle L \rangle_h/R_c$, the ratio of coronal layer thickness to the core size. A disordered phase was observed when ϕ_{PS} was low. By increasing the value of ϕ_{PS}, fcc and bcc crystal structures were formed at low (<1.5) and high (>1.5) $<L>_h/R_c$ values, respectively.

SUMMARY

Here we summarize some basic conclusions related to the formation of amphiphilic block copolymer micelles in organic solvents:

1. Different physical methods have been employed to study many facets of micellization behavior in organic media and more newly developed techniques have extended the characterization capability in this field.
2. The nature of the interaction between the copolymer chains and the solvent plays a dominant role on the micellization behavior of block copolymers. Micelles will not form in a solvent good for all the blocks. In a selective solvent, both close association and open association processes have been reported and they can interchange by varying the solvent quality.
3. For micelles obeying a closed association process, a smaller CMC, a higher association number, and a more negative $\Delta G°$ value can be expected by decreasing the solution temperature, increasing the length of less soluble block, decreasing the length of soluble block, or choosing a worse solvent for the less soluble block. Micelles usually have spherical core-shell structures, but other shapes and shape transitions have been reported. The interface between the micellar core and the shell could become more diffuse at temperatures which improves the solvent quality.
4. For AB diblock copolymers and ABA triblock copolymers in a solvent selective for A blocks, core-shell micelles will form. For ABA triblock copolymers in a solvent selective for the middle B block, flower-like micelles, branched open structures or intermediate structures may form. The association ability of triblock copolymers is much weaker then that of diblock copolymers, with a much higher CMC value and lower association number under the same conditions.
5. Block ionomers show similar association behavior to that of nonionic copolymers. However, it is strongly affected by the neutralization degree and the length of ionic blocks. The effect of counter ions is also important for short ionic blocks.
6. Several hypothesis have been introduced to explain the anomalous micellization. Although the origin of this undesirable phenomenon is still under discussion, more results seem to support the observation that chemical heterogeneity of block copolymers is responsible for the phenomenon.
7. The newly synthesized heteroarm star-block copolymers have similar micellization behavior as that of simple block copolymers.

II. AGGREGATION OF AMPHIPHILIC BLOCK COPOLYMERS IN SUPERCRITICAL CARBONDIOXIDE

In supercritical fluids (SCFs), where the fluid temperature and pressure are above those of the corresponding critical point, the properties of the fluid are uniquely different from either the conventional gas or liquid states. The physical properties of near-critical fluids ($T/T_c \geq 0.75$), where the temperature is just slightly below the critical point temperature, and of SCFs are intermediate between those of a normal liquid and a gas[118,119]. Diffusion coefficients are 10 to 100 times higher in SCFs than in liquids, which is a significant advantage for applications (e. g., reactions or separations) that depend upon transport processes. The density of near-critical fluids and of SCFs is strongly dependent on the pressure of the system, which is an advantage to forming micelles in SCFs since the solvent quality of the continuous phase can be readily controlled by manipulation of the system pressure. Micelles and microemulsions having a SCF as the continuous phase have been reported to exist in several SCFs [120-124]. Most of the studies were about the AOT surfactant (sodium bis(2-ethylhexyl) sulfosuccinate). Supercritical fluids are becoming increasingly important solvent systems for use in polymer science and engineering [125]. Supercritical CO_2, in particular, is a widely used solvent due to its low cost, moderate critical conditions (T_c = 31 °C, P_c = 73.8 bar), and environmentally benign nature, which makes it a good replacement for some toxic organic solvents. It has been shown that only a few classes of polymeric materials are significantly soluble (tens of percent) in supercritical CO_2 at relatively "mild" (T < 100 °C, P < 350 bar) conditions, these being amorphous fluoropolymers and silicones [125-128]. We categorize [129] these materials as being "CO_2-philic", while conventional polymers, having either hydrophilic *or* lipophilic character, are relatively insoluble in CO_2 and are termed "CO_2-phobic":

Polymeric Materials

"CO_2-philic"	"CO_2-phobic"
1. Fluoropolymers	1. Hydrophilic
2. Silicon-polymers	2. Lipophilic

Polymerization in supercritical CO_2 is an environmentally friendly way to make polymers. However, the polymers, which could be made in supercritical CO_2 by homogeneous polymerization [125, 130-132], are very limited due to the insolubility of most polymers in this solvent. Thus it is necessary to pursue heterogeneous polymerization processes to extend the advantages of polymerization in supercritical CO_2 to other systems. CO_2-based emulsion polymerization has received much attention. The experiments performed by DeSimone *et. al.* about the precipitation polymerization of acrylic acid in supercritical CO_2 and the polymerization of the same polymer in supercritical CO_2 with surfactants as stabilizers indicated that no significant increase in product molecular weight was achieved with any of the six surfactants employed [133]. However, amphiphilic block copolymer surfactants could be very effective in emulsifying CO_2-insoluble materials and stabilizing growing polymer colloids dispersed in CO_2 during dispersion polymerization processes[134-135], resulting in the formation of high molar mass polymers with high rates of polymerization. The polymerization in supercritical CO_2 with polymeric surfactants also makes it easier to control the product particle size and morphology. The approach could be used to synthesize new materials with special condensed structure. It is obvious that the study of micellar behavior of block copolymers in supercritical CO_2 is important, i. e., an improved understanding of the self-assembly of amphiphilic block copolymers in supercritical CO_2 will also aid in the development of new routes for polymer synthesis in this relatively benign solvent. Although only preliminary studies on the self-assembly behavior of amphiphilic block copolymers in supercritical CO_2 have been carried out, the importance of this study is obvious.

1. AGGREGATION OF DIBLOCK COPOLYMERS IN SUPERCRITICAL CO_2

The micellar behavior of a block copolymer in a supercritical fluid could be more complex than in other systems. Besides the dependence of the phase behavior on concentration, temperature and the nature of the solvent employed, the phase behavior could also be dependent on the system pressure, which governs the density of the supercritical fluid and the solubility of a given substance in the supercritical CO_2. The association of semifluorinated diblock alkane $F(CF_2)_{10}(CH_2)_{10}H$ and perflouroalkylpoly(ethylene oxide) diblock surfactant, $F(CF_2)_{6\text{-}10}(CH_2CH_2O)_{3\text{-}8}H$, has been reported by Fulton *et al.* [120]. Very recently a SANS study on the micellar behavior of 1,1-dihydroperfluorooctyl acrylate-b-styrene diblock copolymer (PFOA-b-PS) in supercritical CO_2 was carried out by DeSimone *et al.*[135].

1.1 Semifluorinated diblock alkane $F(CF_2)_{10}(CH_2)_{10}H$

Semifluorinated diblock or triblock n-alkanes consist of either one or two fluorinated chains covalently bonded to a hydrocarbon segment. The chemical formula for the diblock alkanes is $F_3C(CF_2)_{m-1}(CH_2)_{n-1}CH_3$, abbreviated as F_mH_n. These alkanes can be regarded as semiflexible, low-molecular-weight copolymers of linear perfluorocarbons and hydrocarbons [136]. There have been several studies about the micellar behavior of semifluorinated n-alkanes [137-139] in selective solvents under normal pressure (1 atmosphere). The light-scattering study [138] of $F_{12}H_{10}$ in octane which is selective to the hydrocarbon block suggests that the material forms micelles of approximately 130 unimers at 35 °C and low concentrations (0.3-1.9 wt %). The degree of association decreases sharply at higher temperatures (60 °C) and in less selective solvents such as toluene. The F_8H_{12}/perfluorotributylamine and F_8H_{16}/perfluorooctane solutions, where the solvent is selective to the fluorocarbon segment, were reported to form premicellar aggregates on the order of four to six molecules per aggregate according to dynamic light-scattering and fluorescence probe solubilization studies [139]. Recently, the phase behavior of solutions of semifluorinated n-alkanes, heptadecafluorotetracosane F_8H_{16} and pentacosafluorooctacosane $F_{12}H_{16}$, in perfluorooctane and isooctane was studied in detail by Nostro and Chen [137] using light-scattering and small-angle neutron-scattering (SANS) techniques. A clear liquid solution of F_8H_{16} or $F_{12}H_{16}$ in perfluorooctane (PFO) or isooctane (i-OCT) is formed when it is heated up. However, at room temperature the solution becomes a white soft gel due to the strong interactions between the outer corona and the solvent. The results show the presence of fairly monodisperse small micelles in F_8H_{16}/perfluorooctane solutions in the liquid phase above the critical micelle concentration. The micelles contain a core consisting of hydrogenated segments and a shell made up of the fluorinated chains and the solvent. The SANS study shows that the aggregates are almost spherical micelles with a core radius of about 1.3 nm.

The semifluorinated alkane $F_{10}H_{10}$ has been shown to be highly soluble in CO_2 and to form gels from 20 wt % liquid CO_2 solutions at room temperature [140]. Very recently Fulton *et al.* [120] studied the micelle behavior of $F_{10}H_{10}$ in supercritical CO_2 by means of small-angle X-ray scattering (SAXS). The experiments focused on 5-6 wt % concentrations of the semifluorinated alkanes in supercritical CO_2 at 65 °C. The study shows that as the pressure of the system is reduced, there is an increase in the scattered intensity across the entire q range. This is in part due to decreases in the density of the supercritical fluid. The solvent density decrease increases the scattering contrast between the particle and the solvent. The association behavior of the $F_{10}H_{10}$ in supercritical CO_2 was analyzed by using the Guinier approximation for the particle form factor given by

$$P(q) = P(0)\exp(-R_g^2q^2/3) \tag{10}$$

where R_g and q are the radius of gyration of the particle and the scattering vector, respectively. As shown in Figure 7, the scattering data under the three different pressures are fitted to a Guinier analysis. The slope $(-R_g^2/3)$ of the line from a plot of ln I(q) versus q^2 yields the radius of gyration for the micelle. For spherical particles, $R_g = (0.6)^2 r$, where r is the radius of the

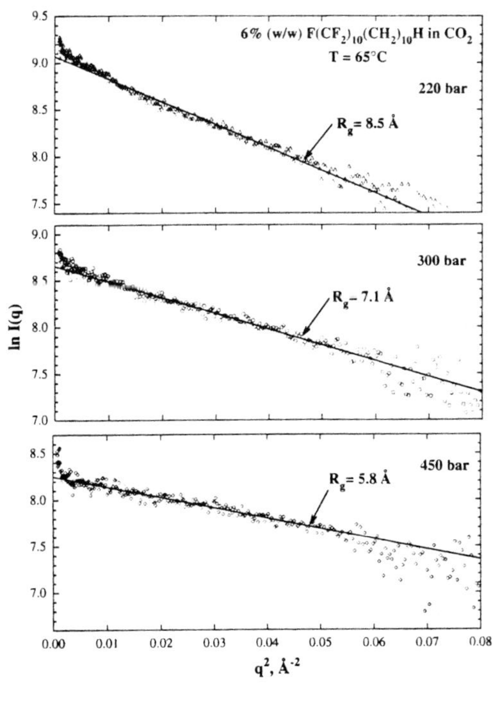

Figure 7. Guinier analysis of the $F_{10}H_{10}$ in supercritical CO_2 at 65 °C and three different pressures, 220, 300 and 450 bar.
Reprinted from Reference 120, with permission.

particle. As shown in Figure 7, the radii of gyration of the micelles under pressures 220, 300 and 450 bar are 5.8, 7.1 and 8.5 Å, respectively. Since the electron density of the perfluorinated portion of the $F_{10}H_{10}$ is approximately 3 times higher than that of the protonated portion, the perfluorinated portion dominates the scattering of the solution and the above calculated R_g values actually corresponds to the domain size consisting of fluorinated segments. For a rigid rod having a length h equivalent to the 10 carbon perfluorinated segment, the radius of gyration is $R_g=(h^2/12)^{1/2}$, or 4.2 Å. The fluorinated segment is, in fact, semirigid, i.e. its R_g is less than 4.2 Å, and the R_g of the $F_{10}H_{10}$ chain should be larger than that of the fluorinated segment due to the contributions from the protonated portion of the chain. The predicted size is just slightly smaller than the 5.8 Å measured at 450 bar. This may indicate that there is a slight amount of aggregation but certainly nothing larger than three to four unimers per aggregate under the present experimental conditions (65 °C and 220 to 450 bar). The extent of aggregation would be enhanced at temperatures well below 65 °C.

1.2 Perflouroalkylpoly(ethylene oxide) $F(CF_2)_{6\text{-}10}(CH_2CH_2O)_{3\text{-}8}H$

$F(CF_2)_{6\text{-}10}(CH_2CH_2O)_{3\text{-}8}H$ is a commercial nonionic surfactant, with a trade name of Zonyl FSO-100 manufactured by Du Pont. It is a small diblock copolymer consisting of a perfluoroalkane chain covalently linked to a poly(ethylene oxide) segment. The solubility of Zonyl FSO-100 in supercritical CO_2 shows a strong dependence on system pressure. Below about 400 bar, a small amount of the sample, most likely the fraction containing higher molecular weight blocks of PEO, is insoluble in the supercritical CO_2. The SAXS curve (Figure 8) of the solution of Zonyl FSO-100 in supercritical CO_2 at 530 bar, where it is a single phase, suggests that the polydisperse material forms aggregates. The fitted scattering curve is also shown in Figure 8. The behavior represented by the SAXS spectra of the FSO-100 surfactant is characteristic of a system having a broad distribution of sizes. An inversion technique of ReFiorentin has been utilized to extract size and particle distribution information

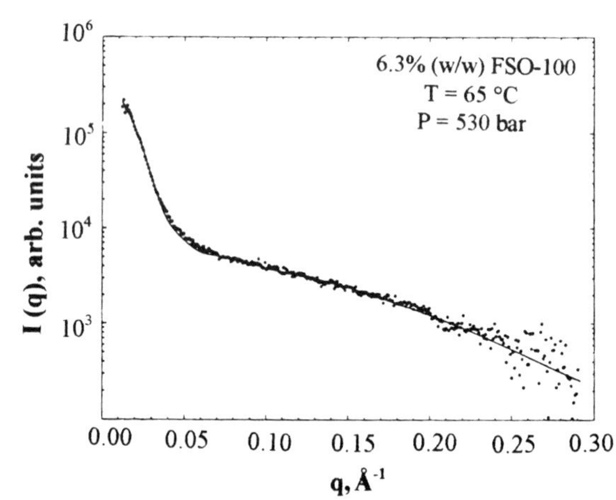

Figure 8. SAXS spectra (points) for 6.3% (w/w) of FSO-100 in supercritical CO_2 at 65 °C and 530 bar, and the distribution fit (line) of the scattering curve.
Reprinted from Reference 120, with permission.

from the 530 bar single phase system, under the assumptions that the micelles represent a polydisperse system of homogeneous spheres and that the interparticle interference is negligible. The result suggested that a small number of large aggregates were present in the sample, and were responsible for the large scattering at q less than about 0.08 $Å^{-1}$ as shown in Figure 8. Their mean radius was 84 Å, with a polydispersity index of 0.31. Such analysis is preliminary at best. Further study should be interesting.

2. MICELLE BEHAVIOR OF BLOCK COPOLYMER IN SUPERCRITICAL CO_2 AND WATER MIXTURES

Fulton *et al.* [120] also used SAXS to study the micellar behavior of a poly(1,1-dihydroperfluorooctylacrylate-g-ethylene oxide) copolymer or PFOA-g-PEO in supercritical CO_2 and water mixtures. The graft copolymer consisting of a CO_2-philic backbone of PFOA and CO_2-phobic PEO grafts. The number average molecular weight of the CO_2-phobic PEO grafts is 5 kg/mol with a molecular weight distribution of 1.10 by GPC analysis, and the weight fraction of ethylene oxide of the copolymer is 15% by proton NMR analysis. Due to complete insolubility of 5 kg/mol PEO in CO_2, the CO_2-phobic PEO grafts are supposed to associate into a micellar structure in the presence of water to promote hydrogen bonding. That is, the PEO grafts are supposed to exist in the core of the micelle, with the PFOA backbone forming a shell to limit the CO_2-PEO interaction. The supercritical CO_2 solutions of 0.6 wt % and 1.9 wt % of PFOA-g-PEO copolymer, with a water-to-surfactant weight ratio of 0.32 (i. e., $[H_2O]/[EO]$ is 5.2), were studied at 60 °C and three different pressures, 255, 300 and 470 bar. The lower pressure data sets are very near the phase boundary that occurs below 238 bar for the 0.6% solution and at slightly lower pressure, 234 bar, for the 1.9% solution. Like the SAXS data of the $F_{10}H_{10}$ solution, the scattered intensity of the two solutions also increases across the entire q range with decreasing system pressure, as shown in Figure 9 for the 0.6% (w/w) solution. This is in part due to decreases in the density of the supercritical CO_2. The solvent density decrease increases the scattering contrast between the particle and the solvent. All the scattering data for PFOA-g-PEO show a characteristic peak at about q = 0.041 $Å^{-1}$, and shows an evidence of a second scattering peak at approximately q = 0.06 $Å^{-1}$. The oscillations in the region from $0.012 < q < 0.07$ $Å^{-1}$ are consistent with scattering from spherical micelles having a small degree of polydispersity. There is also qualitative evidence of a small increase in the size of the particles as the pressure is reduced as shown by the shift in the scattering peaks to lower q. A nonlinear, least-squares fit of spherical core-with-shell model to the low q region of the SAXS spectra of the 0.6% and 1.9% (w/w) solutions was carried out. Table 10 gives the size and polydispersity parameters derived from the fitting, where $\rho_{solvent}$ is the scattering length density of the solvent, $\overline{R}_{core}$ is the mean core radius of the micelles, $\overline{R}_{shell}$ is the mean outer radius based on an assumed Gaussian distribution of sizes, σ is the standard deviation, $l_{tail} = \overline{R}_{shell} - \overline{R}_{core}$ is the length of the PFOA tail and $\sigma/\overline{R}_{shell}$ represents the polydispersity of the system. It is seen from Table 10 that the outer radii of the aggregates are about 125 Å with relatively low polydispersities of about $\sigma/\overline{R}_{shell} = 0.16$, suggesting a very uniform association process. The radii of the micelles clearly increase either as the concentration of the copolymer is increased or as the pressure of the system is reduced. The radius of the PEO core is approximately 105 Å. This is much larger than the diameter of a single, collapsed 5K PEO chain. The calculation shows that these aggregates contain approximately 600 grafts in the core, and that the estimated degree of aggregation is 120 unimers.

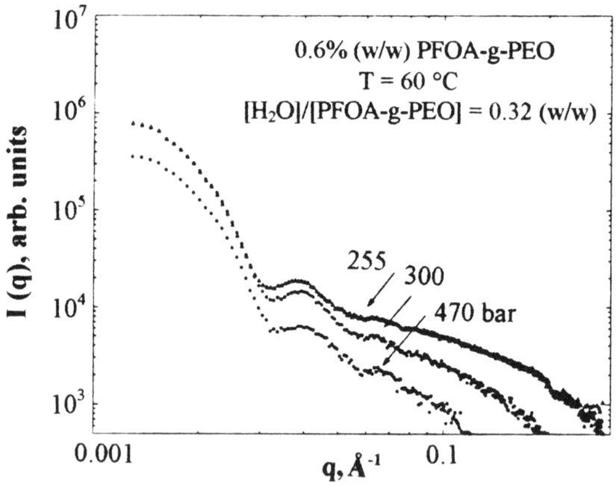

Figure 9. SAXS spectra for 0.6% PFOA-g-PEO in supercritical CO_2 at 60 °C and three different pressure, 255, 300, and 470 bar.
Reprinted from Reference 120, with permission.

Table 10
Size and polydispersity Parameters from core with shell model fit of the scattering data for PFOA-*g*-PEO in supercritical CO_2

% PEO (w/w)	pressure bar	$\rho_{solvent}$ $e^-/Å^3$	$\overline{R}_{shell}$ Å	l_{tail} Å	$\overline{R}_{core}$ Å	polydispersity $\sigma/\overline{R}_{shell}$
0.6	470	0.275	117	15.2	102	0.162
0.6	300	0.244	122	19.8	102	0.155
0.6	255	0.230	124	35.8	88	0.159
1.9	470	0.275	119	14.6	104	0.180
1.9	300	0.244	129	16.5	113	0.165
1.9	255	0.230	148	22.4	125	0.153

Reprinted from Reference 120, with permission.

The above preliminary study shows that a micellar phase could be formed by amphiphilic block copolymers in supercritical CO_2 or its mixture with water. Besides the micellar dependence on copolymer composition, molecular weight of the copolymer and the nature of the CO_2-phobic" block, as it is observed in a classic solvent or a solvent mixture, the micellar behavior of block copolymers in supercritical CO_2 also shows a dependence on the system pressure. The preliminary study is helpful to our understanding of the micellar behavior of amphiphilic block copolymers in supercritical fluids and to develop applications of polymer synthesis in supercritical fluids. Obviously, much more studies on the micellization of amphiphilic block copolymers in supercritical CO_2 and other supercritical fluids are needed.

III. MICELLES IN DILUTE BLOCK COPOLYMER/HOMOPOLYMER SOLUTIONS

Analogous to micellar behavior of block copolymers in a selective organic solvent, blending of a AB block copolymer with a large amount of its corresponding low-molecular-weight homopolymer A can result in the formation of a micellar phase, due to a net repulsive interaction between the two different monomers since most polymers are not miscible. The micelles are characterized by a core that is rich in B block of the copolymer and an outer shell composed of A homopolymer chains mixed with A blocks of the copolymer. Both theoretical and experimental studies have been made on micellar formation in such dilute solid solutions, including, theories [141-145] presented by different research groups and experimental (SAXS, TEM and NMR) investigations [15,146-149]. Micelles of spherical geometry were assumed in the theories developed by Leibler, Orland, and Wheeler[141] and by Whitmore and Noolandi[142]. Rigby and Roe [146-147] used SAXS to study spherical micelle formation in mixtures of styrene-butadiene diblock copolymers with the butadiene homopolymer by varying temperature, copolymer concentration, and relative block lengths. The NMR self-diffusion study of the micellar behavior of ethylene oxide-b-dimethylsiloxane diblock copolymer in poly(ethylene oxide) melts was made by Kiraly, Cosgrove and Vincent [15]. A theory [145] presented by Mayes and Olvera de La Cruz analysed the effects of varying copolymer length and homopolymer molecular weight on micelle morphology. Kinning, Winey and Thomas [148] studied dilute poly(styrene-b-butadiene)/polystyrene systems via transmission electron microscopy(TEM). They observed changes in micelle morphology from spherical to nonspherical micelles (i.e. cylindrical micelle, vesicles, and other lamellar type structures) by varying diblock copolymer concentration, asymmetry, and homopolymer molecular weight. The micellar behavior of amphiphilic block copolymers in this kind of dilute solid solution is not reviewed in this chapter. Some of the studies are presented here in order to show the similarity of micelle formation of block copolymers in polymer blends when compared with that in more conventional nonaqueous organic solvents.

REFERENCES

1. Z. Tuzar and P. Kratochvil, Micelles of Block and Graft copolymers in solutions in Surface and Colloid Science, Vol. 15, E. Matijevic Ed., Plenum Press, New York, 1993.
2. K. Kon-no, Properties and Applications of Reversed Micelles in Surface and Colliod Science, Vol. 15, E. Matijievic Ed., Plenum Press, New York, 1993.
3. B. Chu and Z. Zhou, Physical Chemistry of Polyoxyalkylene Block Copolymer Surfactants in Nonionic Surfactants, V. M. Nace Ed., Marcel Dekker, Inc., New York, 1996.
4. A.J.I. Ward and C. Du Reau, Surfactant Association in Nonaqueous Media in Surface and Colloid Science, Vol. 15, E. Matijevic Ed., Plenum Press, New York, 1993.
5. A. Desjardins and A. Eisenberg, Macromolecules, 24 (1991) 5779.
6. Z. Gao, X.F. Zhong and A. Eisenberg, Macromolecules, 27 (1994) 794.
7. C. Honda, K. Sakaki and T. Nose, Polymer, Vol. 35, Number 24 (1994) 5309.

8. C. Honda, Y. Hasegawa, R. Hirunuma and T. Nose, Macromolecules, 27 (1994) 7660.
9. C. Honda, K. Yamamoto and T. Nose, Polymer, in press.
10. M. Pacovska, K. Prochazka, Z. Tuzar and P. Munk, Polymer, Vol. 34, Number 21 (1993) 4585.
11. X. He, J. Herz and J.-M. Guenet, Macromolecules, 21, (1988) 1757.
12. Z. Zhou, B. Chu, and D. Peiffer, Macromolecules, 26 (1993) 1876.
13. A.S. Yeung and C.W. Frank, Polymer, Vol. 31, November (1990) 2089.
14. J. Godward, F. Heatley and C. Price, J. Chem. Soc. Faraday Trans., 89(18) (1993) 3471.
15. Z. Kiraly, T. Cosgrave and B. Vincent, Langmuir, 9 (1993) 1258.
16. W. Brown, R. Johnsen, P. Stilbs and B. Lindman, J. Phys. Chem., 87 (1983) 4548.
17. P.G. Nilsson, H. Wennerstrom and B. Lindman, B. Chem. Scr., 25 (1985) 67.
18. M. Jonstromer, M. Sjoberg and T. Warnheim, J. Phys. Chem., 94 (1990) 7549.
19. C. Chachaty, Prog. NMR Spectrosc., 19 (1987) 183.
20. K. Prochazka, B. Medhage, E. Mukhtar, M. Almgren, P. Svoboda, J. Trnena and B. Bednar, Polymer, Vol. 34, Number 1 (1993) 103.
21. D. Kiserow, J. Chan, C. Ramireddy, P. Munk and S.E. Webber, Macromolecules, 25 (1992) 2338.
22. A.S. Yeung and C.W. Frank, Polymer, Vol. 31, November (1990) 2101.
23. J. Duhamel, A. Yekta, S. Ni, Y. Khaykin and M.A. Winnik, Macromolecules, 26 (1993) 6255.
24. F. Calderara, Z. Hruska, G. Hurtrez, J.-P. Lerch, T. Nugay and G. Riess, Macromolecules, 27 (1994) 1210.
25. G. Liu, C.K. Smith, N. Hu and J. Tao, Macromolecules, 29 (1996) 220.
26. G. Wu, Z. Zhou and B. Chu, J. Polymer Sci., Part B, Vol. 31 (1993) 2035.
27. Z. Zhou, B. Chu and D.G. Peiffer, Langmuir, 11 (1995) 1956.
28. G. Wu, B. Chu and D.K. Schneider, J. Phys. Chem., 98 (1994) 12018.
29. Z.B. Alfassi and W.G. Filby, Chem. Phys. Lett., 144 (1988) 83.
30. Y.-C. Jean and H. J. Ache, J. Am. Chem. Soc., 99 (1977) 7504.
31. Y.-C. Jean and H. J. Ache, J. Phys. Chem., 82 (1978) 811.
32. Y.-C. Jean and H. J. Ache, J. Am. Chem. Soc., 100 (1978) 6320.
33. J.C. Ravey, M. Buzier and C. Picot, J. Colloid Inter. Sci., 97, No. 1, Jan. (1984) 9.
34. G. Wu, Z. Zhou and B. Chu, Macromolecules, 26 (1993) 2117.
35. A. Desjardins, T.G.M. Van de Ven, A. Eisenberg, Macromolecules, 25 (1992) 2412.
36. J.R. Quintana, M. Villacampa and I.A. Katime Macromolecules, 26 (1993) 601.
37. J.R. Quintana, M. Villacampa and I.A. Katime, Polym. Inter., 36 (1995) 325.
38. B. Chu, Z. Zhou and G. Wu, J. Non-crystal. Solids., 172-174 (1994) 1094.
39. R. Hilfiker, B. Chu and J. Xu, J. Colliod Inter. Sci., 133 (1989) 176.
40. J.R. Quintana, R.A. Salazar and I.A. Katime, J. Phys. Chem., 99 (1995) 3723.
41. J.R. Quintana, R.A. Salazar and I.A. Katime, Macromolecules, 27 (1994) 665.
42. J.R. Quintana, M.D. Janez and I.A. Katime, Macro. Rapid Common., 16 (1995) 607.
43. J.R. Quintana, R.A. Salazar and I.A. Katime, Macromol. Chem. Phys., 196 (1995) 1625.
44. J.R. Quintana, M. Villacampa, M. Munoz, A. Andrio and I.A. Katime, Macromolecules, 25 (1992) 3129.
45. M. Villacampa, J.R. Quintana, R.A. Salazar and I.A. Katime, Macromolecules, 28 (1995) 1025.
46. J.R. Quintana, M. Villacampa and I.A. Katime, Polymer,34 (1993) 2380.

47. J.R. Quintana, M. Villacampa and I.A. Katime, Macromolecules, 26 (1993) 606.
48. D. Ngugen, C.E. Williams and A. Eisenberg, Macromolecules, 27 (1994) 5090.
49. M. Moffitt, K. Khougaz and A. Eisenberg, Acc. Chem. Res., 29 (1996) 95.
50. H. G. Elias in Light Scattering from Polymer Solutions; M.B. Huglin Ed.; Academic Press: New York, (1972) Chapter 9.
51. K. Khougaz, Z. Gao and A. Eisenberg, Macromolecules, 27 (1994) 6341.
52. Z. Gao and A. Eisenberg, Macromolecules, 26 (1993) 7353.
53. Z. Zhou, B. Chu, G. Wu and D. Peiffer, Macromolecules, 26 (1993) 2968.
54. K. Khougaz, D. Nguyen, C.E. Williams and A. Eisenberg, Can. J. Chem., 73 (1995) 2086.
55. X.F. Zhong and A. Eisenberg, Macromolecules, 27 (1994) 1751.
56. X.F. Zhong, S.K. Varshney and A. Eisenberg, Macromolecules, 25 (1992) 7160.
57. J.R. Quintana, M. Villacampa, R.A. Salazar and I.A. Katime, J. Chem. Soc. Faraday Trans., 88(18) (1992) 2739.
58. C. Price, K.D. Kendall, R.B. Stubbersfield and B. Wright, Eur. Polym. J., 20 (1984) 685.
59. C. Price, A.L. Hudd and R.B. Stubbersfield, Polymer, 21 (1980) 9.
60. M.N. Jones, G. Pilcher and L. Espada, J. Chem. Thermodyn., 2 (1970) 333.
61. C. Price, E.K.M. Chan, G. Pilcher and R.B. Stubbersfield, Eur. Polym. J., 25 (1987) 170.
62. C. Price, C. Booth, P.A. Canham, T. de V. Naylor and R.B. Stubersfield, Br. Polym. J., 16 (1984) 311.
63. C. Price, E.K.M. Chan, A.L.Hudd and R.B. Stubbersfield, Polym. Commun., 27 (1986) 196.
64. C. Price, R.B. Stubbersfield, S. El-Kafrawy and K.D. Kendall, Br. Polym. J., 21 (1989) 391.
65. J.R. Quintana, M. Villacampa, M. Munoz, A. Andrio, and I.A. Katime, Macromolecules, 25 (1992) 3125.
66. J.R. Quintana, M.D. Janez, M. Villacampa and I.A. Katime, Macromolecules, 28 (1995) 4139.
67. M. Villacampa, E.D. De Apodaca, J.R. Quintana and I.A. Katime, Macromolecules, 28 (1995) 4144.
68. I.A. Katime, J.R. Quintana and L.C. Cesteros, Eur. Polym. J., 26 (1990) 1121.
69. A. Guinier and G. Fournet, Small Angle Scattering of X-rays (in English), C. Walker trans., Jonh Wiley & Sons, Inc., New York; Chapman & Hall, Ltd., London, 1955.
70. G.T. Oostergetel, F.J. Esselink and G. Hadziioannou, Langmuir, 11 (1995) 3721.
71. L. Pu, G. Yuan, B. Li and Z. Xu, Makromol. Chem. Macromol. Symp., 58 (1992) 189.
72. J.-P. Gouin, A. Eisenberg and C.E. Williams, Macromolecules, 25 (1992) 1368.
73. K. Khougaz X.F. Zhong and A. Eisenberg, Macromolecules, in press.
74. C.L. Phoon, J.S. Higgins, W. Burchard and D.G. Peiffer, Macro. Rep., A2 (suppl. 2) (1992) 179.
75. M. Antonietti, S. Heinz, M. Schmidt and C. Rosenawer, Macromolecules, 27 (1994) 3276.
76. J.R. Quintana, M. Villacampa and I.A. Katime, Makromol. Chem., 194 (1993) 983.
77. Z. Tuzar, C. Konak, P. Stepanek, J. Plestil, P. Kratochvil and K. Prochazka, Polymer, Vol. 31, November (1990) 2118.
78. J. Aime, D. Reibel and C. Mathis, Syn. Met., 55-57 (1993) 127.
79. D.A. Ylitalo, A.S. Yeung and C.W. Frank, Photochemical Processes in organized

molecular system, Editor-in-chief: K. Honda.
80. D.F. Siqueira, S.P. Nunes and B.A. Wolf, Macromolecules, 27 (1994) 4561.
81. H. Utiyama, K. Takenaka, M. Mizumori, M. Fukuda, Y. Tsunashima and M. Kurata, Macromolecules, 7 (1974) 515.
82. T.P. Lally and C. Price, Polymer, 15 (1974) 325.
83. P.A. Canham, T.P. Lally, C. Price and R.B. Stubbersfield, J. Chem. Soc., Faraday Trans., 1, 76 (1980) 1857.
84. C. Price, N. Briggs, J.R. Quintana, R.B. Stubbersfield and I. Robb, Polym. Commum., 27 (1996) 292.
85. C. Price, in Developments in Block Copolymers, Vol. 1, I. Goodman (ed.), Applied Science Publishers, London and New York, 1982, pp. 39.
86. Z. Tuzar and P. Kratochvil, Macromol. Chem., 170 (1973) 177.
87. Z. Tuzar, P. Bahadur and P. Kratochvil, Makromol. Chem., 182 (1981) 1751.
88. Z. Zhou and B. Chu, Macromolecules, 20 (1987) 3089.
89. J.R. Quintana, R.A. Salazar, M. Villacampa and I.A. Katime, Macro. Chem., 194 (1993) 2497.
90. T.N. Khan, R.H. Mobbs, C. Price, J.R. Quintana and R.B. Stubbersfield, Eur. Polym. J., 23 (1987) 191.
91. J. Selb and Y. Gallot, Makromol. Chem., 181 (1980) 809.
92. Z. Tuzar, J. Stehlicek, C. Konak and F. Lednicky, Makromol. Chem., 189 (1988) 221.
93. K.A. Cogan and A.P. Gast, Macromolecules, 23 (1990) 745.
94. L.J.M. Vagberg, K.A. Cogan and A.P. Gast, Macromolecules, 24 (1991) 6512.
95. R. Hilfiker, D. Wu and B. Chu, J. Colloid Inter. Sci., 135, No.2 (1990) 573.
96. N.P. Balsara, M. Tirrell and T.P. Lodge, Macromolecules, 24 (1991) 1975.
97. G. Wu and B. Chu, Macromolecules, 27 (1994) 1766.
98. E. Raspaud, D. Lairez, M. Adam and J.-P. Carton, Macromolecules, 27 (1994) 2956.
99. J. Plestil, D. Hlavata, J. Hrouz and Z. Tuzar, Polymer, Vol. 31, November (1990) 2117.
100. A. Hamnett, J.C.H. Kerr, J.W. White and L. Dai, J. Chem. Soc. Faraday Trans., 89(2) (1993) 277.
101. G. ten Brinke and G. Hadziioannou, Macromolecules, 20 (1987) 486.
102. K. Rodrigues and W.L. Mattice, Langmuir, 8 (1992) 456.
103. R. Saito, K. Ishizu and T. Fukutomi, Polymer, Vol. 32, Number 3 (1991) 531.
104. C. Tsitsilianis, P. Chaumont and P. Rempp, Makromol. Chem., 191 (1990) 2319.
105. C. Tsitsilianis, C. Graff and P. Rempp, Eur. Polym. J., 27 (1991) 243.
106. D. Rein, J.P. Lamps, P. Rempp, P. Lutz, D. Papanagopoulos and C. Tsitsilianis, Acta Polym., 44 (1993) 225.
107. K. Ishizu and S. Yukimasa, Polymer Vol. 34, Number 17 (1993) 3753.
108. C. Tsitsilianis and O. Kouli, Macromol. Rapid Commun., 16 (1995) 591.
109. A.A. Samii, B. Lindman and G. Karlstrom, Prog. in Col. Polym. Sci., 82 (1990) 280.
110. A. Samii, G. Karlstrom and B. Lindman, Langmuir, 7 (1991) 1067.
111. R. Kjellander and E. Florin, J. Chem. Soc., Faraday Trans. 1, 77 (1981) 2053.
112. R.E. Goldstein, J. Chem. Phys., 80 (1984) 5340.
113. J. Hirschfelder, D. Stevenson and H. Eyring, J. Chem. Phys., 5 (1937) 896.
114. G. Karlstrom, J. Phys. Chem., 89 (1985) 4962.
115. M. Schott, A. E. Royce and S. K. Han, J. Colliod Interface Sci., 98 (1984) 196.
116. D.F. Siqueira, S.P. Nunes and B.A. Wolf, Macromolecules, 27 (1994) 1045.

117. G.A. McCornell, A.P. Gast, J.S. Huang and S.D. Smith, Phys. Rev. Lett., September 27, Vol. 71, Number 13 (1993) 2102.
118. M.A. McHugh and V.J. Krukonis (eds.), Supercritical Fluid Extraction, Butterworths, Boston, 1986.
119. M.E. Paulaitis, J.M.L. Penninger, R.D. Gary, and R. Davidson (eds.) Chemical Engineering at Supercritical Fluid Conditions, Ann Arbor Science Publisher, Ann Arbor, MI, 1984.
120. J.L. Fulton, D.M. Pfund, J.B. McClain, T.J. Romack, E.E. Maury, J.R. Combes, E.T. Samulski, J.M. DeSimone and M. Capel, Langmuir, 11 (1995) 4241.
121. R.W. Gale, J.L. Fulton and R.D. Smith, J. Amer. Chem. Soc., 109 (1987) 920.
122. J.L. Fulton, J.P. Blitz, J.M. Tingey and R.D. Smith, J. Phys. Chem., 93 (1989) 4198.
123. J.M. Tingey, J.L. Fulton and R.D. Smith, J. Phys. Chem., 94(1990), 1997.
124. R.D. Smith, J.L. Fulton, J.P. Blitz and J.M. Tingey, J. Phys. Chem., 94 (1990) 781.
125. J.M. DeSimone, Z. Guan and C.S. Elsbernd, Science, 257 (1992) 945.
126. Z. Guan, J.R. Combes, C.S. Elsbernd and J.M. DeSimone, Polym. Prepr., 34 (1993) 447.
127. T. Hoefling, D. Sofesky, M. Reid, E. Beckman and R.M. Enick, J. Supercrit. Fluids, 5 (1992) 237.
128. K.A. Consani and R.D. Smith, J. Supercrit. Fluids, 3 (1990) 51.
129. J.M. DeSimone, E.E. Maury, Y.Z. Menceloglu, J.B. McClain, T.J. Romack and J.R. Combes, Science, 265 (1994) 356.
130. Z. Guan, J.R. Combes, Y.Z. Menceloglu and J.M. DeSimone, Macromolecules, 26 (1993) 2663.
131. J.R. Combes, Z. Guan and J.M. DeSimone, Macromolecules, 27 (1994) 865.
132. J.M. DeSimone, E.E. Maury, R.M. Lemert and J.R. Combes, Makromol. Chem., Makromol. Symp., 67 (1993) 251.
133. T.J. Romack, E.E. Maury and J.M. DeSimone, Polym. Prepr., 35 (1994) 741.
134. K.K. Kapellen, C.D. Mistele and J.M. DeSimone, Macromolecules, 29 (1996) 495.
135. D.A. Canelas, D.E. Betts and J.M. DeSimone, Macromolecules, 29 (1996) 2818.
136. G.L. Gaincs, Langmuir, 7 (1991) 3054.
137. P. Lo Nostro and S.-H. Chen, J. Phys. Chem., 97 (1993) 6535.
138. J. Hopken, C. Pugh, W. Richtering and M. Moller, Makromol. Chem., 189 (1988) 911.
139. M.P. Turberg and J.E. Brady, J. Am. Chem. Soc., 110 (1988), 7797.
140. A. Iezzi, P. Bendale, R.M. Enick, M. Turburg, J. Brady, Fluid Phase Equilib.,52 (1989) 307.
141. L. Leibler, H. Orland and J.C. Wheeler, J. Chem. Phys., 79 (1983) 3550.
142. M.D. Whitmore and J. Noolandi, Macromolecules, 18 (1985), 657.
143. L. Leibler and P.A. Pincus, Macromolecules, 17 (1984) 2922.
144. R.J. Roe, Macromolecules, 19 (1986) 728.
145. A.M. Mayes and M. Olvera de le Cruz, Macromolecules, 21 (1988) 2543.
146. D. Rigby and R.J. Roe, Macromolecules, 17 (1984) 1778.
147. D. Rigby and R.J. Roe, Macromolecules, 19 (1986), 19, 721.
148. D.J. Kinning, K.I. Winey and E.L. Thomas, Macromolecules, 21 (1988) 3502.
149. F.S. Bates, C.V. Berney and R.E. Cohen, Macromolecules, 16 (1983)1101.

Structures of amphiphilic block copolymers in their liquid and solid states.

Anthony J. Ryan[a,b], Shao-Min Mai[c], J. Patrick A. Fairclough[a], Ian W. Hamley[d]

[a]Manchester Materials Science Centre, UMIST, Grosvenor Street, Manchester M1 7HS, UK.
[b]CLRC Daresbury Laboratory, Warrington, WA4 4AD, UK.
[c]Department of Chemistry, University of Manchester, Manchester M13 9PL UK.
[d]School of Chemistry, University of Leeds, Leeds LS2 9JT UK.

NOTATION

E is used to denote an oxyethylene unit, $-OCH_2CH_2-$, and B is used to denote an oxybutylene unit, $-OCH_2CH(CH_2CH_3)-$. For example, a diblock is denoted E_mB_n. In the same spirit we use P to denote an oxypropylene unit, C to denote a methylene unit, S to denote styrene, I to denote isoprene and O to denote a poly(alkylene) formed by hydrogenation of a diene.

1. INTRODUCTION

Block copolymers are macromolecules comprised of two or more chemically different chains joined together by covalent bonds. The development of this field originated with the discovery of termination-free anionic polymerisation, which remains the best method for producing monodisperse block copolymers with well-defined architecture. A variety of molecular architectures are possible, e.g. AB diblock, ABA triblock, starblock copolymers. Current theories deal almost exclusively with diblock and starblock copolymers, with diblocks receiving the most comprehensive treatment [1].

The different component chains in block copolymers usually mix endothermically, and while short blocks may mix at high temperatures, long blocks are incompatible (i.e. thermodynamically immiscible). Consequently, the material (bulk) properties of many block copolymers are dominated by their tendency to spontaneously separate into microphases when the temperature is lowered. Similar effects are found in concentrated solutions of block copolymers. In fact the systematic study of the phase separation of block copolymers began with Skoulios et al. [2] who discovered well-organized microphase structures in SE-block copolymers in various solvents. It should also be mentioned that at the same time microphase-separated gels were described for poly(oxyalkylene) block copolymers in water, including block copolymers of ethylene oxide and 1,2-butylene oxide closely related to those investigated in this work, though the gels were not structurally characterised at that time [3]. Following these initial studies, the microphases exhibited by many block copolymers, in solution and in bulk, have been studied by many different groups. Particularly over the last twenty years, theoretical and experimental scientists have combined to make the phase behaviour of block copolymer in the liquid state well understood [1]. However, phase separation of block copolymers in the solid (semi-crystalline) state is less well understood [4-6].

This chapter introduces the theory of phase separation in block copolymers, then describes experimental developments in the microphase structures and behaviours of block copolymer melts, particularly recent work on I_mS_n and E_mB_n. Finally, the phase behaviour of crystallisable block copolymers is briefly reviewed, with emphasis on E_mB_n studied in our laboratory.

2. THEORY

2.1 Block copolymer melts

Theoretical aspects of microphase separation have been reviewed [1,7], and relevant references can be found therein. Starting from the pioneering studies of Helfand and Wasserman [8] and

Leibler [9], major contributions have come from self-consistent mean-field theories. Considering an AB block copolymer comprising $N = N_A + N_B$ chain units, each occupying the same volume (v) and having the same statistical length (b), theory predicts equilibrium phase behaviour dependent only on the product χN and the composition defined by the mole fraction of (say) A chain units, $x_A = N_A/N$. For this special case, parameter b is uniquely determined by the radius of gyration of the copolymer,

$$R_g^2 = Nb^2/6 \tag{1}$$

and parameter χ by the non-combinatorial Gibbs energy of mixing

$$\Delta_{mix}G_{nc} = kT\chi\, N_A x_B = kT(\chi N)\, x_A x_B \tag{2}$$

For the copolymers of interest here, the unlike segments mix endothermically, and the temperature dependence of χ will be given approximately by

$$\chi = \alpha + \beta/T \tag{3}$$

where α and β are constants dependent on composition. Microphase separation is favoured by lowering the temperature.

The main features of microphase separation in such a block copolymer system are clear. At low values of χN the block copolymer comprises a single disordered phase in which the chains have unperturbed dimensions [1,9]. Connectivity leads to a characteristic correlation length scaling as the radius of gyration, i.e. as $N^{1/2}$. Microphase separation is predicted at $\chi N \approx 10$ (the weak segregation limit, WSL). Free-energy minimisation results in the formation of ordered structures which are dependent on the composition. The surface curvature and domain size is effectively set by the ratio of the radii of gyration of the polymer chains forming the two blocks. The transition, analogous to freezing, is from a disordered state (liquid-like) to an ordered state (crystal-like): although the polymer chains locally are amorphous the microstructure has long-range order. The phase diagram shown in Figure 1, for the I_mS_n system, is determined experimentally *via* a combination of SAXS, SANS, TEM and rheology [10], in comparison with the phase boundary which is predicted theoretically [9]. The phase diagram is presented in terms of the product χN and block copolymer composition, ϕ. The major phase-separated structures of spheres (bcc), hexagonally-packed cylinders (hex) and lamellae (lam) are observed along with the complex hexagonally perforated lamellar (hpl) and cubic Ia3d (G) phases which are intermediate to the predicted structures.

As the WSL is approached, theory [11-16] predicts fluctuations in the disordered phase, with a characteristic dimension scaling as $N^{4/5}$. Immediately beyond the WSL, limited demixing of A and B occurs to form a microphase-separated phase with a characteristic dimension variously predicted [12-16] to scale approximately as $N^{0.8}$ to $N^{0.95}$. In the strong segregation limit (SSL, $\chi N >> 10$), AB contacts are limited to a narrow interfacial region, the microphases are essentially pure A and pure B, and the characteristic length is predicted [1,7] to scale as $N^{2/3}$. In the SSL, any decrease in Gibbs energy resulting from minimisation of interfacial area must be balanced against the resulting increase in Gibbs energy from the more extended chain conformation. Hence, at equilibrium in the SSL, the major phase-separated structures of bcc spheres, hexagonally-packed cylinders and lamellae depend only on composition [1,7]. In the WSL, theory [9] accounts for the same structures. The complex hexagonally perforated lamellar and cubic Ia3d mesophases (phases intermediate between the predicted structures) recently observed in weakly segregated diblocks [10,17-22] are not generally found to be the lowest energy structures by theory. However, the most recent theoretical studies of Matsen and coworkers [7] predict the cubic Ia3d (also known as gyroid, G) to be stable.

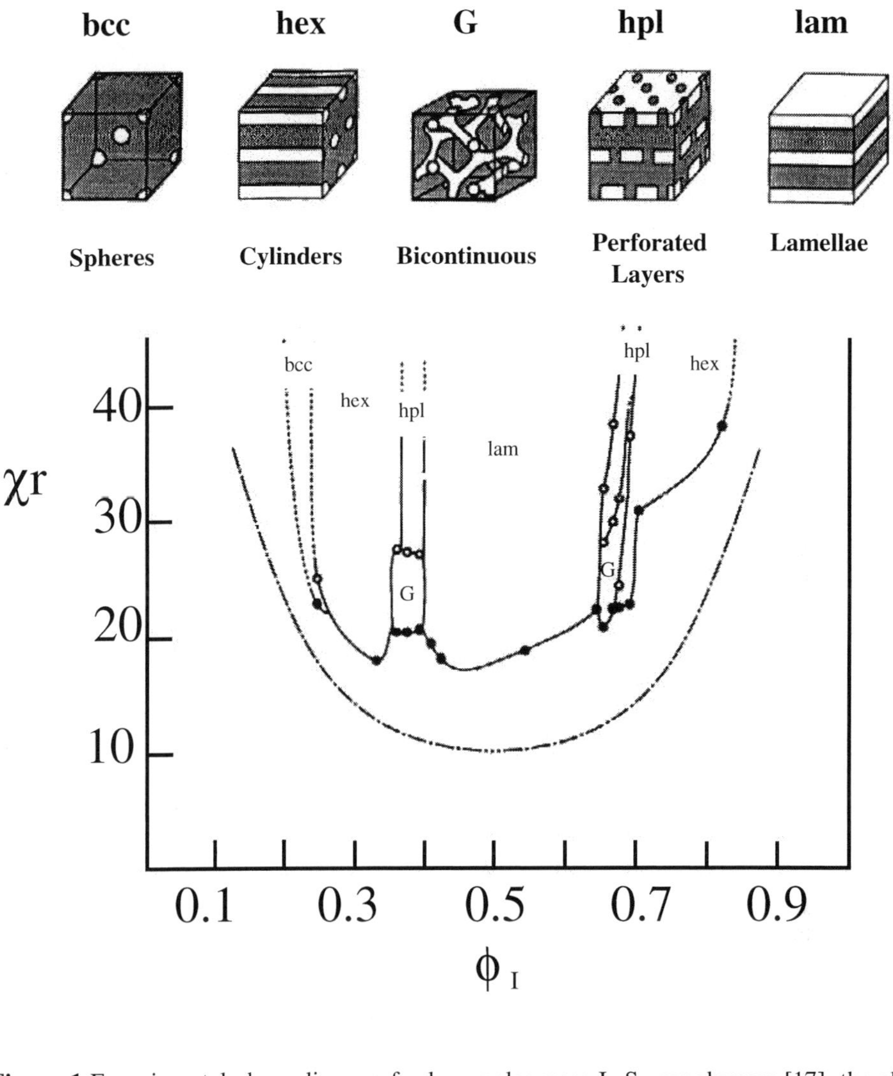

Figure 1 Experimental phase diagram for low molar mass I_mS_n copolymers [17], the close circles represent order-disorder transitions (ODT) and the open circles order-order-transitions (OOT), the solid curve is a guide to the eye. The ODT and OOTs were determined by rheology and the structures by a combination of TEM, SANS and SAXS [10]. The dotted curve is the mean field phase boundary from Liebler [9]

The assumption of equal volume and equal statistical length are overly restrictive in the application of theory to many block copolymers, including the E_mB_n and E_mP_n copolymers of present interest. The problem has been discussed in a number of recent papers. [23-26] In the Flory-Huggins treatment of mixtures of two polymers, eqn. (2) is recast in terms of volume fraction (ϕ) and number of segments (r_v). The two types of segments (E and B or P) each have the same reference volume, v_o. For systems comprising E and B units, it is convenient to use the volume occupied by an E unit (v_E) as reference, whence

$$r_v = r_{vE} + r_{vB} = N_E + N_B(v_B/v_E) \qquad (4)$$

where v_B is the volume occupied by a B unit. Room temperature values of the densities of liquid poly(oxyethylene) and poly(oxybutylene) [26] lead to a value of $v_B/v_E \approx 1.89$. With values of χ based on the same reference volume, the equation for the non-combinatorial part of the free energy of mixing per chain, analogous to eqn. (2), is

$$\Delta_{mix}G_{nc} = kT(\chi r_v)\phi_E\phi_B \qquad (5)$$

where ϕ_E and ϕ_B are volume fractions. This formulation requires a phase diagram plotted as χr_v *versus* ϕ_E. If the two types of segment defined by volume have equal statistical lengths, and if parameter χ is independent of concentration, then this phase diagram will be symmetrical in ϕ_E. Generally the segment with the smallest volume is chosen for normalisation in the case of unequal segment volumes.

Asymmetry in the phase diagram is caused by difference in statistical lengths, which enters through the radius of gyration. The unperturbed square radius of gyration is referenced to r_s segments of identical statistical length (b_o), in the present case put equal to that of an E unit: i.e.

$$R_g^2 = r_s b_E^2/6 \qquad (6)$$

where

$$r_s = r_{sE} + r_{sB} = N_E + N_B(b_B^2/b_E^2) \qquad (7)$$

With the chain length expressed in segments defined by volume,

$$R_g^2 = [r_{vE} + r_{vB}(v_E b_B^2/v_B b_E^2)]\, b_E^2 = [r_{vE} + \varepsilon r_{vB}]\,(b_E^2/6) \qquad (8)$$

where parameter $\varepsilon = (v_E b_B^2/v_B b_E^2)$ is equivalent to that described by Vavasour and Whitmore [24], and similar to that described by Almdal et al. [22,25]. Reported [27,28] values of the unperturbed dimensions of poly(oxyethylene) and poly(oxybutylene) lead to $b_B^2/b_E^2 \approx 1.22$, whence $\varepsilon \approx 0.65$. Following that work, it would be predicted that the microphase separation (disorder-order) boundary in the present system would be little changed from the hypothetical symmetrical case, but that the order-order boundaries would be affected, being significantly shifted towards higher volume fractions of oxyethylene (E). In Figure 2 the phase boundaries have been calculated [29] using the incompressible Gaussian chain model and the self-consistent field theory (SCFT) of Helfand [30]. By using the numerical procedure introduced by Matsen and Schick [29], the mean-field phase diagram for $\varepsilon = 0.6$ has been calculated without invoking any of the traditional approximations; e.g., the narrow interface and unit-cell approximations of Helfand and Wasserman [8], or the Landau expansion and single-harmonic approximations of Liebler [9]. The phase diagram presented improves upon the Vavasour and Whitmore version [24] (also calculated for $\varepsilon = 0.6$), which used approximations for the Wigner-Seitz unit cells of the hex and bcc phases, and because higher order basis functions were used a region of stability of the Ia3d (G) is predicted[7].

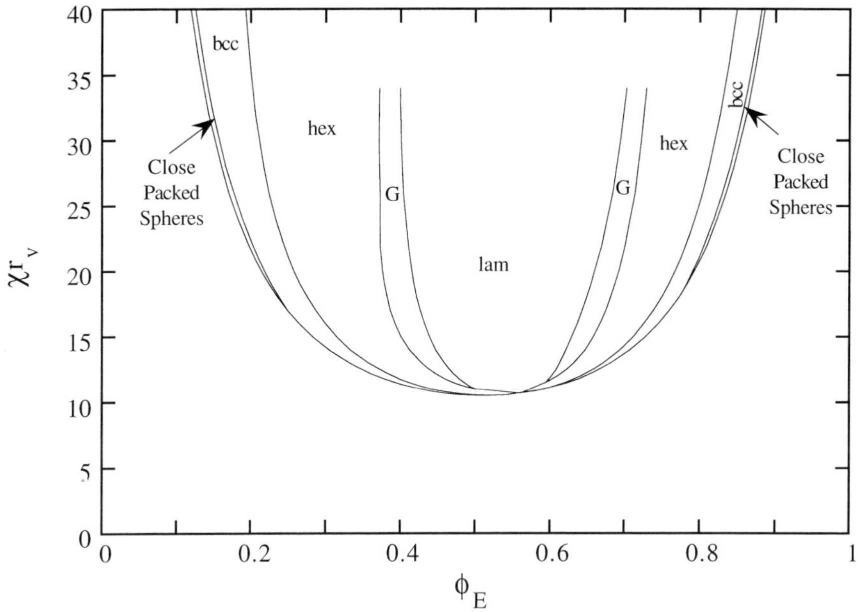

Figure 2 The phase diagram for ε = 0.6 calculated [31], using the numerical procedure introduced by Matsen and Schick [29], from the incompressible Gaussian chain model and the self-consistent field theory (SCFT) of Helfand [30].

2.2 Block copolymers with one semicrystalline block

If one or more of the blocks can crystallise the microphase-separation behaviour becomes more complicated. For non-glassy block copolymers, it may be taken for granted that crystallisation of one block will result in separation into lamellar microphases, as the negative Gibbs energy of crystallisation will greatly outweigh the Gibbs energy of dislocation of any microphase-separated structure in the melt. There are, however, important effects due to the crystallising chain being covalently bonded to a non-crystallisable chain. The ongoing flux of non-crystallisable chains of the second block has to be accommodated, and the ratio of cross-sectional areas of the two types of block is an important consideration if space is to be filled at normal densities. Published theories [4-6] recognise this competition between preferred low-Gibbs-energy conformations, i.e. unfolded chains for the crystallised component and randomly-coiled chains for the non-crystallised component, and minimise the Gibbs energy by introducing an equilibrium degree of chain folding in the crystalline layer. Thus there is an important distinction to be drawn between homopolymers, where chain folds are metastable and are introduced by crystallisation kinetics, and block copolymers where an equilibrium degree of chain folding exists in the crystalline layer. If coil dimensions increase with increase in temperature, then the number of folds at equilibrium should increase with temperature. However, the behaviour generally observed is that more folds form as the crystallisation temperature is decreased, indicating that kinetic (non-equilibrium) effects may dominate.

3. Experimental methods

A decade ago, transmission electron microscopy (TEM) was the pre-eminent experimental technique for studying block copolymer structure. The combination of relatively large monodisperse microphases and efficient heavy-metal staining techniques (e.g. osmium tetroxide) produced truly spectacular electron micrographs of ordered phases in polystyrene-polydiene block copolymers [32,33]. During the past decade, the analytical capabilities have been greatly enhanced by the development of small-angle scattering techiques that complement advances in TEM and provide access to new equilibrium and fluctuation quantities. Hashimoto and coworkers [33,34] have played a leading role in the application of small-angle X-ray scattering (SAXS) to investigation of microdomain size, packing and interfacial mixing profile (interfacial thickness), whereas Bates and coworkers [10,12,19-22,25,35] have exploited labelling techniques and shear orientation combined with small-angle neutron scattering (SANS). Furthermore with SANS the direct determination of the conformation of blocks is possible, however, there is a contrast matching problem, and the method is not widely used [1, 33].

Within the weak segregation region the most significant feature is the order-disorder transition (ODT). Identification of the ODT temperature is often complicated by its weak-first-order character, and by significant composition fluctuations above and below the transition temperature. Rheological and scattering (X-ray, neutron) measurements during a temperature ramp are often used [34,35] and some workers have used form birefringence in a similar way [36,37]. Scattering from oriented samples in a temperature dependent SAXS or SANS experiment has been used, the loss of orientation at the ODT resulting in a very marked change in scattering pattern [38]. Recently high resolution SAXS has been used to measure the ODT from the change in peak shape between the ordered and fluctuating liquid structures [33,39,40].

For semicrystalline block copolymers a combination of SAXS, for the ≈ 100 Å block copolymer structure, and WAXS, for the ≈ 4 Å crystal structure, is required. The use of modern synchrotron X-ray sources has allowed the combination of SAXS/WAXS to be developed for time-resolved experiments [41]. Studies of crystallisable block copolymers (structures and kinetics) have been performed for polyolefins [42-46] and polyethers [26,47]. If the copolymers crystallise to form spherulites optical microscopy can be used to follow the crystallisation kinetics in exactly the same manner as semicrystalline homopolymers.

4. Liquid structures

For historical reasons the majority of amphiphilic copolymers have been designed and synthesised to take advantage of their forming mesophase structures in water. By necessity the molar mass of the hydrophobic part of the molecule must be kept small and, therefore, the majority of amphiphilic polymers are of low molar mass. We have already seen that forming an ordered structure in a block copolymer requires a critical value of χN to be exceeded. Low molar masses (typically < 5000 g mol^{-1}) and a small interaction parameter, particularly in the case of EP systems, means that the literature for EC and EP copolymers contains no reports of ordered melts as far as we are aware. Furthermore most of these polymers crystallise from a disordered phase. The literature on ES copolymers [2,48,49] is concentrated on the solution and solid states even though the polymers would be expected to form ordered melts.

4.1 Ordered liquid structures

The EB [39]and OE [50] systems have had their ordered melt structures studied in most detail. Generally the structures are the same as those found in polyolefin block copolymers. The periodic arrangement in space of the segregated microdomains give rise to well-defined, and often very sharp, Bragg reflections in SAXS experiments. Their angular positions and intensities are related to the crystallographic structure and the sharpness of the diffraction lines provides valuable information about the perfection of the crystal lattices involved and their coherence. Four kinds of ordered phase structures have been found for E_nB_m: lam, hex, G and bcc [39,50]. The are reports of hexagonally-perforated packed lamellae (hpl) structures for OE systems [50] but there is still controversy as to whether this phase is a metastable transition state, and even if the structure is correctly assigned. The type of lattice can be determined by SAXS from unoriented melts by comparing the observed sequences of reflection positions (d-spacings) with characteristic lattices, as given in Table 1.

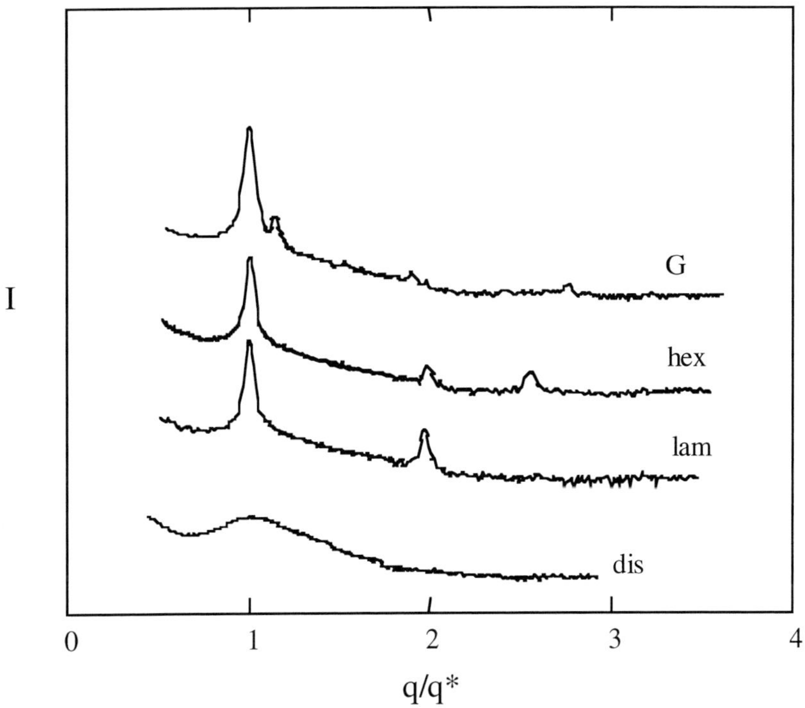

Figure 3 A composite diagram of the SAXS patterns from the G, hex, lam and dis structures for the E_mB_n system. q* is the value of q at the peak of the first order reflection. The intensity scale is logarithmic and the patterns have been shifted for vertically for clarity.

The scattering patterns in Figure 3 are representative of the lam, G, hex and disordered structures of E_mB_n. The number of Bragg reflections observed is important. Usually, the X-ray patterns of ordered-liquid block copolymers contains just a few diffraction lines. Occasionally, however, they contain up to fifteen lines [10], indicating a high degree of ordering and very sharp interfaces. The existence of high-order reflections is an important observation, because it shows that the spatial concentration fluctuations of the block copolymer segments deviate from the sinusoidal-wave-type fluctuation characteristic of the weak segregation limit [1,7] which leads to a single narrow peak, but rather have a square-wave-type fluctuations characteristic of the well-defined structures of the strong segregation limit [1,8,39].

Table 1 Ratios of consecutive reciprocal Bragg spacings for observed melt structures.

Morphologies	Ratios of q_n/q^* where $q = 2\pi/d$
lam (lamellae)	1 : 2 : 3 : 4 : 5 : 6 ......
G (Ia3d - gyroid)	1: √(4/3) : √(7/3) : √(8/3) : √(10/3) : √(11/3) ··
hex (packed cylinders)	1 : √3 : √4 : √7 : √9 : √11 ......
bcc (body centred cubic spheres)	1 : √2 : √3 : √4 : √5 : √6 ......

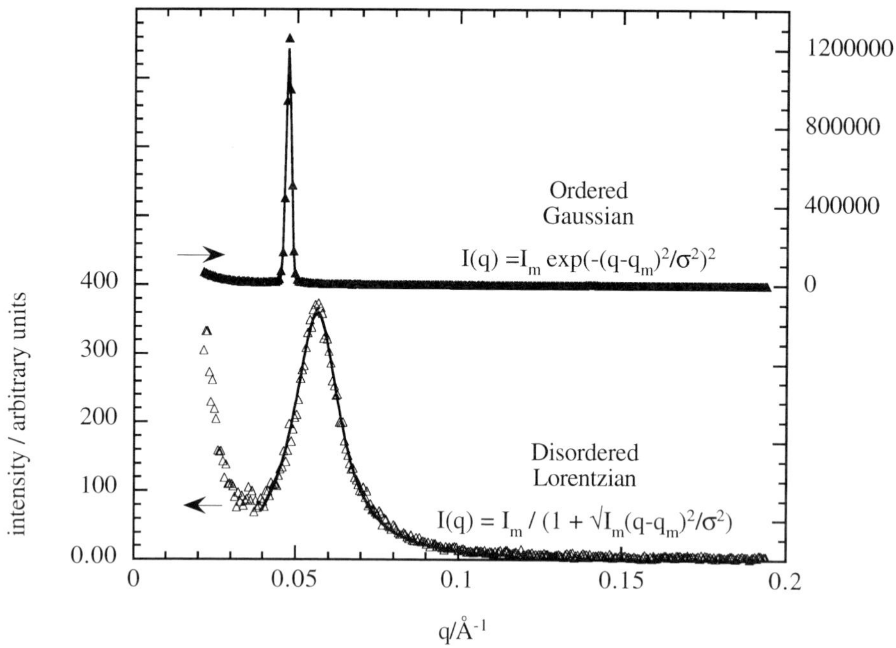

Figure 4 SAXS patterns for ordered and disorded melts of copolymer $E_{96}B_{47}$, the symbols are data points and the continuous lines are fits [39]. The data for the disordered melt were fitted with a Lorentzian and those for the ordered melt with a Gaussian.

4.2 Fluctuations in disordered melts

It is well established [11,12,19-21,39,40,51] that the order-disorder transition in block copolymer melts is weakly first order, on cooling from a high temperature the segment density continuously transforms from being homogeneous (mean field) to having fluctuations of a characteristic length prior to the ODT. The fluctuations are a signal that the ODT is being approached and the amplitude of the segment desity fluctuations increases as the temperature is reduced. The ODT is the transition from a system with time-dependent sinusoidal density fluctations to a system with a time-independent sinusoidal density pattern where the characteristic length of the fluctuations and the pattern does not change.

The general phase and scattering behaviour of lamellar E_mB_n is illustrated by the specific example of copolymer $E_{96}B_{47}$ which has an ODT intermediate between its melting point and the high temperature limit of the experiments as determined by the thermal degradation of polyethers. SAXS patterns for $E_{96}B_{47}$ are shown in Figure 4. In order to illustrate clearly the difference between the scattering from ordered and disorder material, the patterns shown are those obtained from samples at temperatures 30 K apart on either side of the ODT. The peak obtained for the disordered material is best fitted by a Lorentzian function with $\sigma/q^* \approx 1/3$. This pattern is very similar to others in the literature [40,51] covering a wide range of copolymer chemical structure and molecular weight. It can also be fitted with the full mean-field scattering profile given by

$$S^{-1}(q) = N^{-1}F(x, f) - 2\chi_{eff} \qquad (9)$$

where χ_{eff} is an effective interaction parameter, F(x, f) is a function relating to the correlation functions of a non-interacting (Gaussian) copolymer chain [7], and $x = (qR_g)^2$. In fact [51], equation (9) is only discriminating if absolute intensities and the instrument resolution function are well known. Otherwise the values obtained for the fitting parameters (including χ_{eff}) are meaningless [39,40,51]. The peak for the ordered material is best fitted by a Gaussian function with $\sigma/q^* \approx 1/50$. This narrow peak is resolution limited, and is in fact the point-spread function of the instrument (combination of beam profile and detector) . Just below the ODT the sample probably has a sinsusoidal density profile [1,9,11,39,40,51] with only a few Fourier modes overlapping around the mode with wave number q*, giving a very narrow peak. Most other workers have observed another Lorentzian for their ordered copolymers, probably due to wavelength dispersity [39,40,51], and possibly with a contribution to spreading from copolymer polydispersity.

4.3 The ODT

The transition from disorder to order (DOT) can be observed by scattering and rheological techniques. In SAXS a change in line shape from Gaussian to Lorentzian on heating (and *vice versa* on cooling) occurrs over a 3 K interval for E_mB_n, and the intermediate patterns can be fitted by a weighted average of the two functions in a manner similar to that described by Sakamoto [40]. There is significant hysteresis on heating and cooling, for example the difference in temperature was 8 K for $E_{96}B_{47}$ at a temperature ramp rate of 10 K min^{-1}.

In order to measure the ODT, the conventional method of data analysis is [12,21,25,33,34,38-40,51] to plot the reciprocal peak position, $1/(q^*/\text{Å}^{-1})$, and the reciprocal peak-maximum intensity, $1/I(q^*)$, as functions of reciprocal temperature, $1/(T/K)$: see Figure 5. The data shown were recorded during heating and cooling copolymer $E_{96}B_{47}$ in its liquid state between 25 °C and 190 °C [39]. The rapid changes in $1/I(q^*)$ on heating or cooling copolymer $E_{96}B_{47}$ at temperatures above 150 °C are associated with the order-disorder transition. There is some scatter in the data above the ODT if $1/I(q^*)$ is plotted, but nevertheless it is possible to characterise the effect as a step change in $I(q^*)$. The step is obvious if $I(q^*)$ is plotted against T, since this avoids compression of the abscissa in the region of interest and the noise in the data above the ODT is not accentuated [39]. The hysteresis (described above) is clearly seen in Figure 5, the ODT (heating) occurring at a temperature some 8 K higher than the DOT (cooling). In this the order-disorder transition resembles the crystal/melt transition, though the undercooling is larger for crystallisation. As would be expected, the extent of hysteresis increased with an increase in temperature ramp rate. Near-perfect reversibility of $I(q^*)$ could be found if copolymers were cycled between 10 °C above T_m and 10 °C below the ODT.

The melting temperature ($T_m \approx 60$ °C) and crystallisation temperature ($T_c \approx 25$ °C) are marked on the inset. The near reversibility of $1/q^*$ on heating and cooling is obvious from the inset to Figure 5, particularly in the high temperature range (T > 130 °C), and with small differences between values in the low temperature range. The value of $1/q^*$ decreased with increasing temperature, which suggests that R_g decreases with temperature, as found for polyolefin [12,21,51] and polystyrene-polydiene diblock copolymers [34, 40]. Discontinuities related to those in $I(q^*)$ could be found at the ODT for the other parameters used in fitting the diffraction pattern, i.e. the peak width and the peak shape [39].

The ODT in E_nB_m has also been determined [39] by the step in the low-frequency (1 rad s^{-1}), in-phase shear modulus, G', during a temperature ramp. The ODT is determined with a prescision of ± 1 K from the step function in G' which is associated with the loss of rigidity in the interfaces [35].

4.4 Temperature dependence of χ

In order to locate the copolymers on the phase diagram, the temperature dependence of χ is required, i.e. the coefficients in equation (3). The values of χ at the ODTs of symmetrical E_nB_m have been calculated from the mean-field approximation of its critical value, $\chi_c = 10.5/r_v$, assuming χ to be independent of molecular weight [39]

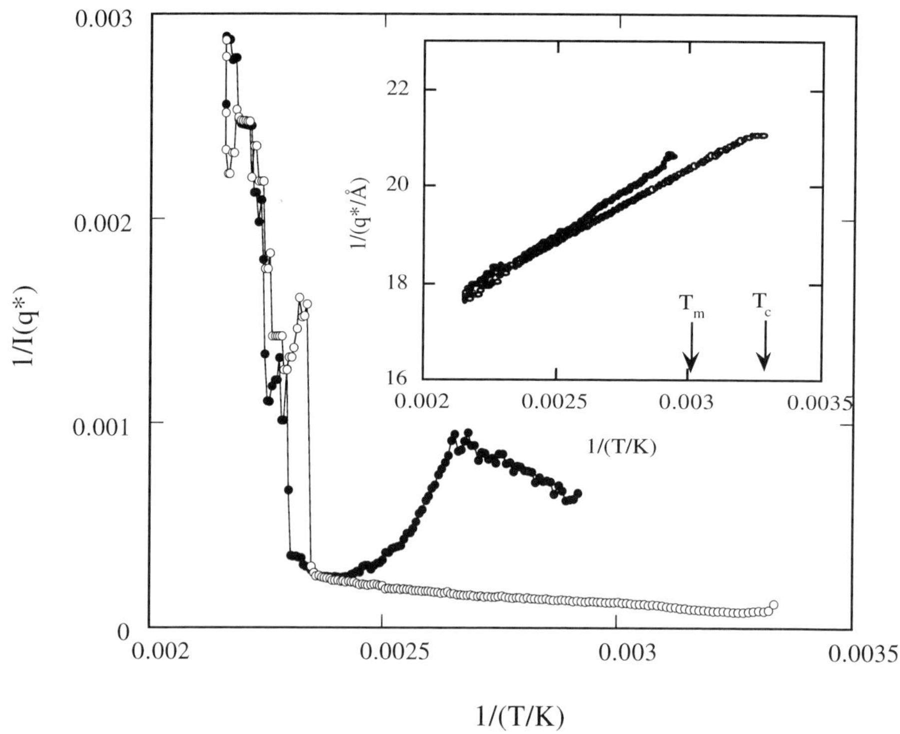

Figure 5 Determination of the ODT from SAXS data, the filled symbols for the heating ramp and the closed symbols for cooling ramp. The main graph is the reciprocal peak intensity, $1/I(q^*,T)$ as a function of reciprocal temperature, $1/(T/K)$. A plot of the reciprocal peak position, $1/(q^*/\text{Å}^{-1})$ as a function of reciprocal temperature is included as an inset [39].

$$\chi = 48.5/T - 0.0504 \qquad (T \text{ in K}) \tag{10}$$

This equation can be compared with $\chi = 34.1/T - 0.02$ found for I_mS_n [34] and $\chi = 15.0/T - 0.0055$ found for PE-PEE copolymers [51]. We have already discussed the ODT being a transition from an ordered melt into a fluctuating melt and that using $\chi_c = 10.5/r_v$ underestimates the value of χ. In order to correct for fluctuation effects the finite molar mass correction, $\tilde{N}$ (= Nb^6v^{-2}) should be calculated, but even so fluctuation theory [11] does not strictly apply for $\tilde{N} \leq 10^4$. It should be noted that the fluctuation regime in these low molar mass copolymers is quite large [39,40,50,51], up to ≈ 50 °C above the ODT.

4.5 Domain size scaling in the melt

The scaling of the lamellar spacing ($d=2\pi/q^*$) versus the molecular size (r_v) has been determined for symmetric ($\phi_E \approx 0.5$) E_nB_m at 70 °C. The experimental data [39] give

$$d = 3.5 r_v^{0.69\pm0.03} \tag{11}$$

measured in Å. The theoretically calculated domain size increases most rapidly with r_v near the ODT. Away from the ODT at $r_v \sim 300$, the theoretical result [7] is approximately

$$d = 2.4\, r_v^{\,0.73} \tag{12}$$

Moreover as the melts become increasingly segregated, the exponent is expected to approach a value of 2/3 [7,52]. The difference between experiment and theory can be attributed to underestimating χ, as a result of not accounting for fluctuations.

5. SOLID STRUCTURES

5.1 Literature Review

As mentioned in section 2.2, the usual result of crystallisation of one block is separation into lamellar microphases. Copolymers with one glass-forming block are exceptions to this rule, since a structure formed in the melt in the temperature interval between the glass transition temperature of one block and the crystallization temperature of the other may persist in a metastable microphase-separated solid state: e.g. diblock copolymers of ethylene oxide and styrene [2,48,49,53], ethylene and styrene [54] or ethylene and vinyl-cyclohexane [46, 55]. However, for copolymers with low glass transition temperatures, the crystalline structure should be independent of microphase structure in the melt. Recent experiments on mobile systems with microphase-separated melts, e.g. butadiene/ε-caprolactone copolymers [56], have shown that this is generally the case, although cylindrical morphologies have been reported [57] for crystalline poly(dimethylsiloxane)/poly(oxyethylene) copolymers with lengthy E blocks. Any possibility of crystal morphology being influenced by liquid morphology can be avoided by studying block copolymers with homogeneous melts, e.g. ethylene and ethylene-alt-propylene [42-44].

Some understanding of the possible equilibrium morphologies of crystallizable block copolymers comes from studies of uniform oligomers with two crystallizable components. In uniform C_mE_n systems, it has been shown that initial crystallization of the oxyethylene block in helical conformation (area of cross-section per chain = 21.4 Å^2) allows subsequent crystallization of trans-planar alkyl blocks, whereas initial crystallization of trans-planar alkyl blocks (area of cross-section per chain = 18.4 Å^2) precludes crystallisation of the oxyethylene block [58,59]. The layer crystals formed have their oligomeric blocks locally crystalline or liquid-crystalline, as appropriate, and also locally normal or tilted relative to the layer end-group plane. Related structures have been found for nonuniform C_mE_n ($M_n < 3000$ g mol^{-1}, $M_w/M_n < 1.10$) [60], but in that case the initial crystallization of either component essentially precluded crystallization of the other, this being a direct result of disorder at the block interface. This effect in nonuniform block copolymers with two crystallizable components was first recorded many years ago by Perret and Skoulios in their pioneering studies of diblock copolymers of ethylene oxide and ε-caprolactone [61].

The copolymers of interest here have one crystallisable block (E block) and one non-crystallisable rubbery block (atactic B or P block). The morphologies and crystallisation rates of such copolymers are more difficult to interpret than are those of corresponding homopolymers. Because the crystallising chain is covalently bonded to a non-crystallisable chain, the ongoing flux of non-crystallisable chains of the second block has to be accommodated, and the ratio of cross-sectional areas of the two types of block is an important consideration if space is to be filled at normal densities. There is an important distinction to be drawn between low-molar-mass homopolymers (e.g. PEG4000) where chain folds introduced as a result of rapid crystallisation may unfold on annealing, and block copolymers where an equilibrium degree of chain folding may exist in the crystalline layer.

Most published [42-46,53,56,57] experimental studies of the crystallisation of block copolymers have been on fairly long chains, where it did not prove possible to quantify low extents of chain folding. Yang et al. [26] recently studied a range of low-molar-mass E_mB_n copolymers which formed either unfolded or once-folded crystals on crystallisation from their disordered melts. The methods used were simultaneous WAXS/SAXS/DSC, low-frequency Raman spectroscopy

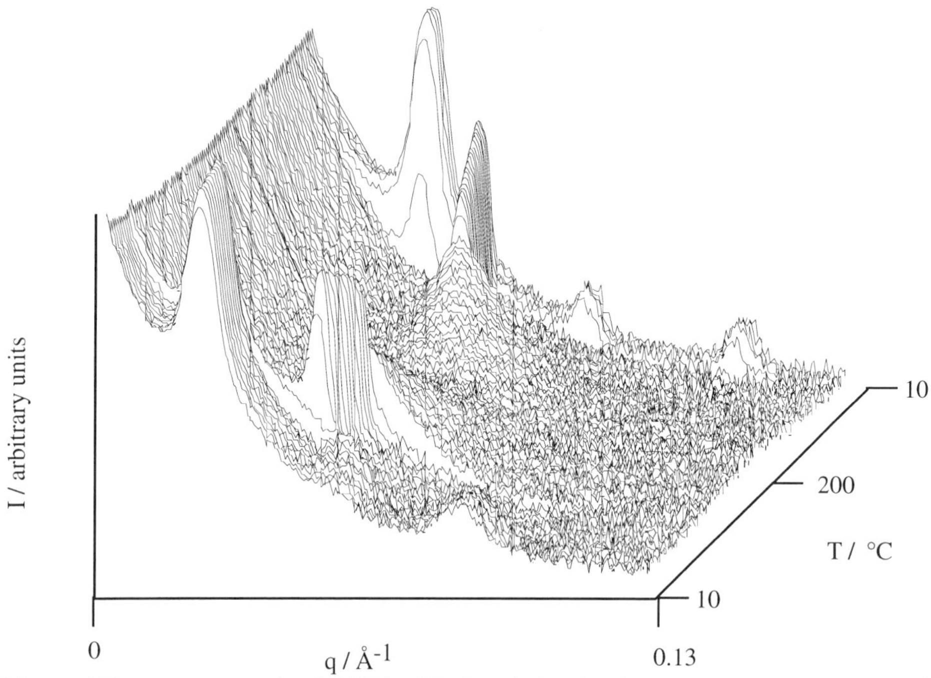

Figure 6 Temperature resolved SAXS of $E_{74}B_{37}$ during heating and cooling at 10 °C min^{-1} between 10 °C to 200 °C. The length scale changes continuously at the ODT but the discontinuous change in length scale on melting and recrystallisation is obvious [47].

(i.e. determination of LAM-1 frequency) and conventional DSC. The influence of the non-crystallisable block was illustrated by the fact that the E block of copolymer $E_{31}B_{30}$ folded on crystallisation at room temperature, whereas the E blocks of copolymers with shorter B blocks, in the range $E_{29}B_{20}$ to $E_{32}B_{3}$, were unfolded.

Like oxybutylene blocks, oxypropylene blocks formed by anionic polymerization are atactic. WAXS, SAXS, low-frequency Raman spectroscopy and DSC was used by Viras et al. [62] to show that $P_nE_mP_n$ with short E blocks (i.e. m = 39 or 48, $M_n \leq 2100$ g mol^{-1}) and n in the range 1-10 crystallized with their E blocks in unfolded-chain conformation, but with the chains (or parts of the chains) significantly titled with respect to the lamellar end plane for copolymers with lengthy P blocks. Chain folding was important for $P_nE_mP_n$ with longer E blocks: e.g. 75 units, $M_n \approx 3300$ g mol^{-1}.

Published theories [4-6] recognise competition between preferred low-Gibbs-energy conformations of the two components in semi-crystallisable block copolymers, i.e. unfolded chains for the crystallised component and randomly-coiled chains for the non-crystallisable component, and minimise the Gibbs energy by introducing an equilibrium degree of chain folding in the crystalline layer.

5.2 Integral chain folding in E_nB_m

Time-resolved SAXS is shown in Figure 6 for $E_{74}B_{37}$ during heating and cooling at 10 °C min^{-1} between 10 °C and 200 °C [47]. There are thermally reversible transitions between metastable many-folded crystals (d_c = 160-170 Å) and ordered melt (T ≈ 51 °C) and between ordered melt and disordered melt (T ≈ 114 °C). The initially-formed melt was stretched compared to equilibrium melt, the discrepancy in length scale being ≈ 15 %. Relaxation to the equilibrium value occurred within 100 s. Post-melt stretch has also been observed in an EO block copolymer [50] and is due to the difference in length scales between the crystalline and molten states which imposes a packing constraint on melting because of the need to conserve density. There was a difference of 4 degrees between the ODT (heating) and the DOT (cooling), but a difference of 23 degrees between melting and crystallisation temperatures. This last value is significantly larger than that observed (7 °C) for a poly(oxyethylene) of similar molar mass to the E block (M_n = 3000 g mol^{-1}, E_{68}) which was melted and crystallised under similar conditions.

The change in length scale that needs to be accommodated is a significant contribution to the barrier to crystallisation. If the copolymer were to recrystallise in its once-folded condition (with once folded length l_1 = 219 Å) both E and B blocks would need to stretch by a factor of 2 from their equilibrium values in the melt (d = 114 Å). In order to overcome the consequent Gibbs energy barrier to nucleation the polymer must be undercooled to such an extent that the formation of a more-folded metastable state is kinetically preferred. The undercooling needed for the first order process of crystallisation, which is accompanied by discontinuous changes in length scale, enthalpy and density, is clearly much larger than that required for the weakly first order process at the DOT, where the length scale changes continuously.

The SAXS pattern from the melt of copolymer $E_{74}B_{37}$ at 80 °C (see Figure 7) shows two orders of reflection consistent with a lamellar structure with $d = 2\pi/q^* \approx 114$ Å. Assuming that the block copolymer is stretched from its Guassian conformation in the melt [1,7,9], the radius of gyration (R_g) of this polymer is estimated to be 54 Å using the relation $R_g = r_v^{2/3}b/\sqrt{6}$ [1], with the statistical segment length b = 5.0 Å [39]. The measured spacing is twice the R_g as expected for a lamellar structure. The SAXS patterns of the solid copolymer show three or four orders of reflection, and indicate lamellar structures with long range order and little polydispersity in lamellar thickness. Estimates of the lamellar spacing to be expected for a structure with unfolded E blocks having their helices normal to the lamellar end plane, l_0, were obtained from

$$l_0 / \text{Å} \approx 0.95 z_E / \phi_E \tag{13}$$

where z_E is the number of chain atoms (C and O) in the E block, and 0.95 Å is the length per chain atom along the poly(oxyethylene) helix [63]. It is simple, therefore, to estimate the repeat length in a system in which the E blocks have n folds as

$$l_n = l_0 / (n + 1) \tag{14}$$

As indicated in Figure 7, the structure grown at 50 °C by a self-seeding process (see below for details) had d_c = 220±4 Å compared with the spacing calculated for a once-folded structure, l_1 = 219±5 Å. By contrast, the structures grown in the range 38 °C < T_c < 45 °C had 160±4 Å ≤ d_c ≤ 170±4 Å, and as l_2 = 146±3 Å there was no direct match to integer folding. The experimental result could be rationalised for a structure with twice-folded E blocks if either some of the E blocks did not crystallise and contributed to the non-crystalline layer, or if the longer E blocks were preferentially crystallised (the polydispersity of the E blocks was also 1.05) and the molecules with shorter E blocks were excluded from the lamellar stacks. An alternative explanation is that the E blocks were once-folded but were tilted in the lamellar stack.

To obtain the once-folded structure of $E_{74}B_{37}$ a self-seeding technique was required, similar to that used by Buckley and Kovacs [64]. This involved crystallising the copolymer at 38 °C, holding it for a short time at 52 °C so that most of the twice-folded crystals melted, and quickly dropping the temperature to 50 °C when once-folded crystals were observed by SAXS to grow on the nucleating surface provided. This melting and recrystallising process was required for the transformation: direct crystallisation from the melt did not occur over a period of one hour

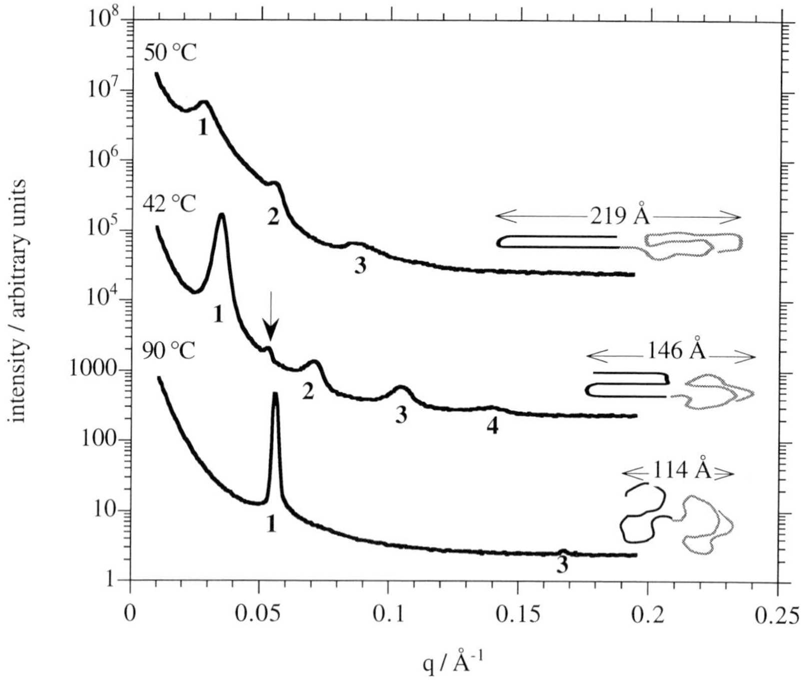

Figure 7 SAXS patterns for copolymer $E_{74}B_{37}$ showing the ordered melt structure (T = 90 °C), the metastable structure (T_c = 42 °C) and the equilibrium, once-folded structure grown at 50 °C by a self-seeding process. Numerals indicate the positions of the reflections from the stacked lamellae, and the arrow indicates the position of the peak arising from the coexisting ordered melt. The insets show calculated repeat lengths for possible conformations [47].

at 50 °C, whereas twice-folded crystals grew from the melt at a reasonable rate at 48 °C. Primary crystallisation to form metastable, multiply-folded structures occurred in many E_mB_n copolymers, and the stable less-folded structures could only be formed when there was already a crystal surface present to reduce the nucleation barrier. This effect in a lengthy copolymer is understandable in terms of a nucleation barrier dominated by the difference in length scale between the ordered melt and the stable solid. Since $d_m \sim r_v^{2/3}$ [1,7,8] and (to a fair approximation) $l_n \sim r_v$, the activation energy for nucleation of a given n-folded structure should scale as $\Delta E \sim r_v^{1/3}\phi_E/n$.

The quantisation in length scale found for the crystallised copolymers is similar to that observed in low-molar-mass poly(oxyethylene)s (i.e. polyethylene glycol, PEG, $3000 \leq M_n \leq 10000$ g mol^{-1}, chain length E_{68} to E_{230} [64-68]), where crystallisation occurs with integer numbers of folds depending on the chain length and crystallisation temperature. The effect occurs in the homopolymers because the Gibbs energy finds a metastable minimum in a system where the chain ends and folds are located at the lamellar surface. Slow crystallisation of the lower PEGs (E_{136} and lower) resulted in crystals with unfolded chains [68]. Annealing the lower PEGs (E_{90} and lower) at temperatures below their melting points resulted in rapid unfolding of any folded crystals [64]. Initially-formed once-folded (and possibly twice-folded) crystals of the higher PEGs (E_{136} and higher) were stable to annealing: unfolded structures were formed by self-seeding [65]. In the copolymer $E_{74}B_{37}$ the twice-folded structure could not be unfolded by

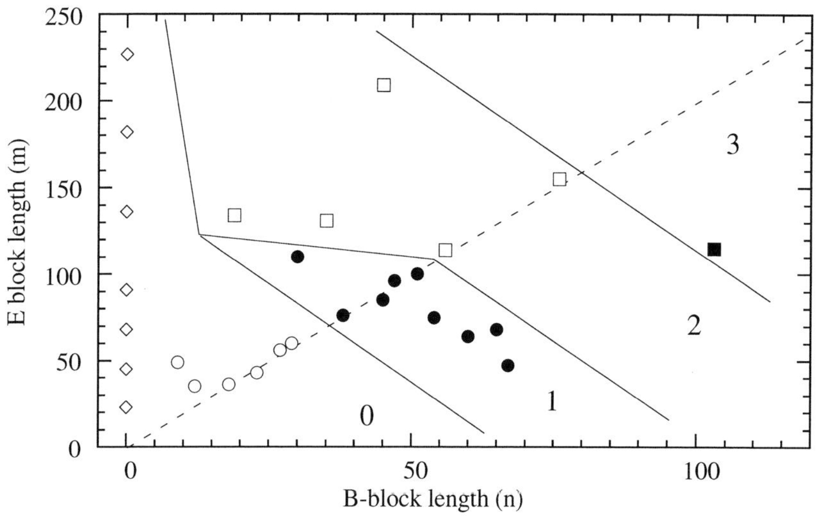

Figure 8 The number of folds per E-block at equilibrium as a function of E-block (m) and B-Block length. The dotted line represents the composition with $\phi_E = 0.5$ and the solid lines are guides to the eye separating regions of 0, 1, 2 and 3 folds.

heating below T_m, and only the once-folded structure could be obtained by self-seeding. In view of the relatively short length of the E blocks of copolymer $E_{74}B_{37}$ compared to most of the PEGs,the inability to unfold an E block (either by annealing or self-seeding) is seen as as an indication of an equilibrium state. In fact the structure with a once folded E-block is the equilibrium structure for $E_{74}B_{37}$ and other E_nB_m where $\phi_E \sim 1/2$. This is because the contour length of the B-block is much shorter than the fully extended crystal length of the E-block and an unfolded layer structure could not form at normal densities.

There appears to be a balance between the number of folds at equilibrium and the lengths of the two blocks. Figure 8 presents the number of folds as a function of the E-block and B-block lengths. The homopolymer E has no equilibrium folds whereas the number of folds at equilibrium increases with the total chain length (m+n) and the volume fraction of B (n/m). The lines on the "phase diagram" are guides to the eye and separate the regions of 0, 1, 2 and 3 folds. The dotted line represents the compositions with $\phi_E = 0.5$. The simple model which conserves density would not allow equilibrium structures to have zero folds along this line, that zero-fold equilibrium structures are observed illustrates that the molecules have methods of overcoming the frustration of packing constraints to adopt unfolded conformations with structure that has liquid crystalline order in the B-blocks [26].

Polymers that have disordered or lamellar melt phases show an increase in the characteristic length scale on crystallisation of the E block which implies stretching of the amorphous B block above that experienced in the melt. Higher molar mass copolymers have folded stems whose length was determined by the balance between the free-energy of E chain folding and B stretching and the crystallisation kinetics. Metastable structures formed during rapid crystallisation that could be melted to grow equilibrium folded structures. There is a particular case of epitaxy between the hexagonal melt and lamellar crystalline phases that involves the formation of metastable many-folded structures with no change in length scale, these metastable structures are difficult to anneal out as they coexist with the melt. The kinetically induced folded structures observed in homo-E are much larger than in the equilibrium folded structures observed in copolymers due to the balance of energetics for folding E and stretching B.

Acknowledgements
We thank Dr. Colin Booth for his helpful advice during the preparation of this review and for getting us all involved in this subject.

REFERENCES

1 F. S. Bates and G. H. Fredrickson, Annu. Rev. Phys. Chem., 41 (1990) 525.
2 A. Skoulios, G. Finaz and J. Parrod, C.R. Hebd. Seances Acad. Sci., 251 (1960) 739; A. Skoulios and G. Finaz , C.R. Hebd. Seances Acad. Sci., 252 (1961) 3467; 250 (1961) 265.
3 I. R. Schmolka and A. J. Raymond, J. Am. Oil Chem. Soc., 42 (1965) 1088; I. R. Schmolka, J. Am. Oil Chem. Soc., 51 (1977) 110.
4 E. A. DiMarzio, C. M. Guttman and J. D. Hoffmann, Macromolecules, 13 (1980) 1194.
5 M. D. Whitmore and J. Noolandi, Macromolecules, 21 (1988) 1482.
6 T. Viglis and A. Halperin, Macromolecules, 24 (1991) 2090.
7 M. W. Matsen and F. S. Bates, Macromolecules, 29 (1996) 1091.
8 E. Helfand and Z. R. Wasserman, Macromolecules, 9 (1976) 879; 11 (1978) 960; 13 (1980) 994.
9 L. Leibler, Macromolecules, 13 (1980) 1602.
10 A. K. Khandpur, S. Förster, F. S. Bates, I. W. Hamley, A. J. Ryan, W. Bras, K. Almdal and K. Mortensen, Macromolecules, 28 (1995) 8796.
11 G. H. Fredrickson and E. J. Helfand, J. Chem. Phys., 87 (1987) 697.
12 K. Almdal, J. H. Rosedale, F. S. Bates, G. D. Wignall and G.H.Fredrickson, Phys. Rev. Lett., 65 (1990) 1112.
13 M. Banaszak and M. D. Whitmore, Macromolecules, 25 (1992) 249.
14 K. R. Shull, Macromolecules, 25 (1992) 2122.
15 J. D. Vavasour and M. D. Whitmore, Macromolecules, 25 (1992) 5177.
16 M. Muthukumar, Macromolecules, 26 (1993) 5259.
17 S. Förster, A. K. Khandpur, J. Zhao, F. S. Bates, I. W. Hamley, A. J. Ryan and W. Bras, Macromolecules, 27 (1994) 6922.
18 D. A. Hajduk, P. E. Harper, S. M. Gruner, C. C. Honeker, G. Kim and E. L. Thomas, Macromolecules, 27 (1994) 4063.
19 M. F. Schulz, F. S. Bates, K. Almdal and K. Mortensen, Phys. Rev. Lett., 73 (1994) 86.
20 I. W. Hamley, K. A. Koppi, J. H. Rosedale, F. S. Bates, K. Almdal and K. Mortensen, Macromolecules, 26 (1993) 5959.
21 F. S. Bates, M. F. Schulz, A. K. Khandpur, S. Förster, J. H. Rosedale, K. Almdal and K. Mortensen, Faraday Discuss., 98 (1994) 7.
22 I. W. Hamley, M. D. Gehlsen, A. K. Khandpur, K. Koppi, J. H. Rosedale, M. F. Schulz, F. S. Bates, K. Almdal and K. Mortenson, J. Phys. II France, 4 (1994) 2161.
23 E. Helfand and A. M. Sapse, J. Chem. Phys., 62 (1975) 1327.
24 J. D. Vavasour and M. D. Whitmore, Macromolecules, 26 (1993) 7070.
25 K. Almdal, K.A. Koppi, F. S. Bates and K. Mortensen, Macromolecules, 25 (1992) 1743.
26 Y.-W. Yang, S. Tanodekaew, S.-M. Mai, C. Booth, A. J. Ryan, W. Bras and K. Viras, Macromolecules, 28 (1995) 6029.
27 D. R. Beech and C. Booth, J. Polym. Sci. (A-2), 7 (1969) 545.
28 C. Booth and R. Orme, Polymer, 11 (1970) 626.
29 M. W. Matsen and M. Schick, Phys. Rev. Lett., 72 (1994) 2660
30 E. Helfand, J. Chem. Phys., 62 (1975) 999.
31 Private Communication - Mark Matsen
32 R. A. Brown, A. J. Masters, C. Price and X.-F. Yuan, in Comprehensive Polymer Science, Eds. C. Booth, C. Price, Pergamon, Oxford, 1989, Vol. 1, Ch.6.
33 H. Hasegawa and T. Hashimoto, in Comprehensive Polymer Science, Eds. S.L. Aggarwal and S. Russo, Pergamon, Oxford, 1996, 2nd. Suppl., Ch. 14, and references therein.

34 H. Hasegawa, H. Tanaka, T. Hashimoto and C.C. Han, Macromolecules, 20 (1987) 2120.
35 K. A. Koppi, M. Tirrell, F. S. Bates, K. Almdal and K. Mortensen, J. Rheol., 38 (1994) 999.
36 N. P. Balsara, B. A. Garetz and H. J. Dai, Macromolecules, 25 (1992) 6072.
37 K. R. Amundsen, E. Helfand, S. S. Patel, Z. Quan and S. D. Smith, Macromolecules, 25 (1992) 1934.
38 F. S. Bates, J. H. Rosedale and G. H. Fredrickson, J. Chem. Phys., 92 (1990) 6252.
39 S.-M. Mai, J. P. A Fairclough, I.W. Hamley, R.C. Denny, M. Matsen, B.-X. Liao, C. Booth and A.J. Ryan, Macromolecules, 29 (1996) 6212.
40 M. Sakamoto and T. Hashimoto, Macromolecules, 28 (1995) 6825.
41 W. Bras, G. E. Derbyshire, S. Clarke, A. Devine, B.U. Komanschek, J. Cooke, and A.J. Ryan, J. Appl. Cryst. 28 (1994) 26.
42 P. Rangarajan, R. A. Register and L. J. Fetters, Macromolecules, 26 (1993) 4640.
43 P. Ranjaran, R.A. Register, D.H. Adamson, L.J. Fetters, W. Bras, S. Naylor and A.J. Ryan, Macromolecules, 28 (1995) 1422.
44 A. J. Ryan, I. W. Hamley, W. Bras and F. S. Bates, Macromolecules, 28 (1995) 3860.
45 P. Ranjaran, R.A. Register, D.H. Adamson, L.J. Fetters, W. Bras, S. Naylor and A.J. Ryan, Macromolecules, 28 (1995) 4932.
46 I.W. Hamley, A.J. Ryan, J.P. Fairclough, N.J. Terrill, F.S. Bates, E. Towns-Andrews, Macromolecules, 29 (1996) 8835.
47 A.J. Ryan, J.P.A. Fairclough, I.W. Hamley, S-M Mai and C. Booth, Macromolecules, 30 (1997) Issue 6.
48 C. Tsitsilianis, G. Staikos, A. Dondos, P. Lutz and P Rempp, Polymer, 33 (1992) 3369.
49 A. Dondos and D. Papanagopoulous, J. Polym. Sci., Polym. Phys. Edn., 34 (1996) 1281.
50 M.A. Hillmyer, F.S. Bates, K. Almdal, K. Mortensen, A.J. Ryan and J.P.A. Fairclough, Science, 271 (1996) 976.
51 J.H. Rosedale, F.S. Bates, K. Almdal, K. Mortenson, G.D. Wignall, Macromolecules, 28 (1995) 1429.
52 A.N. Semenov, Sov.Phys.JETP, 61 (1985) 733.
53 M. Gervais and B. Gallot, Makromol. Chem., 171 (1973) 154; 178 (1977) 1547.
54 R. E. Cohen, P. L. Chang, K. Douzinas, P. Kofinas and C. V. Berney, Macromolecules, 23 (1990) 324.
55 I.W. Hamley, A.J. Ryan, J.P.A. Fairclough, F.S. Bates, W. Bras, E. Towns-Andrews, Polymer, 37 (1996) 4425.
56 S. Nojima, K. Kato, S. Yamamoto and T. Ashida, Macromolecules, 25 (1992) 2237.
57 M. Galin and A. Mathis, Macromolecules, 14 (1981) 677.
58 C. Campbell, K. Viras, M. J. Richardson, A. J. Masters and C. Booth, Makromol. Chem. 194 (1993) 799.
59 S. G. Yeates and C. Booth, Eur. Polym. J., 21 (1985) 217.
60 R. C. Domszy, R. H. Mobbs, Y. K. Leung, F. Heatly and C. Booth, Polymer, 20 (1979) 1204; 21 (1980) 588.
61 R. Perret and A. Skoulios, Makromol. Chem., 162 (1972) 163.
62 F. Viras, Y.-Z. Luo, K. Viras, R. H. Mobbs, T. A. King and C. Booth, Makromol. Chem., 189 (1988) 459.
63 J.R. Craven,H. Zhang,C. Booth, J. Chem. Soc., Faraday Trans., 87 (1991) 1183.
64 C. P. Buckley and A. J. Kovacs, Prog. Colloid Polym. Sci., 58 (1975) 44.
65 C. P. Buckley and A. J. Kovacs, Colloid Polym. Sci., 251 (1976) 695.
66 J. P. Arlie, P. Spegt and A. Skoulios, Makromol. Chem., 104 (1967) 212.
67 P. Spegt, Makromol. Chem., 140 (1970) 167
68 D.R. Beech, C. Booth, D.V. Dodgson, R.R. Sharpe and J. R. Waring, Polymer, 13 (1972) 73.

Structural Polymorphism of Amphiphilic Block Copolymers in Mixtures with Water and Oil: Comparison with Solvent-Free Block Copolymers and Surfactant Systems

Paschalis Alexandridis,[a,b] Ulf Olsson,[a] Per Linse[a] and Björn Lindman[a]

[a] Physical Chemistry 1, Center for Chemistry and Chemical Engineering, Lund University, P.O. Box 124, Lund S-22100, Sweden

[b] Department of Chemical Engineering, State University of New York at Buffalo, Buffalo NY 14260-4200, U. S. A.

Binary or ternary mixtures of amphiphilic poly(ethylene oxide)-poly(propylene oxide) (PEO-PPO) block copolymers and one or two solvents selective for the PEO and/or the PPO blocks exhibit a very rich structural polymorphism. The microstructure resulting from the self-assembly of the solvated block copolymers can vary from "normal" (PPO+oil core) micelles in solution, through all types of normal and "reverse" (PEO+water core) lyotropic liquid crystals (normal micellar cubic, hexagonal, and bicontinuous cubic, lamellar, reverse bicontinuous cubic, hexagonal, and micellar cubic), to reverse micelles in solution, as the relative volume fraction of the "PPO"-like (i.e., PPO plus its selective solvent) components increases over that of the "PEO"-like components, and the polymer blocks swell to different extents and thus modulate the interfacial "curvature". Recent findings on the phase behavior and structure of solvated block copolymers are reviewed. These results are placed in the context of the morphology observed in block copolymer blends and in surfactant-water-oil systems. The morphological versatility of block copolymers, which is promoted by the addition of selective solvents, has repercussions on aspects both fundamental and applied/practical.

1. INTRODUCTION

Amphiphilic (αμφιφιλος) are called molecules such as surfactants and lipids, which have an affinity (φιλος) for both (αμφι) "water" and "oil", two solvents which are immiscible with each-other. Amphiphiles play a fundamental role in biology (e.g., as constituents of the cell membranes) and find widespread technological applications (e.g., as ingredients and/or processing aids in pharmaceuticals, foods, detergents, personal care products, plastics) because of their unique ability to self-organize at interfaces and in solution and thus modify interfacial properties and enhance compatibility or partition [1,2]. The tendency of surfactants and lipids to self-organize in (usually aqueous) solutions is reflected in the formation of thermodynamically stable (spontaneous self-assembly) micelles [3], microemulsions [4] (in multi-component systems with water, oil, and, often, cosurfactant), and lyotropic liquid crystals (i.e., planar, cylindrical and spherical micelles, with one-, two-, or three-dimensional order, respectively) [5], as well as in the formation of kinetically stabilized vesicles, emulsions and dispersions [6]. Block copolymers, macromolecules consisting of chemically different contiguous parts (blocks), can also express amphiphilic character when the different blocks prefer to segregate rather than to mix. Block copolymers are known to associate into micelles when dissolved in a solvent selective for one of the blocks [7,8], and to attain a variety of ordered microstructures, such as lamellae, cylinders, and spheres [9,10] in the absence of any solvent ("dry") or in mixtures with homopolymers and/or other block copolymers.

An interesting class of amphiphiles, developed originally for use as surface active agents in dilute aqueous solutions [11], but shown to also form structures in the absence of any solvent (similarly to "dry" block copolymers) [12], are block copolymers consisting of poly(ethylene oxide) (PEO) and poly(propylene oxide) (PPO). From the recent spur of activity in this field came the realization that ternary isothermal systems consisting of a PEO/PPO block copolymer, "water", and "oil" (where "water" and "oil" are selective solvents for the PEO and PPO blocks, respectively) can exhibit a very rich structural polymorphism [13-26]. A number of PEO-PPO-PEO triblock copolymers have been identified which self-assemble under equilibrium conditions into a variety of lyotropic liquid crystalline microstructures consisting of spherical, cylindrical, or planar microdomains, of normal ("oil"-in-"water") or reverse ("water"-in-"oil") morphology and discrete or interconnected (bicontinuous) topology [16-19,22,23]. Such an ability to form many normal and reverse structures at the same temperature has never before been observed in ternary systems of common surfactants with water and oil. Moreover, the variety of structures formed by a PEO/PPO block copolymer in the presence of solvents is much greater than that of a block copolymer of a given composition in the absence of solvents.

In this review we present the main features of the self-assembly behavior of amphiphilic PEO-PPO-PEO block copolymers in mixtures with selective solvents, "water" and "oil", and we place such features in the context of the molecular ordering observed in solvent-free block copolymers and in surfactant systems in the presence of water and oil.

2. BLOCK COPOLYMER PHASE BEHAVIOR

We first examine the tendency of block copolymers to form ordered structures in the "melt" state. Compared to the self-assembly of surfactants and lipids in solution, this is a relatively recent, but very active, topic of investigation. One-component block copolymer systems are simpler than multicomponent surfactant systems, and their understanding has advanced to the point where theories are available to qualitatively predict their phase behavior.

2.1 One-component block copolymer systems

While macroscopic phase separation is common occurrence in homopolymer blends owing to the usually very low entropy of mixing and unfavorable enthalpy of mixing of homopolymers, the different blocks in a block copolymer cannot phase separate in the same manner because they are joined together by covalent bonds. Instead, the different blocks of a block copolymer are able to segregate (microphase separation) and form domains with different morphologies. The stability of different morphologies in block copolymers results from an interplay between enthalpic (e.g., contact between chemically different blocks) and entropic (e.g., chain stretching, confinement, and frustration) contributions to the system free energy, and is often described in terms of a χN vs f phase diagram [27,28]. A χN vs f diagram, predicted from mean field theory is shown in Figure 1. The term χN, where χ is the Flory-Huggins interaction parameter between the different blocks and N is proportional to the block copolymer molecular weight, expresses the tendency for block segregation, with χ reflecting enthalpic contributions and N entropic contributions. f, the volume fraction of one block with respect to the whole block copolymer, is the composition variable. The higher the χN, the higher the degree of segregation between the different blocks and the higher the tendency for forming ordered microstructures. Typically χ increases with decreasing temperature. In one-component symmetric block copolymers, an order-disorder transition is observed at $\chi N \approx 10.5$ [28,29]. For $\chi N < 10.5$, the block copolymers are in a disordered state ("solution"). Upon an increase in the block segregation (by, e.g., increasing the block length, or decreasing the temperature), the block copolymers form ordered microstructures with domains enriched in a given block. The characteristic dimensions of the domains are on the order of nanometers, and their morphology (e.g., spheres, cylinders, and lamellae) depends on the composition f of the block copolymer, as shown in Figure 1. The degree of segregation and the corresponding "sharpness" of the interfaces between adjacent domains increase with increasing χN. The degree of segregation is characterized as "weak" and "strong", respectively, in the vicinity of $\chi N \approx 11$ and for $\chi N > 60$. While a variation in temperature may induce a change in microstructure (thermotropic polymorphism), f is the parameter most often varied (by synthesizing a new block copolymer) in order to attain a different morphology. Experimentally determined phase diagrams are now available for a variety of single-component block copolymer systems [27]. Recent advances in self-consistent mean-field theory permit calculation of the free energy of ordered block copolymer phases possessing any symmetry, and show reasonable agreement with experiment [28,29].

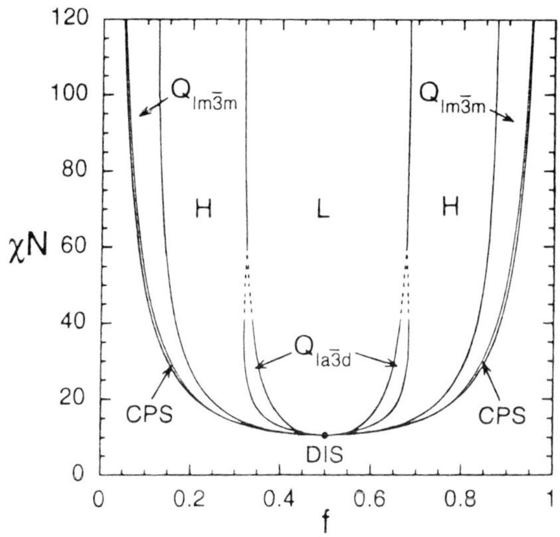

Figure 1. χN vs f phase diagram for conformationally symmetric AB block copolymer melts (one-component system), predicted from mean-field theory. The different regions are labelled according to the block copolymer morphology: L (lamellar), H (hexagonally-packed cylinders), Q_{Ia3d} (bicontinuous cubic of the crystallographic spacegroup Q_{Ia3d}, gyroid), Q_{Im3m} (bcc spheres), CPS (close-packed spheres), and DIS (disordered). Dashed lines denote extrapolated phase boundaries, and the dot denotes the mean-field critical point. (Reproduced from ref [28]; © 1996 American Chemical Society)

2.2 Two- and three-component block copolymer - homopolymer systems

As discussed above, the interfacial curvature in solvent-free block copolymers is set by the relative composition f of the different blocks [27]. Consequently, a given block composition can typically result in only one type of microstructure. For different structures to be obtained, different relative block compositions (resulting from the synthesis of different polymers) are required. Multicomponent systems, wherein other homopolymers or block copolymers are added to a block copolymer of a given f, can result into a wide variety of morphologies [30,31]. For example, addition of an A-homopolymer to an AB-block copolymer modifies the ratio of A and B in the binary system and increases the amount of A relative to that of B; it is thus expected to shift the phase stability in the direction which corresponds to a block copolymer of lower effective B content than the original one [30]. Self-consistent field theory has proved useful in predicting the phase stability of a symmetric AB-block copolymer in the presence of an A-homopolymer [30] to form a binary system, or both A- and B-homopolymers to form a ternary system [31]. Figure 2 presents the "isothermal" (constant χN) phase diagram for a symmetric AB-diblock copolymer of length N with an A-homopolymer of length α_A = 0.3N and a B-homopolymer of length α_B = 0.3N (three-component system), predicted from mean-field theory. The symmetric AB-diblock copolymer on its own forms a lamellar microstructure, but, upon the addition of A-homopolymer [B-homopolymer], the morphology changes first to that of cylinders and then to that of spheres with B [A] interior. At a given block copolymer content (e.g., 50%), a change in the content of A- and B-homopolymers can result in microstructures ranging from B-in-A spheres, to B-in-A cylinders, to lamellae, to A-in-B cylinders, and to A-in-B spheres. Note that the results of Figure 2 have been obtained in the weak segregation regime, and that the length of the added homopolymer is smaller than that of the block copolymers (longer homopolymers cannot be accommodated in the domains formed by the block copolymer). These theoretical studies demonstrate that the addition of short homopolymer(s) or selective solvent(s) can dramatically increases the structural (lyotropic) polymorphism, thereby reducing the need to synthesize a new block copolymer every time a specific morphology is required. The experimental findings on the ternary block copolymer - water - oil systems presented below confirm the expectations raised by the theory.

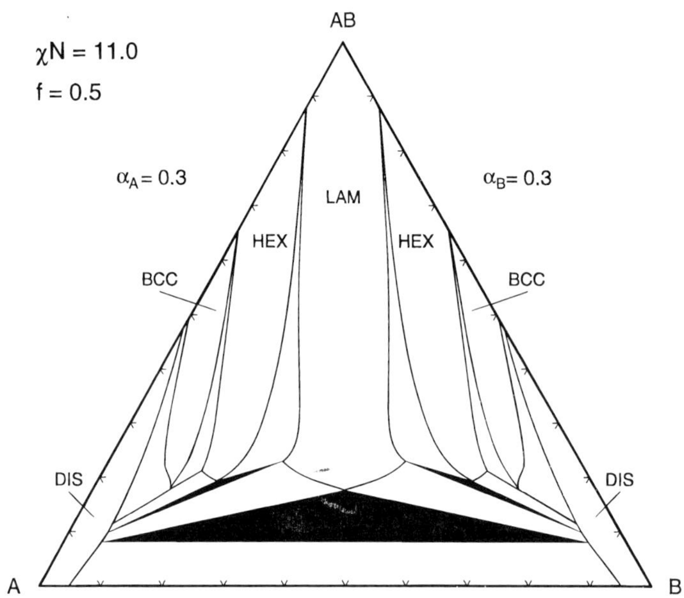

Figure 2. Phase diagram for a symmetric AB-diblock copolymer of length N with an A-homopolymer of length α_A = 0.3N and a B-homopolymer of length α_B = 0.3N (three-component system), predicted from mean-field theory. LAM denotes a lamellar phase; HEX, hexagonally arranged cylinders; BCC spheres on a body-centered-cubic lattice; and DIS, homogeneous phases. Regions of three-phase coexistence are shaded, biphasic regions are unlabeled. (Reproduced from ref [31]; © 1997 American Chemical Society)

3. AMPHIPHILIC BLOCK COPOLYMERS IN MIXTURES WITH WATER AND OIL

A well-studied example of amphiphilic block copolymers, which have been found to exhibit a structural polymorphism similar to that predicted for block copolymer-homopolymer blends (see Figure 2) in mixtures with water and oil, are those consisting of poly(ethylene oxide) as the hydrophilic block and poly(propylene oxide) or poly(butylene oxide) as the hydrophobic block. Below we present the main features of the self-assembly of such block copolymers in binary and ternary systems with solvents, and discuss the effects on self-assembly of the block copolymer molecular weight and block composition, and of the solvent quality.

3.1 Binary amphiphilic block copolymer - water systems

A number of PEO-PPO-PEO block copolymers have been shown to associate in aqueous solutions to form micelles consisting of a (mainly) PPO core and a corona dominated by hydrated PEO segments [32-34]. The formation of micelles and their structure have been investigated over the last decade with numerous techniques [32-34], and the micellization process is now relatively well understood. In addition to their association into micelles, some PEO-PPO-PEO block copolymers have been known to form "gels" in water [13,35]. Some of the "gel" phases reported in the literature are of micellar cubic structure but most others still lack proper structural identification; terms such as "hard gel" and "soft gel" are sometimes used to describe the texture (but not the microstructure) of various lyotropic liquid crystalline (LLC) phases. Such LLC "gel" phases are frequently encountered during the processing and formulation of amphiphiles, however they are much less studied than the micellar solutions.

We have studied the phase behavior and structure of binary amphiphilic polymer - water systems as a function of polymer concentration and temperature for three PEO-PPO-PEO block copolymers of different composition, $EO_6PO_{34}EO_6$, $EO_{13}PO_{30}EO_{13}$, and $EO_{37}PO_{58}EO_{37}$ (Pluronics L62, L64, and P105, respectively) [15]. The choice of these copolymers allowed the observation of the effects on the self-assembly of both the block copolymer molecular weight and its PEO/PPO composition. All three block copolymers form lamellar structures at low solvent (water) contents; with increasing solvent content, different microstructures (based on cylindrical or spherical micelles) can be observed. The number of LLC phases formed increases with the PEO content and the molecular weight of the polymers in the order L62 < L64 < P105. Only a lamellar LLC phase is formed by $EO_6PO_{34}EO_6$, while $EO_{37}PO_{58}EO_{37}$ is capable of self-assembling with increasing solvent (and decreasing block copolymer) concentration into lamellar, hexagonal, and cubic LLC phases as seen in Figure 3.

Upon heating, the LLC phases of the $EO_6PO_{34}EO_6$-water and $EO_{13}PO_{30}EO_{13}$-water systems swell with water; no such swelling is detected for the $EO_{37}PO_{58}EO_{37}$-water system. The effects of temperature on the phase behavior of the PEO-PPO-PEO copolymers can be understood by invoking the reverse solubility (i.e., decrease in solubility with increasing temperature) of PEO and PPO in water. Upon a temperature increase, the PEO-water and PPO-water χ interaction parameters increase [36] and the PEO-PPO interaction parameter decreases [37]. Overall, the "hydrophobicity" of the polymer is enhanced and the swelling of the PEO blocks by water is decreased. Thus, the stability of the various LLC regions in the binary polymer-water phase diagrams shifts to lower polymer concentrations as the temperature increases. The thermal stability of the LLC regions increases in the order cubic < hexagonal < lamellar and $EO_6PO_{34}EO_6 < EO_{13}PO_{30}EO_{13} < EO_{37}PO_{58}EO_{37}$. An increase of the temperature results in a decrease in the interfacial area and an increase in the periodicity of the $EO_6PO_{34}EO_6$ and $EO_{13}PO_{30}EO_{13}$ lamellae. Both the lamellar periodicity and the average interfacial area per block copolymer molecule decrease with increasing polymer content.

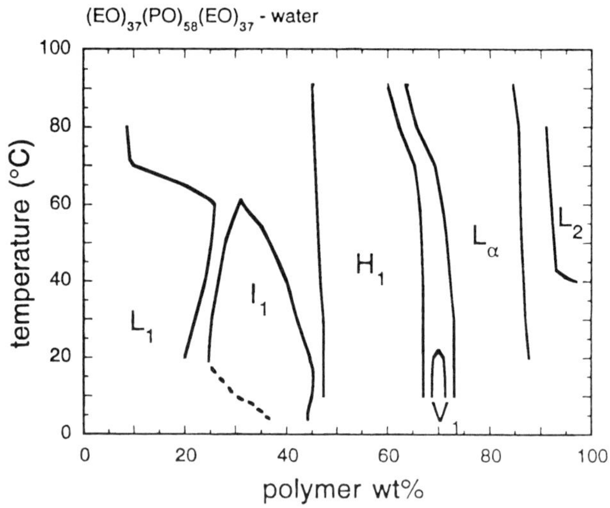

Figure 3. The concentration-temperature phase diagram of the $EO_{37}PO_{58}EO_{37}$ (Pluronic P105) block copolymer – ^2H_2O binary system (the concentrations are expressed in wt%). The block copolymer has 50% PEO and can be considered symmetric. The phase boundaries of the one-phase regions are drawn with solid lines. I_1, H_1, V_1, and L_α, denote micellar cubic, hexagonal (cylindrical), bicontinuous cubic, and lamellar (smectic) lyotropic liquid crystalline phases, respectively, while L_1 and L_2 denote water-rich and polymer-rich solutions. The samples whose compositions fall outside the one-phase regions are dispersions of two different phases. The tie-lines in the two-phase regions are parallel to the concentration axis (isothermal). (Adapted from ref [15]; © 1996 American Chemical Society)

3.2 Ternary amphiphilic block copolymer - water - oil systems

Over the last five years we have examined the phase behavior and microstructure of a number of PEO-PPO-PEO block copolymers (covering a range of molecular weights and PEO/PPO composition ratios) in the presence of water and xylene as selective solvents for the PEO and PPO blocks, respectively [16,19,21,23]. Recently we have also established the generality of the features observed in the PEO-PPO-PEO - water - xylene systems by looking into other types of block copolymers such as PEO-PBO, where PBO: poly(butylene oxide), and PEO-PS, where PS: polystyrene [18,20]. Here we discuss the phase behavior and microstructure features of ternary amphiphilic block copolymer - water - oil systems in the context of the $EO_{19}PO_{43}EO_{19}$ – water – p-xylene system which exhibits a very rich structural polymorphism [19]. We believe that such polymorphism transcends this particular system and is instead a general feature of amphiphilic block copolymer - selective solvent systems.

Figure 4 shows the isothermal (25 °C) phase diagram obtained for the ternary system $EO_{19}PO_{43}EO_{19}$ – water – p-xylene. The phase boundaries of the one-phase regions are drawn with solid lines. The samples whose compositions fall outside the one-phase regions are dispersions of two or three (depending on the location in the phase diagram) different phases [16,19]. A record number of nine regions of different microstructure are present in this phase diagram. Seven different lyotropic liquid crystalline phases have been identified (in terms of location in the phase diagram) and characterized in terms of microstructure (see references [16,19] for details): lamellar (L_α), normal (H_1) and reverse hexagonal (H_2), normal (V_1) and reverse bicontinuous cubic (V_2), and normal (I_1) and reverse micellar cubic (I_2). Two isotropic solution phases, one (normal micellar) rich in water (L_1) and one with high xylene-to-water ratio (L_2), have also been identified. The different modes of self-organization of the block copolymers in the presence of water and oil are illustrated in the schematics of Figure 4.

This modulation of the interfacial curvature (and the resulting microstructure) can be attributed to the ability of the macromolecular blocks to swell with solvent to a varying extent depending on the amount (and quality) of solvent(s) available [19]. For a PEO-PPO block copolymer of a given block composition and molecular weight, the types of structures obtained in the presence of selective (water and oil) solvents appear to be a function of the volume fraction of the polar (PEO + water) / apolar (PPO + xylene) components (note, however, that, for certain microstructures, e.g., cubic, to form, both the water/oil ratio and the total block copolymer content are important). This is similar to the structure variation observed in one-component AB block copolymer melts, at temperatures below the order-disorder transition, upon changing the A/B block composition (see Figure 1), and also similar to the ternary AB-block copolymer - A-homopolymer - B-homopolymer phase behavior presented in Figure 2. Thus the structural polymorphism in the amphiphilic copolymer - water - oil systems appears to be governed by similar mechanisms as that in solvent-free block copolymers [28,38,39].

We note that the different regions in the ternary phase diagram of Figure 4 are arranged around the L_α region. The lack of "mirror image" symmetry in the present phase diagram is primarily due to the different nature of the PEO-water and the PPO-xylene interactions. Differences in the conformational and volume-filling characteristics of each block, as well as fluctuation effects, could also account for the lack of symmetry around the lamellar region, as discussed recently for diblock copolymer melts [26].

3.3 Progression of microstructure in block copolymer - water - oil systems

The progression of structure in the block copolymer - water - oil phase diagrams of the type discussed above can be discussed at two levels, (i) at varying water/oil ratio and constant total copolymer content (along composition lines parallel to the basis of the phase triangle of Figure 4) and (ii) at increasing total copolymer content and constant copolymer/oil [copolymer/water] ratio (along dilution lines originating from the water [oil] corner of the phase triangle of Figure 4, or along composition lines parallel to the basis of the Figure 5a [5b] plot).

An example of structure progression for the case of varying the water/oil ratio at constant copolymer content, is the phase sequence $I_1 \rightarrow H_1 \rightarrow L_\alpha \rightarrow H_2 \rightarrow I_2 \rightarrow L_2$, observed at ≈40 wt% copolymer. This phase/structure sequence is consistent with a decreasing interfacial mean curvature (H) from the left (H>0) to the right (H<0) side of the phase diagram, with H=0 in the lamellar phase [40,41]. Note that higher (>65 wt%) copolymer concentrations are required for the V_1 and V_2 structures (expected to occur between H_1–L_α and L_α–H_2) respectively, to form; at such copolymer concentrations the I_1 and I_2 structures are no longer stable (as discussed below, the microstructure is also affected by the block copolymer concentration).

The progression of structure with increasing total block copolymer content at the copolymer-water (oil-lean) and the copolymer-oil (water-lean) sides of the ternary phase diagram is shown in Figure 5, where the degree that the different phases/structures can accommodate oil [water] is plotted as a function of the copolymer weight fraction in water (Figure 5a) [oil (Figure 5b)]. In the oil-lean part of the phase diagram (at a constant copolymer/oil ratio), the $EO_{19}PO_{43}EO_{19}$ amphiphilic block copolymer self-assembles upon increase of its concentration (and the relative contribution of PPO to the ternary composition) into the following structures: $L_1 \rightarrow I_1 \rightarrow H_1 \rightarrow V_1 \rightarrow L_\alpha \rightarrow L_2$. Note that this sequence does not include any of the reverse (water-in-oil) liquid crystalline structures (V_2, H_2, I_2); the PPO content of the copolymer is such (60%) that it does not favor such structures in the absence of apolar solvent. The sphere → cylinder → plane sequence of structures, this time of the reverse (water-in-oil) morphology, is also observed in the water-lean part of the phase diagram with increasing block copolymer concentration (and increasing PEO volume fraction in the ternary copolymer-water-oil system) at a constant copolymer/water weight ratio (of about 1÷4, corresponding to ≈ 1.4 water molecules per EO segment).

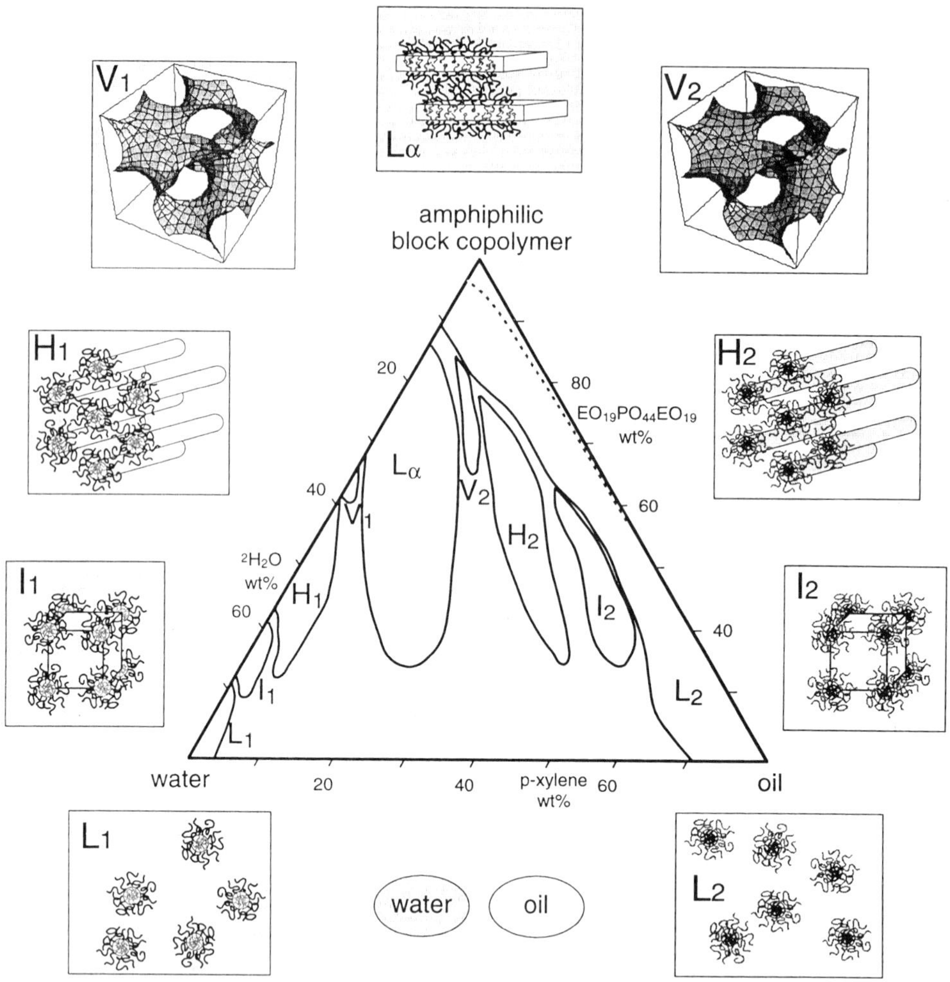

Figure 4. The phase diagram of the $EO_{19}PO_{43}EO_{19}$ (Pluronic P84) – ^2H_2O – p-xylene ternary system at 25°C. The phase boundaries of the one-phase regions are drawn with solid lines. I_1, H_1, V_1, L_α, V_2, H_2, and I_2, denote normal (oil-in-water) micellar cubic, normal hexagonal, normal bicontinuous cubic, lamellar, reverse (water-in-oil) bicontinuous cubic, reverse hexagonal, and reverse micellar cubic lyotropic liquid crystalline phases, respectively, while L_1 and L_2 denote water-rich (normal micellar) and water-lean (reverse micellar) solutions. The concentrations are expressed in wt%. Schematics of the different modes of self-organization of the amphiphilic block copolymers in the presence of solvents ("water" and "oil") are shown adjacent to the respective phases in the phase diagram. The amphiphiles are localized at the interfaces between the water and oil domains (shaded). The Ia3d/Gyroid minimal surface is used as a representation of the microstructure in the V_1 and V_2 phases. (Reproduced from ref [19]; © 1998 American Chemical Society)

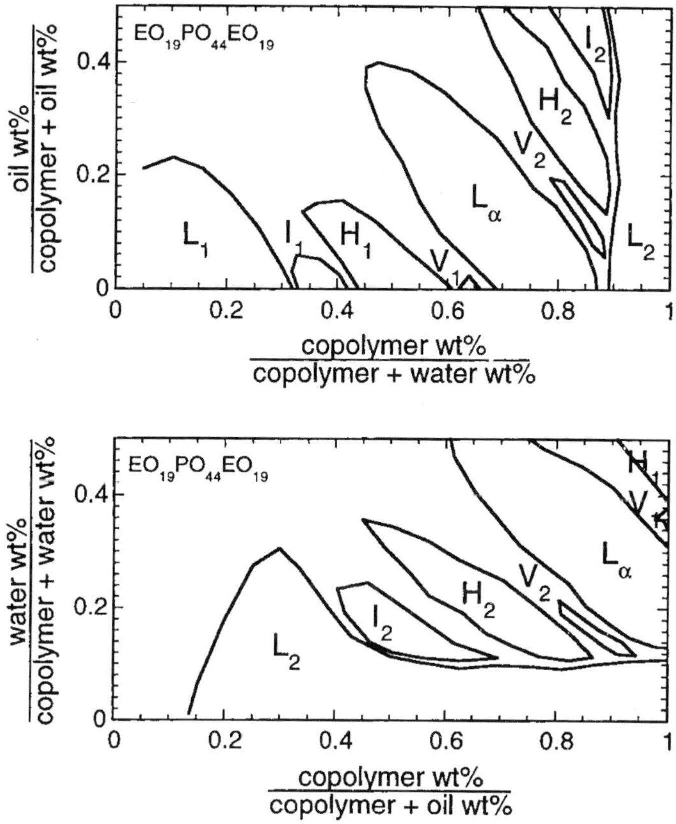

Figure 5. (top) Boundaries of the various phases along the copolymer-water side of the ternary phase diagram of Figure 4 plotted in the "reduced" coordinates X: block copolymer weight fraction in its mixture with water, and Y: oil weight fraction in its mixture with the block copolymer. (bottom) Boundaries of the various phases along the copolymer-oil side of the ternary phase diagram plotted in the "reduced" coordinates X: block copolymer weight fraction in its mixture with oil, and Y: water weight fraction in its mixture with the block copolymer. (Reproduced from ref [19]; © 1998 American Chemical Society)

The decrease [increase] in the interfacial curvature (and the interrelated progression from a normal [reverse] spherical to a normal [reverse] cylindrical and then to a lamellar arrangement) upon an increase of the copolymer concentration at constant copolymer/oil [copolymer/water] ratio (Figure 5a [5b]) is a consequence of the relative increase of the PPO [PEO] block volume fraction in the system. The tilt to the left, observed in the phase boundaries as plotted in Figure 5, is a reflection of the curvature decrease [increase] upon the increase of the oil/water [water/oil] ratio. This resembles the observed increased swelling with water of the lyotropic phases in PEO/PPO copolymer - water systems upon increasing temperature (see Figure 3 and related discussion), which has been interpreted in terms of decreasing effective interfacial curvature with increasing temperature (due to dehydration of the PEO block) [15].

As discussed above, for a PEO/PPO block copolymer of a given block composition and molecular weight, the types of structures obtained in the presence of selective solvents appear to be a function of the volume fraction of the polar ("water"-like) / apolar ("oil"-like) components. The important point here is that the microstructure in such systems is not tied up to a specific block copolymer molecular weight and block composition, which define a point in the χN vs f phase diagram of one-component "dry" block copolymers, but can vary significantly under the influence of the selective solvents. Of course, the block copolymer molecular weight and block composition will still affect the phase behavior and microstructure in systems of block copolymers and selective solvents. Below we present some of our recent findings on this topic. We also present some evidence on how the quality of the solvent and its ability to swell the different blocks can also affect the phase behavior and structure.

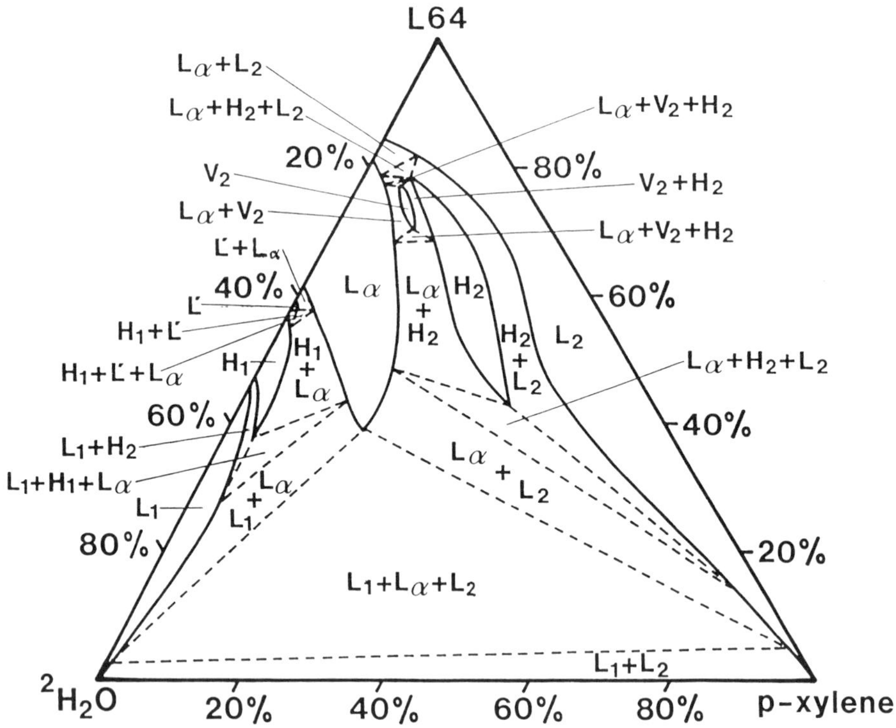

Figure 6. Isothermal (25 °C) phase diagram of the ternary system $EO_{13}PO_{30}EO_{13}$ (Pluronic L64) - water (^2H_2O) - p-xylene, showing the phase boundaries (solid lines) of the one-phase regions, the two-phase regions and the three-phase triangles (dotted lines). The concentrations are expressed in wt%. The notation of the different phases is the same as the one used in Figure 4. (Reproduced from ref [16]; © 1998 American Chemical Society)

3.4 Effect of block copolymer molecular weight

In block copolymer melts, an increase in the copolymer molecular weight for a given A/B block composition increases the block segregation and the tendency for organization. Recall that the degree of segregation is proportional to N (see Figure 1 and related discussion) which is a function of the polymer molecular weight. In order to examine the effects of copolymer molecular weight on the self-assembly of amphiphilic block copolymers in the presence of selective solvents, we undertook a study of a group of PEO-PPO-PPO block copolymers having the same, roughly symmetric (40% PEO), block composition and varying molecular weight, under the same solvent and temperature conditions. Figure 6 shows the $EO_{13}PO_{30}EO_{13}$ (Pluronic L64) - water - xylene ternary isothermal phase diagram.

Comparison of the phase diagrams of Figures 4 and 6 confirms the trend of increasing block segregation and tendency for self-organization with increasing copolymer molecular weight, and establishes the validity of this trend in the case of PEO/PPO block copolymer self-assembly in selective solvents. The block copolymer examined in Figure 4, $EO_{19}PO_{43}EO_{19}$ (Pluronic P84), has a 45% higher molecular weight than, but is of the same composition as $EO_{13}PO_{30}EO_{13}$ (Pluronic L64), whose phase diagram with water and xylene is shown in Figure 6 [16]). An increase in the composition stability range of the different structures, and also a formation of more types of lyotropic liquid crystalline structures) is observed with $EO_{19}PO_{43}EO_{19}$, as compared to $EO_{13}PO_{30}EO_{13}$. More details on the effects of copolymer molecular weight on the self-assembly of amphiphilic block copolymers (in the presence of selective solvents) are presented elsewhere for a group of PEO/PPO block copolymers having the same block composition (40% PEO) and varying molecular weight [23].

Another illustration of molecular weight effects comes from the group of Pluronics L62 ($EO_6PO_{34}EO_6$), L92 ($EO_8PO_{47}EO_8$) and L122 ($EO_{11}PO_{70}EO_{11}$), all with 20% PEO, the binary concentration-temperature phase diagrams of which are presented in Figure 7 [39]. These data confirm that a certain minimum molecular weight is required for the PEO and PPO blocks to segregate and to form liquid crystalline structures. For example, Figure 7 shows that the aqueous solution of Pluronic L62 (of lower molecular weight) displays only one LLC phase (a lamellar phase in the 62-77 wt % polymer range) as compared to the three LLC phases which appear in aqueous solutions of Pluronic L92 and L122. Note also that the range of the normal hexagonal regions increases when going from L92 to L122.

An important result from the study of molecular weight effects concerns the dependence of the lamellar characteristic spacing on the polymer molecular weight. The data shown in Figure 7 (obtained such data for a set of three copolymers having 20% PEO [39]) suggest a $d \sim N^{1/2}$ scaling, which is characteristic of a random (Gaussian) coil. This is an indication that the block copolymer chains are in the weak segregation regime. The presence of the selective solvent, however, makes the segregation more pronounced and gives rise to well defined (judging from small-angle scattering patterns and from the high viscosity exhibited by the ordered samples) microstructures.

3.5 Effect of block copolymer composition

Apart from the ternary composition and block copolymer molecular weight effects (outlined in the previous paragraph), the block composition of the amphiphilic block copolymer is expected to play a major role in the microstructure attained in the ternary copolymer-water-oil systems. After all, the block composition is the main determinant of the microstructure observed in solvent-free block copolymers [27], and the chemical composition of typical surfactants ("headgroup" and "tail") affects their hydrophile/lipophile ratio and self-assembly-in-solution properties [2,5,42,43]. We have addressed the effects of block composition on the self-assembly (and resulting microstructure) of amphiphilic block copolymers in the presence of selective solvents ("water" and "oil": butyl acetate or butanol), by examining the ternary phase behavior and structure of two copolymers, $EO_{20}PO_{70}EO_{20}$ and $EO_{100}PO_{70}EO_{100}$, having the same block architecture ($EO_XPO_YEO_X$) and PPO middle-block size, but different PEO end-block sizes (Pluronic P123: $EO_{20}PO_{70}EO_{20}$ contains 30% EO and Pluronic F127: $EO_{100}PO_{70}EO_{100}$ 70% EO) [20-22]. The ternary phase diagrams are presented in Figure 8.

In the $EO_{20}PO_{70}EO_{20}$ phase diagrams (left column of Figure 8), the lamellar structure (of zero interfacial curvature) is the most extensive. The high PEO (hydrophilic) content of $EO_{100}PO_{70}EO_{100}$, however, favors oil-in-water LLC structures with high interfacial curvature.

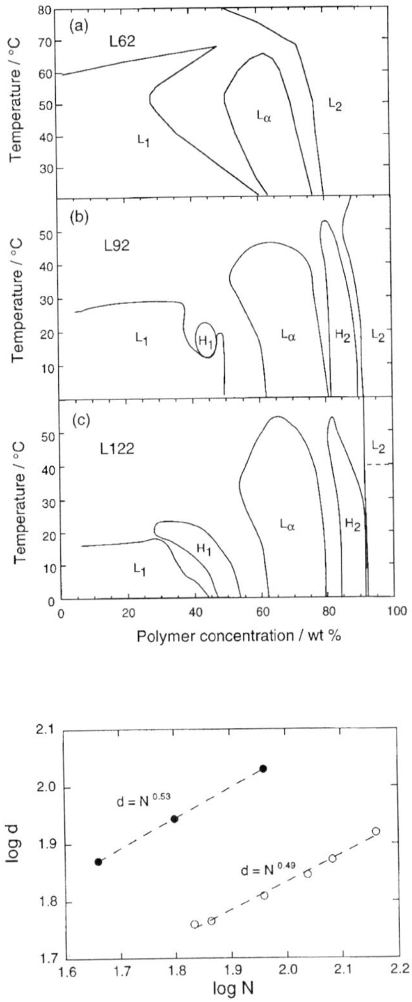

Figure 7. (top) Concentration-temperature phase diagrams for three PEO-PPO-PEO block copolymers having the same block architecture and composition (20% PEO) but different molecular weights (increasing in the order L62<L92<L122, where Pluronics L62, L92 and L122 can be represented by the formulas $EO_6PO_{34}EO_6$, $EO_8PO_{47}EO_8$, and $EO_{11}PO_{70}EO_{11}$, respectively). The notation used for the different phases is the same one used in Figure 4. (bottom) Logarithm of the lamellar domain spacing, d, vs the logarithm of number of polymer monomers, N, for 75.0/25.0 (wt%) polymer/water samples (filled symbols), where polymer: L122, L92 and L62, and the corresponding quantities (empty symbols) calculated from mean field theory. (Reproduced from ref [39]; © 1999 American Chemical Society)

No reverse LLC phases are formed by $EO_{100}PO_{70}EO_{100}$. Figure 8 shows how the increase of the PEO block size effects the preferred self-organization of the amphiphilic block copolymers. Indeed, a clockwise shift (to the direction of increasing "oil" content) in the type (curvature) of structure can be observed when the block copolymer changes from $EO_{20}PO_{70}EO_{20}$ to $EO_{100}PO_{70}EO_{100}$ for both the butyl acetate and butanol ternary phase diagrams, with the micellar cubic structure in $EO_{100}PO_{70}EO_{100}$ replacing the normal hexagonal (H_1) in $EO_{20}PO_{70}EO_{20}$, the normal hexagonal structure in $EO_{100}PO_{70}EO_{100}$ replacing the lamellar in $EO_{20}PO_{70}EO_{20}$, and the lamellar phase in $EO_{100}PO_{70}EO_{100}$ moving to higher oil contents. It is notable that the normal hexagonal phase in the $EO_{100}PO_{70}EO_{100}$, systems occurs at approximately the same composition range as the lamellar phase in the $EO_{100}PO_{70}EO_{100}$ systems [20-22]. The copolymer-water side of the $EO_{100}PO_{70}EO_{100}$ ternary phase diagrams is dominated by the micellar cubic LLC structure; significant amounts of oil are needed in order for cylindrical (H_1) and lamellar microstructures structures to form.

The effects of block asymmetry are also evident in the phase diagrams of low PEO content block copolymers shown in Figure 7. These systems are notable because they show both normal hexagonal (H_1) and reverse hexagonal (H_2) microstructures formed by the same block copolymer in a binary system with water. The low PEO content block copolymers favor reverse structures such as H_2, but in the presence of sufficient water the block copolymers can also attain normal structures like H_1. This feature has not been observed in other surfactants before, and is another reflection of the flexibility of the amphiphilic block copolymers in attaining different microstructures aided by the solvent.

Many of the PEO-PPO-PEO block copolymer - water - oil phase behavior features observed experimentally can be captured qualitatively by mean-field polymer models as demonstrated recently [39,40]. In addition to providing help in rapid screening through the large number of degrees of freedom afforded by polymeric amphiphiles, such models can provide detailed structural information which is very difficult to obtain from experiments. The free energy of different possible microstructures of binary and ternary block copolymer - water - oil systems was calculated using mean-field lattice theories, based on an extension of the Flory-Huggins theory to heterogeneous systems originally developed by Scheutjens and Fleer. Thereafter phase diagrams were constructed employing the relative stability of the microstructures. In order to describe the decreased solubility in water exhibited by PEO-and PPO-containing polymers with increasing temperature, the EO and PO segments were assigned in the model with two conformational states each, one polar and one apolar [39].

A ternary phase diagram for the $EO_{27}PO_{61}EO_{27}$ - water - xylene system predicted from the theory described above [40] is shown in Figure 9a. Several different ordered phases as well as water-rich and water-poor disordered phases were obtained. Note the good qualitative agreement between this phase diagram and that of Figure 4. The effect on the phase behavior of extending the PEO block length is strong. Parts b and c of Figure 9 show the phase diagrams for the $EO_{50}PO_{61}EO_{50}$ - water - xylene and the $EO_{107}PO_{61}EO_{107}$ - water - xylene systems, respectively, and confirm the effects of block asymmetry shown in Figure 8. As the PEO block length increases, the normal ordered phases are favored, as predicted by the model.

3.6 Effect of solvent quality

An important issue in our understanding and control of self-assembly has to do with how the nature of the oil present affects the phase behavior and structure afforded by the amphiphilic block copolymers. The oil can swell to a varying degree the hydrophobic blocks of the macromolecular amphiphile, and thus its selection will be of great importance in modulating the interfacial "curvature". We examined a PEO-PPO-PEO triblock copolymer (Pluronic F127 $EO_{100}PO_{70}EO_{100}$) in ternary isothermal systems with water and each of three "oils" (of varying polarity), p-xylene, n-butyl acetate, and butan-1-ol [21]. A large variety of microstructures is found in all three ternary systems, but the overall phase behavior (e.g., the number of phases and the composition range of their stability) depends on the "oil" used.

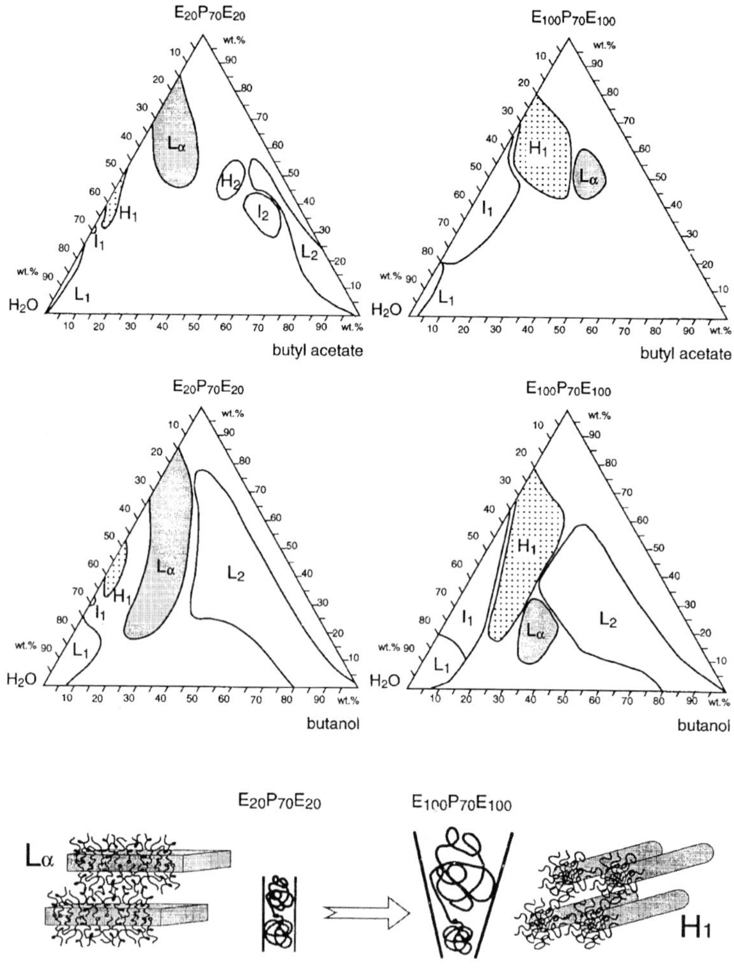

Figure 8. Phase diagrams of the (top left) $EO_{20}PO_{70}EO_{20}$ (Pluronic P123) - n-butyl acetate - water, (bottom left) $EO_{20}PO_{70}EO_{20}$ - butan-1-ol - water, (top right) $EO_{100}PO_{70}EO_{100}$ (Pluronic F127) - n-butyl acetate - water, and (bottom right) $EO_{100}PO_{70}EO_{100}$ - butan-1-ol - water ternary systems at 25°C. The notation is the same as in Figure 4. The four phase diagrams presented here illustrate the change in phase behavior when (i) the size of the copolymer PEO block changes from EO_{20} to EO_{100} (left to right) and (ii) the "oil" changes from butyl acetate to butanol (top to bottom). The schematic represents the change of the self-organization of the amphiphilic block copolymers from the lamellar (L_α) to the normal hexagonal (H_1) structures upon an increase of the PEO block size and the corresponding increase in the interfacial curvature. The shaded areas represent the apolar domains. For reasons of clarity, only the copolymer molecules at the cross-section of the (planar and cylindrical) assemblies are depicted, and only small sections of the lyotropic liquid crystals are shown. (Reproduced from ref [22]; © 1998 American Chemical Society)

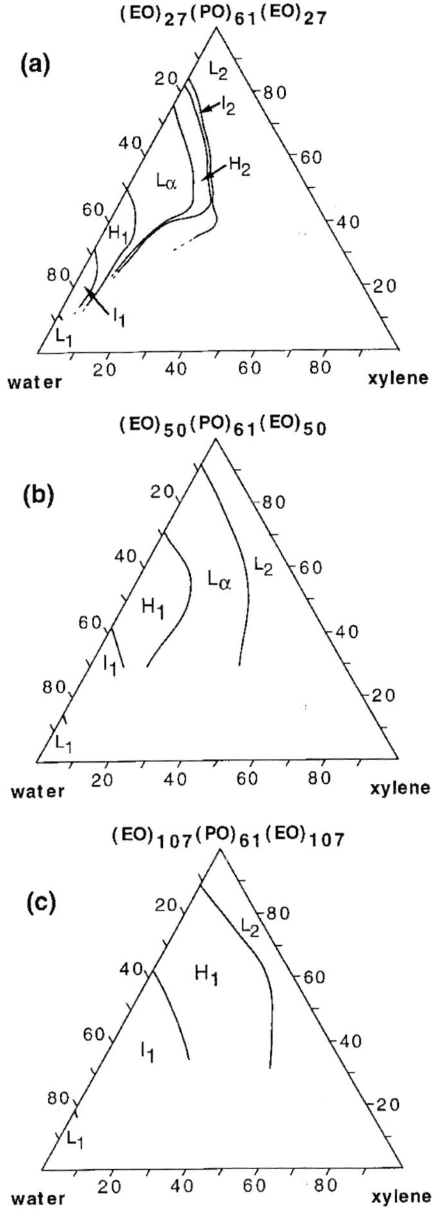

Figure 9. Ternary phase diagrams of (a) $EO_{27}PO_{61}EO_{27}$ - water - xylene system, (b) $EO_{50}PO_{61}EO_{50}$ - water - xylene system, (c) $EO_{107}PO_{61}EO_{107}$ - water - xylene system, predicted from mean-field theory as discussed in ref [40]. The middle PPO block is the same in all three block copolymers, but the PEO blocks increase in size from (a) to (c). The notation is the same as in Figure 4. (Reproduced from ref [40]; © 1999 American Chemical Society)

Comparison of the ternary phase diagrams prepared with the three different organic solvents (the phase diagrams for butyl acetate and butanol are shown in Figure 8) leads to the following main observations. The extent of the regions where the samples are of L_1, I_1 and H_1 microstructure, all of the "oil"-in-water topology, increases to accommodate increasing amounts of organic solvent when butyl acetate or butanol are used instead of xylene. This can only be partly attributed to the finite solubility of butyl acetate or butanol in water, because, as we discuss below, there are additional effects by these solvents on the copolymer swelling. Such effects are, e.g., evident in the pronounced swelling of the H_1 phase with water and in the stability of the L_α region at considerably lower block copolymer contents when changing solvent from butyl acetate to butanol. Butanol is unique in that it can stabilize lamellar and hexagonal phases at low (down to 12 wt. %) block copolymer concentrations, and also the L_2 solution phase at high copolymer concentrations (up to 60 wt %). It thus acts in a manner comparable to cosurfactants known from systems containing low-molecular-weight amphiphiles [2]. The normal hexagonal and the lamellar regions in the cases of butyl acetate and xylene are stable at copolymer concentrations above 45 w%. Up to 30 wt % butyl acetate but only 10 wt. % xylene can be accommodated in the normal hexagonal structure. The stability of the cubic structure from 20 up to 65 wt % copolymer concentration range, and its ability to solubilize significant amounts of butanol and butyl acetate are notable, given the practical applications that such cubic "gels" are finding [2,5].

The analysis of small-angle X-ray scattering (SAXS) measurements, performed in the H_1 and L_α regions for all three $EO_{100}PO_{70}EO_{100}$ phase diagrams, enabled us to get a firmer grip on the role of the organic solvent [21]. For the same ternary composition and structure, the lattice spacing was found to decrease as the "oil" became less hydrophobic (more polar), i.e., from xylene to butyl acetate to butanol. This trend was observed in both the normal hexagonal and the lamellar regions, and suggested a different location of the "oils". The location of the "oil" in the different microstructures has been deduced from the structural dimensions obtained from SAXS and the constraints imposed by the total volume fraction of the copolymer and oil, and the consideration that the interfacial area per copolymer molecule be independent of the overall composition. We thus showed that all xylene is located in the interior (apolar core) of the oil-in-water microdomains, butanol is located at the interfacial region between PPO and PEO, while butyl acetate contributes to both the core and the interfacial domains.

4. SURFACTANT PHASE BEHAVIOR

The PEO-PPO-PEO block copolymers are similar (in terms of the PEO headgroup and the block architecture) to the nonionic alkyl-oligo(ethylene oxide) surfactants. In this respect, it is instructive to compare the self-assembly features of these two groups of amphiphiles.

4.1 Binary surfactant - water systems

The alkyl-oligo(ethylene oxide) surfactants form microdomains with curvature towards oil ("normal" structures) at lower temperatures, while at higher temperatures curvature towards water is favored ("reverse" structures) [41,42]. In binary systems with water, this property is manifested by the formation of dilute lamellar and L_3 ("sponge") phases at higher temperatures in addition to micellar and the usual LLC phases at lower temperatures [40,41]. It is instructive to consider the series of $C_{12}EO_4$ - $C_{12}EO_6$ surfactants [43], where the size of the alkyl part is the same and that of the EO part increases (similar to the series Pluronic L62 - L64 examined in ref [15]), and that of $C_{12}EO_6$ - $C_{16}EO_8$, where the ratio of the alkyl and the EO part is the same but the molecular weight increases (similar to the series Pluronic L64 - P105 examined in ref [15]). The only LLC phase in the $C_{12}EO_4$-water system is the lamellar; an increase in the number of EO segments allows the formation of hexagonal and bicontinuous cubic LLC phases in the $C_{12}EO_6$-water system, in addition to the lamellar [43]. The hexagonal structure in the $C_{12}EO_6$-water system melts at ~40°C and the lamellar at ~75°C; on the other hand, the

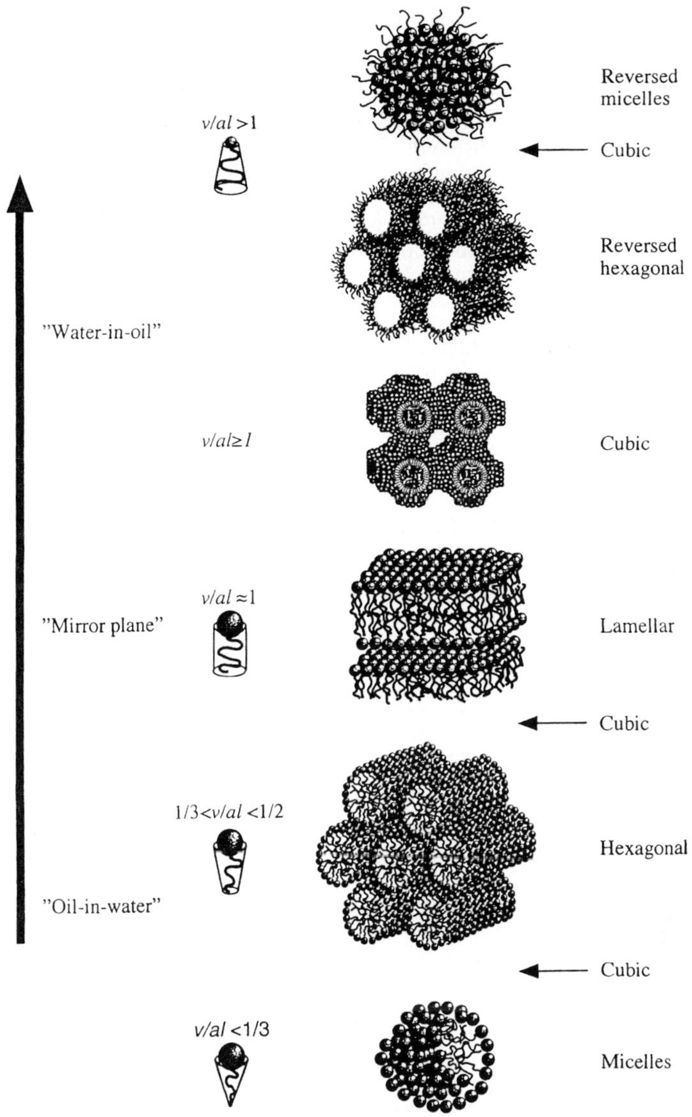

Figure 10. Packing parameters of surfactant molecules and preferred aggregate structures for geometrical packing reasons. (Reproduced from ref [2]; © 1998 J. Wiley & Sons)

hexagonal structure formed by the higher molecular weight $C_{16}EO_8$ melts at ~60°C and the lamellae are stable beyond 100°C. In addition to the hexagonal and lamellar LLC phases, a micellar cubic LLC region is formed in the $C_{16}EO_8$-water system at temperatures <10°C.

The trends in the phase behavior of the C_iEO_j surfactants in water outlined above are very similar to the trends exhibited by the PEO-PPO-PEO block copolymers in water. However the two classes of amphiphiles are different. Typical surfactants, such as the C_iEO_j molecules, are characterized by a strong segregation between the polar and apolar domains. In the PEO-PPO-PEO block copolymer systems, the segregation between the hydrophilic PEO and the "hydrophobic" PPO is weaker (PPO is relatively polar because of the ether oxygen), the packing of the block copolymer molecules at the PEO/PPO interface is less ordered and there may be some intermixing between the PEO and PPO blocks at the polar-apolar interface, and finally water is not as completely excluded from the PPO domains as it is from the alkyl domains of the C_iEO_j surfactants [39,40]. In order to counteract the relatively weak segregation between PEO and PPO, high molecular weights are required for the PEO-PPO-PEO block copolymers to arrange into ordered structures (i.e., larger N is required to compensate for a smaller χ). This weak segregation, however, is one of the reasons leading to the structural polymorphism observed in PEO-PPO-PEO block copolymers in ternary systems with water and oil; discussion on this follows.

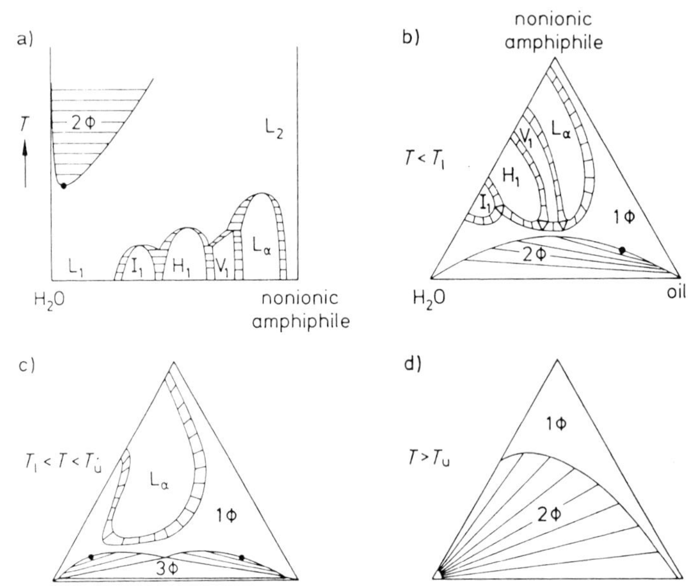

Figure 11. (a) Lyotropic mesophases in the binary system H_2O - C_iE_j, where C_iE_j denotes a nonionic alkyl ether surfactant with a hydrophobe consisting of i carbon atoms and a hydrophile of j PEO segments. (b) Schematic of the extension of the lyotropic mesophases into the Gibbs triangle (ternary phase diagram) below T_l, where T_l is the "lower critical tie line" temperature (see ref [41] for details). (c) Schematic of the extension of the lyotropic mesophases into the Gibbs triangle in the three-phase temperature interval (T_u-T_l), where T_u is the "upper critical tie line" temperature (see ref [41] for details). (d) Schematic of the phase diagram above T_u. (Reproduced from ref [41]; © 1985 VCH Verlagsgesellschaft mbH)

4.2 Ternary surfactant - water - oil systems

The surfactant phase behavior can to a first approximation be related to the shape of the surfactant molecule which is expressed in terms of the surfactant "packing" parameter, $N_S = \upsilon / (l\,\alpha)$ (where υ stands for the volume of the hydrophobic part of the surfactant molecule, l the length of the hydrocarbon chain, and α the effective area per polar headgroup) [44]. Small values of the surfactant parameter imply highly curved assemblies (spherical micelles for $N_S \approx 0.33$ and infinite cylinders for $N_S \approx 0.5$) but when N_S is close to unity, planar bilayers usually form [44], as illustrated in Figure 10. The packing parameter of ionic surfactants can be modulated to some extent by the presence of salts which screen the electrostatic repulsion between the surfactant headgroups and decrease α [41,44]. Temperature can be used to alter the packing parameter in nonionic poly(ethylene oxide)-containing surfactants because of the "dehydration" of PEO upon heating and the subsequent reduction in α; lower temperatures thus favor structures with curvature towards oil, while higher temperatures are needed for the curvature to be towards water [41,44]. However, a structural polymorphism such as that exhibited by the amphiphilic block copolymers presented here has never been observed in ternary isothermal systems formulated with low-molecular weight surfactants [19].

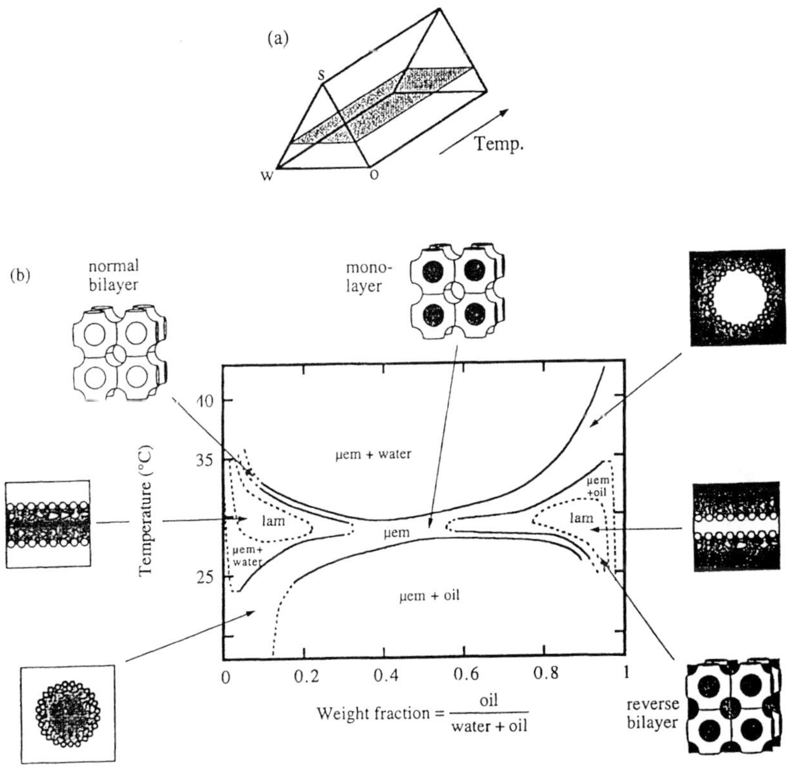

Figure 12. Temperature evolution of the microstructures (from oil-in-water micelles, to bilayer structures, to water-in-oil micelles/microemulsions upon heating) in different regions of a constant-surfactant-content cut of a ternary nonionic surfactant - water - oil system. (Reproduced from ref [2]; © 1998 J. Wiley & Sons)

The "preferred" interfacial curvature constrains the phase behavior of surfactants so that a ternary isothermal surfactant-oil-water phase diagram exhibits either normal (oil-in-water) or reverse (water-in-oil) phases, but usually not both types at the same temperature [5,41,42]. Figures 11 and 12 illustrate the evolution with temperature of the phase behavior expected for a C_iEO_j surfactant in oil and water.

For amphiphilic block copolymers, the curvature is set initially by the relative length of the different blocks in the copolymer, but can be readily tuned by the varying degree of swelling of the different blocks by the respective selective solvent. The flexibility of the PEO chain as well as the great difference in size between the solvent molecules and the copolymer blocks facilitate the aforementioned swelling. The effective persistence length of the hydrocarbon chain in typical surfactants is of the order of the assembly characteristic dimension. On the contrary, in the system under consideration here the persistence length of the copolymer is expected to be much smaller than the characteristic dimension of the domain, which, in turn, is much smaller than the stretched polymer length. This extra degree of freedom afforded by the solvents is responsible for the unique structural polymorphism observed: both "normal" and "reverse" structures can be formed at the same temperature with the same block copolymer in the presence of two solvents selective for each block. We can conjecture from the comparison to block copolymers (Section 2) and to surfactants (Section 4) that the self-assembly of amphiphilic block copolymers in selective solvents is a link between the self-assembly of dry block copolymers and the self-assembly of surfactants in water (and oil/cosurfactant).

5. SUMMARY AND OUTLOOK

A number of significant contributions emerge from our investigation on the phase behavior and microstructure of PEO-PPO-PEO block copolymers in water and oil (where water is a selective solvent for PEO and xylene a selective solvent for the PPO blocks): (i) the remarkable structural polymorphism afforded by these particular class of block copolymers in the presence of selective solvents, (ii) the relevance/connection of the self-assembly behavior of amphiphilic block copolymers in the presence of selective solvents to the self-assembly of solvent-free block copolymers and the self-assembly of surfactants in water/oil, (iii) the good agreement between experiments and mean-field polymer theory in predicting the phase behavior, and (iv) the repercussions of (i) - (iii) on aspects both fundamental (e.g., inter-molecular and inter-assembly interactions, thermodynamics of self-assembly) and applied/practical (e.g., formulation for drugs, design of novel nanoporous materials - recall that PEO-PPO-PEO block copolymers are commercially available). PEO-PPO-PPO block copolymers emerge as model systems, which can help clarify both the thermodynamics and the dynamics of self-assembly phenomena that are analogous to those in a wide range of synthetic and biological systems.

The structural polymorphism exhibited by amphiphilic block copolymers in the presence of selective solvents is remarkable and cannot be attained by, e.g., low-molecular weight surfactants. It can be attributed to the macromolecular nature of the amphiphilic block copolymers which allows for fine-tuning of the molecular packing (and interfacial curvature) by the varying degree of swelling of the hydrophilic and hydrophobic blocks. The location of the different phases in the ternary phase diagram depends primarily on the relative amount of hydrophobic-hydrophilic material. Hence, we can identify the volume fraction of apolar components, f, as the major parameter governing the phase behavior. The extend (particularly at low copolymer concentrations) of the phases depends also on the copolymer effective length (affected by both the block architecture and the molecular weight) which (in the absence of long-range interactions) dictates the range of interaction between the structural elements (micelles, cylinders, lamellae) of the assemblies.

Acknowledgements. Financial support from the Swedish Natural Science Research Council (NFR) is gratefully acknowledged. We thank Drs Ali Khan, Peter Holmqvist, Birgitta Svensson, and Mårten Svensson for productive collaboration and helpful discussions.

REFERENCES

1. D.F. Evans and H. Wennerström, The Colloidal Domain: Where Physics, Chemistry, Biology, and Technology Meet, VCH Publishers, New York, 1994.
2. B. Jönsson, B. Lindman, K. Holmberg and B. Kronberg, Surfactants and Polymers in Aqueous Solutions, J. Wiley & Sons, Chichester, 1998.
3. B. Lindman and H. Wennerström, Top. Curr. Chem., 87 (1980) 1.
4. J. Sjöblom, R. Lindberg and S.E. Friberg, Adv. Colloid Interface Sci., 95 (1996) 125.
5. R.G. Laughlin, The Aqueous Phase Behavior of Surfactants, Academic Press, London, 1994.
6. D.D. Lacic, Liposomes: from Physics to Applications, Elsevier, Amsterdam 1993.
7. Z. Tuzar and P. Kratochvil, Surface and Colloid Science, E. Matjevic (ed.), Plenum Press, New York, 1993; Vol. 15.
8. P. Alexandridis and T.A. Hatton, The Polymeric Materials Encyclopedia, J.C. Salamone (ed.), CRC Press, Boca Raton, FL, 1996; p. 743.
9. F.S. Bates, Science, 251 (1991) 898.
10. D.J. Lohse and N. Hadjichristidis, Curr. Opin. Colloid Interface Sci., 2 (1997) 171.
11. I.R. Schmolka, Surfactant Sci. Ser., 1 (1967) 300.
12. K. Mortensen, W. Brown and E. Jorgensen, Macromolecules, 28 (1995) 1458.
13. G. Wanka, H. Hoffmann and W. Ulbricht, Macromolecules, 27 (1994) 4145.
14. K. Mortensen, J. Phys.: Condens. Matter, 8 (1996) A103.
15. P. Alexandridis, D. Zhou and A. Khan, Langmuir, 12 (1996) 2690.
16. P. Alexandridis, U. Olsson and B. Lindman, Macromolecules, 28 (1995) 7700.
17. P. Alexandridis, U. Olsson and B. Lindman, J. Phys. Chem., 100 (1996) 280.
18. P. Alexandridis, U. Olsson and B. Lindman, Langmuir, 13 (1997) 23.
19. P. Alexandridis, U. Olsson and B. Lindman, Langmuir, 14 (1998) 2627.
20. P. Alexandridis, P. Holmqvist and B. Lindman, Colloids Surf. A, 129 (1997) 3.
21. P. Holmqvist, P. Alexandridis and B. Lindman, Macromolecules, 30 (1997) 6788.
22. P. Holmqvist, P. Alexandridis and B. Lindman, J. Phys. Chem. B, 102 (1998) 1149.
23. B. Svensson, P. Alexandridis and U. Olsson, J. Phys. Chem. B, 102 (1998) 7541.
24. P. Alexandridis, Curr. Opin. Colloid Interface Sci., 2 (1997) 478.
25. G.W. Wu, L.Z. Liu, V.B. Buu, B. Chu, and D.K. Schneider, Physica A, 231 (1996) 73.
26. A. Caragheorgheopol and S. Schlick, Macromolecules, 31 (1998) 7736.
27. F.S. Bates, M.F. Schulz, A.K. Khandpur, S. Förster, J.H. Rosedale, K. Almdal and K. Mortensen, Faraday Discuss., 98 (1994) 7.
28. M.W. Matsen and F.S. Bates, Macromolecules, 29 (1996) 1091.
29. M.W. Matsen and M. Schick, Curr. Opin. Colloid Interface Sci., 1 (1996) 329.
30. P.K. Janert and M. Schick, Macromolecules, 31 (1998) 1109.
31. P.K. Janert and M. Schick, Macromolecules, 30 (1997) 137.
32. P. Alexandridis and T.A. Hatton, Colloids Surf. A, 96 (1995) 1.
33. M. Almgren, W. Brown and S. Hvidt, Colloid Polym. Sci, 273 (1995) 2.
34. B. Chu and Z. Zhou, Surfactant Sci. Ser., 60 (1996) 67.
35. I. R.Schmolka, J. Am. Oil Chem. Soc., 54 (1977) 110.
36. P. Linse, J. Phys. Chem., 97 (1993) 13896.
37. M. Malmsten, P. Linse and K.-W. Zhang, Macromolecules, 26 (1993) 2905.
38. J. Noolandi, A.-C. Shi and P. Linse, Macromolecules, 29 (1996) 5907.
39. M. Svensson, P. Alexandridis and P. Linse, Macromolecules, 32 (1999) 637.
40. M. Svensson, P. Alexandridis and P. Linse, Macromolecules, 32 (1999) 5435.

41. M. Kahlweit and R. Strey, Angew. Chem. Int. Ed. Engl., 24 (1985) 654.
42. U. Olsson and H. Wennerström, Adv. Colloid Interface Sci., 49 (1994) 113.
43. J.D. Mitchell, G.J.T. Tiddy, L. Waring, T. Bostock and M.P. McDonald, J. Chem. Soc., Faraday Trans. 1, 79 (1983) 975.
44. J.N. Israelachvili, Intermolecular and Surface Forces, 2nd ed., Academic Press, 1991.

Small-Angle Scattering Studies of Block Copolymer Micelles, Micellar Mesophases and Networks

Kell Mortensen[a]

[a]Condensed Matter Physics and Chemistry Department,
Risø National Laboratory, DK-4000 Roskilde, Denmark.

Small-angle scattering is a key technique in relation to experimental studies of complex fluids and polymers. The application of small-angle neutron scattering is reviewed in relation to structural studies of block copolymers and their solutions. It is shown that such scattering experiments provide detailed information on both size and form of the polymer as well as their aggregates, and on the thermodynamics characterizing the self-assembling. The scattering data gives the detailed structure of block copolymer micelles as well as information on the inter-micellar interactions. In dense systems, the small-angle technique is used to study thermodynamics of the disorder-to-order phase transitions, and to study details of the ordered phases. It is shown that aqueous systems of triblock copolymers composed of poly(ethylene oxide) and poly(propylene oxide) have not only interests for applications, but constitute a very good model system in which a variety of thermodynamics can be studied. This include for example the unimer-to-micelle self-assembling, micellar sphere-to-rod shape transformation, micellar hard-sphere crystallization into cubic or hexagonal order, and order-to-bicontinuous phase-transitions.

1. INTRODUCTION

Static and dynamic neutron scattering have been key techniques in studying the basic properties and revealing some of the universal structures and scaling laws for polymers and other complex fluids.

An important property of complex polymer materials is the degree of miscibility between various units in the material, i.e. the miscibility between polymers and a solvent, between different polymers, or between various regimes (blocks) within a single polymer chain. This factor plays a main role for the material properties and phase behavior. As a general rule it is found that polymers do not mix. When low-molecular weight liquids normally makes an ideal mixture, this is a result of the dominating entropic contribution to the Gibbs free energy. Even though the intermolecular forces usually are repulsive, the enthalpic contribution is only small (except for special liquids like water). In polymers, on the other hand, the gain in mixing-entropy is very much reduced due to the large molecular dimensions, whereas the enthalpic parts remain like small molecules. Therefore, it is practically only possibly to make ideal blends of few, special polymers.

The absence of miscibility is, on the other hand, the basis for attractive structural properties. When various parts within a single chain are immiscible, or have different solubility in selective solvents, the molecules may self-assembly into well organized structures to minimize direct contact between the immiscible units. This self-association gives rise to a wide range of phase behavior, including the formation of micelles of various form and sizes, complex structured microemulsions, and liquid crystalline phases on the mesoscopic length scale.

The association of amphiphilic macromolecules into complex structures is analogous to that of small surfactants like soaps and lipids. The polymeric nature of amphiphil macromolecules will, however, give rise to new properties which may be the basis for important extensions of the use of amphiphil molecules.

Diblock and triblock copolymers have been extensively studied both in the melt and in solution. The solutions, which is the main topic of here, are in particular interesting when only one of the blocks are soluble. In such case a variety of aggregates may be formed, depending on the molecular architecture. Often, the polymers then self-associate into microscopic well defined aggregates where the non-soluble parts form a core shielded from the solution by the soluble blocks. Such micellar aggregates may have the form of spherical objects, or they may be elongated stif rods or less flexible 'worms', or have the form of flat discs. More complex structures can also be found, as for example networks of micelles or bi-continuous phases.

A variety of block copolymers have been investigated when mixed with a selective solvent. These include organic solvents with copolymers composed of blocks of poly(styrene), PS; poly(isoprene), PI; poly(ethylene), PE; poly(ethylene propylene), PEP; and poly(ethylene butylene), PEB. Most of the polymers which have been studied in relation to aqueous solutions, are based on poly(ethylene oxide), PEO, as the water soluble block. The non- or less-soluble blocks have f.ex. been poly(propylene oxide), PPO [1–4], poly(dimethyl siloxane) (PDMS) [5] poly(butylene oxide), PBO [6], poly(styrene) [7], and poly(isobutylene), PIB [8]. Most experimental as well as applied studies have concerned triblock copolymers with the center block of poly(propylene oxide), PPO, known as Pluronics or Poloxamers. A large number of papers on such PEO-PPO-PEO tri-block copolymers have been published in the last few years, including extended reviews [1–4]. Only few studies have been published on the reversed Pluronics with molecular architecture PPO-PEO-PPO [9]. The PEO-PPO block copolymers constitute an ideal model system to study the properties as a function of amphiphil character. While PEO is water soluble within the 0-100°C-temperature range, PPO is only soluble at low temperatures and for relative low polymer concentrations. It is therefore possibly to control the degree of the amphiphil character of the block-copolymer, just by varying the temperature.

Below we will focus on di- and tri-block copolymers which self-organize into micellar aggregates. We will discuss both AB, ABA and BAB type of chemical architecture, where A is the soluble block. The AB and ABA type of systems to be discussed organize as a general rule in individual micellar aggregates, which interact through a near hard-sphere type of potential. Such micellar systems are visualized in Fig. 1. BAB polymers may also form individual micelles, but this imply that all polymer chains start and end in the same micelle having the middle A-block dispersed into the liquid. More likely, such micelles form interconnected networks, where cores are connected by the soluble A-polymer block, as shown schematically in Fig. 2.

Scattering experiments give important insight into the assembling characteristics. Fig. 3

and Fig. 4 shows the characteristic scattering pattern, $I(q)$, typically obtained from aggregates of respectively AB(A) and BAB type of block copolymers. The peak is indicative of interacting micelles. The peak position depends on both the (local) micellar volume fraction and the interaction distance.

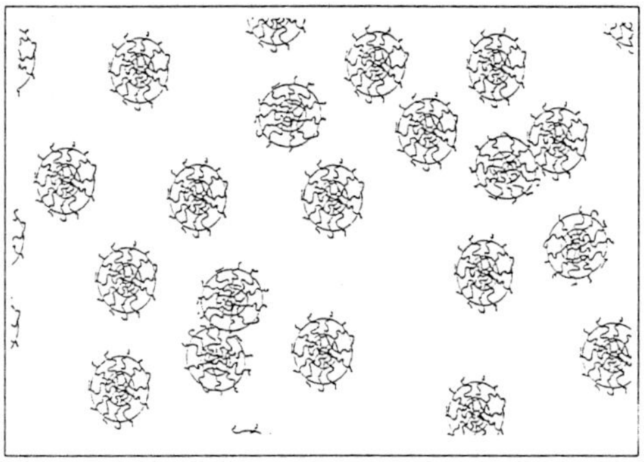

Figure 1. Schematic representation of spherical micelles of AB and/or ABA type of block copolymers in A-solvent, resulting in independent hard-sphere interacting spherical aggregates.

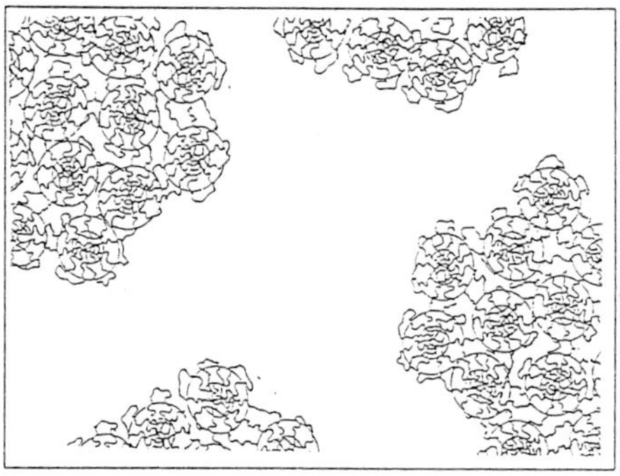

Figure 2. Schematic representation of spherical micelles of BAB type of block copolymers in A-solvent, resulting in domains of interconnected networks of spherical aggregates.

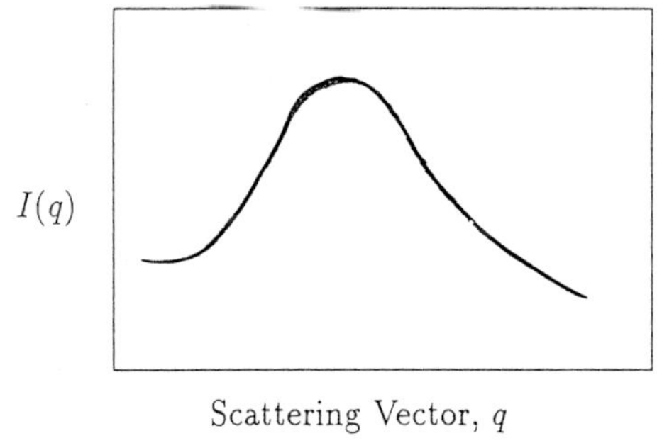

Figure 3. Typical scattering function of individual, hard-sphere interacting micelles.

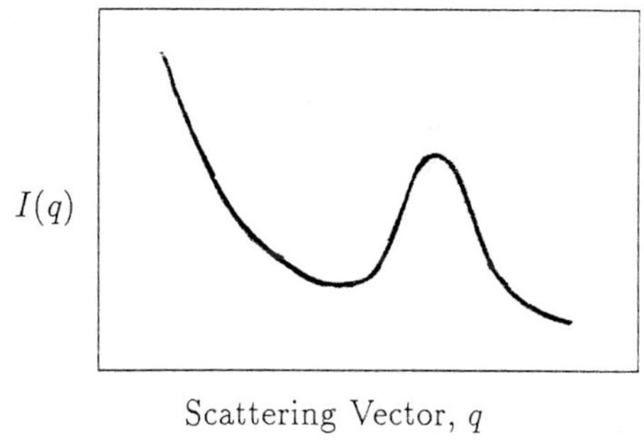

Figure 4. Typical scattering function of networks of connected spherical micelles.

Block copolymer micellar aggregates have wide range of potential and applied uses, as for example viscosity controlling agents, soaps and detergents, lubricant's and units for controlled drug release. An important class of block copolymers are those in which water is a selective solvent.

2. Small Angle Scattering Instrumentation

Small-angle x-ray (SAXS) and neutron scattering (SANS) facilities consists in principle of few, independent units: the radiation source, the monochromating system, the collimator and the detector. In addition, special sample environment are typically an integrated part of the instrument. While most SAXS instruments are laboratory facilities, SANS facilities must due to the need for intense neutron flux always be located at large installations of either thermal research reactors or neutron-spallation sources. The additional possibilities for high resolution and quick data sampling from synchrotron x-ray sources, do, however, make the use of such SAXS-instruments at large installations still more important.

2.1. Small-Angle X-ray Scattering

In instruments using conventional x-ray tubes as the source, the Kratky type of SAXS instrument is the most commonly used [10]. This type of instrument has optimized the flux and resolution using slit geometry for collimation and simple filter for monochromatization. The detector is a simple or a linear sensitive detector. The Kratky-apparatus is often sufficient for studies of solution with not too anisotropic aggregates. The Kratky instrument has also successfully been used in studies of micelles with liquid crystalline order [11], but due to the extreme anisotropic resolution function it may cause significant problems in such samples, with misleading values of scattering momenta. In x-ray sources of high power rotating anodes and synchrotron facilities, pin-hole cameras are usually used. In these systems the collimation consists of two pinhole to define the beam, one near the source, and a second relative near the sample. A third pinhole is usually needed between the defining pinhole and the sample to reduce parasitic scattering and Fraunhofer diffraction from the slits. The monochromatization is obtained using Bragg-scattering from single crystals, and the scattered photon are detected by an area sensitive detector.

2.2. Small-Angle Neutron Scattering

Small-angle neutron scattering instruments are basically always of the pin-hole type, and facilitated with area sensitive detectors. In thermal reactor instruments, the monochromatization is usually obtained using the transmission of a mechanical velocity selector. Few instruments uses the Bragg scattering from a crystal with broad mosaic spread, or super mirrors. Spallation source instruments can also with advance use such monochromatization units, in particular in the case of significant inelastic scattering processes. Usually, however, the full beam is used, and time of flight analysis split the spectra into given neutron wavelengths.

2.3. Auxcilliary in situ devices.

The success of many recent studies utilizing small-angle scattering technique for polymers and liquids, are facilitated by the access to new experimental tools which can be operated *in situ* during the scattering experiment. These include pressure cells [12] and mechanical stretching and shearing devices [13]. Within studies of soft condensed matter the application of shear field has in particular given important new insight into the underlying physics. Hadziioannou et al. [14] showed that highly oriented cylinder and lamellae structures of block copolymer melts could be produced in oscillatory shear flow, whereas Bates and coworkers demonstrated shear oriented single domain cubic phases [15,16]. Ackerson et al. [17] and Lindner et. al. [18], have shown how colloidal crystals can be oriented in the continuous shear flow in a Couette cell. Block copolymer micellar crystals can equivalently be oriented in the Couette cell, as shown by Mortensen et al. [19] and Gast et al. [20].

2.4. Couette shear cell

A simple shear cell used in SANS experiments is the Couette type shown schematically in Fig. 5. Both inner and outer cylinders of the Couette cell can be manufactured of neutron transparent materials like quartz, aluminium or niobium. In the design of Fig. 5 the apparatus is contained in a larger steel cylinder allowing removal of the driving motor and inner Couette cylinder from the top through the open shaft. Conical contact surfaces guarantee that the inner cylinder always is in perfect centricity with the outer cylinder. The outer cylinder is mounted vacuum tight to the bottom of the steel cylinder with a 120° V-shaped inside bottom to prevent air bubbles being caught when the cell is spun. The diameter of the inner cylinder is typically of the order of 30 mm, and the annular gap of 0.5-1 mm. The apparatus is placed in the horizontal neutron beam so that the neutrons pass either radially through the center of the cylinders, or tangential to the cylinder. Couette devices are typically operated with shear gradients in the $\dot{\gamma}=1-3000$ sec^{-1} range.

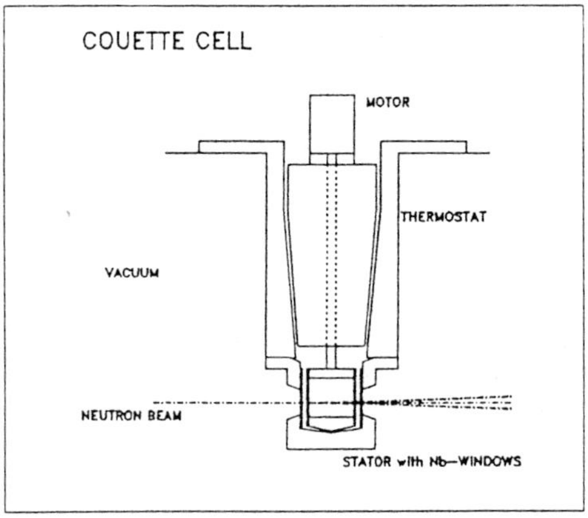

Figure 5. Sketch of Couette cell for SANS.

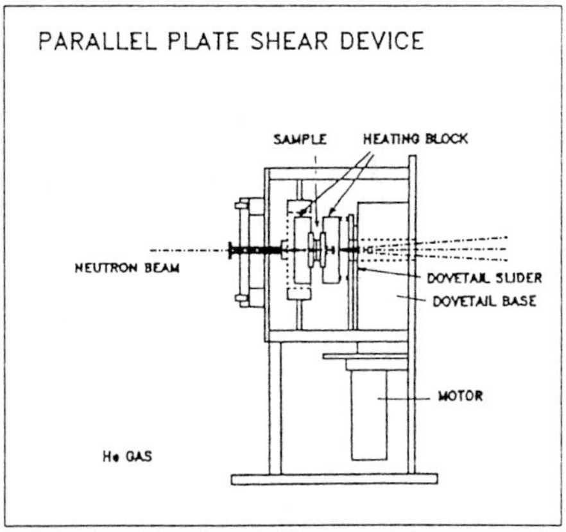

Figure 6. Sketch of parallel plate shear device for SANS.

2.5. Parallel Plate Cell

A parallel-plate device for *in situ* studies of block copolymer melts and more viscous micellar suspensions are shown schematically in Fig. 6. The fist instrument of this type were specially constructed for SANS [13,21], while more recent devises for both neutrons (Risø) and x-ray (Daresbury) facilities have modified a commercial rheology instrument (Rheometrics, RSA2), thus providing the possibility for simultaneous rheological and structural information.

2.6. Pressure Cell

Pressure cells for studies of complex liquids exists for both x-ray and neutrons. The typical range of available pressures are in the 1-2000 bar range.

3. Scattering function from dilute solutions.

The theory of small-angle scattering has been presented in various text books [10,22–24] and will only briefly be reviewed below.

The interaction between matter and radiation is for x-ray scattering dominated by the response of the electrons which is accelerated in the electric field of the beam. To a good approximation each electron scatters the beam as a dipole proportional to the term $b_e = e^2/m_e c^2$, where e is the electronic charge, m_e the electron mass and c the velocity of light. b_e, which has the dimension of length, b_e=0.282·10^{-4}Å, is the 'scattering length'. In the interaction with an atom, one must integrate over all electrons, taking the phases into account:

$$b_z = \int b_e \rho(r) d^3r \tag{1}$$

where $\rho(r)$ is the charge distribution in the atom. The integral, which depends on the scattering angle θ, defines the *atomic form factor.* For small-angle scattering ($\theta \approx 0$), the $\int \rho(r) d^3r$-term approaches the number of electrons, i.e. Z, thus giving the atomic scattering length $b_Z \approx b_e Z$.

The interaction between neutrons and (non-magnetic) matter is dominated by the interaction with the nuclei. The scattering from different isotopes of the same atom may therefore be quite different. This is in particular the case for the two hydrogen isotopes, ^{1}H (abbreviated H) and ^{2}H (abbreviated D for deuterium). Since the neutron is a spin-$\frac{1}{2}$ particle there are two different couplings between the neutrons in the beam and a nucleous, unless the nucleous spin approach null (as in for example ^{16}O and ^{12}C, see Table 1). The average of these two interactions gives coherent scattering, whereas the deviation from average is incoherent and gives a uniform background. Only the coherent scattering term give information on the static structure. In Table 1 is given the scattering lengths of the most relevant atoms for polymer solutions.

The amplitude of x-ray or neutron radiation scattered from a collection of objects, located by vectors $\bar{r}_i$ and characterized by the scattering length b_i is given by

$$f(q) = \sum b_i \cdot \exp[i(\bar{q} \cdot \bar{r}_i)] \tag{2}$$

where $\bar{q}$ is the momentum transfer with length $|\bar{q}| = q = 4\pi/\lambda \sin(\theta/2)$, λ being the radiation wavelength and θ being the scattering angle. The intensity $I(q)$ is proportional to the square of the scattering amplitude

Table 1
Neutron scattering parameters

Nuclei	Coherent Scattering Length	Incoherent Scattering Length
^{1}H	$-0.374 \cdot 10^{-12}$cm	$2.527 \cdot 10^{-12}$cm
^{2}H	$0.667 \cdot 10^{-12}$cm	$0.404 \cdot 10^{-12}$cm
^{16}O	$0.580 \cdot 10^{-12}$cm	$0.0 \cdot 10^{-12}$cm
^{12}C	$0.665 \cdot 10^{-12}$cm	$0.0 \cdot 10^{-12}$cm
^{14}N	$0.937 \cdot 10^{-12}$cm	$0.20 \cdot 10^{-12}$cm
^{28}Si	$0.411 \cdot 10^{-12}$cm	$0.0 \cdot 10^{-12}$cm
^{31}P	$0.513 \cdot 10^{-12}$cm	$0.02 \cdot 10^{-12}$cm
^{35}Cl	$1.165 \cdot 10^{-12}$cm	$0.61 \cdot 10^{-12}$cm

Ref: Neutron News, Vol 3, 1992

$$I(q) = K \cdot |f(q)|^2 \tag{3}$$

where K is given by the geometrical configuration, the intensity of the incident flux, and the sample transmission. Normalizing with respect to these parameters, we get the differential cross section $d\sigma/d\Omega$. From Eq. 2, $d\sigma/d\Omega$ can thus be written

$$\frac{d\sigma(q)}{d\Omega} = \sum\sum b_i b_j \exp[i(\bar{q} \cdot \bar{r}_{ij})] \tag{4}$$

where $\bar{r}_{ij} = \bar{r}_i - \bar{r}_j$. It is useful to replace the scattering length b with the parameter η describing fluctuations from average: $\eta = b - < b >$, and defining the correlation function $\gamma_{ij} = \eta_i \eta_j / < \eta^2 >$. Inserted in Eq. 4, this gives

$$\frac{d\sigma(\bar{q})}{d\Omega} = < \eta^2 > \cdot \sum \gamma_{ij} \exp[i(\bar{q} \cdot \bar{r}_{ij})] \tag{5}$$

In small-angle scattering experiments the resolution is not on atomic distances (~ 1Å). The scattering lengths may therefore be replaced by a continuous function, defined as the scattering length density, ρ, given by $\rho = \sum b_i / V$, where V is an appropriate (molecular) volume. Equivalently, the fluctuation parameter η is redefined as a continuous function: $\eta(\bar{r}) = \rho(\bar{r}) - < \rho >$. The prefactor $< \eta^2 >$ is in a two-component system with scattering length density ρ_1 and ρ_2 easily shown to be equal $(\rho_1 - \rho_2)^2 = \Delta\rho^2$. With the summation in Eq. 5 replaced by an integration, we then have:

$$\frac{d\sigma(\bar{q})}{d\Omega} = \Delta\rho^2 \int_{sample} \gamma(\bar{r}) \exp[i(\bar{q} \cdot \bar{r})] d\bar{r} \tag{6}$$

The integration in Eq. 6 is over the whole sample volume. For a monodisperse ensemble of scattering objects, particles, the integration of the correlation function can be split into the intra-particle self-correlation term and the inter-particle correlation giving

$$\frac{d\sigma(\bar{q})}{d\Omega} = nV^2(\Delta\rho)^2 \cdot \int_{intra} \gamma(\bar{r}) \exp[i(\bar{q} \cdot \bar{r})] d\bar{r} \cdot \int_{inter} (\gamma(\bar{r}) - 1) \exp[i(\bar{q} \cdot \bar{r})] d\bar{r} \tag{7}$$

where n is the number density of particles, and spherical symmetry is assumed. The first integral in Eq. 7 determines the normalized particle form factor, $P(q)$, and the last integration term gives the inter-particle structure factor, $S(q)$. The scattering function can thus be written

$$\frac{d\sigma(\bar{q})}{d\Omega} = n\Delta\rho^2 V^2 P(q) S(q) \tag{8}$$

For very dilute systems, $S(q)=1$ and the particle form factor is determined directly from the scattering function.

For polydisperse systems Pedersen [25] proposed an approximation assuming sub-systems of monodisperse ensembles. The total scattering function is then the sum of these systems weighted by a size distribution function $D(R)$

$$\frac{d\sigma(\bar{q})}{d\Omega} = \Delta\rho^2 \int D(R) V^2(R) P(q) S(q) dR \tag{9}$$

Kotlarchyk and Chen [26] showed that it is possibly formally to write the scattering function of polydisperse systems as Eq. 8, but with the form factor replaced by the weighted mean value:

$$V^2 \tilde{P}(q) = < V^2 P(q) >$$

and the structure factor replaced with

$$\tilde{S}(q) - 1 = \beta(q)[S(q) - 1],$$

where $\beta(q)$ is a correction term determined by the degree of polydispersity. The scattering function of non spherical particles can be treated equivalently, using Eq. 8.

3.1. Contrast Variation

Small-angle neutron scattering experiments are particularly useful for detailed studies of the local structures of block copolymer aggregates. This is done by systematic studies of systems in which the contrast is varied by exchanging hydrogens with deuterium in specific places. This is most easily done simply by changing the composition of the solvent in terms of normal and deuterated material. Alternatively, specific deuterium labeling is chemically build into the block copolymer molecule during the synthesis. While the effect of changing the contrast of a pure two-phase systems is only to change the prefactor in front of the scattering function, more complex structures with varying scattering length density gives significant different form of the scattering function. This is shown in the example given in Fig. 7, showing PEO-PPO-PEO micelles in mixtures of water and oil, with different contrasts.

4. Model Functions

In the following we will assume that the block copolymer aggregates are monodisperse and randomly oriented, so Eq. 8 is valid. We will review some of the most important analytical expressions for the scattering function of typical block copolymer aggregates, eg. micelles with spherical, elliptical or rod-like shapes. Most of the expressions can be found in the book of Guinier and Fournet [22], or in the extended list of known scattering cross sections recently summarized by Pedersen [28].

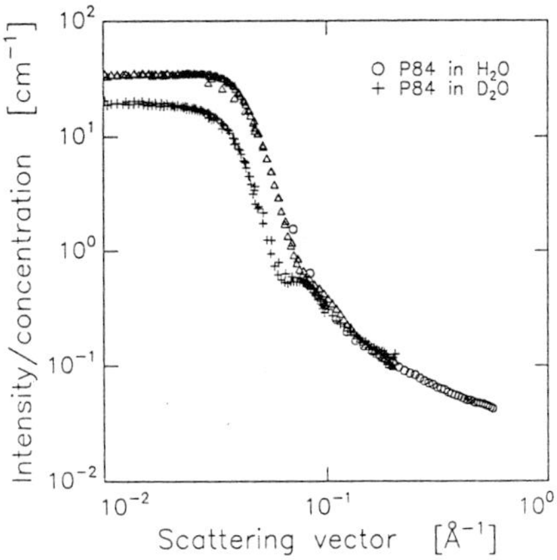

Figure 7. Scattering function of PEO-PPO-PEO block copolymer aggregate in mixture of water and oil, with two different contrasts [27].

4.1. Block Copolymers Dissolved as Gaussian Coils

If the solvent is a theta solvent for both blocks, the copolymers will be dissolved as individual random coils. The scattering function is then determined by the Debye function, F_c [22]:

$$F_c(q) \sim x^{-2}(\exp(-x) + x - 1) \tag{10}$$

where $x=(qR_g)^2$, R_g being the polymer radius of gyration.

4.2. Scattering from Spherical Micelles

The form factor, F_s of a homogeneous spherical particle with radius R_c has the analytical form [22]

$$F_s(q) = [\Phi_s(q)]^2 = [\frac{3}{(qR_c)^3}(\sin(qR_c) - qR_c\cos(qR_c))]^2 \tag{11}$$

which approach the $F_s(q) \propto q^{-4}$ Porod law at large q-values.

The form factor Eq. 11 represents a good approximation to block copolymer micelles if they form a dense spherical aggregate, or if the contrast of the micellar corona has been matched to the solvent. The spectrum shown in Fig. 7 represents the latter situation.

Typically, however, the micelles are of the 'star-type' (Fig. 1) with relative dense cores of the non-soluble blocks and a corona of the soluble part extended into the solvent. At large q-values such block copolymer micelles rather shows a form factor which approach a q^{-2} behavior at large q-values.

4.2.1. Concentric Spherical Shells

A possibly solution for better description of such spherical symmetric micelles is to use the form factor of concentric shells.

The form factor of a single shell with inner and outer radius R_1 and R_2, respectively, is obtained from the form factor of dense sphere, $P = P_{sphere}$ (Eq. 11) by subtracting the empty core with the proper volume weighting [22]:

$$P(q) = \frac{V(R_1)P_{sphere}(q, R_1) - V(R_2)P_{sphere}(q, R_2)}{V(R_1) - V(R_2)} \tag{12}$$

This form factor can be generalized to a form factor of N concentric shells with radii R_i and ρ_i being their respective scattering length densities:

$$P(q) = \frac{\rho_1 V(R_1)P_{sphere}(q, R_1) + \sum(\rho_i - \rho_{i-1}V(R_i)P_{sphere}(q, R_i)}{\rho_1 V(R_1) + \sum V(R_i)(\rho_i - \rho_{i-1}} \tag{13}$$

4.2.2. Micellar Form Factor

A more realistic form factor of spherical block-copolymer micelles including the dense core and corona of Gaussian chains can be approached by an analytical expression [29]. This form factor includes both the self-correlation of the spherical core, self-correlation of the chains in the corona, the cross term between the core and chains and cross term between different chains. The self-correlation term of the sphere (F_s and Φ_s) and chains (F_c) are given by the formula Eq. 11 and Eq. 10, respectively. The cross terms are calculated using the Debye equation, using infinitely thin shells and taken into account the correct weighting functions for, respectively, solid sphere and Gaussian chains. The interference term between core and chains thereby get the form:

$$S_{sc}(q) = \Phi_s(q, R_c)x^{-1}(1 - e^{-x})\sin(qR_c)/(qR_c) \tag{14}$$

and the term between chains attached to the surface:

$$S_{cc}(q) = x^{-2}(1 - e^{-x})^2 \, [sin(qR_c)/(qR_c)]^2 \tag{15}$$

where x=$(R_g q)^2$, as before. The resulting form factor of a block-copolymer micelle with aggregation number N_{agg} can then be expressed as [29]

$$F_{mic}(q) = N_{agg}^2 \rho_s^2 F_s(q, R) + N_{agg}\rho_c^2 F_c(q, L, b) + N_{agg}(N_{agg} - 1)\rho_c^2 S_{cc}(q) + 2N_{agg}\rho_s\rho_c S_{sc}(q) \tag{16}$$

where ρ_s and ρ_c are the excess scattering length densities of blocks in the core and in the chains of the corona, respectively.

4.3. Ellipsoidal Micelle

The form factor of a dense ellipsoid with rotation symmetry is not analytically, but must be calculated numerically. Averaging over orientations gives the form factor for an ellipsoid with axes R, R, ϵR [22]

$$P(q) = \int_0^{\pi/2} [\frac{3(\sin(qR) - qR\cos(qR)}{(qR)^3}] \, \sin\alpha \, d\alpha \tag{17}$$

The form factor for concentric ellipsoidal shells can be calculated in analogy to that of concentric spheres.

4.4. Rod-Like Micelles

The form factor expression for rod-like micelles involves an integration over the first order Bessel function $B_1(x)$. The form factor of rods with radius R and length L has the form [22]

$$P(q) = \int_0^{\pi/2} [\frac{2B_1(qR\sin\alpha)}{qR\sin\alpha} \cdot \frac{\sin(qL\cos\alpha)/2}{qL\cos\alpha/2)}]^2 \cdot \sin\alpha \, d\alpha \qquad (18)$$

4.5. Inter-Micellar Correlations

The scattering function of micellar solutions, Eq. 8, becomes increasingly dominated by the structure factor $S(q)$ as the micellar concentration is increased.

The structure factor can often successfully be determined using the Ornstein-Zernike and Percus-Yevick approximations [30] using a hard-sphere interaction potential, but other approximations are also available. $S(q)$ is in the hard-sphere Percus-Yevick approximation given by the micellar volume fraction ϕ and the hard-sphere interaction distance R_{hs} [31–33]:

$$S(q) = \frac{1}{1 + 24\phi G(qR_{hs}, \phi)/(2qR_{hs})} \qquad (19)$$

where G is a trigonometric function of $y{=}qR_{hs}$ and ϕ

$$\begin{aligned} G(y,\phi) &= \alpha(\phi)/4y^2 \cdot [\sin(2y) - 2y\ \cos(2y)] + \beta(\phi)/8y^3 \cdot [4y\ \sin(2y) + (2-4y^2)\cos(2y) - \\ &\quad + \gamma(\phi)/32y^5 \cdot [-16y^4\ \cos(2y) + 4[(8y^2-6)\cos(2y) + (8y^3-12y)\sin(2y) + 6]] \end{aligned}$$

and α, β, and γ are given by the hard sphere volume fraction ϕ:

$$\alpha = (1+2\phi)^2/(1-\phi)^4$$
$$\beta = -6\phi \cdot (1+\phi/2)^2/(1-\phi)^4$$
$$\gamma = \phi/2 \cdot (1+2\phi)^2/(1-\phi)^4$$

The peak-position q_{max} of $I(q)$ of hard-sphere systems are frequently interpreted simple as given by the mean micellar distance: $2\pi/q_{max} = D = [\frac{4\pi}{3}R_{hs}^3/\phi]^{1/3}$, i.e. given by the micellar number density only. This is not true, the position of q_{max} is determined by a complex function of R_{hs} and ϕ.

5. Examples:

Aqueous solutions of block copolymers of PEO and PPO constitutes an ideal class of systems in which self assembling and correlations can be studied in great details. The solubility of PPO in water changes markedly by relative small changes in thermodynamic parameters like temperature and pressure. It is therefore possibly in details to study phenomena like the micelle-formation as a function of the amphiphilic character of the molecules continuously by for example changing the temperature. Likewise, it is possibly to study form-transformations of the aggregates, as well as correlations between aggregates and possibly crystalline mesophases. Below, we will show examples of the use of small angle scattering technique for studies of micellar systems. The examples will mainly be taken from studies of Pluronics, but some other systems are mentioned as well.

5.1. Unimers

The miscibility of polymers in solvents may vary substantially with polymer concentration and thermodynamic parameters like temperature and pressure. This is for example the situation for poly(propylene oxide), PPO, in water. PPO is soluble in water only at low temperatures and relative low polymer concentrations. Block copolymers of water soluble poly(ethylene oxide), PEO, and PPO therefore appear as independent polymer chains, called *unimers*. Within statistical error, the scattering function of these unimers is in agreement with the polymers obeying Gaussian conformation. This is shown by the experimental scattering data of 5% $EO_{99}PO_{65}EO_{99}$ (F127) obtained at T=5°C, given in Fig. 8. The solid line represents the best fit to the Debye function of Gaussian chains, Eq. 10, including instrumental smearing and a constant I_{ic} which represents incoherent background from the sample. The resulting radius of gyration for $EO_{99}PO_{65}EO_{99}$ is R_g=22Å. Equivalent good agreements to the Debye function have within statistical error been observed in other Pluronics [33–35].

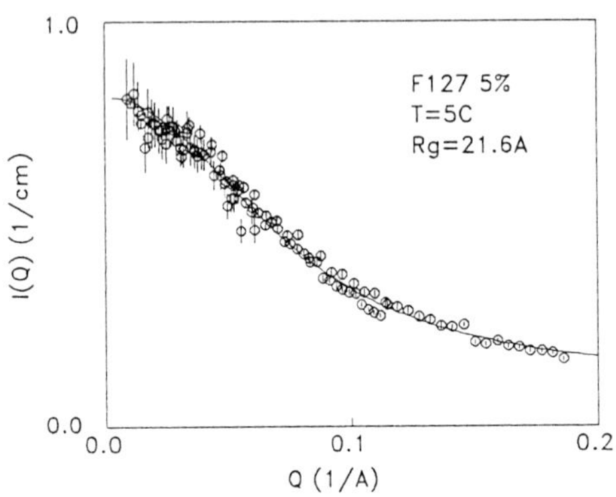

Figure 8. Scattering function of an aqueous solution of $EO_{99}PO_{65}EO_{99}$ unimers (5% solution at 5°C). The solid line represents the best fit to the Debye-function.

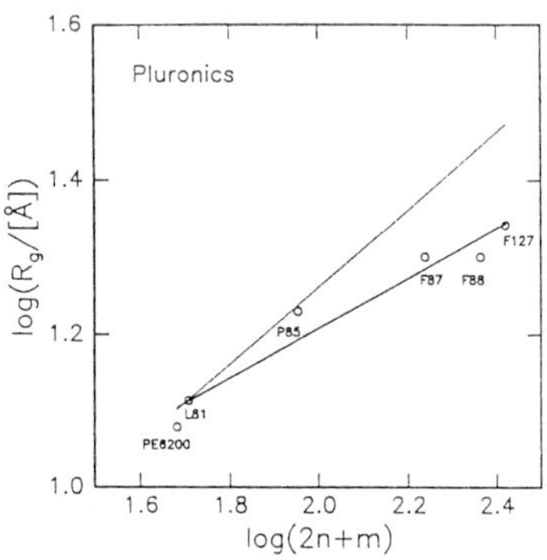

Figure 9. The radius of gyration R_g as obtained from SANS data of a variety of PEO-PPO-PEO unimers in aqueous solutions. The dotted line represents the expected R_g-values expected for Gaussian conformation with monomer size 2Å and Kuhn segment length 10Å.

Fig. 9 shows the experimental R_g values of different Pluronics, as plotted against the total number of monomers ($2m+n$). The experimental R_g values appear rather small if the copolymers really are dissolved as Gaussian chains. It has been argued that acqueous solution of PEO-chains with more than 10 units are predominantly in the helical meander configuration [36,37] with monomer length l of the order of 2Å. The Kuhn segment length b is of the order of 10Å [38,29]. If we assume that the PPO chain have the same characteristics as the PEO chain (which is a lower limit) this leads to a radius of gyration

$R_g = \sqrt{Nlb/6} = 30$Å for $EO_{99}PO_{65}EO_{99}$. The broken line in Fig. 4 represents this calculation as a function of $N = 2m + n$. The reduced experimental value of R_g may indicate that the less soluble PPO-block is significant more compact than Gaussian chain. The unimer thus resemble that of a uni-molecular micelle.

5.2. Spherical Micelles

The poly(propylene oxide) block becomes more hydrophobic upon increasing temperature. For a wide range of Pluronics, this leads to a well defined temperature upon which spherical micelles are formed. These micelles are composed of a core dominated by propylene oxide blocks, and surrounded by a corona of hydrated ethylene oxide subchains. Model calculation have been able to describe adequately the micellation process and proved that the PPO blocks are responsible for the entropy-driven micellation process [39,40].

The scattering function of $EO_{99}PO_{65}EO_{99}$-micelles in a 1%, 2% and 5% polymer solutions are shown in Fig. 10 [41]. The data have been normalized to polymer concentration. The scattering functions clearly show a side maximum as expected from the formfactor of dense spherical objects with sharp interface

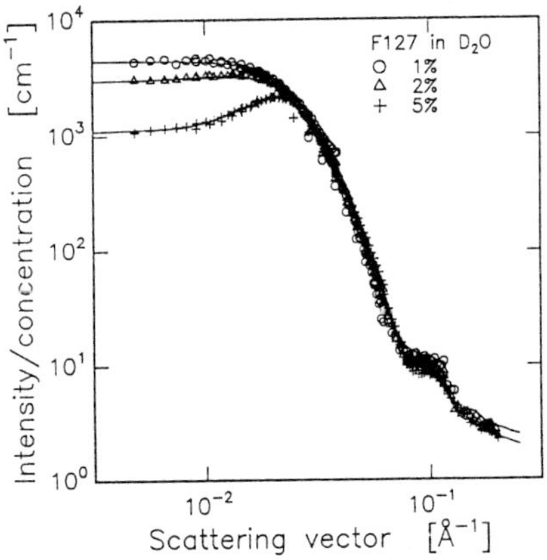

Figure 10. Experimental scattering function of $EO_{99}PO_{65}EO_{99}$-micelles in a 1%, 2% and 5% aqueous solutions. The data have been normalized to polymer concentration. The solid line represents best fits to the micellar scattering function, Eq. 16, including hard-sphere inter-micellar correlations, Eq. 19.

It appear from Fig. 10 that the position of the side maximum is effectively unaffected by the polymer concentrations, showing that the micellar size, and thereby the aggregation number, is rather independent of the polymer concentration. At low q-values, on the other hand, a significant reduction in scattering intensity is seen resulting from inter-micellar correlations of higher concentrations. The temperature dependence (not shown) of the side-maxima is also weak, but significant, reflecting an increase in micellar size upon raising temperatures.

The significance of the side maximum is different for different Pluronics. For low molecular weight systems, the side maximum is significantly reduced, if visible at all. The side-maximum observed in neutron scattering of Pluronic micelles is generally weak due to the rather gradual change in scattering length density at the interface between the micellar core and corona. Since there is quite some difference in electron density from the PPO-melt core to the hydrated PEO corona, x-ray scattering has a larger change in scattering length density. x-ray scattering shows accordingly significantly more pronounced side maxima[42]. The factor which reduces the side-maxima in both x-ray and neutron scattering experiments, is possibly dispersity in size or shape. It thus seems that the high-molecular weight Pluronic micelles have significantly more well characterized spherical shape and size than low molecular systems.

Excellent fits are obtained to the experimental scattering function as represented by the representative $EO_{99}PO_{65}EO_{99}$ micellar data shown in Fig. 10. The solid lines are best fit using Eq. 16, and including instrumental smearing.

In the attempt to reduce the number of fitting parameters when analyzing extended number of scattering data, one may simple use the formfactor of dense spheres (Eq. 11) rather than the more complex (Eq. 16). Good fits are then obtained to the experimental scattering function for q-values up to approximately 2-3 times the peak-position q_o. The scattering function is thereby expressed in an analytical form determined by only three parameters: the hard-sphere volume fraction ϕ, and the two radii characterizing the micelles: the core-radius R_c and the hard-sphere interaction radius R_{hs}. In extended data sets one can then obtain detailed information on the temperature and concentration dependence of the micellar structure.

It is important to note, that in spite of the dispersed polymer corona a near-hard sphere interaction potential is expected as a result of entropic repulsion.

5.2.1. Micellar Size and Aggregation Number

As already concluded based on the position of the side maximum observed in the formfactor, the Pluronic micellar core-radius, R_c, appear to be basically independent of polymer concentration (Fig. 10). Likewise the hard-sphere interaction radius is relative independent of concentration, showing that the thermodynamics controlling the copolymer aggregation first of all is determined by temperature.

An increase in both R_c and R_{hs} of the PEO-PPO-PEO micelles upon increasing temperatures reflects changes in aggregation number. The core radius of $EO_{25}PO_{40}EO_{25}$ micelles, as an example, changes from approximately 40Å at 20°C to 50Å at 50°C. Simultaneously, the hard-sphere interaction radius changes from 50 to 70Å. The changes in the micellar sizes with temperature show that the micelles are dynamic aggregates in which the individual copolymer chains constantly move from one micelle to another. This allows optimizing the aggregation number according to thermodynamic parameters.

The change in micellar radius for different Pluronics shows very similar characteristics. Fig. 11 shows in a double logarithmic plot R_c versus reduced temperature for the series of PEO-PPO-PEO micelles with similar size PPO block: $EO_{25}PO_{40}EO_{25}$, $EO_{65}EO_{39}EO_{65}$ and $EO_{96}PO_{39}EO_{96}$ for 9% and 20% copolymer concentrations. It appears that, when R_c is plotted against the reduced temperature T-T_{cm1}, the data follow a common master curve with the empiric scaling relation:

$$R_c \propto (T - T_{cm1})^{0.2} \tag{20}$$

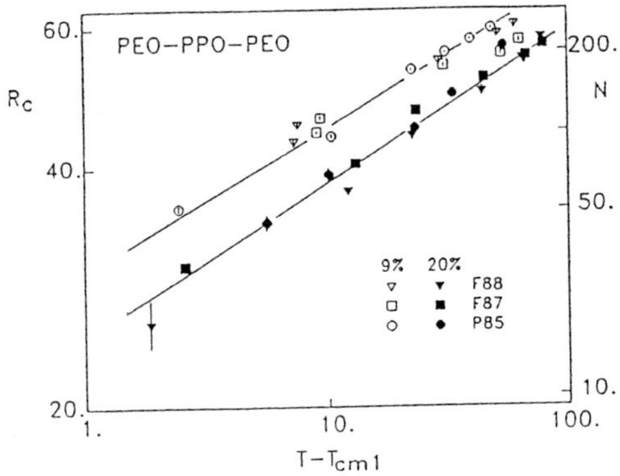

Figure 11. The micellar core-radius of various PEO-PPO-PEO block copolymer micelles plotted versus the relative temperature $T - T_{cm1}$, T_{cm1} being the critical micellation temperature [34]. The scale on the left axis gives the aggregation number N.

The aggregation number N_{agg} can be calculated from the the core dimension. If we assume that the core consist only of propylene oxide, we have

$$4\pi R_c^3/3 = N_{agg} \cdot n \cdot V_{PO} \tag{21}$$

where n=40 is the poly(propylene oxide) degree of polymerization. $V_{PO}=M_{PO}/(\rho_{PO}N_A)$ is the dry propylene oxide volume. With the molar weight M_{PO}=58 and the mass density ρ_{PO}=1.01g/cm^3 we get V_{PO}=95.4Å^3 This leads to an aggregation number of $N_{agg} \approx 58$ at T=20°C and $N_{agg} \approx 116$ at T=40°C.

5.2.2. Critical Micellar Values and Micellar Volume Fraction

One of the important parameters obtained from fits to the experimental scattering function is the micellar volume fraction, ϕ.

In Fig. 13 and Fig. 12 are shown the micellar volume fraction of a 28% aquous solution of $EO_{96}PO_{39}EO_{96}$ (F88) as a function of respectively temperature and pressure. One sec that at low temperatures, the micellar volume fraction is zero corresponding to that all copolymers is dissolved as independent unimers.

Above a well defined critical micellation temperature, T_{cm1}, there appear a broad range where micelles and unimers coexists thermodynamically. The micellar volume fraction increases roughly linearly with temperature and reaches an approximately constant level above a second characteristic temperature, T_{cm2}. Above T_{cm2} 'all' copolymers have aggregated in micelles. The broad range with linear increase in micelle density is rather different from low-molecular surfactants, which at the critical micellar temperature typically shows a step-like function in micelle-concentration. The origin of the broad range where significant concentrations of both micelles and unimers coexist may be a result of the polymeric nature, including polydispersity.

The application of hydrostatic pressure, Fig. 12 has an effect corresponding to lowering the temperature, i.e. PPO get more hydrophilic upon application of pressure.

For polymer concentrations below approximately 20%, the limiting micellar volume fraction, ϕ_o varies linearly with concentration [54,33]. This is in agreement with the

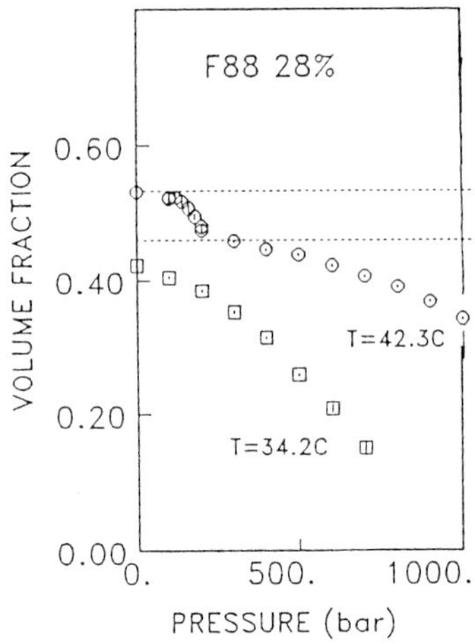

Figure 12. The micellar volume fraction of an aqueous solution of 28% $EO_{96}PO_{39}EO_{96}$ block copolymer, as obtained versus pressure [12].

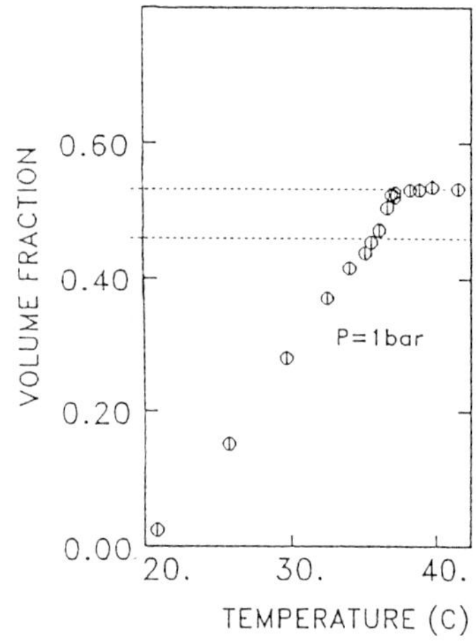

Figure 13. The micellar volume fraction of an aqueous solution of 28% $EO_{96}PO_{39}EO_{96}$ block copolymer, as obtained versus temperature [19].

postulate that all polymers have aggregated in micelles above T_{cm2}. Above $c \sim 20\%$, on the other hand, the micelle volume fraction reach a limit of the order of $\phi_c \sim 0.53$. In this regime there therefore remain some unimers present. The ϕ=0.53 value is the critical value for hard-sphere crystal formation [43,19], and will be discuss further below.

The aggregation number, which was discussed above on the basis of the size of the micellar core, can be independently calculated on the basis of the micellar volume fraction. In the regime where ϕ increases linearly with temperature, the system is thermodynamic stable with both micelles and polymers present. The number of micelles per volume, ϕ/V_{hs}, ($V_{hs}=4\pi/3R_{hs}^3$ being the hard-sphere micelle volume) and the number of free chains per volume, n_p, is related through

$$n_p + N_{agg} \cdot \phi/V_{hs} = c/V_p \tag{22}$$

where c/V_p represents the total concentration of polymer chains, V_p being the dry volume of the polymer. Above T_{cm2} there are basically no free polymer chains, i.e. n_p=0. This leads to the aggregation number:

$$N_{agg} = c/\phi_o \cdot 4\pi R_{hs}^3/3V_p \tag{23}$$

where $\phi_o=\phi_o(c)$ is the saturating value, linearly dependent on polymer-concentration c. The dry polymer volume is given as the sum of the volume of monomers, $V_p=2mV_{EO}+nV_{PO}$, where $V_i=M_i/(\rho_i N_A)$, i=PO or EO. $V_{PO}=95.4\text{Å}^3$ as given above, and with M_{EO}=44 and

ρ_{EO}=1.01 g/cm^3 the ethylene oxide volume is V_{EO}=73.1Å^3. Using the fitted value for R_{hs} we then get an aggregation number of $N_{agg}\approx 37$ at 20°C, increasing to 78 at 40°C for the $EO_{25}PO_{40}EO_{25}$ copolymer micelles. These values are slightly smaller than those obtained from the core-dimension. If we, on the other hand, use a more realistic model where the core dimensions are given by a dense propylene oxide core covered by a dense monolayer shell of ethylene oxide we get, within the whole temperature-regime, perfect agreement between the two routes of calculating the aggregation number. The effective thickness of the dense EO-shell is then 2.8Å.

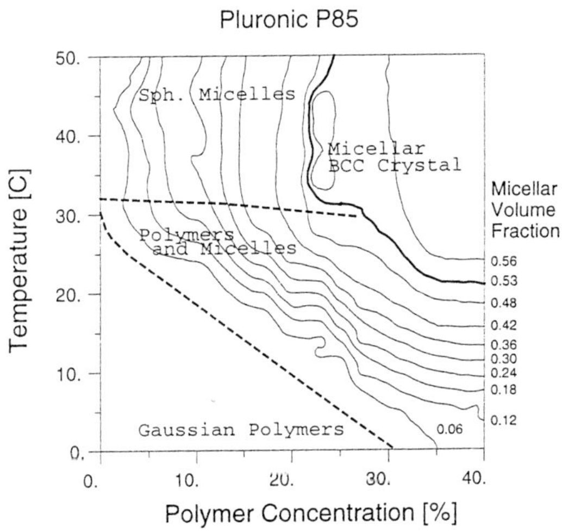

Figure 14. Contour plot of the micellar volume fraction of the $EO_{25}PO_{40}EO_{25}$ system as a function of polymer concentration and temperature. The broken lines separates pure unimer phase (I), pure spherical micellar phase (III) and coexisting micellar and unimer phase (II). The solid line represents ϕ=0.53 and separates liquid and para-crystalline phase (IV).

In Fig. 14 is shown in a contour plot the micellar volume fraction as resulting from neutron scattering data of $EO_{25}PO_{40}EO_{25}$ obtained for polymer concentrations in the range 0-40% and at temperatures up to 50°C, i.e. the range of spherical micelles. The variation in ϕ separates into 4 regimes as indicated in the figure.

At low temperatures and concentrations (regime I in the figure), all polymers are dissolved as the unimers discussed above. Above a line of critical micellation temperatures (T_{cm1}) and concentrations (cmc_1), regime II, micelles and unimers coexists. The solution is totally dominated by micelles in regime III. The line separating phase II and III is well defined and reproducible characterized by the relative concentration independent temperature T_{cm2}.

For polymer concentrations below approximately 20%, the limiting micellar volume fraction, ϕ_o, varies linearly with concentration (phase-III), as discussed above. Above c ~20%, on the other hand, the micellar volume fraction reach a limit of the order of ϕ_c ~0.53 (phase-IV). In this regime the micelles form an ordered structure, as discussed below.

For constant polymer concentration below 20%, Fig. 14 shows a minor decrease in micellar volume fraction in the high temperatures phase-III. This reflects the increasing aggregation number, resulting in larger micelles which occupy less volume.

5.3. Rod-like Micelles

At high temperatures, the scattering pattern of many PEO-PPO-PEO copolymer micelles changes in character. For low concentrations of $EO_{25}PO_{40}EO_{25}$, this appears at approximately 70°C. Above $T \approx 70°$ the low angle scattering intensity increases and in the low concentration regime the characteristic correlation peak totally disappears.

Fig. 15 shows the scattering function of 1% $EO_{25}PO_{40}EO_{25}$ obtained at 80°C. The solid line is a fit to cylinder with axes 50Å and length 500Å. There are systematic differences which can be associated with contribution to the scattering from the PEO-chains and possibly polydispersity or deviation from rod-like (worm-like) form.

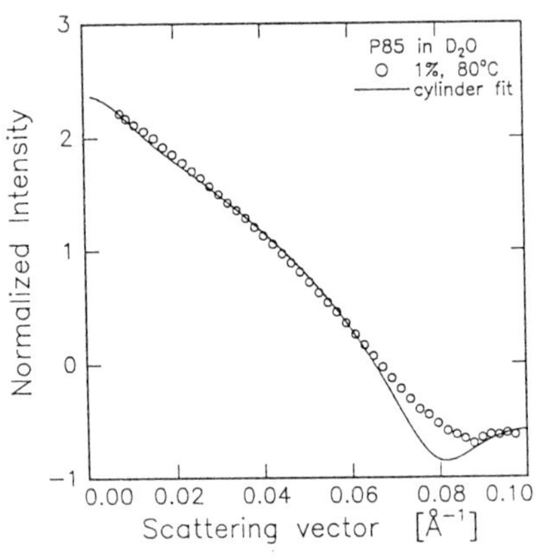

Figure 15. The scattering pattern of $EO_{25}PO_{40}EO_{25}$ in the rod phase. The solid line represents scattering from a cylinder with radius 50Å and length 500Å.

The low-temperature spherical micelles have apparently within a small temperature regime transformed into rod- or worm-like micelles. Based on Fourier transform of the scattering function it was anticipated [33] that the micelles within a narrow temperature range continuously change from spherical via prolate ellipsoids to worm- or rod-like micelles. The scattering data would, however, also be consistent with coexistence of spherical and worm-like micelles within the given temperature regime. Such behavior has recently been observed in a study combining neutron scattering and cryo-electron microscopy measurements on the closely related aqueous system of PEO-PIB-PEO, where PIB is poly(isobutylene) [44].

In block copolymer micelles of higher PEO-content, the phase lines (unimer-to-micelle and spherical-to-rod) are shifted. In $EO_{99}PO_{65}EO_{99}$ and $EO_{96}PO_{39}EO_{96}$, f.ex., the sphere-to-rod transition for low polymer concentrations only appear at $T \approx 95°C$ [41,34], whereas the $EO_{96}PO_{39}EO_{96}$ appear at approximately 90°. On increasing polymer concentration, the sphere-to-rod transition temperature is reduced.

The origin of the sphere-to-rod transition is attributed to the size of the spherical aggregates. As seen from the plot Fig. 11 the core-radius of the spherical micelles increases markedly with increasing temperature. Close to T=70°C for low concentration $EO_{25}PO_{40}EO_{25}$, the core-radius approach 50Å giving on average only 2.5Å per PO-monomers for those chains crossing the micelle-center. This can be accommodated in two ways. Either the PPO-blocks are highly stretched which entropically is very costly, or there is a large mixing of EO and PO inside the core with a resulting increase in chemical potential [33]. Linse has in a lattice model calculated the chemical potential [45] and thereby theoretically verified that the rod-like structure due to these properties are favored at elevated temperatures.

5.4. Disc-like Micelles

At high polymer concentrations, many aqueous systems of PEO-PPO-PEO block copolymers form assemblies of disc like or lamellae structure. Individual micelles of disc like form have been proposed in block copolymers of low PEO-content, eg. $EO_6PO_{36}EO_6$ [35]. The reason for disc like shape is in these system obviously resulting from the optimum way to shield the large core of non-soluble PPO-blocks.

A quite different reason for disc-like micelles can be found in for example micelles of polyethylene, PE, in which the form is governed by polymer crystallization. Diblock copolymers of PE and poly(ethylene propylene), PEP, form in decane large disc-like micelles, in which the PE-blocks have crystallized into extended lamellar sheets [46]. The sheets are surrounded on both sides by PEP brushes.

6. Micellar Liquid Crystalline Phase

In the contour plot of the micellar volume fraction, ϕ, shown in Fig. 14, we see that ϕ saturates at a limiting value of the order of $\phi_c \approx 0.53$. When this ϕ_c boarder-line is crossed the micellar liquid undergoes a first order phase transition to a cubic crystal [19] which constitute a transparent, paste-like material with elastic shear modulus of the order of 10^4-10^5Pa [47,48]. The liquid and crystalline domains coexist in the range between ϕ_c=0.47 and ϕ_c=0.53 [12] (see also Fig. 13 and Fig. 12). These characteristics are in agreement with simple hard-sphere crystallization [43].

The polycrystalline paste-like powder-sample is easily aligned into one macroscopic monodomain crystal upon application of simple continuous or oscillatory shear. In spite of the fact that the individual micellar aggregates are dynamic in the sense that copolymers on a short time scale move from one micelle to another, the mono domain crystalline phase remains for more than weeks after the shear has been stopped.

While the shear thus have dramatic influence on the texture, it does not have marked effects on the phase behavior of the PEO-PPO-PEO systems. There has been no observations of any shear dependence in the cubic phase for shear rates up to 6000 sec^{-1}. This is different from observations on crystalline ordered PS-PI diblock copolymer micelles in decane [20]. In the BCC phase of these PS-PI micelles, Gast *et al.* observe at moderate shear rates a continuous deformation of the lattice into a twin structure with sliding crystalline layers. At higher shear rates the crystalline long range order is lost, and an amorphous or liquid-like structure develops. In micellar networks of triblock copolymers of PS-PEB-PS, the order-disorder phase transition temperature changes significantly by shear rate [49], as has been seen in pure melts of both di-block [50] and tri-block copolymer melts [51].

6.1. Single Crystal Crystallography

The poly-crystalline phase of PEO-PPO-PEO micelles can effectively be oriented in both the static shear of a Couette cell and in the oscillatory shear of plan-parallel devices, as shown in Fig. 5 and Fig. 6, respectively. In fact, the sample is so easy to shear align that mounting the material between two parallel plates (eg. quartz or aluminum) and by hand just move one plate relative to the other, results in a well oriented single phase crystal with mosaicity typically of the order of 10°. Such shear-oriented crystals are free to be rotated, and are accordingly suitable for crystallographic studies in the neutron beam, and thereby index the observed Bragg-reflections. Fig. 16 shows an example of such studies performed on $EO_{96}PO_{39}EO_{96}$-micelles. The figure shows two two-dimensional scattering pattern, obtained with the beam along the shear gradient and when the sample is rotated 35° around the primary [110]-reflection, respectively. The scattering conclude body centered cubic lattice, bcc, as indicated by the associated Miller indexes.

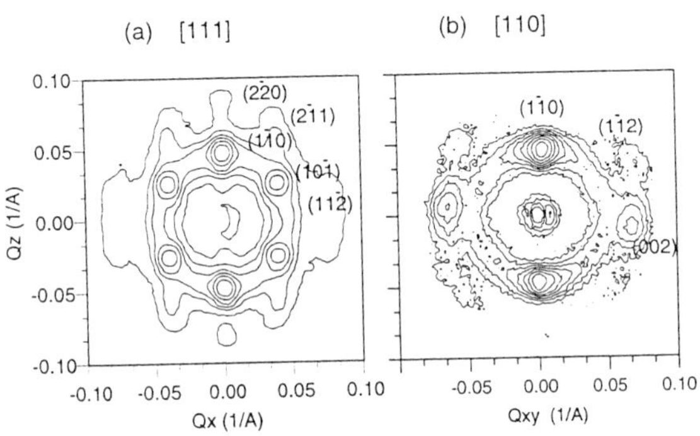

Figure 16. Two-dimensional scattering pattern of $EO_{96}PO_{39}EO_{96}$-micelles in the cubic bcc phase. The scattering pattern are obtained with the shear axis horizontal and with shear gradient parallel to the beam (left) and the with the sample rotated by 35° around the vertical axis (right), respectively.

Most classical hard-sphere systems crystallize on face-centered cubic, fcc, lattice, rather than bcc. The reason for bcc-phase in many micellar system is related to the influence of the polymer-chains dispersed into the solvent, given rise to a softening of the interaction potential. Only in the limit of a thin corona, the micelles are true hard-sphere systems and should form a fcc solid. As the corona and thereby the length scale of the repulsion increases, the micellar system favors bcc symmetry. This has been nicely demonstrated experimentally in PS-PI diblock copolymer micelles [52].

Experiments on the cubic mesophase of PEO-PPO-PEO micelles have always revealed the shear gradient parallel to [111] and with the flow direction parallel to [112]. [110]-types of Bragg reflections dominates therefore the scattering pattern, as shown in the examples given in Fig. 17. In the two-phase regime with $0.47<\phi<0.53$, the orientation appears, however, to depend on both history and applied shear field. While the [111]-axis is still parallel to the shear gradient, both [112] and [110]-axes are found parallel to the flow direction.

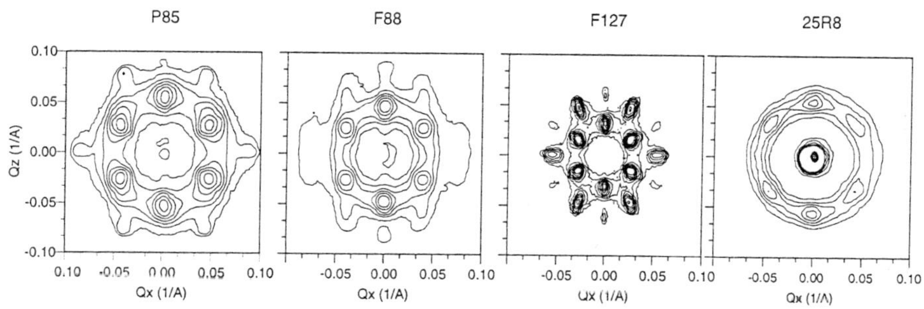

Figure 17. Two-dimensional scattering pattern of ABA-type of block copolymer micelles: $EO_{25}PO_{40}EO_{25}$, $EO_{96}PO_{39}EO_{39}$, and $EO_{99}PO_{65}EO_{99}$; and BAB-type of block copolymer micelles: $PO_{15}EO_{165}PO_{15}$ in the cubic phases. The scattering pattern are obtained with the shear axis horizontal and shear gradient parallel to the beam.

In ordered phases of PPO-PEO-PPO triblock copolymer micellar networks, as discussed further below, equivalent shear orientation have been found in the cubic phase. In a triblock copolymer gel of polystyrene and poly(ethylene buthylene), PS-PEB-PS, on the other hand, the bcc cubic phase orient in a twinned structure arranged with a [111] direction parallel to the shear, and [100] direction parallel to the shear gradient, Fig. 18 [49]. This is the orientation typically observed in the ordered phase of amorphous diblock copolymer melts (see eg. [53,13]) as well as classical systems like metals.

In the limit of high temperature and/or high polymer concentration, the cubic phase

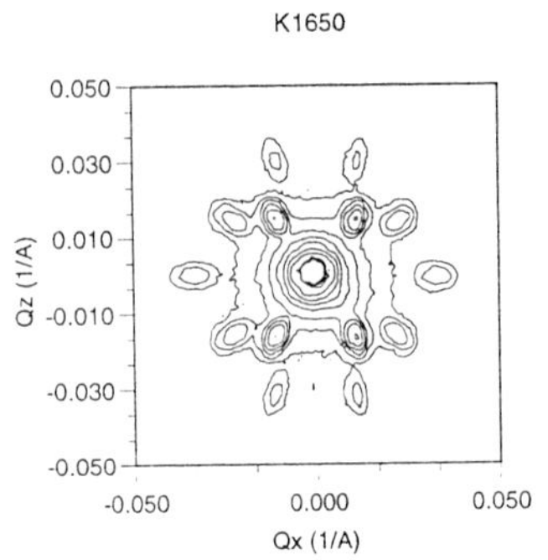

Figure 18. Two-dimensional scattering pattern of the ordered state of an interconnected network of PS-PEB-PS micelles in oil, revealing twinned bcc-structure. The scattering pattern are obtained with the shear axis horizontal and shear gradient parallel to the beam [49].

melts at the point where the spherical micelles transform into rod-like form, as discussed above. In some systems, eg. $EO_{25}PO_{40}EO_{25}$ [54], there are a separate phase between the cubic and the hexagonal rod-phase. Other materials, e.g. ($EO_{27}PO_{61}EO_{27}$) [55], shows a direct transition from the cubic to the hexagonal. The cubic phase of $EO_{25}PO_{40}EO_{25}$ remains to concentrations of the order of 45-50 % copolymers. At 45 % the cubic phase is observed from the lowest temperature measured (5°C) up to T=17°C. In the 50 % copolymer sample no six-fold por other atterns characteristic of the bcc phase are observed.

6.2. Cubatic Order

The Bragg reflections of the micellar crystals of typical PEO-PPO-PEO systems are far from resolution limited. Analysis, including instrumental resolution, shows that the line-shapes are Lorentzian rather than Gaussian [19]. The correlation length (ξ) as obtained from Lorentzian fits to the first order reflections

$$\frac{d\sigma(q)}{d\Omega} = \frac{I_o}{1 + [(q - q_o) * \xi]^2} + I_i \tag{24}$$

shows a small discontinuity at T_m, in agreement with the first order phase transition. In the ordered phase, however, the correlation length remains typically only of the order of a few hundred Å, i.e. the bond-length correlation is only a few lattice constants (roughly 2, 6 and 10 lattice constants for $EO_{25}PO_{40}EO_{25}$, $EO_{96}PO_{39}EO_{96}$, and $EO_{99}PO_{65}EO_{99}$, respectively). The orientational order, on the other hand, is macroscopic. This indicate *cubatic* classification of the crystalline state[19]. This para-crystalline nature markedly reduce the intensity of higher order reflections.

The reason for the cubatic nature of the ordered phase of PEO-PPO-PEO-micelles is unclear. It is possibly that the polydispersity in the micellar size play a role. Pluronics with large PEO-blocks tend to make more well characterized spherical micelles, as evident from the more pronounced side peaks in the form factor. Accordingly one should expect both more narrow peaks, and better resolved higher order reflections. This is consistent with experiments on different Pluronics. Fig. 17 shows the shear oriented scattering pattern of $EO_{25}PO_{40}EO_{25}$, $EO_{96}PO_{39}EO_{96}$, $EO_{99}PO_{65}EO_{99}$, and $PO_{15}EO_{156}PO_{15}$. Note that the latter is of the reverse architecture PPO-PEO-PPO, making a network of interconnected micelles. Another effect which however also affect the higher order reflections in the Pluronics series is the ratio between core size and inter-micellar distance. The low PEO-content polymers form relative larger cores. Their form factor will accordingly reduce the intensity of higher order reflections.

7. Nematic Phase

Upon application of steady shear, rod or worm-like micelles align to various degree depending on polymer concentration and shear-rate. In the rod-phase of 1% $EO_{25}PO_{40}EP_{25}$ copolymer solution, shear has only a relative small effect, possibly as a result of both relative small rods and because of the low viscosity, i.e. high relaxation rate. In samples of higher polymer concentration, on the other hand, shear has a marked effect as displayed in Fig. 19.

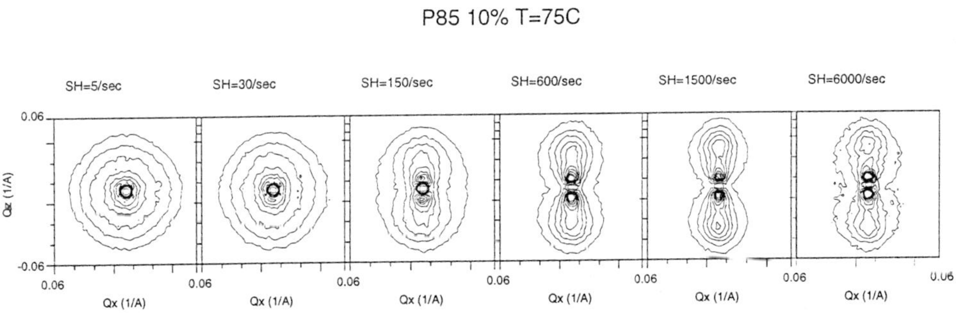

Figure 19. Two-dimensional scattering pattern of a 10% aqueous solutions of $EO_{25}PO_{40}EO_{25}$ in the rod-phase, as a function of external shear rate.

Fig. 19 shows the shear dependence of a 10% solution of $EO_{25}PO_{40}EP_{25}$ as obtained approximately in the middle of the rod-phase (T=76°C). Although the data still have to be finally analyzed, it is clear that the micelles align in the shear field, forming a nematic phase. The degree of alignment can be quantified by the ratio between scattered intensity parallel and perpendicular to the shear. For shear-rates below approximately 1000/sec a near linear increase in this order parameter is observed. For higher shear-rates, the value saturates at a finite number.

The anisotropic scattering pattern relaxes instantaneously back to an azimuthally isotropic structure when the shear is stopped. In the neutron scattering experiment we observe isotropic scattering within the time resolution of a few seconds after the shear has been stopped.

Other systems of rod-like micelles have shown a hexagonal phase even at low concentrations [56,57]. This is not the case for low concentration Pluronics. In an experiment of 10% $EO_{25}PO_{40}EP_{25}$ where the Couette cell was displaced from a 'through center' to a tangential configuration, there were no sixfold pattern as characteristic of the hexagonal phase. In higher concentration, on the other hand, the PEO-PPO-PEO block copolymer rod-like micelles clearly form a permanent hexagonal rod-structure, as discussed below. The exact transition between nematic and hexagonal order has not been identified, but seems to be somewhere between 10 and 20 % copolymer concentration.

8. Hexagonal Rod-Phase

In aqueous systems of $EO_{25}PO_{40}EP_{25}$ and related Pluronics, the rod-like micellar phase form a solid high viscous suspension for polymer concentrations approaching 15-20%. This solid-like phase appear to be hexagonally ordered rod structure.

As in the cubic phase, this phase can be oriented by shear in a Couette cell, or between two parallel plates allowing crystallographic studies. In Fig. 20 is shown the pattern with the beam respectively parallel and perpendicular to the shear gradient. Two reflections clearly appear in the parallel configuration, while the perpendicular configuration shows the six fold pattern characteristic of the hexagonal rod-pattern, i.e. the beam parallel to the flow and to the rods. No other reflections were observed at any sample orientation.

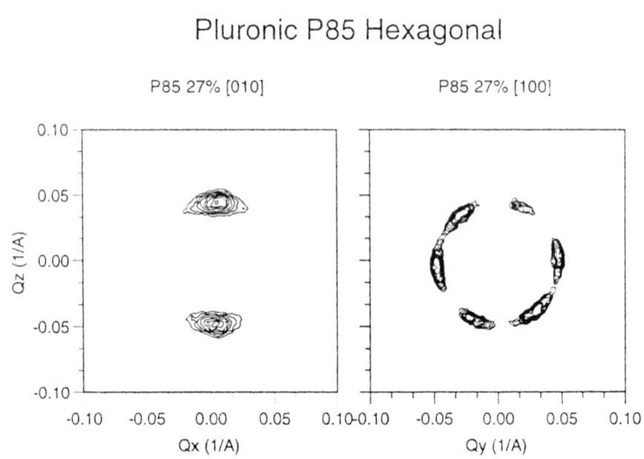

Figure 20. Two-dimensional scattering pattern of a 27% aqueous solutions of $EO_{25}PO_{40}EO_{25}$ in the hexagonal rod-phase. Left: xz-plane; right: yz-plane, where the x-axis is parallel to the flow and the y-axis is parallel to the shear gradient.

The two patterns shown in Fig. 20 are to some extend not in agreement with each other. Rotating the sample 90° we should not expect to see the two strong 1st order reflections at q_o, but only higher-order terms like those at $\sqrt{3} \times q_o$ (which is however not visible in these samples). The reason for first order peaks in the $q_x - q_z$-plane is that the shear oriented sample always consists of two types of domains. This becomes evident from a detailed study of scattering pattern in three dimensions. The major part of the sample have the $(1\bar{1}0)$-axis parallel to the shear gradient, but a minor part of the sample is aligned with the (110)-axis parallel to the shear gradient. The (001) rod-axis is always parallel to the shear flow. In studies using the Couette cell for shear alignment, it is actually scattering from the minor part of the sample which dominates, since the major part have no first order peaks in this scattering plane and there is no higher order reflections. In the 90°-orientation displayed in Fig.14b, however, the major part of the sample dominates. Similar observation of two-domain structure from hexagonally ordered rods have also been done in block-copolymer mesophases [58].

9. Lamellar Phases

Aqueous solutions of typical PEO-PPO-PEO block copolymers form lamellar mesophases within wide ranges of the temperature-concentration phase diagram. In the $EO_{25}PO_{40}EO_{25}$-system, the lamellar phase is observed over the 0-90°C temperature range [59], for polymer concentrations of the order of 50% or more. In aqueous systems the low-PEO content copolymer, $EO_6PO_{36}EO_6$-system, the lamellar phase is stable already at 10% polymer concentration [35]. The scattering function of the lamellar phase is dominated by a single correlation peak with only weak or no higher order reflections.

In Fig. 21 is shown an example of the scattering function of a typical lamellar phase: a narrow first order Bragg peak and a shoulder indicating the second order reflection. The data represent that of a 30% solution of $EO_6PO_{36}EO_6$ [35]. The low-q scattering is typical for swollen lamellaes with major in-plane fluctuations. The lamellar periodicity, d, increases linearly with polymer concentration as expected for ideal one-dimensional swelling

$$d = \delta/\phi \tag{25}$$

with bilayer thickness δ and polymer volume fraction ϕ.

The scattering pattern of the lamellar phase becomes usually anisotropic when the sample is exposed to shear. The result is a characteristic pattern with two symmetrical broad arcs [59]. Studies of shear alignment of block-copolymer lamellar phases have revealed that such materials typically orient into one of two kind of crystal orientations: either in *parallel lamellar* (lamellae parallel to xz plane) or in *perpendicular lamellar* (lamellae parallel to xy-plane of shear). In the lamellar phase of PEO-PPO-PEO aqueous solutions one find usually lamellar domains in all directions which have the lamellaes parallel to the direction of flow (x-axis).

10. Bi-continuous phases

While bi-continuous microemulsions frequently appear in ternary phase diagrams of oil, water and low-molecular surfactants, there has only recently been observations of

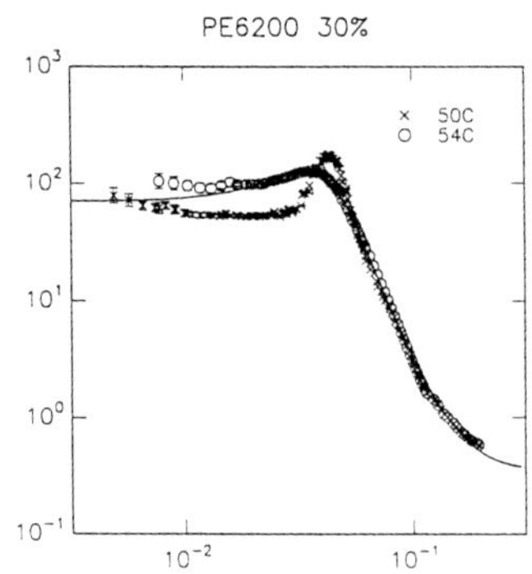

Figure 21. Scattering function of 30% $EO_6PO_{36}EO_6$ aqueous solution in the lamellar L_α-phase and in the bi-continuous L_3-phase [35].

such phases in binary systems of block copolymers and solvent. The first observation was made in an aqueous solution of the low PEO-content PEO-PPO-PEO triblock copolymer, $EO_6PO_{36}EO_6$ previously mentioned [35]. In Fig. 21 are shown scattering functions of 30% $EO_6PO_{36}EO_6$ solutions in the lamellar L_α-phase described above, and in the bi-continuous L_3 sponge-phase with the characteristic scattering function as described by Teubner and Strey [60].

More recently, the microemulsion sponge phases have been observed in systems of ABA tri-block copolymers dissolved in A-homopolymers [61] and in a ternary systems of AB-diblock copolymer and A- and B-homopolymers [62]

11. Block Copolymer Micellar Networks

While the ABA-kind of block copolymers form individual micelles which interact through hard-sphere like repulsions, quite different behavior can be expected for BAB-kind of aggregates.

The PPO-PEO-PPO BAB-kind of copolymers associate also in spherical micelles. Since, however, the water soluble block is in the middle, aqueous solutions form opaque materials because of interconnected domain structures as schematically visualized in Fig. 2. In the scattering pattern such domains of micellar networks are characterized by a major low-angle scattering on top of the pattern observed for individual micelles. Such behavior is observed, for example, in the system, $PO_{15}EO_{156}PO_{15}$. Only for polymer concentrations above approximately 50%, the network constitutes the whole sample. Within this 'isotropic' phase, single domain cubic ordered mesophases have been observed in analogy with the ABA kind of micellar systems, as shown in Fig. 17 [9]. Other interconnected micellar systems expand the whole material even at only few percents of copolymers [63].

The BAB-micellar systems have potentially interesting elastic properties as a result of the network structure. One may imagine that very well defined networks can be designed with controllable sizes of both the rubber-like middle blocks and the knots formed by the micellar cores. One may even be able to control the position of the knots since these can be organized on a crystalline lattice. The dynamics of the individual molecules within PEO-PPO kind of polymers are very fast, which makes the elastic properties of this particular network beyond the frequency range of practical interests.

In triblock copolymer systems of crystalline or glass forming polymers, permanent stable networks are possibly. Tri-block copolymers of polystyrene, PS, and rubber like materials like poly(ethylene propylene), PEP, form such stable networks when dissolved in oil [63]. Already with a few percent of such polymers in the oil, a permanent network of interconnected glassy PS spheres is formed throughout the sample, resulting in rubber-like elastic material. In some parts of the phase diagram, these materials form solid like ordered structures in analogy with the independent ABA-micelles (Fig. 18) [49].

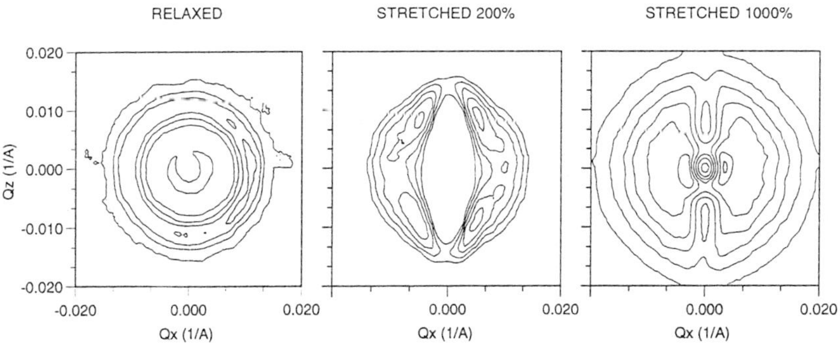

Figure 22. Two-dimensional scattering function of micellar network of PS-PEP-PS tri-block copolymers in oil, as obtained when stretched to various degree [64,65].

Stretching these materials gives rice to interesting structures, as shown in the contour plot of Fig. 22 [65]. The scattering pattern of samples stretched up to roughly 100% shows markedly enhanced scattering intensity in specific positions: *pseudo Bragg reflections* revealing stretching induced positional order of the PS-cores. At higher stretching ratios butterfly shaped patterns appear indicating inhomogeneity in the distribution of micellar cores and links [64].

12. Phase behavior of bulk block copolymers

Bulk block copolymers form also ordered mesophases, as revealed through intense studies during the last decade. The ordered phases includes bcc-structure of spherical micelles, hexagonal ordering of rod-like micelles and lamellar phases. Moreover, the tri-continuous cubic Ia3d-phase has been observed as well as complex modulated lamellar phases [66].

The pristine PEO-PPO-PEO materials form lamellar phases at low temperatures [59]. In opposition to the mesophases of both most block copolymers studied as well as the block copolymer solutions, where the self-assembling is controlled by the Flory-Huggins solubility parameter, the PEO-PPO-PEO lamellar phase is formed as a consequence of the crystallization of PEO.

In $EO_{25}PO_{40}EO_{25}$ the scattering momentum of the [10]-Bragg reflection is q_{10}=0.0272 Å^{-1}, giving a lamellar periodicity of $d=2\pi/q_{10}$=230 Å. To accomodate this large periodicity both the PEO and the PPO blocks have to be highly stretched.

At the melting point of PEO the block copolymer material undergoes an order-to-disorder transition. In the disordered phase there are still some composition fluctuations, but the Flory-Huggins interaction parameter characterizing the PEO-PPO-PEO system is relative small: $\chi = 0.092$. For the typical polymer sizes N studied, this χ-value is too small to accomodate the χN-value needed for ordering.

Other related PEO-block copolymer melts based on comparable molecular weights have significant larger Flory-Huggins interaction parameters. This is the case for poly(ethyl ethylene)-poly(ethylene oxide) (PEE-PEO) which show a variety of χ-driven mesophases above the PEO-melting temperature, and a conventional order-disorder transition at higher temperature [67].

13. Concluding Remarks

In the present review we have described the use of small-angle neutron scattering to study block copolymer micellar phases. The technique provide experimental values for size and form of the polymers as well as their aggregates. It gives information on aggregation number and aggregation thermodynamics. The scattering experiments also provide information on the interactions between polymeric aggregates. It gives detailes on positional correlations between these structures, including detailed information on ordered structures.

It is shown that the aqueous systems of PEO-PPO-PEO triblock copolymers is a very good model system for experimental thermodynamic studies on the block copolymer self-assembly, as the amphiphilic character can be varied continously by changing either temperature or pressure.

The information obtained from scattering experiments can be used directly in the design and optimization of new polymeric materials. Moreover, experimental scattering studies of complex polymer systems provides a challenge for exploring new physical phenomena.

REFERENCES

1. B. Chu, Langmuir 11 (1995) 414.
2. M. Almgren, W. Brown and S. Hvidt, Colloid Polym. Sci. 273 (1995) 2.
3. P. Alexandriris, V. Athanassiou and T.A. Hatton, Langmuir 11 (1995) 2442.
4. K. Mortensen, J. Phys. Conden. Matter 25 (1996) A103..
5. J. Yang and G. Wegner 1992 Macromolecules 25 (1992) 1786.
6. Y.-Z. Luo, C.V. Nikolas, D. Attwood, J.H. Collett, C. Price, C. Booth, B. Chu and Z.-K. Zhou J. Chem. Soc. Faraday Trans. 89 (1993) 539.
7. C. Tsitsilianis, G. Staikos, A. Dondos, P. Lutz and P. Rempp, Polymer 33 (1992) 3369.
8. B. Gao and J. Kops, Polym. Bulletin 34 (1995) 279.
9. K. Mortensen, W. Brown, E. Jørgensen, Macromolecules 27 (1994) 5654.

10. O. Glatter and O. Kratky, Small-Angle X-Ray Scattering, Academic Press, 1982.
11. P. Alexandridis, U. Olsson and B. Lindman Langmuir 12 (1996) 1419.
12. K. Mortensen, D. Schwahn and S. Janssen, Phys. Rev. Lett. 71 (1993) 1728.
13. K. Koppi, F. Bates, K. Almdal, and K. Mortensen J. Rheology, 38 (1994) 999.
14. G. Hadziioannou, A. Mathis, and A. Skoulios, Colloid Polym. Sci. 257 (1979) 136.
15. K. Almdal, K. Koppi, and F.S. Bates, Macromolecules 26 (1993) 4058.
16. F.S. Bates, K. Koppi, M. Tirell, K. Almdal and K. Mortensen Macromolecules 27 (1994) 5934.
17. B. Ackerson, J.B. Hayter, N. Clark and L. Cotter, J.Chem.Phys. 84 (1986) 2344.
18. se f.ex. H.M. Lang et al, J. Rheol. 36 (1992) 743.
19. K. Mortensen, W. Brown, and B. Nordén Phys.Rev.Lett. 68 (1992) 2340.
20. G.A. McConnell, M.Y. Lin and A. Gast, Macromolecules 28 (1995) 6754.
21. K, Mortensen, K. Almdal, F.S. Bates, K. Koppi, M. Tirell and B. Nordén, Physica B213 (1995) 682.
22. A. Guinier and G. Fournet, Small-Angle Scattering of X-Ray, J. Willey & Sons, 1955.
23. R. Hosemann and S. N. Bagchi, Direct Analysis of Diffraction by Matter, North-Holland, Amsterdam, 1962.
24. J.S. Higgins and H.C. Benoit, Polymers and Neutron Scattering, Oxford Science Publ., Oxford 1994.
25. J.S. Pedersen, J. Appl. Cryst. 27 (1994) 595.
26. M. Kotlarchyk and S.-H. Chen, J. Chem. Phys. 79 (1983) 2461.
27. B. Svensson, P. Alexandridis, U. Olsson and K. Mortensen, To be published (1997).
28. J.S. Pedersen, J. Colloid. Polymer Science. To be published (1997).
29. J.S. Pedersen and M. Gerstenberg, Macromolecules 29 (1996) 1363.
30. J.K. Percus and G.J. Yevick, Phys. Rev. 110 (1958) 1.
31. N.W. Ashcroft and J. Lekner, Phys. Rev. 145 (1966) 83.
32. D.J. Kinning and E.L. Thomas, Macromolecules 17 (1984) 1712.
33. K. Mortensen and J.S. Pedersen, Macromolecules 26 (1993) 805.
34. K. Mortensen and W. Brown, Macromolecules 26 (1993) 4128.
35. E. Hecht, K. Mortensen and H. Hoffmann, Macromolecules 28 (1995) 5465.
36. H. Rosch, *in Nonionic Surfactants*, Ed. M.J.Schick, Dekker, New York (1967).
37. Z. Zhou and B. Chu, J. Colloid. Interface Sci. 126 (1988) 171.
38. S.M. Aharoni, Macromolecules 16 (1983) 1722.
39. P. Linse, Macromolecules 26 (1993) 4437.
40. P. Alexandridis, J.F. Holzwarth and T.A. Hatton, Macromolecules 27 (1994) 2414.
41. K. Mortensen and Y. Talmon Macromolecules 28 (1995) 8829.
42. O. Glatter, G. Scherf, K. Shillen and W. Brown, Macromolecules 27 (1994) 6046.
43. M.O. Robbins, K. Kremer and G.S. Grest, J. Chem. Phys. 88 (1988) 3286.
44. K. Mortensen, Y. Talmon, B. Gao and J. Kops, To be published (1997).
45. P. Linse, J.Phys.Chem. 97 (1993) 13896.
46. D. Richter, D. Schneiders, M. Monkenbusch, L. Willner, L.J. Fetters, J.S. Huang, M. Lin, K. Mortensen and B. Farago, Macromolecules, (1996) in press.
47. W. Brown, K. Schillen, M. Almgren, S. Hvidt and P. Bahadur, J. Phys. Chem. 95 (1991) 1850.
48. G. Wanka, H. Hoffmann and W. Ulbricht, Colloid Polym. Science 268 (1990) 101.

49. R. Kleppinger, K. Mortensen, N. Mischenko, K. Almdal and H. Reynaers To be published (1997).
50. K. Almdal, K. Mortensen, K. Koppi, M. Tirell and F.S. Bates, J. Physique II (France) 617 (1996).
51. T. Tepe D. Hajduk, M.A. Hillmyer, P.A. Weimann, M. Tirell, F.S. Bates, K. Almdal, and K. Mortensen, To be published (1997).
52. McConnell GA, Gast AP, Huang JS, and Smith DJ 1993 Phys. Rev. Lett. **71** 2102
53. K. Almdal, K.A. Koppi and F.S. Bates, Macromolecules 26 (1993) 4058.
54. Mortensen K 1992 Europhys. Lett. **19** 599
55. K. Mortensen, Unpublished.
56. W.A. Hamilton, P.D. Butler, S.M. Baker, G.S. Smith, J.B. Hayter, L.J. Magid and R. Pynn, Phys.Rev.Lett. 72 (1994) 2219.
57. W.A. Hamilton, P.D. Butler, S.M. Baker, G.S. Smith, J.B. Hayter, L.J. Magid and R. Pynn, Phys.Rev.Lett. 74 (1995) 335.
58. F.S. Bates, K.A. Koppi, M. Tirrell, K. Almdal and K Mortensen, Macromolecules 27 (1994) 5934.
59. K. Mortensen, W. Brown and E. Jørgensen, Macromolecules 28 (1995) 1458.
60. M. Teubner and R. Strey, J. Chem. Phys. 87 (1987) 3195.
61. J.H. Laurer, J.C. Fung, J.W. Sedat, D.A. Agard, S.D. Schmit, J. Samseth, K. Mortensen, R.J. Spontak, Langmuir, (1997) submitted.
62. F.S. Bates, W.W. Maurer, P.M. Lipic, M.A. Hillmyer, K. Almdal, K. Mortensen and T.P. Lodge To be published 1997.
63. N. Mischenko, K. Reynders, R. Scherrenberg, K. Mortensen, F. Fontaine, R. Graulus, and H. Reynaers, Macromolecules, 27 (1994) 2345.
64. K. Reynders, N. Mischenko, K. Mortensen, N. Overbergh and H. Reynaers, Macromolecules 28 (1995) 8699.
65. N. Mischenko, K. Reynders, K. Mortensen, and H. Reynaers. J. Polymer Science, Polymer Physics (1996)
66. F.S. Bates, M.F. Schultz, A.K. Khandpur, S. Förster, J.H. Rosedale, K. Almdal and K. Mortensen Trans. Faraday Soc. 98 (1994) 7.
67. M. Hillmyer, F.S. Bates, K. Almdal, K. Mortensen, and A. Ryan, Science 271 (1996) 976.

Fluorescence studies of amphiphilic block copolymers in solution

R. Zana

Institut C. Sadron, CNRS-ULP, 6 rue Boussingault, 67000 Strasbourg, France

Block copolymers are made up of two or more polymeric sequences (blocks). The most common copolymers are of the A_nB_m or $A_nB_mA_n$ type, where A and B are the repeat units (monomeric units) and n and m the corresponding degrees of polymerization. Because of the large values of n and m, usually tens to hundreds of units, block copolymers are intrinsically amphiphilic whenever A and B are of sufficiently different nature. Consider for instance poly(styrene-*co*-isoprene), where the two blocks are hydrophobic or lipophilic, in the usual sense of these words. Nevertheless, this copolymer forms micelles in a hydrophobic solvent such as cyclohexane, which is a "good solvent" for the polyisobutylene block but a "bad solvent" for the polystyrene block above 34.5 °C, the so-called "theta temperature" of the polystyrene-cyclohexane system. This example can be generalized and it may be stated that block copolymers greatly extend the concept of amphiphilicity with respect to small amphiphilic molecules. For the latter the difference in nature between A and B must be much more pronounced. Another important fact is that small amphiphilic molecules are generally surface-active, while a block copolymer characterized by very large values of n and m may display little surface activity. This behavior is observed when the isolated copolymer molecules can form monomolecular copolymer micelles where the block with the least affinity for the solvent forms a compact coil protected from contacts with the solvent by a corona constituted by the block with the most affinity for the solvent [1,2]. Monomolecular micelles form however low the copolymer concentration and in this conformation the copolymer may not adsorb at interfaces, being in a state of lower energy with respect to the adsorbed state which may involve a conformational reorganization costly in free energy. Also, the adsorbed state may be of somewhat lower energy but separated by a high energy barrier from the monomolecular micelle state, which results in kinetic stability of the latter. This situation can give rise to a surface or interfacial tension slowly varying with time, owing to the copolymer conformational reorganization at the interface.

Thus, block copolymers can form micelles containing from one to a very large number of copolymer molecules, both in water and in organic solvents, pure or mixed, depending on the chemical nature of the repeat units A and B. The concentration where polymolecular polymer micelles form can be referred to as critical micelle concentration (cmc), as for small amphiphilic molecules. The cmc takes very different values depending on the experimental conditions, essentially the quality of the solvent, which can be modulated by changing the solvent nature (more or less polar), for instance. Likewise, the dynamics of block copolymer micelles will stretch over an enormously wide range of time. It must be emphasized that these characteristics are not very different from a qualitative viewpoint from those of small amphiphilic molecules for which changes of solvent quality can also result in large changes of equilibrium properties (cmc, for instance) and dynamic properties of micelles. Cetyltrimethylammonium bromide (CTAB) provides with a good example as it has been investigated in both water and formamide. The reported cmcs and aggregation numbers are respectively 0.9 mM and about 100 in water [3], and 100 mM and less than 10 in formamide [4]. The residence time of a cetyltrimethylammonium ion in CTAB micelles can be estimated to be of the order of 10^{-4} in water and 10^{-7} s in formamide [5]. Block copolymers are expected to show even larger changes.

Amphiphilic block copolymers, particularly the low molecular weight diblock and triblock copolymers with A = ethylene oxide (EO) and B = propylene oxide (PO), are currently arising much interest in view of their increasing range of applications and their peculiar properties. The trademarks Pluronic®, Synperionic®, or Poloxamer® are often used. Much information on amphiphilic block copolymers has been gained using fluorescence probing methods. The next paragraph briefly examines these methods, focussing on the type of information they can provide. The following five paragraphs review the main results obtained using these methods and concerning cmc and cmT (critical micellization temperature), micropolarity and microviscosity of block copolymers micelles, and micelle aggregation numbers and dynamics. A last paragraph deals with miscelleanous studies.

1. METHODS USING FLUORESCENCE FOR THE STUDY OF AMPHIPHILIC BLOCK COPOLYMERS IN SOLUTION

This Section focusses on the type of information that several methods based on fluorescence can provide on block copolymer micelles. The basic principles of fluorescence are not discussed as they are detailed in several textbooks [6-10]. The use of fluorescence probing methods for the study of micellar solutions of conventional surfactants has been reviewed [11-17]. The reader can refer to these comprehensive papers as the same methods apply to conventional surfactants and amphiphilic block copolymers.

1.1. Labeling of the systems

Most block copolymers of interest, particularly the $EO_nPO_mEO_n$, (also denoted as PEO-PPO-PEO in the following) are not fluorescent and their study by means of fluorescence methods requires the introduction in the block copolymer system of a fluorescent molecule (referred to as probe, thus fluorescence probing). This probe is added to the system and, in the ideal situation, is completely insoluble in the intermicellar solution and fully solubilized in the micelles. The probe can also be covalently attached to the block copolymer. In both instances, its static and dynamic fluorescence properties can then be used to obtain information on the micelles, the probe being alone or in the presence of quenchers, fully or partially solubilized in the micelles. Copolymers containing an intrinsically fluorescent block, like polystyrene, can be studied without addition of probe.

1.1.1. Aqueous systems

Both free probes and probes covalently attached to copolymers have been used for the study of aqueous systems. Pyrene is the most frequently used free probe owing to its very low solubility in water (about 5×10^{-7} M); its long lifetime (400 ns in hydrophobic media against 200 ns in water); the sensitivity of its emission and excitation spectra to the polarity of its environment; the formation of pyrene excimer, and the large number of compounds or ions able to quench its fluorescence [18, 19].

Other often used free probes are the pyrene derivatives pyrenecarboxaldehyde (PyCA) [20,21], 1,1'-dipyrenylpropane (DPyP) [22], and 1,1'-dipyrenylmethylether (dipyme) [23], and the diphenylhexatriene (DPH) [24]. Octadecylrhodamine B (ORB) [25], 1,3-dinaphthylpropane (DNP) [26], 1,3-bis(N-carbazolyl)propane (BCP) [26], [11-(3-hexyl-1-indolyl)-undecyltrimethylammonium bromide (6In11, a fluorescent surfactant) [26], 1-anilino-naphthalene (AN) [27], 8-anilino-1-naphthalenesulfonate (ANS) [27] and 4-bromo-1-acetonaphthone (4BAN) [28] are other free probes which have been used to a lesser extent. The spectroscopic probes used for the study of amphiphilic systems have been recently reviewed [29]. Basically, any of the probes discussed can be used for the study of block copolymer micelles.

Block copolymers have been labeled with pyrenyl [30], naphthyl or dansyl [31] groups, for instance, for studies in aqueous solutions. At this stage it must be recalled that a covalently attached probe may not yield the same information as the same probe in the free state. Indeed it may considerably perturb that part of the polymer in the vicinity of its point of attachment, whereas the effect of a free probe is the same over the whole copolymer, on a time average basis [32,33].

1.1.2. Systems in organic solvents.

The studies generally used covalently attached fluorescent groups such as carbazolyl [34], anthracenyl [35], pyrenyl [36] and naphthyl [37].

1.1.3. Preparation of the systems for fluorescence probing studies

The probe is usually only sparingly soluble in water and used at such a low concentration that it cannot be directly introduced in the system under investigation. In most instances, a stock solution of the probe in a volatile solvent (ethanol, tetrahydrofuran) is prepared and the appropriate amount of this solution is injected into the copolymer solution under stirring. The amount of solvent thus introduced must represent less than 0.5 % of the volume of the system in order that its effect be completely negligible. The solution is then stirred for several hours at 30-40°C.

The required amount of probe stock solution can also be introduced in a volumetric flask, spread over the flask wall and the solvent evaporated under a gentle stream of air. The copolymer solution is then added and the probe solubilized by vortex stirring for a long time. We experienced difficulties when using this last procedure for very hydrophobic probes like DPyP. The fluorescence intensities were always lower and the results less reproducible than with the first method suggesting that solubilization of DPyP may require an extremely long time (several days or weeks). Kabanov et al [38] reported similar effects for the solubilization of DPH (see Figure 3 below). Winnik et al [23] reported on artefacts that may arise from an incompletely solubilized probe and suggested to use the probe excitation spectrum to detect the possible presence in the system of probe microcrystals arising from the method of preparation of the solutions. Also, for those probes capable of forming excimers, the excimer peak is somewhat red-shifted.

Another method consists in preparing a saturated aqueous solution of probe and use it as solvent for preparing the solutions to be investigated.

For water-soluble probes such as AN, ANS or ORB the required amount of an aqueous solution of probe is directly added to the amphiphile solution.

The labeled solutions can be studied as they are or, better, after deoxygenation, as molecular oxygen is a strong quencher of the fluorescence emission of aromatic probes. Deoxygenation by successive freeze-pump-thaw cycles (3 cycles for aqueous systems, 4 cycles for systems using organic solvents) is recommended in time-resolved fluorescence quenching studies (see below) for determination of aggregation numbers.

1.2. Spectrofluorometry

1.2.1. Fluorescence emission spectra

These spectra have been used to estimate the polarity and viscosity of the immediate environment of the probe, referred to as *micropolarity* and *microviscosity*, respectively. The probe is introduced in the system at very low concentration, in the micromolar range. The required information is obtained by procedures that are illustrated by the following examples.

The ratio I_1/I_3 of the intensities of the first and third vibronic peaks in the fluorescence emission spectrum of pyrene [18,19] and the wavelength λ_{max} of the emission maximum of PyCA [20,21] and ANS and AN [27] are strongly polarity-dependent. Calibration curves have been prepared by plotting the variation of I_1/I_3 or λ_{max} for a series of solvents or solvent mixtures of varying polarity, characterized by the dielectric constant or another appropriate property. The polarity of the system investigated is then obtained from the measured value of I_1/I_3 or λ_{max}, using the calibration curve.

The fluorescence emission intensity of DPH [24] and ORB [25] is strongly polarity-dependent. This property can be used to estimate the polarity of the surrounding medium when these probes are included in amphiphilic aggregates.

The emission spectrum of DPyP [22], dipyme [23] and other similar probes [26] consists of an emission arising from the probe in a state where one fluorescent moiety is in the excited state, between 370 and 395 nm for pyrenyl moieties (monomer emission), and a wide band around 480 nm arising from the excimer that pyrene can form intramolecularly (excimer emission). The ratio I_M/I_E of the intensities of monomer and excimer emissions increases with the viscosity sensed by the two pyrenyl moieties in their motion required for excimer formation. As above, the microviscosity value is obtained using a calibration curve.

The above methods provide only relative values of the polarity or viscosity of the probe environment. Indeed these values depend on the compounds used to prepare the calibration curves. It is better to compare directly the measured quantities (I_1/I_3, or λ_{max}, or I_M/I_E) characterizing various amphiphilic systems. Mixtures of an alcohol and a polyalcohol of high viscosity, such as glycerol, and water-alcohol mixtures are recommended for preparing calibration curves for microviscosity and micropolarity, respectively. Indeed, aromatic probes are preferentially located in the micelle palisade layer, separating bulk water and hydrophobic core, where their environment is believed to resemble short-chain alcohols [11]. Figures 1 and 2 give exemples of calibration curves for microviscosity from DPyP [39] and micropolarity from PyCA [20]. The results for DPyP show that large differences can be found between calibration curves prepared using different solvent mixtures, which result in differences in microviscosity values. Thus, the microviscosity of sodium dodecylsulfate (SDS) micelles was found to be 10.1 and 15.9 when using the apolar mixture and the polar mixture respectively (see legend of Figure 1), an acceptable difference. However the corresponding values for CTAB, 19.1 and 111, and for the nonoionic surfactant Triton X-100, 29.7 and 250, show unacceptable differences. Still different values were obtained when replacing DPyP by dipyrenyldecane, probably because different probes reside at different micelle sites, on a time average basis [39]. The same problems exist for micropolarity as can be seen in Figure 2

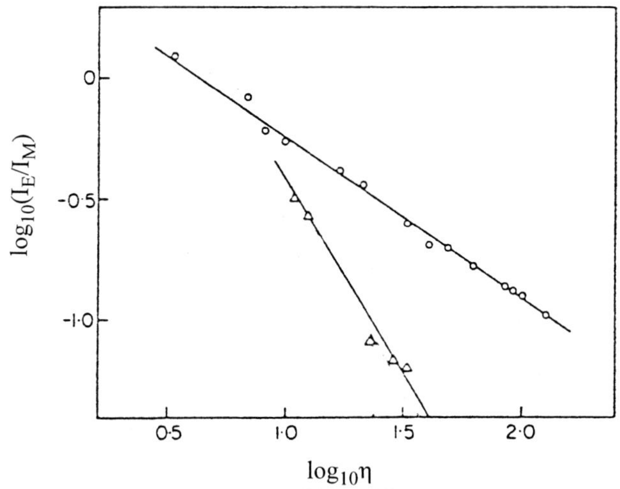

Figure 1. DPyP: variation of the I_E/I_M ratio in (O) n-hexadecane-liquid paraffin mixtures and in (Δ) ethyleneglycol-triethyleneglycol mixtures. Adapted with permission from [39].

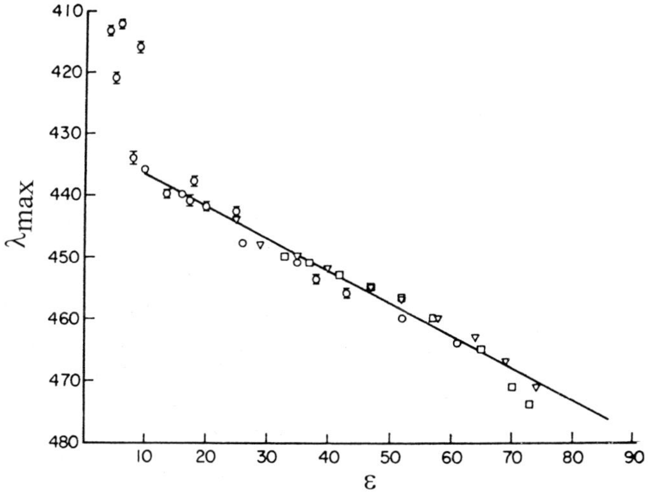

Figure 2. PyCA: variation of λ_{max} (in nm) with solvent dielectric constant in mixtures of water with (O) methanol; (□) ethanol; and (∇) dioxane. Symbols ($\bar{\text{Q}}$) refer to pure solvents. Reproduced from [20] with permission of the American Chemical Society.

where the values in organic solvent clearly fall off the curve defined by aqueous mixtures. Absolute values of microviscosities and micropolarities may not have much meaning unless one compares values for a series of homologous compounds as for instance a series of surfactants of differing chain length, or a series of block copolymers like Pluronics.

The *critical micelle concentration* (cmc) of amphiphilic compounds can be obtained from the concentration dependence of any of the above properties.

The *partition coefficient* of solubilizates between micelles and bulk phase can be determined from the variation of emission intensity of a micelle-solubilized probe upon addition of solubilizate to the aerated amphiphile solution [40]. This variation occurs because the solubilizate affects the quenching of the probe by molecular oxygen solubilized in the system.

Last, variations of fluorescence intensity with time can be used to monitor *dynamic events* taking place in amphiphilic systems. For instance two populations of identical micelles one with a solubilized probe (or an energy donor), the other with a solubilized quencher (or an energy acceptor) can be rapidly mixed. The monitoring of the variation of fluorescence intensity with time yields information on how fast micelles can exchange their components and form and breakdown [34].

1.2.2. Fluorescence excitation spectra

The excitation spectrum of pyrene shows a shift of the (0,0) band from 332.5 nm in water to 339 nm in the hydrophobic environment of amphiphilic micelles. This shift can be used to estimate cmc and micelle polarity [41-43]. The excitation spectra can also used to check that the probes are fully solubilized, as they show peculiar bands when the system still contains probe microcrystals [23].

1.2.3. Static fluorescence quenching

The decrease of fluorescence intensity associated with the quenching of a fluorescence probe by an appropriate quencher can be used to determine aggregation numbers of micelles of conventional surfactants [44]. In principle, this method can be applied to aqueous block copolymer micelles. The fluorescence intensities in the absence of quencher, I_0, and in the presence of quencher, I_Q, are related to the quencher concentration [Q] and micelle molar concentration [M] according to:

$$\log (I_0/I_Q) = -[Q]/[M] \qquad (1)$$

The slope of the $\log (I_0/I_Q)$ vs [Q] plot yields [M]. The micelle aggregation number, N, is then calculated from

$$N = (C\text{-cmc})/[M] \qquad (2)$$

where C is the amphiphile concentration.

The following conditions must be met, however, for this method to yield reliable results:

(i) probe and quencher must be fully solubilized in the micelles with $[\text{probe}]/[M] << 1$ and $[Q]/[M] < 2\text{-}3$;

(ii) probe and quencher distributions between micelles must be frozen on the fluorescence time scale (their residence times in micelles must be long with respect to the fluorescence probe lifetime);

(iii) the rate constant for intramicellar quenching, k_Q, must be much larger than the probe decay rate constant, k.

These conditions are best fulfilled by using probes and quenchers insoluble in water or strongly partitioned in micelles, the probes having a very long lifetime (low k) and the quenchers being efficient (large k_Q). The problem in using static quenching is that one cannot know whether condition (iii) is fulfilled without performing preliminary time-resolved fluorescence quenching (TRFQ) experiments on the system under investigation. Since TRFQ can also be used to determine aggregation numbers without the above restricting conditions, it is not recommended to use static quenching because condition (iii) may be rarely fulfilled in block copolymer systems owing to the high micellar microviscosity (see below).

1.3. Fluorescence anisotropy or polarization

The system is illuminated by a polarized light and the emission intensity is measured in directions perpendicular ($I_\perp$) and parallel ($I_{//}$) to the incident beam. The fluorescence polarization p and anisotropy r are defined as:

$$p = (I_{//} - I_\perp)/(I_{//} + I_\perp) \qquad \text{and} \qquad r = (I_{//} - I_\perp)/(I_{//} + 2I_\perp) \tag{3}$$

Perrin's equation relates p and r to the viscosity of the probe environment (microviscosity), η_e, to the probe lifetime, τ, and volume, V, according to:

$$\frac{r_0}{r} = \left(\frac{1}{p} - \frac{1}{3}\right) \Big/ \left(\frac{1}{p_0} - \frac{1}{3}\right) = \left(1 + \tau\frac{RT}{\eta_e V}\right) \tag{4}$$

where p_0 and r_0 are the fluorescence polarization and anisotropy of the probe in a solvent of infinite viscosity (frozen motion). The value of the microviscosity is obtained from the values of p and p_0, provided τ is known or has been measured. This method does not requires a calibration curve. Equations 3 and 4 show that an increase of p corresponds to an increase of microviscosity. Note that the changes of p with any parameter are directly related to microviscosity changes, contrary to the variations of the I_E/I_M ratio (see above). Indeed the latter are also sensitive to changes of micelle size, with I_E/I_M decreasing upon increasing size, as a result of decreasing efficiency in excimer formation.

ORB [25] and 6In11 [26, 45] have been used as probes for polarization measurements.

1.4. Time-resolved fluorescence quenching

In homogeneous solutions containing a probe and a quencher, the fluorescence intensity I(t) at time t decays with time according to:

$$I(t) = I(0) \exp[(-k - k_Q[Q])t] \tag{5}$$

where I(0) is the fluorescence intensity at time t = 0, and k_Q is the quenching rate constant.

In micellar solutions, where the probe and quencher are completely solubilized in the micelles and where the probe, quencher and micelle molar concentrations are such that [probe]/[M] << 1 and [Q]/[M] $\approx$ 1, the fluorescence decay equation becomes [11,14-17,46,47]:

$$I(t) = I(0) \exp\{-A_2t - A_3[1 - \exp(-A_4t)]\} \quad (6)$$

where the A_i's are time-independent. The decay is thus biphasic, its fast and slow components arising from micelles containing one probe and one or more quencher, and from micelles containing only one probe, respectively.

When the probe and quencher distributions are frozen on the fluorescence time scale (time for probe or quencher redistribution >> probe lifetime $\tau = 1/k$) the following expressions hold:

$$A_2 = k = 1/\tau \qquad A_3 = [Q]/[M] \qquad \text{and} \qquad A_4 = k_Q \quad (7)$$

where k_Q is now the intramicellar quenching rate constant. The long time part of the decay (slow decay) is not affected by the introduction of the quencher. The micelle aggregation number is obtained from the value of [M], using Equation 2.

Theory suggests to restrict the use of TRFQ to systems with $k_Q/k > 1$, and $[Q]/[M] < 2\text{-}3$ [48-50]. This last condition can probably be somewhat relaxed in the case of block copolymers as they generally have a much larger size than quenchers.

More complex expressions of the A_i's must be used when the distributions are not frozen on the fluorescence time scale [15-17] and in the case of polydisperse micelles [51,52]. From the A_i values obtained by fitting Equations (5) and (6) to the decay curves in the absence and presence of quencher one can calculate the values of N, k, k_Q, and of the rate constants of the intermicellar exchange processes responsible for probe and/or quencher redistribution. Information on micelle polydispersity (weight average aggregation number, standard deviation and skewness of the distribution) can also be obtained from measurements of N at different quencher concentrations. The existence of exchange processes is revealed by the fact that $A_2 > k$. Note that the theory for fluorescence decay in polydisperse systems with intermicellar exchanges of probe and quencher is still being worked out [53].

The main advantage of TRFQ with respect to other methods used for the determination of micelle aggregation numbers is that TRFQ is completely insensitive to intermicellar interactions and to complications arising from the micelle shape. Indeed, Equations 6 and 7 hold irrespective of micelle shape: spherical, cylindrical or disk-like. However, the expression of k_Q depends on micelle shape and microviscosity of the medium in which probe and quencher diffuse prior to reacting. In fact, the values of k_Q can be analyzed to yield information on micelle shape and microviscosity.

For conventional amphiphiles (surfactants) in aqueous solutions the TRFQ method applies for values of N between 15-20 and several hundreds, using pyrene as fluorescent probe and various quenchers, most notably alkylpyridinium chlorides with 12 to 16 carbon atoms in the alkyl chain, and dimethylbenzophenone (DMBP). This range is more restricted in the case of amphiphilic block copolymers because of the high micelle microviscosity (see below).

1.5. Non radiative energy transfer

This method uses donor and acceptor molecules, the latter having an absorption spectrum that overlaps the donor fluorescence emission spectrum. When the donor is excited, its emission is partly absorbed by the acceptor which then emits light with its own emission spectrum. This technique has been mostly used to study the dynamics of block copolymer micelles in water and organic solvents [34,54] and to obtain information on the internal organisation of micelles using labels that are covalently bound to the copolymer [55,56].

2. CRITICAL MICELLE CONCENTRATION AND CRITICAL MICELLE TEMPERATURE

The critical micelle concentration (cmc) of an amphiphile is the concentration at which micelles appear in the solution at a given temperature. The critical micelle temperature (cmT) is the temperature at which micelles appear in an amphiphile solution of a given concentration. In principle, this concentration represents the cmc value at temperature T.

Many fluorescence probing studies have reported cmc and cmT values. When comparisons can be made the results are not always in agreeement, contrary to what is usually observed for conventional amphiphiles. There are several possible reasons for this situation. The first one is that most studies used samples of copolymers "as received". These raw products often contain impurities in the form of more hydrophobic homopolymers if one is dealing with diblock copolymers, or of homopolymers and diblock copolymers when dealing with triblock copolymers. These impurities can affect the behavior of the copolymers investigated, particularly at interfaces, giving rise to a minimum in the surface tension plots [57], and have been reported to affect the association behavior in solution [58]. Also, variations of composition can occur from batch to batch and from manufacturer to manufacturer for a given copolymer. Thus, large differences in cmc values measured by means of surface tension and in shapes of some surface tension versus concentration plots in the reports of Alexandridis et al (PEO-PPO-PEO samples from BASF Corp., NY) [59] and Kabanov et al (PEO-PPO-PEO samples from Serva) [38] may originate from such variations. Moreover, even in the absence of impurities, the copolymers are always polydisperse in composition and block lengths and the different methods of measurements weigh differently this polydispersity, yielding different results. A last reason is that the probe used must be fully solubilized in the micelles as soon as they form in the solution. This may not be the case however when the cmc is extremely low, because the micelle volume fraction is very small just above cmc and the probe continues to sense a partly aqueous environment even at concentrations above cmc, being partitioned between micelles and bulk phase. In this case, the variation of fluorescence property has a sigmoidal shape and stretches over about two decades in concentration without being directly related to the cmc [42,60].

Figure 3 shows a typical example of pyrene partitioning between the polystyrene (PS) core of PEO-*b*-PS-*b*-PEO copolymers and the intermicellar solution [61]. The variation of the pyrene intensity ratio I_1/I_3 is sigmoidal and progressive. The cmc may be thought to be located between 10^{-3} and 10^{-2} g/dm^3, where the variation of I_1/I_3 is the steepest. However the cmc would then be nearly independent of the hydrophobe content of the copolymer, a physically unreasonable result. Besides X-ray scattering showed that micellization occurred below 10^{-5} g/dm^3. Other studies [42,43] showed similar slow variations of the pyrene intensity ratios I_1/I_3 (emission spectrum) and I_{338}/I_{333} (excitation spectrum) and of the average pyrene fluorescence lifetime [41,42]. Pyrene partition was taken into account and the concentration where the experimental curve separated from that calculated on the basis of pyrene partition at low copolymer concentration was interpreted as the cmc. This interpretation may need reexamination as the calculations assumed a single partition constant. However, at low copolymer concentration, under the conditions used in the studies, the micelles contained several pyrene molecules and the partition coefficient is then a decreasing function of the mole fraction of pyrene in the micelles [60]. This effect is sufficient to explain the observed differences. These systems should be reexamined by means of scattering

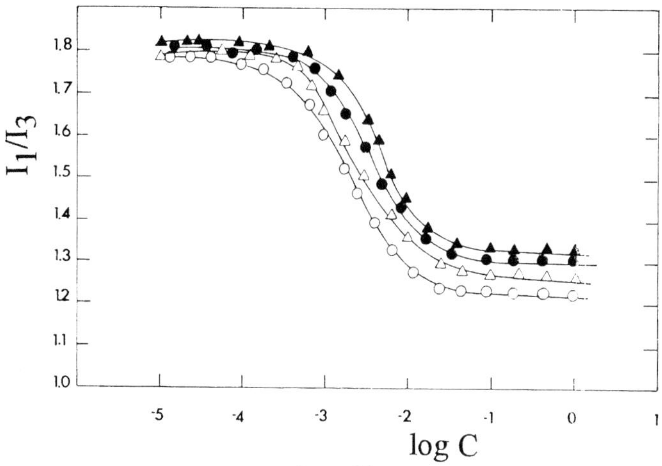

Figure 3. Variation of the pyrene fluorescence intensity ratio with the concentration (g/dm^3) of solutions of PEO-*b*-PS-*b*-PEO block copolymers of increasing PS weight fraction: (▲) 0.17; (●) 0.20; (Δ) 0.24; and (O) 0.28. Reproduced with permission from [61].

techniques to detect the onset of micelle formation as in [61].

Figures 4 and 5 show typical results for cmc determinations by means of various probes. The study in Figure 4 used the intensity of fluorescence, I_f, and the fluorescence polarization, p, of ORB [25]. The cmc of the $EO_{78}PO_{30}EO_{78}$ (Pluronic L68) shows very well at 50 °C as a one-step and steep increase of I_f and p. The study in Figure 5, which refers to $EO_{38}PO_{51}EO_{38}$ (Pluronic P85) shows a one-step increase of pyrene fluorescence intensity [38]. On the contrary, in Figure 6 the changes of λ_{max} of PyCA and of the microviscosity measured using DNP and 6In11 occurs in two steps for $EO_{13}PO_{30}EO_{13}$ (Pluronic L64) [26]. The first step was attributed to the formation of monomolecular copolymer micelles, and the second step to the formation of polymolecular polymer micelles with an aggregation number of 52 [26]. The authors point out, however "that some of the effects might be related to the presence of impurities, e.g., diblock copolymers." Similar two steps variations of λ_{max} with copolymer concentration for the probes ANS and AN in aqueous solutions of PEO-PHEMA-PEO (PHEMA = poly(2-hydroxyethylmethacrylate)) were given [27c] the same explanation. In this study also, one cannot discard the presence of diblock impurities. The results in Figures 4 and 5 confirmed the strong decrease of cmc of Pluronics upon increasing temperature.

The examples above suggest that fluorescence probing alone is not sufficient to establish unambiguously the micellization of amphiphilic block copolymers. The presence of hydrophobic polymeric impurities capable of binding the probe and/or forming monomolecular polymer micelles can give rise to results seemingly indicating micellization. This behavior was referred to as "anomalous micellization" of block copolymers and was discussed in detail by Tuzar and Kratochvil [62]. The impurities are solubilized in the micelles proper when these occur.

Whenever possible fluorescence studies should be backed by scattering studies.

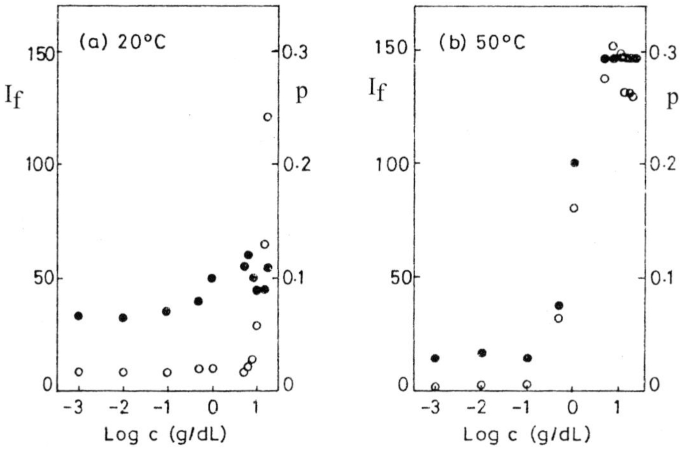

Figure 4. Pluronic F68 at 20 and 50 °C: variations of the fluorescence intensity, I_f (O),and polarization, p (●), with the copolymer concentration (ORD,1 μM). Reproduced from [25] with permission of the American Chemical Society.

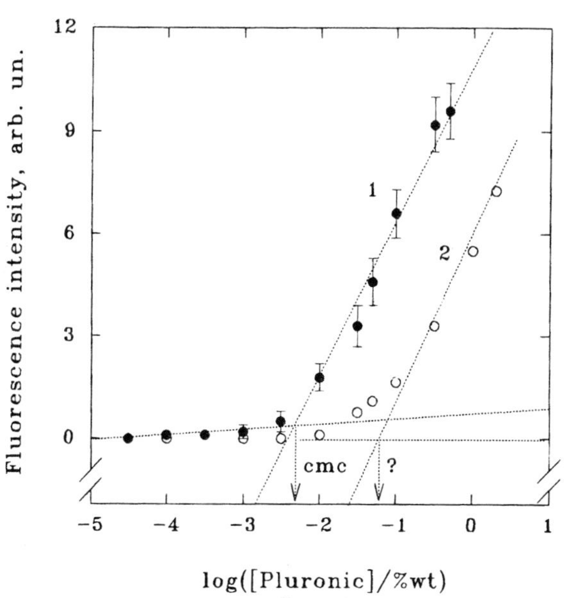

Figure 5. Pluronic P85 at 37 °C: variation of the fluorescence emission intensity of DPH (0.5 μM, λ_{exc} = 380 nm, λ_{em} = 430 nm) with concentration, determined after allowing 1h (O) and 16 h (●) for DPH solubilization. Reproduced from [38] with permission of the American Chemical Society.

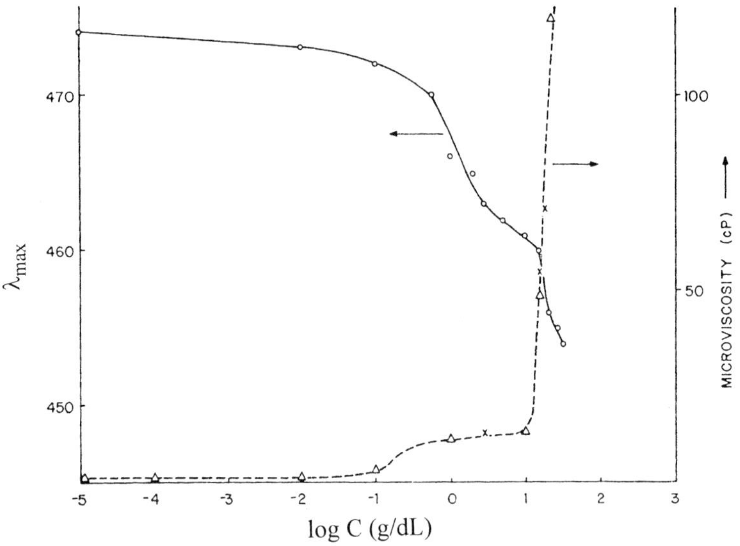

Figure 6. Effect of the Pluronic L64 concentration on λ_{max} (in nm) of PyCA fluorescence (O) and microviscosity measured from the 6In11 fluorescence polarization (Δ) and DNP I_E/I_M ratio ($\times$). Reproduced from [26] with permission of the American Chemical Society.

The cmT have also been much investigated. Figure 7 illustrates the determination of the cmT of Pluronics F108 and P104 (see their structure in Table 1) from the variations of the pyrene polarity ratio I_1/I_3 and also from the DPH absorption intensity with the Pluronic concentration. The cmT was taken as the temperature at midpoint of the sharp decrease in I_1/I_3 as it agrees well with the onset of rapid increase of DPH absorption. Note that the range of "rapid change" of I_1/I_3 stretches over about 10 °C. This is probably in part due to the heterogeneity of, and the presence of impurities in, the samples (used "as received") [63,64]. Figure 8 shows the I_1/I_3 vs concentration plots for several Pluronics, whose characteristics are listed in Table 1. These results give the essential features of the effect of several parameters on the cmT:

(i) a random copolymer of EO and PO does not micellize, contrary to block copolymers (Figure 8a, no effect on I_1/I_3);

(ii) the cmT increases as the copolymer molecular weight decreases at constant composition (Figure 8b);

(iii) the cmT increases with the length of the PEO block at nearly constant length of the PPO block (Figure 8c) but decreases when the length of the PPO block is increased at nearly constant length of the PEO block (Figure 8d);

(iv) the cmT decreases upon increasing copolymer concentration (Figure 8e).

There have been a number of other studies of block copolymers relating to micellization and cmc determinations. The formation of micelles in solutions of poly(maleic acid-*b*-p-

Table 1
Characteristics of some Pluronics and composition of the micelle core (adapted from [63,64]).

Pluronic	PO weight%	OP units	OE units	(PO weight%)$_{core}$
P65	50	29	2×17	63
P85	50	40	2×26	67
P103	70	60	2×17	73
P104	60	61	2×27	---(74/26/0)[a]
P105	50	56	2×37	72
F108	20	50	2×132	69 (68/28/4)[a]
P123	70	69	2×19	74

[a] Values calculated taking into account the possible presence of water in the micelle core.

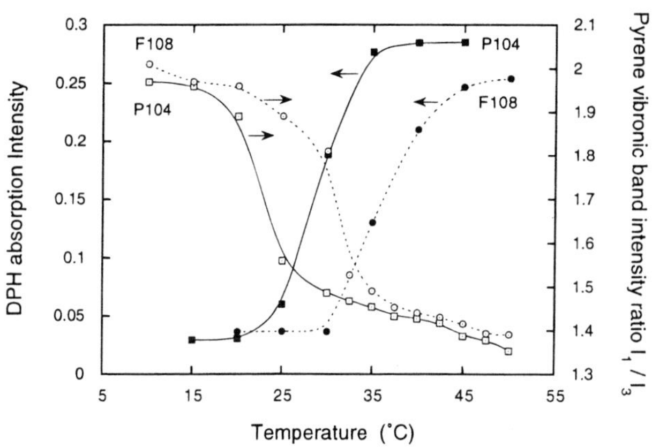

Figure 7. Temperature effect on the pyrene fluorescence intensity ratio I_1/I_3 in P104 and F108 aqueous solutions (open symbols). The filled symbols are the DPH absorption intensity data for the same copolymer solutions. Reproduced from [63] with permission of the American Chemical Society.

N,N-dimethylaminostyrene) [65] and of poly(vinylphenanthrene-*b*-methacrylic acid) [66] was evidenced from the ANS fluorescence intensity enhancement and also by studying the quenching of phenanthrene by various quenchers. The study did not specify whether the micelles formed intra or intermolecularly.

The cmc's of a series of poly(ethyleneoxide-*b*-benzylaspartate) have been determined using pyrene fluorescence probing (variations of I_1/I_3, I_{339}/I_{334} (excitation spectra) and total fluorescence intensity with the polymer concentration) [67]. The results showed little change of the measured cmc's with sample composition and molecular weights. This study used the methods described in [42] and discussed at the beginning of paragraph 2. The observed

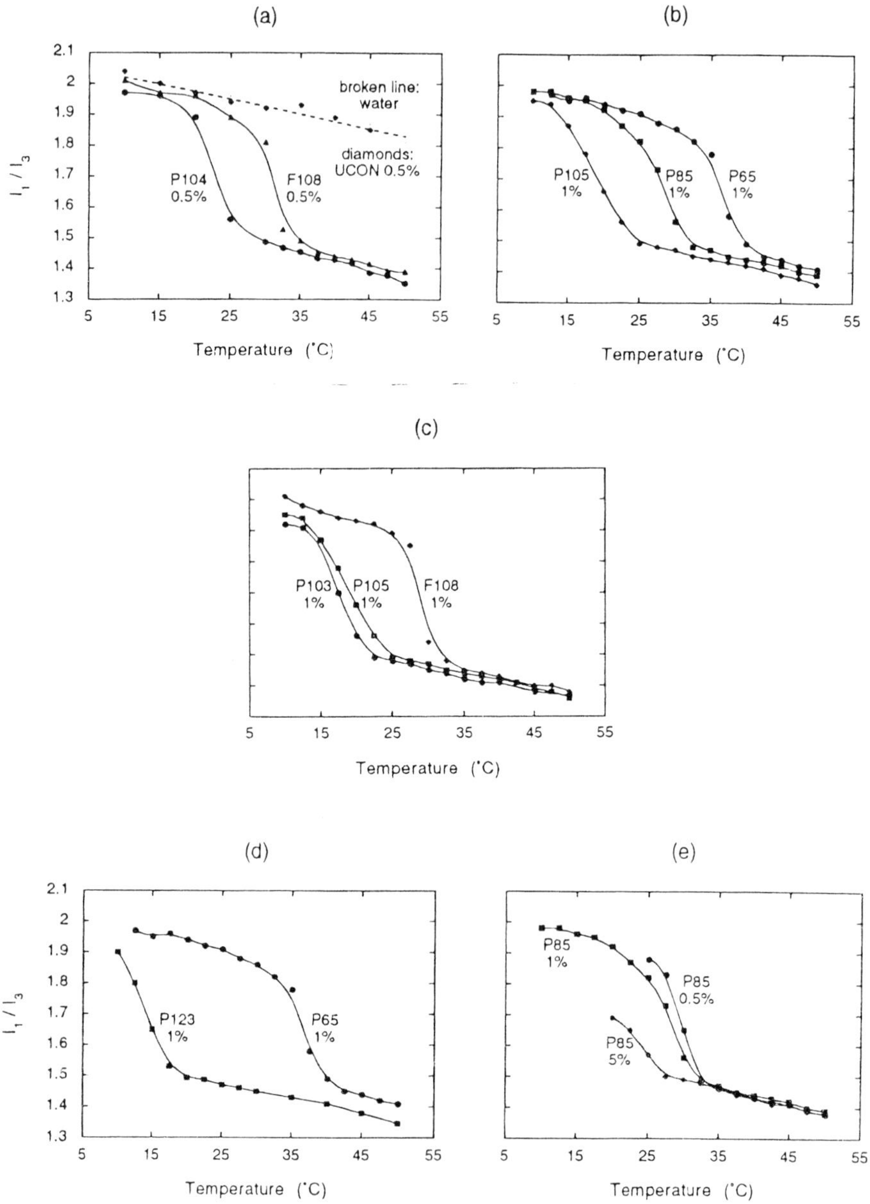

Figure 8. Effect of various structural parameters of Pluronics on the variation of the pyrene intensity ratio I_1/I_3 with temperature for the determination of the cmT of some Pluronics listed in Table 1. Ucon denotes a random copolymer of EO and PO. Reproduced from [64] with permission of the American Chemical Society.

changes of properties were very progressive. They may reflect a partition of pyrene and not really the onset of micelle formation.

Astafieva et al [43,68] have used pyrene fluorescence probing as described in [42] to investigate the micellization behavior of poly(styrene-*b*-sodium acrylate) as a function of block lengths and copolymer composition. For a constant PS block length, the cmc was observed to go through a maximum as the poly(sodium acrylate) block length was increased (Figure 9), a behavior very different from that of nonionic block copolymers. Also the cmc decreased linearly as a function of the square root of the concentration of added NaCl. These results were explained in terms of changes of copolymer conformation.

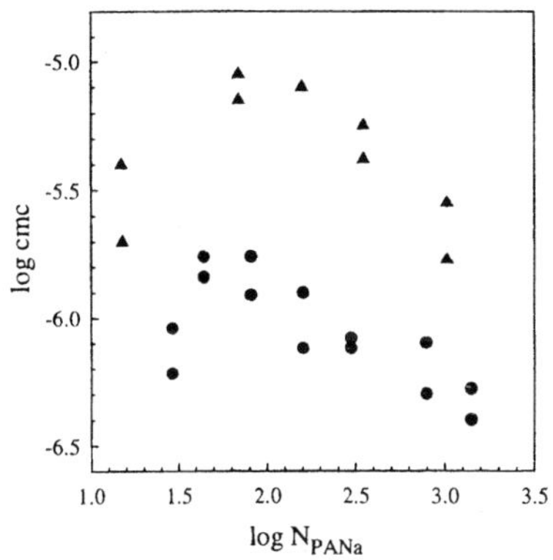

Figure 9. Effect of the poly(sodium acrylate) block length on the cmc values for PS-*b*-PANa series with PS block lengths of 11 (▲) and 23 (●). Reproduced from [68] with permission of the American Chemical Society.

The I_1/I_3 pyrene ratio was used to show that the cmc of P105 increases upon addition of urea [69], a result similar to that found for many conventional surfactants.

An examination of the cmc data for both nonionic and ionic copolymers reveals an unexpectedly weak dependence on the hydrophobic block molecular weight with respect to comparable results for conventional nonionic and ionic surfactants [42,69]. This may indicate the existence of unimolecular polymer micelles [1,62]. Whether these micelles arise from the block copolymer or from some more hydrophobic impurities is open to question. The answer requires comparative studies of purified and raw samples of block copolymers.

All of the results above concerned aqueous solutions of block copolymers. To our knowledge, there has been no reported cmc determinations of block copolymers in organic solvents by means of fluorescence techniques based on free probes. The main reason for this situation is that the most usual aromatic probes (pyrene, DPyP, etc) remain in the organic intermicellar phase for which they have much affinity rather than solubilizing in the solvophobic core, or are partitioned between core, solvophilic shell and intermicellar phase, making the interpretation of the data difficult. Copolymers labeled with acceptor and donor molecules and the use of the nonradiative energy transfer (NRET) method have permitted cmc and cmT determinations for block copolymers in organic solvents and solvent mixtures even in situations where the cmc is so small that more simple techniques such as osmometry or light scattering are not sensitive enough. Thus, the micellization of PS-*b*-PI (polyisoprene) labelled with carbazolyl or naphthyl (donor D) or anthryl (acceptor A) was investigated by NRET in n-heptane/cyclohexane mixtures [70]. Very large changes of $I_f(A)/I_f(D)$ accompanied the formation of micelles, with the cmc decreasing as the quality of the solvent mixture worsened upon decreasing temperature or enriching the solvent mixture in heptane (non solvent) as can be seen in Figures 10. Multimolecular micelles occurred at concentrations as low as 10^{-3} g/L where turbidity is not sensitive to micelle formation. Similar NRET results had been previously reported for graft PS-PMMA (poly(methylmethacrylate)) copolymers in THF/acetonitrile and THF/cyclohexane [71]. NRET was used to investigate the micellization of PS-*b*-PEO copolymers labeled by either phenanthrene (donor) or anthracene (acceptor) groups at the block junctions in mixtures of methanol or water with THF and of methanol with 1,2-dichloroethane [72]. Last, NRET was used for cmc determinations in PMMA-*b*-poly(methacrylic acid) copolymer solutions in methanol/ethylacetate mixtures. The cmc was shown to increase with the methanol content [30].

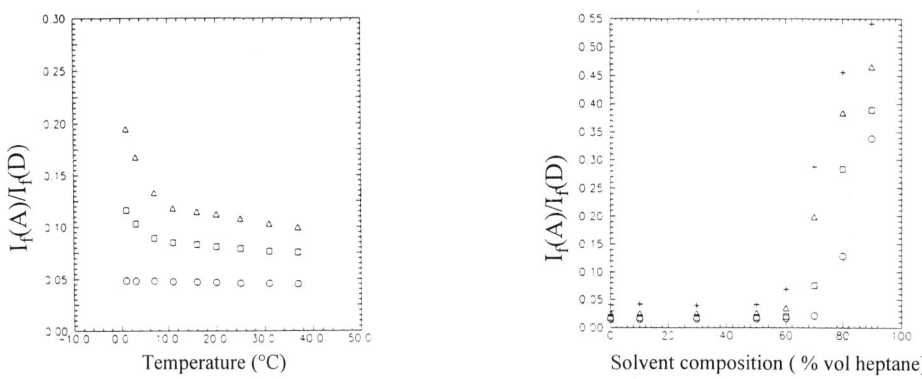

Figure 10. Variations of the intensity ratio $I_f(A)/I_f(D)$ in labeled PS-*b*-PI copolymer solutions in heptane as a function of temperature (left, copolymer concentrations in g/L: (Δ) 1.41; (□) 1.01; and (O) 0.23, no micellization) and in cyclohexane/heptane mixtures as a function of the heptane volume fraction (right, copolymer concentrations in g/L: (O) 0.10; (□) 0.50; (Δ) 1.09; and (+) 2.23). The onset of increase of $I_f(A)/I_f(D)$ corresponds to the cmc. Reproduced from [70] with permission of the American Chemical Society.

3. MICROPOLARITY

In most instances, the micropolarity was measured by the value of the pyrene polarity ratio I_1/I_3 at amphiphile concentrations well above the cmc and at a given temperature. Other properties have been used, however, as for instance the polarity dependence of the wavelength of emission maximum of AN, ANS and PyCA.

Figures 7 and 8 show examples of variations of I_1/I_3 with temperature. There is a fairly large decrease of I_1/I_3, that is of polarity, upon increasing T above the cmT. Figure 11 shows

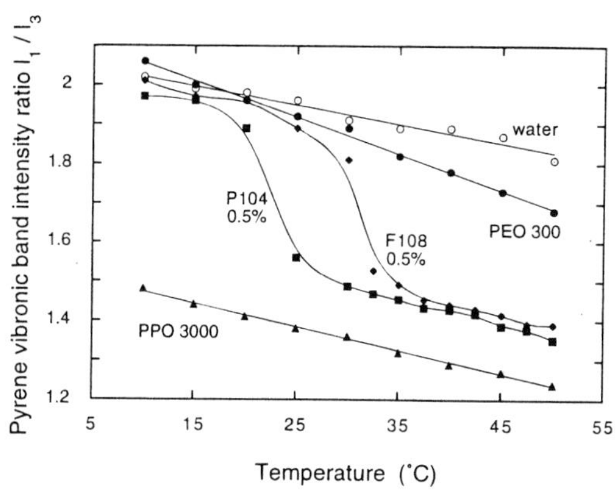

Figure 11. Temperature dependence of the pyrene I_1/I_3 ratio in water, in P104 and F108 solutions and in bulk PEO (molecular weight 300), and PPO (molecular weight 3000). Reproduced from [63] with permission of the American Chemical Society.

that the constituting polymers, that is PEO and PPO, are also characterized by decreasing I_1/I_3 values upon increasing T. The plot for PPO runs nearly parallel to those for the Pluronics. The value of I_1/I_3 at a given temperature above the cmT depends on the copolymer composition. These results have been used to evaluate the composition of the core of Pluronic micelles assuming that the measured value of I_1/I_3 for the copolymer solution is a weighted average of the I_1/I_3 values for PEO and PPO [63]. This assumption was checked on mixtures of these two polymers: I_1/I_3 was found to increase linearly with the PEO weight fraction. Using the results represented in Figure 11 led to the PO weight fractions in the micelle core listed in Table 1. These calculations were later improved to take into account a possible presence of water in the core, for Pluronics F108 and P104 [64]. The results listed in Table 1 show that the micelle core PO content decreases and that the water content increases as the copolymer becomes more hydrophilic. The compositions thus determined did not depend on temperature, suggesting that the linear decrease in I_1/I_3 above cmT arises from changes of polarity of the components of the micelle core and not from changes in its composition.

Turro and Kuo [45] also concluded to a decrease of polarity upon increasing T for Pluronics L64, F68, L92, and F127, using pyrene (I_1/I_3 ratio) and PyCA (λ_{max}) and investigated the effect of pressure in the range between 1 and 1500 bar, using the same probes. Typical results are represented in Figure 12 for L64. The polarity sensed by pyrene and PyCA increased with pressure, an effect opposite to that of temperature. The effect of pressure is more important for copolymers with a higher PO content.

In constrast, a study of the temperature dependence of λ_{max} of AN in solutions of PEO-*b*-poly(2-hydroxyethylmethacrylate) (HEMA) at concentration above cmc (see Figure 13) concluded that the hydrophobic core polarity increased with temperature, owing to an increased mixing of PEO and PHEMA blocks [27c]. Indeed, results for DMF/water mixtures showed that λ_{max} increases with the polarity of the AN environment. Note that the plots in Figure 13 show two steps, attributed to intramolecular polymer micellization

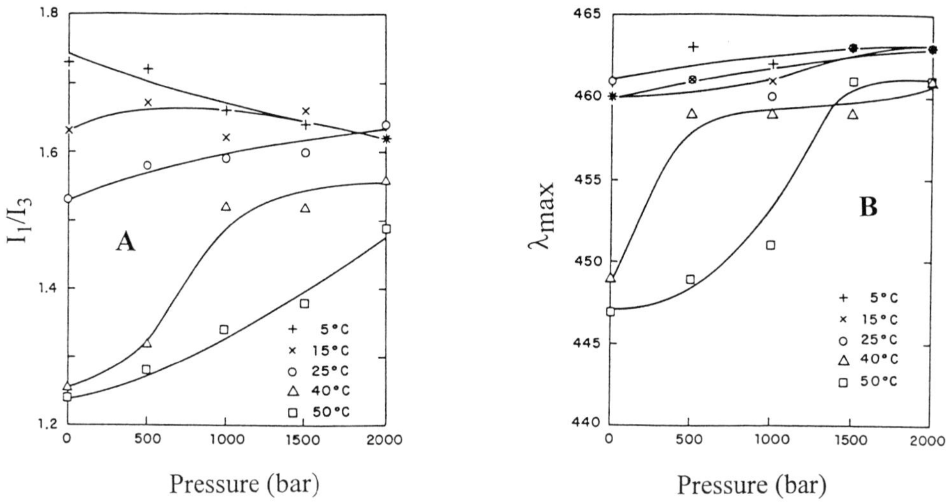

Figure 12. Pluronic L64: variations of I_1/I_3 (A) and λ_{max} (B) with pressure at the indicated temperatures. Reproduced from [45] with permission of the American Chemical Society.

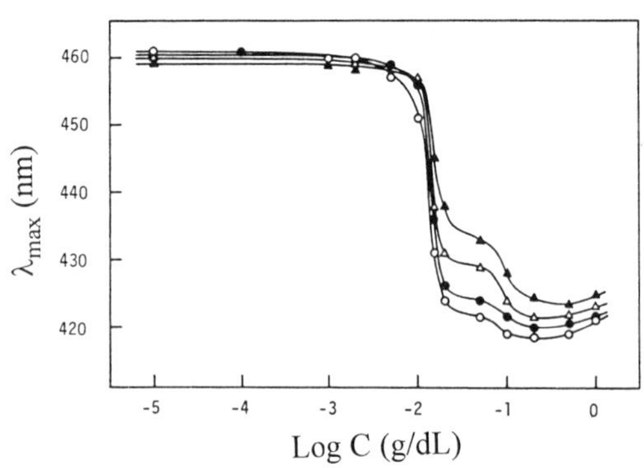

Figure 13. Variation of λ_{max} for AN in solutions of PEO-*b*-PHEMA with the copolymer concentration at (O) 10 °C; (●) 20 °C; (Δ) 30 °C and (▲) 40 °C. Reproduced from [27c] with permission the American Chemical Society.

(monomolecular micelle) and to intermolecular polymer micellization [27c]. This behavior may also be due to the presence of hydrophobic impurities. However, the conclusion concerning the effect of temperature is not affected by this behavior as only the qualitative changes of λ_{max} above cmc matter. It thus appears that the change of micelle polarity is strongly dependent on the nature of the polymer making up the core. The core polarity is expected to decrease when the polymer block constituting the core becomes less soluble in water upon increasing temperature, and conversely.

Additions of urea have very little effect on the micelle polarity sensed by the pyrene I_1/I_3 intensity ratio [69].

4. MICROVISCOSITY

The microviscosity sensed by ORB [25] and 6In11 [26] (fluorescence depolarization) and DNP [26] (excimer formation) solubilized in block copolymer micelles increases steeply when the copolymer concentration becomes larger than the cmc, at constant temperature (see Figures 4 and 6). These increases were attributed to micelle formation and the probe immobilization in the micelle core where it experiences a higher viscosity than in bulk phase, the polymer chains being constrained.

Evidence for a large microviscosity of poly(styrene-*b*-acrylonitrile) micelle cores was presented on the basis of fluorescence polarization of 9-MeA [73].

Nivaggioli et al [74] reported a systematic study of the effect of temperature on the microviscosity of Pluronics, as monitored by the ratio I_M/I_E of the intensities of the monomer and excimer emissions in the fluorescence spectra of dipyme. This ratio was showed to decrease nearly linearly upon increasing viscosity in the range 34 to 70 cP by using a liquid PPO in the temperature range 34 - 50 °C [74]. Several Pluronics were investigated at temperatures above their cmT. In all instances I_M/I_E, that is the microviscosity, decreased upon increasing temperature. However, the decreases were smaller than for bulk PPO, as seen in Figure 14, indicating that the relative microviscosity increases with T. A similar conclusion was previously reached on the basis of the temperature dependence of the I_M/I_E ratio for DNP in L64 solutions: the microviscosity was reported to increase from 46 cP at 10 °C to 87 cP at 40 °C [75]. The microviscosity increased with the Pluronic molecular weight at constant composition, and with the PPO block molecular weight, for PEO blocks of constant molecular weight, respectively [74]. Also, as expected for an intramicellar property, the microviscosity was independent of the copolymer concentration [74] and not affected by the presence of urea up to 4 M [69]. Figure 14 shows that the microviscosity of F108 micelles (PPO block molecular weight: 2920) has the value of the viscosity of bulk PPO of molecular weight 725 at low T and that of PPO of molecular weight 2000 at high T. This behavior reveals that the PPO block conformation in the micelle core becomes more compact as T is increased, probably as a result of dehydration. The results yielded microviscosities ranging between 50 cP for P65 and 130 cP for P123 and P105, at 40 °C. These values are about five times larger than for conventional surfactants. The large microviscosity values found for systems where polymeric structures are involved: oligomeric surfactants, polysoaps, polymer-surfactant complexes and amphiphilic block copolymers, were attributed to the presence in the micelle core of copolymer backbones which hinder the motion of reactants, probes and quenchers [76].

A four block copolymer, the Tetronic T704, $(EO_{12}PO_{14})_2NCH_2CH_2N(EO_{12}PO_{14})_2$, was also investigated [74]. The micelles were found to have a lower microviscosity than that observed in equivalent Pluronic micelles, a behavior attributed to a larger free volume in the micelle core as a result of the peculiar structure of T704.

The complex relationship between temperature and concentration in determining the microviscosity is illustrated in Figure 15. The 0.01 g/dL P68 solution contains no micelles in

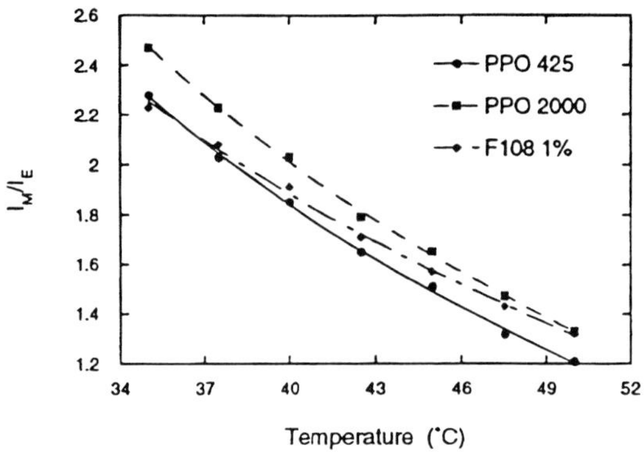

Figure 14. Variation of the monomer to excimer intensity ratio with temperature for a F108 micellar solution above the cmT and two bulk PPOs. Reproduced from [74] with permission of the American Chemical Society.

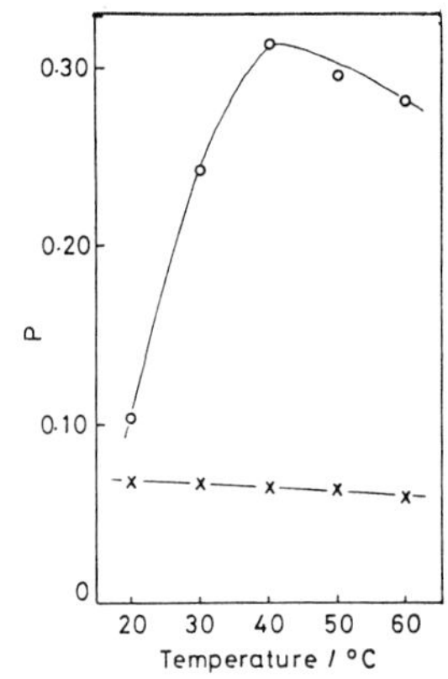

Figure 15. Effect of temperature on ORD fluorescence polarization in 0.01 g/dL (×) and 18 g/dL (O) F68 solutions. Reproduced from [25] with permission of the American Chemical Society.

the whole range of temperature, thus a low value of the fluorescence polarization p of ORB [25]. In the T range investigated the 18 g/dL solution is above the cmT and a maximum in p is observed. The increasing part of the plot was given the same explanation as in Nivaggioli's study [74]. The slowly decreasing part of the plot, at higher T, was attributed to a real decrease of microviscosity reflecting the fact that at about 40 °C the micelles have reached their stage of full compactness and their microviscosity then decreases as in homogeneous solutions [25].

Pyrene excimer formation has been used to study the effect of temperature on microviscosity [75]. Figure 16 shows the presence of a maximum in the variation of the I_E/I_M intensity ratio, which suggests a minimum of microviscosity, not observed with nonylphenolethoxylates [75]. The decreasing part of the plot, at high T, was assigned to the increase of microviscosity with T, as in [74] and [25]. The increasing part of the plot at low T was attributed to a growth of L64 micelles with T, resulting in a larger average number of pyrene per micelle, and thus a larger probability of excimer formation at a given concentration of pyrene. It should be noted that the cmT of the 10 % L64 solution is given as 23.5 °C [77], a value close to that where I_E/I_M is a maximum. A mixed cluster of pyrene and L64 molecules may then be responsible for the initial increase in I_E/I_M. NaCl additions resulted in a significant shift of the maximum in the I_E/I_M curve to lower temperature [75], as expected if this maximum corresponds to a cmT [58].

The effect of pressure on microviscosity was investigated [75]. Figure 17 shows that the microviscosity decreases upon increasing pressure, an effect opposite to that caused by an increase of T. This behavior parallels the increase of micropolarity with pressure [45] and was attributed to an increasing hydration of the Pluronic with pressure.

Several studies using excimer formation or NRET and labeled copolymers showed that the

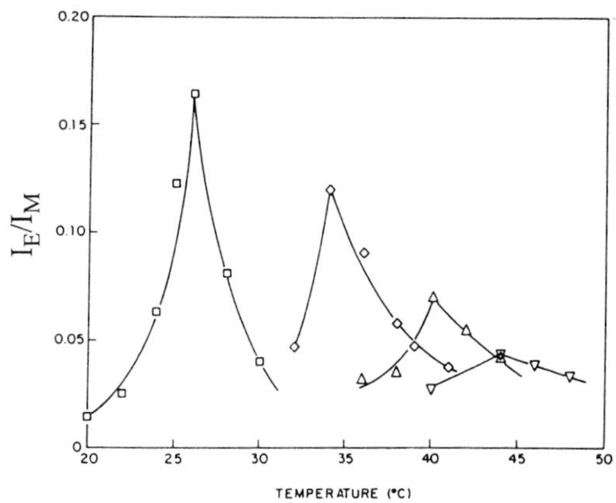

Figure 16. Effect of temperature on the pyrene I_E/I_M ratio in 1 g/dL L64 solution at (□) 1 bar; (◊) 0.5 kbar;(Δ) 1 kbar; (∇) 1.5 kbar. Reproduced from [75] with permission of the American Chemical Society.

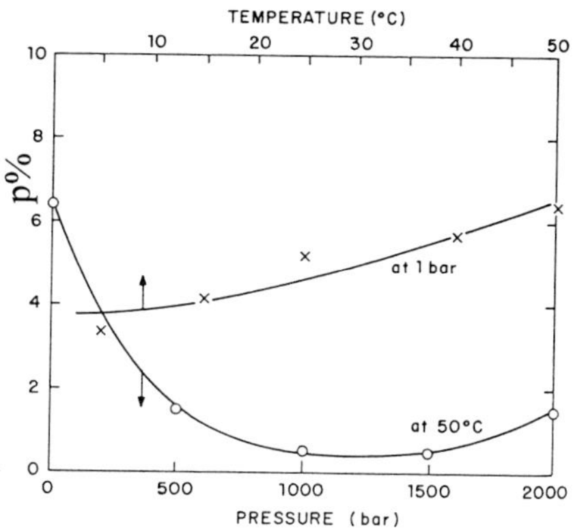

Figure 17. Effect of pressure and temperature on the polarization of 6In11 in a 10 % L64 solution. Reproduced from [45] with with permission of the American Chemical Society.

core of block copolymer micelles in organic solvents or solvent mixtures is characterized by a high microviscosity [31,35,78,79].

5. MICELLE AGGREGATION NUMBERS

As is shown below the reported studies of aggregation numbers of block copolymer micelles by means of fluorescence quenching are still rather fragmentary and the values uncertain mainly because the high micelle microviscosity lowers the quenching rate constants making the analysis of the decay curves less accurate, and also because of the partition of the quencher between micelles and bulk phase.

The first value of an aggregation number obtained by fluorescence probing was reported by Turro and Chung [26], presumably for a 20 g % Pluronic L64 aqueous solution, from pyrene excimer formation. N was estimated to be 52 from the pyrene concentration for which the value of the ratio I_E/I_M was close to 1.

The micelle aggregation numbers in solutions of poly(styrene-*b*-acrylamide), prepared by copolymerization in CTAB micelles, were determined by pyrene fluorescence quenching [73]. Static quenching by nitromethane and time-resolved fluorescence quenching by dibutylaniline yielded results in good agreement for copolymers of different compositions. The styrene degrees of polymerization in the block copolymers were thus determined and found to be were very close to the average styrene occupancy numbers of the CTAB micelles prior to polymerization, suggesting that the polymerization involved all styrene molecules contained in one CTAB micelle. The results however are open to questions. Thus, the CTAB micelle aggregation number used to calculate styrene occupancy numbers were much too low, 60 instead of 100-120. This would substantially increase the styrene occupancy numbers and thus suggest than the styrene block does not necessarily involve all styrene molecules contained in one micelle. Also the reported values of the product $\tau \times k_Q$ (lifetime×intramicellar quenching rate constant) were well below 1, which renders doubtful the use of the static fluorescence quenching method, and introduces large errors in N values from TRFQ [48-50]. The low quenching rate constants reported in this study result from the high microviscosity of the micelle core.

The aggregation of Pluronic L64 and the effect of additions of SDS and KF on the aggregation state of this copolymer have been investigated by Almgren et al [58,80,81]. The study which used the probe-quencher pair pyrene/dimethylbenzophenone yielded an aggregation number of 145 for a 1% L64 solution at 40 °C [80]. In a later report [81] the authors stated that this value should be reduced by a factor of about 2, because the free L64 present in the system was not properly accounted for and noted that, even so, the TRFQ value was still more than twice larger than that estimated from static light scattering. Nevertheless, from a qualitative viewpoint, the fluorescence probing results showed that additions of SDS at concentrations in the millimolar range, i.e., well below the cmc of SDS alone, to L64 or F68 solutions led to the formation of mixed aggregates even at temperatures where the copolymers do not form micelles when alone (case of F68 at 20 and 40 °C) [80]. The aggregation numbers of the Pluronics and SDS in the mixed micelles were evaluated, as well as the micelle polydispersity. In conditions where the Pluronic forms micelles on its own, SDS additions rapidly reduced the copolymer aggregation number, to values as low as 2 in the presence of 0.1 M SDS, and the mixed micelle polydispersity was important [80]. Additions of KF were shown [58] to decrease somewhat the aggregation number of L64 micelles. However the values of N from TRFQ using pyrene excimer formation and pyrene quenching by guaiazulene differed by a factor as large as 5 and were again much larger than from static light scattering. These inconsistent results were attributed to the presence in the sample of L64 (used as received) of about 3 % diblock impurity.

Pyrene excimer formation was used in two other TRFQ studies. A study of a 18.7 g % Pluronic P85 solution showed that the pyrene lifetime is a maximum at about 25 °C, and yielded aggregation numbers of 38 and 58 at 10 and 15 °C, respectively. At these temperatures the fit of the conventional fluorescence decay Equation (6) to the decay curves was good and the product $N \times k_Q$ was found to be nearly constant, as expected for monodisperse spherical micelles [82]. At higher temperatures, Equation (6) did not fit the data which were then analyzed using a fractal approach that implied a large micelle polydispersity. The excimer method was also used to investigate the aggregation behavior of a 1 g% Pluronic P104 solution [64]. The results in Figure 18 show a large and nearly linear increase of N with temperature. Two points must be made at this stage. The first one is that the P104 study reported a self-quenching rate constant $k_Q \approx 1 \times 10^6$ s^{-1} at 30 °C [64], whereas the P85 study [82] reported $k_Q \approx 1.5 \times 10^7$ s^{-1} at 10 °C. This large discrepancy is not easy to understand even when taking into account the differences of copolymer structure. The second point is that the low k_Q value for P104 [64] results in a value of $\tau \times k_Q$ of about 0.25, that is in principle too low for allowing determinations of N values [48-50]. In fact the aggregation numbers were obtained on the assumption of frozen probe and quencher distributions among micelles [64].

In conclusion of this section, it appears that time-resolved fluorescence quenching has been only little successful in determinations of aggregation numbers of block copolymer micelles.

Much remains to be done in this field and future progress requires samples containing no hydrophobic impurities and probe-quencher pairs more efficient than those used thus far, in order to overcome the reduction in quenching efficiency due to the high microviscosity of block copolymer micelles.

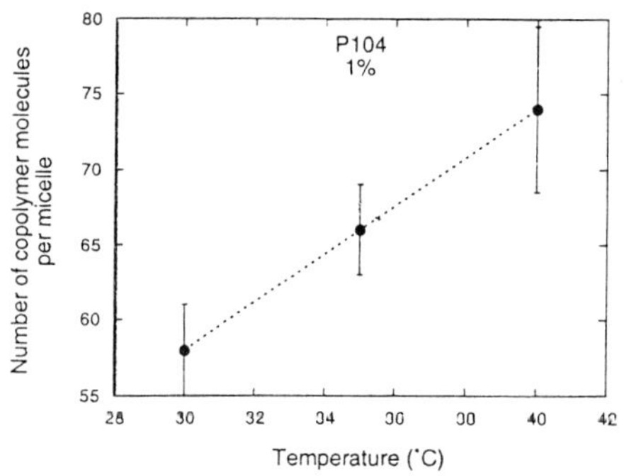

Figure 18. Effect of temperature on the micelle aggregation numbers in a 1 weight % P104 aqueous solution. Reproduced from [64] with permission of the American Chemical Society.

6. DYNAMICS OF BLOCK COPOLYMER MICELLES.

As pointed out in the Introduction, the dynamics of block copolymer micelles may stretch over a very wide range of times, depending on the copolymer molecular weight and quality of the solvent. The temperature at which the measurements are performed may also have an enormous effect, depending on whether it is above or below the glass transition temperature of the polymer constituting the micelle core.

Recall that in micellar solutions of conventional surfactants, free and micellized surfactants are in dynamic equilibrium, constantly exchanging between these two states (surfactant exchange process) [83]. Also the micelles are not frozen objects, they constantly form and breakdown by two well documented mechanisms:

(i) step-wise association/dissociation of one surfactant at a time to/from micelles [83];

(ii) fragmentation/coagulation reactions where the aggregate A_{i+j} breaks into aggregates A_i and A_j, and conversely [84]

The theory of Halperin and Alexander [85] for the dynamics of micellar solutions of block copolymers indicates that these features should be essentially retained, with the micelle formation/breakdown predominantly occurring via the stepwise process. Thus, formation/breakdowns should be characterized by a relaxation time increasing with the copolymer concentration [83], whereas a decrease should be found if fragmentation/coagulation reactions predominate [84]. However, recent results [54] and theoretical calculations [86] suggest that exchanges of copolymer may occur simultaneously via copolymers exiting the micelles and also through contacts between micelles accompanied by hopping of copolymer(s) from one micelle to the other (reactions of the type $A_i + A_j \leftrightarrow A_{i+x} + A_{j-x}$, with x being a small number). The time scale for the second mechanism would be at least one order of magnitude longer than for the first one.

Fluorescence probing techniques are capable of yielding information on the dynamics of block copolymer micelles in different ways. First, the mixing of differently labelled but otherwise identical populations of block copolymer micelles can yield information on the rate at which these micelles mix by exchanging monomeric copolymers, by recording the variations of fluorescence intensity with time. Likewise, mixing experiments invoving either a micellar solution in a given solvent and another solvent capable of dissociating the micelles or a non micellar copolymer solution and a solvent favoring micellization can yield quantitative information on the rate at which micelles breakdown or form, respectively. Also, time-resolved fluorescence probing can inform on the dynamics of local motions within micelles.

Surprinsingly, there are no report of direct studies of the dynamics of PEO-PPO-PEO micelles using fluorescence methods. This is probably because these copolymers are obtained from manufacturers and no attempt was made to label them. However, studies using GPC (gel permeation chromatography) [87], pulsed field gradient NMR [88], and temperature-jump [89] have led to values of the time characterizing the exchange of a Pluronic molecule between micelles and intermicellar solution ranging between hours and milliseconds. Such large discrepancies cannot be attributed only to differences in structure of the Pluronics studied, and more studies are required in order to explain them.

Dynamics studies are more numerous with other block copolymers in water or in selective solvents or solvent mixtures, some of them involving water. Preliminary studies using GPC had suggested that the equilibrium between free and micellized copolymer molecules in organic solvents may be extremely slow under certain conditions [90,91]. The stopped-flow method with light scattering detection was used to study the rate of PS-b-PHI (poly(hydrogenated isoprene)) micelle formation/breakdown in dioxane/heptane mixtures [92]. The results showed that the decay/increase of the light intensity with time were characterized by two time constants in the 1-100 ms range. These fast relaxations were later attributed to the fact that macroperturbations of the micellar equilibria were used in this study [92]. Indeed subsequent studies, by sedimentation [93] or using NRET [34,54,94] showed that the kinetics of exchange was much slower than inferred from stopped-flow. The sedimentation study involved PS-*b*-PMAa (polymethacrylic acid) in water/1,4-dioxane mixtures and showed that the unimer-micelle equilibrium was kinetically frozen for water volume fractions above 30 % [93]. Prochazka et al [34] used PS-*b*-PHI labeled by either carbazole (donor) or anthracene (acceptor) in heptane/dioxane mixtures (the micelles have a PS core in alkanes). The mixing of two identical micelle populations differing only by the label yielded at sufficiently high copolymer content results such as those in Figure 19. The occurrence of energy transfer reflecting exchanges of copolymer molecules between differently labeled micelles shows as an increase of emission intensity from the acceptor and a decrease of emission intensity from the donor. Note that the transfer is slowed down when increasing the content of heptane which renders the PS core of the micelles more compact, and that the time scale is in thousands of seconds (indicating rate constants smaller than 10^{-4} s^{-1}). Similar NRET experiments performed using PS-*b*-PEO copolymers labeled with either naphthalene or pyrene covalently attached to the block junctions, in methanol/water mixtures, also yielded changes of intensity stretching over thousands of seconds and rate constants of the order of 10^{-5} s^{-1} [54,94]. In all NRET studies the variations of the fluorescence intensity with time were well fitted by a sum of two exponential terms with characteristic times in a ratio of 10-100 [34,54,94]. However, the data were not well fitted when assuming that copolymer exchanges took place only via unimer stepwise association/dissociation [54].

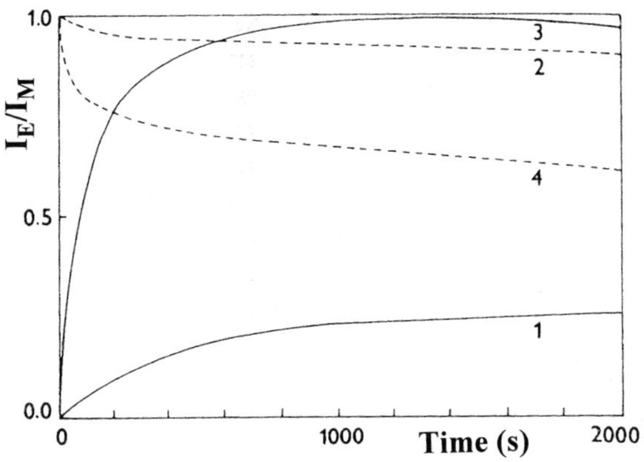

Figure 19. Normalized increase in anthracene emission at 413 nm (curves 1 and 3) and normalized decrease in carbazole emission at 364 nm (curves 2 and 4) after mixing labeled and unlabeled PS-*b*-PHI in 95/5 (v/v) heptane/dioxane (curves 1 and 2) and in 90/10 (v/v) heptane/dioxane (curves 3 and 4) at 25 °C. Reproduced from [34] with permission of the American Chemical Society.

Intermicellar collisions can apparently contribute to this process, as pointed above.

More recently, a PMMA-*b*-PMMa copolymer labeled by pyrene attached at the PMMa end and fluorene attached at the PMMA end and an unlabeled PMMA-*b*-PMMa copolymer, were investigated in ethyl acetate/methanol mixtures [30]. Micellization took place at ethyl acetate volume fractions above 80%. The analysis of the increase of the pyrene fluorescence intensity with time upon mixing of a micellar solution of unlabeled copolymer with a unimer solution of the labeled copolymer, according to a model developped by Liu [95], yielded the first values of the rate constants k^+, for the insertion of a copolymer chain in the micelle, and k^-, for the exit of a chain from a micelle. These values were found to be of the order of $2\text{-}4\times10^3$ $M^{-1}s^{-1}$ and $5\text{-}10\times10^{-4}$ s^{-1} for k^+ and k^-, respectively. Both k^+ and k^- increased with the methanol content, as expected if the PMMA micelle cores become less compact in the presence of a higher content of methanol. The value of k^- is in good agreement with those given above [34,54,94]. The value of k^+ is much smaller than that for a diffusion-controlled reaction, by a factor of up to 10^5, contrary to what is usually found for conventional surfactants [83]. This reveals the existence of a large energy barrier to association, possibly arising from the fact that free copolymer molecules are under the form of unimolecular polymer micelles [1,2,62].

More studies should be performed, preferably using copolymers in aqueous solutions, in order to check further the above results. Such low k^+ values may not be found with other copolymers, Pluronics for instance, because of their lower molecular weights and also because the PPO block is not so hydrophobic. Thus, in terms of free energy of transfer from water to the micellar pseudo-phase one methylene group is equivalent to four to five propylene oxide units [96]. The expression of k^- given by Aniansson et al [83] leads one to anticipate that the value of this rate constant for a Pluronic with about 50 propylene oxide units may be of the same order as that of SDS. Nevertheless, very small exit rate constants are expected for block copolymer micelles where one block is more hydrocarbon-like than POP, as for instance polystyrene, in water. Thus, a PS block made up of 10 repeat units is equivalent to an alkyl chain with 30 carbon atoms, on the assumption that each phenyl group adds one carbon atom to the main chain. The value of the residence time of sodium dodecylsulfate (SDS) in its micelles is about 10^{-5} s [83]. Assuming that the prefactor in the expression of the surfactant residence time has about the same value for SDS and for the copolymer containing the PS block of 10 units, a lower bound value for the residence time of such a copolymer in its micelles will be exp[1.1(30-12)] $\approx 4\times10^8$ times larger than for SDS, that is about 4000 s or one hour (The factor 1.1 comes from the fact that the free energy of transfer from the micelle to the aqueous phase is of about 1.1 RT per mole of methylene group [83]). The residence time is calculated to be 6×10^4 times larger, that is close to one year, if each phenyl groups is taken as equivalent to two carbon atoms. However, such a strongly hydrophobic alkyl chain may not remain fully in contact with water and the copolymer molecule may reduce its free energy by forming an intramolecular micelle. These simple calculations emphasize the interest in performing dynamic studies of block copolymer micelles.

Fluorescence has been used to study the kinetics of exchange of molecules solubilized in block copolymer micelles between micelles and intermicellar solution. Cao et al [97] have investigated the solubilization of pyrene in PMMa-*b*-PS-*b*-PMMa triblock copolymers in water/1,4-dioxane mixtures where the micelles have a PS core. Identical micelle populations one with solubilized pyrene to the extent that excimer formed, the other without pyrene were mixed and the time dependence of the intensity ratio I_E/I_M determined. Figure 20 shows an example of such variation, clearly characterized by two time constants. The results were interpreted as indicating that the rate determining step for the release and exchange of large hydrophobic molecules, such as pyrene, between micelles is the diffusion of the molecules out of the micelles. This process becomes very slow at long times after mixing because of the slow diffusion of the pyrene from the PS micelle core towards the surface [97].

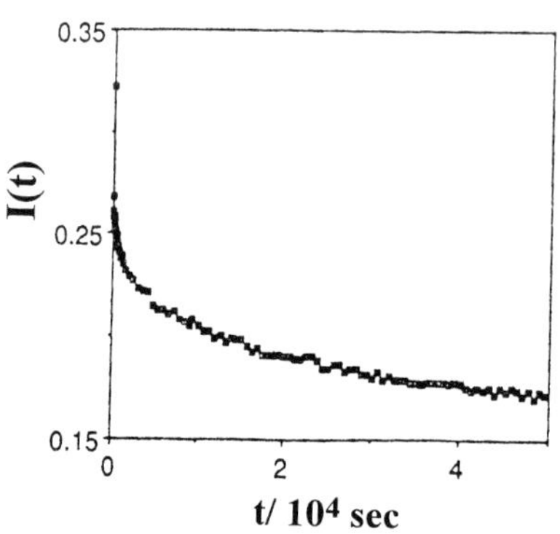

Figure 20. Time dependence of the I_E/I_M intensity ratio after mixing identical PMMa-*b*-PS-*b*-PMMa copolymer micellar solutions one with solubilized pyrene and the other without pyrene in water/1,4-dioxane mixtures. Reproduced from [97] with permission of the American Chemical Society.

Evidence of fast intermicellar exchange of pyrene molecules (time scale: 1 μs) in a 18.7 % w/w Pluronic P85 solution at above 20 °C, was obtained from time-resolved pyrene fluorescence emission at high pyrene content, where excimer occur [82]. This process was not detected at lower temperature. Its occurrence suggested that micelles cluster above a certain temperature, and that within these clusters intermicellar exchanges of pyrene molecules are greatly facilitated. This situation is reminiscent to that prevailing in micellar solutions of nonionic surfactants at the approach of the cloud temperature [98] and in certain microemulsion systems [99].

Time-resolved phosphorescence quenching of 4-bromo-1-acetonaphthone by the water-soluble quencher $NaNO_2$ permitted Hruska et al [28] to determine the exit rate constant of this probe from PS-*b*-POE micelles in water. The reported value, $5\text{-}9\times10^3$ s^{-1}, depends little on the composition and molecular weight of the copolymer and is relatively close to that found for the exit rate of the chemically similar compound 4-bromonaphthalene from micelles of the conventional surfactant CTAB: $2.5\text{-}4\times10^4$ s^{-1} [100]. However the association rate constant was found to be larger than for a diffusion-controlled process, a behavior interpreted as indicating that whenever a probe left a micelle it reentered the same micelle.

PS-*b*-PHI was labeled by anthracene attached at the PS block and investigated in 1,4-dioxane-heptane mixtures using time-resolved fluorescence anisotropy in order to gain information on the mobility of the pendant labels and polymer segments both in the micelle cores and shells and in the unimers [35]. Reorientational motions were quite fast in the unimers, but were much hindered in selective solvents for one of the blocks. The authors concluded that the label reorientational motion is extremely complex and that "the anthracene probe should be used with great care in fluorescence anisotropy studies".

7. MISCELLEANOUS STUDIES

7.1. Internal organization of block copolymer micelles

Polystyrene-*b*-poly(ethylenepropylene) copolymers have been tagged at both ends with pyrene and the cyclization of this polymer in the micelles formed in heptane, a preferentially good solvent for the poly(ethylenepropylene) block, studied by measuring the variations of the I_E/I_M ratio with molecular weight and concentration [36]. The results suggested that cyclization occurs only in unimers and that excitation spectroscopy can be used to study the partition of the copolymer between unimers and micelles.

NRET has been used to show that the free ends of PS-*b*-POE copolymers micellized in methanol/dichloromethane mixtures are distributed throughout the micelles and not accumulated at the micelle center of mass [55], in agreement with theoretical predictions [101]. The polymer were labeled with pyrene (acceptor) at the block junctions or with naphthalene (donor) at the junctions or at the ends. Note that the conclusion reached in this study is the same as for conventional surfactants where the terminal methyl groups have been shown to be distributed throughout the SDS micelle core volume [102].

Steady-state and time-resolved fluorescence anisotropy clearly showed the collapse of the cores of micelles of PS-*b*-PMMA and P*t*BMA-*b*-PS-*b*-P*t*BMA (poly *t*-butylmethacrylate) in methanol/dioxane mixtures upon increasing methanol content [37].

7.2.Solubilization

Block copolymer micelles can be used to solubilize compounds sparingly soluble in water [103]. Cao et al [97] showed that aqueous micellar solutions of PMAa-*b*-PS-*b*-PMAa copolymers can solubilize aromatic compounds such as pyrene and diphenylanthracene. The partition coefficien between micelles and water were reported to be 2.2 and 1.1×10^5, respectively. The micelle-solubilized pyrene was shown to be distributed between the PS core and the PMAa corona and its solubility decreased at higher pH.

The partition coefficient of pyrene between PS-*b*-POE micelles and water was reported to be 3×10^5 [42], on the basis of pyrene lifetime and intensity ratio I_1/I_3, and $I_{338}/I_{332.5}$ (from the excitation spectra) measurements. This value is rather close to that for PMAa-*b*-PS-*b*-PMAa copolymers. Note that the values of the partition coefficients of pyrene between micelles of conventional surfactants and water are around 10^5 [42,104]. The use of the I_1/I_3 ratio for determining pyrene partition coefficients has been criticized [105]. Nakashima et al [106] showed that the method was valid under the conditions used in refs [42].

The solubilization of benzene in aqueous micellar solutions of PS-*b*-PMAa copolymers has been investigated using the pyrene I_E/I_M ratio [31,107]. The value of this ratio levels off when the micelles become saturated with benzene.

The partition coefficients of pyrene and DPH between Pluronic P85 and F108 micelles and water has been determined from fluorescence intensity measurements [38]. The partition coefficient was shown to increase with temperature as expected from the decrease of cmc and increasing hydrophobicity of the copolymers.

7.3. Possible effect of the initiator on the fluorescence properties of block copolymers

Hruska et al [108] showed that the initiator (cumyl potassium) used for preparing PS-*b*-POE copolymers can, under certain conditions, form fluorescent species by fusion of two cycles. These species are attached to the synthesized copolymers and result in unwanted fluorescence emission. This may apply to other initiators and other copolymers as well.

8. CONCLUSIONS

This Chapter reviewed the kind of information that fluorescence probing methods can provide on the micellization of block copolymers in water and other solvents. Much has been achieved concerning micelle formation (cmc and cmT), micropolarity and microviscosity of the micelles, as well as for solubilization by, and internal organization of, block copolymer micelles in water and organic solvents. However, much remain to be done concerning measurements of micelle aggregation numbers and studies of micelle dynamics. For the latter, methods using non radiative energy transfer appear to be most promising. The high microviscosity of block copolymer micelles constitutes a major difficulty in using time-resolved fluorescence quenching methods for the determination of micelle aggregation numbers. Additional measurements on samples of block copolymers from which hydrophobic impurities (homopolymers in the case of diblock copolymers, or diblock copolymers in the case of triblock copolymers) have been removed appear to be required if the changes of micellar properties with the block copolymer molecular weight and composition are to be evaluated more accurately.

REFERENCES

1 C. Sadron, Pure Appl. Chem., 4 (1962) 347.
2 P. Marie, J. Selb and Y. Gallot, C. R.Acad.Sci. (Paris), 282C (1976) 703.
3 Zana, Y. Muto, K. Esumi and K. Meguro, J. Colloid Interface Sci., 123 (1988) 502.
4 X. Auvray, C. Petipas, R. Anthore, I. Rico, A. Lattes, A. Ahmad-Zadeh Samii and A. de Savignac, Colloid Polym. Sci., 265 (1987) 925.
5 J. Lang and R. Zana, Surfactant Solutions: New Methods of Investigation, R. Zana (ed.), M. Dekker, New-York, Chapter 8, p. 405, 1987.
6 C.A. Parker, Photoluminescence of Solutions, Elsevier, Amsterdam, 1968.
7 J.B. Birks, Photophysics of Aromatic Molecules, Wiley-Interscience, New-York, 1970; Organic Molecular Photophysics, Wiley-Interscience, New-York, 1976.
8 N.J. Turro, Modern Molecular Photochemistry, Benjamin-Cummings, Menlo-Park, 1978.
9 J.N. Demas, Excited State Lifetime Measurements, Academic Press, New-York, 1983.
10 J. Lakowicz, Principles of Photochemistry in Microheterogeneous Systems, Academic Press, New-York, 1987.
11 R. Zana, Surfactant Solutions: New Methods of Investigation, R. Zana (ed.), M. Dekker, New-York, 1987, Chapter 5.
12 N.J. Turro, M. Grätzel and A.M. Braun, Angew. Chem. Int. Ed., 19 (1980) 675.
13 J.K. Thomas, Chem. Rev., 80 (1980) 283.
14 J. Lang, The Structure, Dynamics, and Equilibrium Properties of Colloidal Systems, D.M. Bloor and E. Wyn-Jones (eds.), NATO ASI Series, Kluwer Academic Press, 1990, p. 1.
15 M. Gehlen and F.C. De Schryver, Chem. Rev., 93 (1993) 199.
16 M. Almgren, Adv. Colloid Interface Sci., 41 (1992) 9.
17 A.V. Barzykin and M. Tachiya, Heter. Chem. Rev., 3 (1996) 105.
18 K. Kalyanasundaram and J.K. Thomas, J. Am. Chem. Soc., 99 (1977) 2039.
19 D.C. Dong and M.A. Winnik, Photochem. Photobiol., 35 (1982) 17.
20 K. Kalyanasundaram and J.K. Thomas, J. Phys. Chem., 81 (1977) 2176.
21 J.C. Dederen, L. Coosemans, F.C. De Schryver and A. Dederen, Photochem. Photobiol., 30 (1979) 443.
22 K. Zachariasse, Chem. Phys. Lett., 57 (1978) 429.
23 F.M. Winnik, M.A. Winnik, H. Ringsdorf and J. Venzmer, J. Phys. Chem., 95 (1991) 2583.
24 F. Hare, J. Amiell and C. Lussan, Biochem. Biophys. Acta, 555 (1979) 388.
25 K. Nakashima, T. Anzai and Y. Fujimoto, Langmuir, 10 (1994) 658.
26 N.J. Turro and C. Chung, Macromolecules, 17 (1984) 2123.
27 (a) M. Ikemi, N. Odagari and I. Shinohara, Polym. J., 11 (1980) 777; (b) M. Ikemi, N. Odagari, S. Tanaka, I. Shinohara and A. Chiba, Macromolecules, 14 (1981) 34; (c) M. Ikemi, N. Odagari, S. Tanaka, I. Shinohara and A. Chiba, Macromolecules, 15 (1982) 281.
28 Z. Hruska, M. Piton, A. Yekta, J. Duhamel, M.A. Winnik, G. Riess and M.D. Croucher, Macromolecules, 26 (1993) 1825.
29 F. Grieser and C.J. Drummond, J. Phys. Chem., 92 (1988) 5580.
30 C.K. Smith and G. Liu, Macromolecules, 29 (1996) 2060.

31 D. Kiserow, K. Prochazka, C. Ramireddy, Z. Tuzar, P. Munk and S.E. Webber, Macromolecules, 25 (1992) 461
32 F. Quina, E. Abuin and E. Lissi, Macromolecules, 23 (1990) 5173.
33 H. Morawetz, Macromolecules, 29 (1996) 2689.
34 K. Prochazka, B. Bednar, E. Mukhtar, P. Svoboda, J. Trnena and M. Almgren, J. Phys. Chem., 95 (1991) 4563.
35 K. Prochazka, B. Medhage, E. Mukhtar, M. Almgren, P. Svoboda, J. Trnena and B. Bednar, Polymer, 34 (1993) 103.
36 D.A. Ylitalo, A.S. Yeung and C.W. Frank, Polym. Prep. 30(2) (1989) 561.
37 D. Kiserow, J. Chan, C. Ramireddy, P. Munk and S.E. Webber, Macromolecules, 25 (1992) 5338.
38 A.V. Kabanov, I.R. Nazarova, I.V. Astafieva, E.V. Batrakova, V.Y. Alakhov, A.A. Yaroslavov and V.A. Kabanov, Macromolecules, 28 (1995) 2303.
39 C.N. Henderson, B.K. Selinger and A.R. Watkins, J. Photochem., 16(1981) 215.
40 M.V. Encinas and E.A. Lissi, Chem. Phys. Lett., 91 (1982) 55.
41 C.L. Zhao, M.A. Winnik, G. Riess, M.D. Croucher, Langmuir, 6 (1990) 514.
42 M. Wilhelm, C.L. Zhao, Y. Wang, R. Zu, M.A. Winnik, J.L. Mura, G. Riess and M.D. Croucher, Macromolecules, 24 (1991) 1033.
43 I. Astafieva, X. Zhong and A. Eisenberg, Macromolecules, 26 (1993) 7339.
44 N.J. Turro and A. Yekta, J. Am. Chem. Soc., 100 (1978) 5951.
45 N.J. Turro and P.L. Kuo, J. Phys. Chem., 90 (1986) 4205.
46 P. Infelta, Chem. Phys. Lett., 61 (1979) 88.
47 A. Yekta, M. Aikawa and N.J. Turro, Chem. Phys. Lett., 63 (1979) 543.
48 M. Almgren, J.E. Löfroth and J. van Stam, J. Phys. Chem., 90 (1986) 4431.
49 M. Van der Auweraer, S. Reekmans, N. Boens and F.C. De Schryver, Chem. Phys., 132 (1989) 91.
50 M. Almgren, J. Alsins, E. Mukhtar and J. van Stam, J. Phys. Chem., 92 (1988) 4479.
51 M. Almgren and J.E. Löfroth, J. Chem. Phys., 76 (1982) 2734.
52 G. Warr and F. Grieser, J. Chem. Soc., Faraday Trans. 1, 82 (1986) 1813.
53 A.V. Barzykin and M. Tachiya, Het. Chem. Rev., 3 (1996) 105.
54 Y. Wang, C.M. Kausch, M. Chun, R.P. Quirk and W.L. Mattice, Macromolecules, 28 (1995) 904.
55 K. Rodrigues, C.M. Kausch, J. Kim, R.P. Quirk and W.L. Mattice, Polym. Bull., 26 (1991) 695.
56 T.J. Martin and S.E. Webber, Macromolecules, 28 (1995) 8845.
57 N.K. Reddy, P.J. Fordham, D. Attwood and C. Booth, J. Chem. Soc., Faraday Trans., 86 (1990) 1569.
58 M. Almgren, J. Alsins and P. Bahadur, Langmuir, 7 (1991) 446.
59 P. Alexandridis, V. Athanassiou, S. Fukuda and T.A. Hatton, Langmuir, 10 (1994) 2604.
60 O. Anthony and R. Zana, Macromolecules, 27 (1994) 3885.
61 A. Jada, B. Siffert and G. Riess, Colloids Surf. A: Physicochem Eng. Aspects, 75 (1993) 203.
62 Z. Tuzar and P. Kratochvil, Surface and Colloid Science Series, Vol 15, E. Matijevic (ed.), Plenum Press, p.1, 1993.
63 P. Alexandridis, T. Nivaggioli and T.A. Hatton, Langmuir, 11 (1995) 1468.

64 T. Nivaggioli, P. Alexandridis, T.A. Hatton, A. Yekta and M. Winnik, Langmuir, 11 (1995) 730.
65 Y. Morishima, T. Hashimoto, Y. Itoh, M. Kamashi and S.I. Nozakura, J. Polym. Sci., Polym. Chem. Ed., 20 (1982) 299.
66 Y. Morishima, Y. Itoh, T. Hashimoto and S.I. Nozakura, J. Polym. Sci., Polym. Chem. Ed., 20 (1982) 2007.
67 G. Kwon, M. Saito, M. Yokoyama, T. Okano, Y. Sakura and K. Kataoka, Langmuir, 9 (1993) 945.
68 I. Astafieva, K. Khougaz and A. Eisenberg, Macromolecules, 28 (1995) 7127.
69 P. Alexandridis, V. Athanassiou and T.A. Hatton, Langmuir, 11 (1995) 2442.
70 M.D. Major, J.M. Torkelson and A.M. Brealey, Macromolecules, 23 (1990) 1700.
71 A. Watanabe and M. Matsuda, Macromolecules, 18 (1985) 273.
72 F. Calderara, Z. Hruska, G. Hurtrez, J.P. Lerch, T. Nugay and G. Riess, Macromolecules, 27 (1994) 1210.
73 K.C. Dowling and J.K. Thomas, Macromolecules, 23 (1990) 1059.
74 T. Nivaggioli, B. Tsao, P. Alexandridis and T.A. Hatton, Langmuir, 11 (1995) 119.
75 N.J. Turro and P.L. Kuo, Langmuir, 2 (1986) 438.
76 O. Anthony and R. Zana, Langmuir, 12 (1996) 1967 and 3590.
77 P. Alexandridis, J.F. Holzwarth and T.A. Hatton, Macromolecules, 27 (1994) 2414.
78 A. Watanabe and M. Matsuda, Macromolecules, 19 (1986) 2253.
79 K. Prochazka, D. Kiserow, C. Ramireddy, Z. Tuzar, P. Munk and S.E. Webber, Macromolecules, 25 (1992) 454.
80 M. Almgren, J. van Stam, C. Lindblad, P. Li, P. Stilbs and P. Bahadur, J. Phys. Chem. 95 (1991) 5677.
81 M. Almgren, W. Brown and S. Hvidt, Colloid Polym. Sci., 273 (1995) 2.
82 P. Lianos and W. Brown, J. Phys. Chem., 96 (1992) 6439.
83 E.A.G. Aniansson and S. Wall, J. Phys. Chem., 78 (1974) 1024 and 79 (1975) 986; E.A.G. Aniansson, S. Wall, M. Almgren, H. Hoffmann, I. Kielman, W. Ulbricht, R. Zana, J. Lang and C. Tondre, J. Phys. Chem., 80 (1976) 905.
84 M. Kahlweit, J. Colloid Interface Sci., 90 (1981) 92; Pure Appl. Chem. 53 (1981) 2069.
85 A. Halperin and S. Alexander, Macromolecules, 22 (1989) 2403.
86 W.L. Mattice, Macromolecules, 29 (1996) 4761.
87 M. Malmsten and B. Lindman, Macromolecules, 25 (1992) 5440.
88 G. Fleischer, J. Phys. Chem., 97(1993) 517.
89 E. Hecht and H. Hoffmann, Colloids Surf. A: Physicochem. Eng. Asp., 96 (1995) 181.
90 C. Booth, T. Taylor, C. Price, N. Rajab and R. Stubbersfield, J. Chem. Soc., Faraday Trans. 1, 74 (1978) 2352; C. Price, A.L. Hudd, C. Booth and B. Wright, Polymer, 23 (1982) 650.
91 Z. Tuzar, V. Petrus and P. Kratochvil, Makromol. Chem., 175 (1974) 3181.
92 B. Bednar, K. Edwards, M. Almgren, S. Tormod and Z. Tuzar, Makromol. Chem., Rapid Comm., 9 (1988) 785.
93 P. Munk, C. Ramieddy, M. Tian, S.E. Webber, K. Prochazka and Z. Tuzar, Makromol. Chem., Macromol. Symp., 58 (1992) 195.
94 Y. Wang, R. Balaji, R.P. Quirck, and W.L. Mattice, Polym. Bull., 26 (1992) 333.
95 G. Liu, Can. J. Chem., 73 (1995) 1995.
96 G. Wanka, H. Hoffmann and W. Ulbricht, Colloid Polym. Sci., 268 (1990) 101.

97 T. Cao, P. Munk, C Ramireddy, Z. Tuzar and S.E. Webber, Macromolecules, 24 (1991) 6300.
98 W. Binana-Limbélé, N.M. Van Os, L.A.M. Ruppert and R. Zana, J. Colloid Interface Sci., 144 (1991) 458 and references therein.
99 J. Lang, N. Lalem and R. Zana, Colloids Surf. 68 (1992) 199 and references therein.
100 R. Zana, Surfactants in Solution, Vol. 4, K.L. Mittal and P. Bothorel (eds.), Plenum Press, New York, p. 115, 1986 and references therein.
101 K. Rodrigues and W.L. Mattice, J. Chem. Phys., 94 (1991) 761.
102 B. Cabane, R. Duplessix and T. Zemb, J. Phys (Paris), 46 (1985) 2161.
103 R. Nagarajan and S. Ganesh, Macromolecules, 22 (1989) 4312.
104 M. Almgren, F. Grieser and J.K. Thomas, J. Am. Chem. Soc., 101 (1979) 279.
105 W.A. Acree, S.A. Tucker and D.C. Wilkins, Macromolecules, 26 (1993) 7366.
106 K. Nakashima, M.A. Winnik, K.H. Dai, E.J. Kramer and J. Washiyama, Macromolecules, 26 (1993) 7367.
107 K. Prochazka, D. Kiserow, C. Ramireddy, S.E. Webber, P. Munk and Z. Tuzar, Makromol. Chem., Macromol. Symp., 58 (1992) 201.
108 Z. Hruska, M. Winnik, G. Hurtrez and G. Riess, Polym. Comm., 31 (1990) 402.

Direct-Imaging Cryo-Transmission Electron Microscopy in the Study of Colloids and Polymer Solutions

Marganit Goldraich and Yeshayahu Talmon

Department of Chemical Engineering, Technion - Israel institute of Technology, Haifa 32000, Israel.

ABSTRACT

Transmission electron microscopy at cryogenic temperatures is by now a well established technique, applied in the study of a wide range of microstructured liquids, including solutions and dispersion of polymers. In a comprehensive review in the context of this volume centered on amphiphilic block copolymers, we start with the rationale for using cryo-TEM, the basics of the technique, its limitations and its potential. We continue with a survey of the work that has been done on synthetic surfactants in various states of aggregation, emphasizing the type of information one can get from the technique, and how that information is used together with data from non-imaging techniques to completely characterize the microstructure of a liquid or a semi-liquid system. We next survey the work on biologically originated systems, excluding biological systems such as cellular organelles and viruses. Throughout the review we try to elucidate the flexibility of the technique and how its applicability can be extended by proper design of the experiments. Finally, we describe how cryo-TEM has been applied in the study of water-soluble polymers and polymer-surfactant systems, including experiments which involve on-the-grid processing and time-resolved cryo-TEM. Gelling systems are prime candidates to be examined by the latter techniques.

1. INTRODUCTION

It is well accepted now that direct imaging to display structural features with typical dimensions of a few nanometers to several millimeters is essential for complete, unequivocal structural characterization of material systems. The choice of technique is primarily determined by the resolution it provides, but also by the nature of the information it gives, e.g., bulk or surface images, thin films or bulk specimens. Most fluid systems are labile, sensitive to make-up, concentration, and temperature changes, and quite often sensitive to external forces such as shear or compression. Most imaging techniques impose restrictions on the nature of the specimens that can be studied. Thus it takes a careful match of system, imaging technique and specimen preparation to obtain an image that reliably represents the structure of the system.

Imaging techniques may be classified into optical and non-optical ones. While light microscopy and transmission electron microscopy (TEM) are optical techniques, all scanning probe microscopies (scanning tunneling microscopy [STM], atomic force microscopy [AFM], scanning electron microscopy [SEM], etc.) are non-optical, i.e., images are combination of picture elements (pixels), each modulated by a signal characteristic of the technique. While electron microscopy (EM) can be performed only under high vacuum, light microscopy may be carried out in ambient atmosphere. AFM may also be carried out under ambient condition, but specimens need to be anchored to a solid surface.

Specimens for EM must be made compatible with the vacuum in the microscope, and all internal supramolecular motion needs to be arrested, to avoid image blurring. This process is called fixation. Chemical fixation methods that involve adding a compound alien to the system (a fixative/stain) are in most cases not acceptable for soft matter systems (Talmon, 1983; Kilpatrick et al., 1985); thus physical (thermal) fixation is used. Under favorable conditions, such as in thin specimens prepared for cryo-TEM, the liquid phase is vitrified, not frozen, so that the specimen does not undergo phase transformation and component rearrangement, and hence microstructural changes are avoided.

In this review we concentrate on direct-imaging cryo-TEM of thin vitrified specimens of microstructured liquid systems, also known as 'complex fluids' or 'soft condensed matter'. In this technique a thin liquid film, ca 0.3 micrometer thick, of the system to be examined is prepared in a chamber where temperature and humidity are controlled. A small drop is applied onto a perforated polymer film supported on a copper grid, held by a tweezer. Most of the liquid is blotted away, leaving thin liquid films spanning the holes in the polymer film. The grid is then propelled by a spring-loaded mechanism into liquid ethane at its freezing point. The combination of high surface area-to-volume ratio, low temperature and large temperature difference between the freezing and boiling points of the cryogen, and the velocity of entry into the cryogen produces very high heat transfer rates which lead to cooling rates on the order of 100,000 K/s, as has been experimentally shown by Siegel et al. (1994).

Several experimental set-ups are available for controlled-environment. The so-called Controlled Environment Vitrification System (CEVS) was originally developed and described by Bellare et al. (1988). A variant of this has been built and used by Egelhaaf (1995). More recently a flow-thru CEVS (FT-CEVS) was developed by Fink and Talmon (1994), and was described in more detail by Talmon (1996). In all systems evaporation is minimized (except for controlled drying experiments) to avoid concentration changes, and vitrification can be done from room temperature and above. The CEVS allows vitrification of a system initially below room temperature, while the FT-CEVS makes it possible to perform rapid temperature jumps (duration on the order of a second) under fixed relative humidity conditions. Another variant of the CEVS (Siegel et al., 1994) allows us to induce very rapid temperature jumps (tens of degrees in a few millisecond) by shining an intense light beam on the specimens as it travels towards the cryogen. After preparation the vitrified specimens are stored under liquid nitrogen. A vitrified specimen to be examined by cryo-TEM is loaded into a TEM cooling-holder, which keeps it at cryogenic temperatures (typically around

100 K) throughout the examination in the TEM. It should be emphasized that no cryo-protectants (to suppress crystallization) or stains (to enhance contrast) are used.

Vitrified specimens have typically very little inherent contrast and are sensitive to radiolysis by the electron beam. One needs to employ 'minimal-dose techniques' leading to relative low signal-to-noise ratio. Those factors together with the limited mechanical stability of the cooling-holder limit image resolution to about 1 nm. Also, the technique is limited to thin objects, with one dimension at least smaller than about 300 nm. Elastic objects such as liposomes may be flattened during specimen thinning. Quite often one relies on 'phase-contrast' to obtain sufficient contrast for imaging. In the TEM this is achieved by proper defocus of the objective lens, which transforms phase differences to amplitude differences. One needs to be careful, however, as this may affect resolution, or lead to electron-optical artifacts. More about the technique, its applications and limitations may be found in a recent review by Talmon (1996).

While only relatively low-viscosity fluids can be made into thin liquid films, the applicability of the technique has been extended by 'on-the-grid-processing' (OTGP): a thin liquid film of a low-viscosity precursor is prepared, and the more viscous phase is then formed on the grid by inducing a physical or chemical change, such as heating, cooling, controlled drying, pH-change, or mixing of reactants. (Talmon et al., 1990; Siegel et al., 1994; Danino et al., 1997a). Such an on-the-grid process may be stopped before it reaches completion by plunging the grid into the cryogen, to capture intermediate microstructures. A series of such experiments with increasing OTGP durations gives time-resolved cryo-TEM (Siegel et al., 1989; Sein et al., 1995), which provides 'time-sectioning' of the process.

Although in direct-imaging cryo-TEM the possibility of producing image artifacts is minimized, one must always take that possibility into account. Because of that, it is best to compare results of different imaging techniques, or combine imaging techniques with indirect techniques such as small-angle x-ray scattering (SAXS), small-angle neutron scattering (SANS), NMR, fluorescence, and conductimetry. The latter combination is ideal for complete microstructural characterization of a fluid system. Microscopy identifies the building blocks of the system. Those are used to construct a physical model, needed to interpret the quantitative data of the non-imaging techniques. The agreement between the results helps rule out artifacts in the imaging technique.

2. IMAGING SURFACTANT SOLUTIONS

2.1 Micelles

The first report on direct observations by cryo-TEM of micelles involved cylindrical micelles of cetyltrimethylammonium-3,5 dichlorobenzoate (CTA-3,5 DCB) (Bellare et al., 1986). Images of spheroidal micelles, about 9 nm in diameter, were shown by Burns et al. (1990) in the aqueous solutions of HCO-60, a polyethylene castor oil derivative, and linoleic acid. The existence of globular micelles near and well above the critical micellar concentration was evident using cryo-TEM in cetyltrimethylammonium bromide (CTAB) solutions (Vinson et al., 1991).

Spheroidal micelles have been imaged by cryo-TEM in systems of anionic and cationic surfactants in non-aqueous solution as well, despite difficulties of thin-film preparation and low image contrast. Micelles of CTAB in glycerol and sodium dodecyl sulfate (SDS) in formamide were found to resemble closely the appearance of the corresponding structures in aqueous solutions (Lin et al., 1997).

'Rod-like', 'worm-like', 'cylindrical' or 'thread-like' micelles, all synonymous terms for elongated micelles, are found in many surfactant solutions under the proper conditions, and are easily visualized by cryo-TEM. For example, they have been imaged in solutions of the nonionic surfactant hexaethyleneglycol-monohexadecylether ($C_{16}E_6$) at low concentrations (1-3 %wt.), at 30-34 °C, with and without added salt (NaCl or NaSCN). The diameter of those flexible micelles was found to be about 6 nm, and their length up to a micrometer (Lin et al., 1992). That was compared to the SANS study by Cummins et al. (1987), who had assumed the micelles to be rigid cylindrical rods, and deduced that the rods had a diameter of about 6 nm and a length ranging from 160 to 400 nm, depending on temperature. 'Worm-like' micelles of 5 nm diameter were also seen in solutions of the cationic surfactant CTAB with added salt (NaBr) (Vinson et al., 1991). The addition of sodium salicylate to ionic surfactants such as cetyltrimethylammonium chloride (CTAC) led to giant micelle formation. Images of CTAC/sodium salicylate/NaCl solutions showed very long flexible cylindrical micelles and some closed loops formed by those thread-like micelles (Vinson and Talmon, 1989; Clausen et al., 1992).

The evolution of size and shape can be nicely demonstrated by direct imaging. The transition from spherical to cylindrical micelles upon increasing surfactant concentrations was studied in the hexadecyltrimethylammonium dichloro-benzoate system, where two different dichlorobenzoate counterions affect the micelle shape (Magid et al., 1990). With the 3,5-dichlorobenzoate, spheroidal micelles were imaged at low concentrations, while cylindrical ones were imaged at higher concentrations. At intermediate concentrations images of mixed spheroidal and cylindrical structures were recorded.

Many mixed surfactant systems have been studied using cryo-TEM. An example where direct visualization by cryo-TEM confirmed the results of model-dependent methods (self-diffusion and specific electrodes measurements) was presented in the microstructural investigation of SDS and dodecyltrimethylammonium bromide (DTAB) mixed micelles. Mostly spheroidal or slightly elongated micelles were seen in both the SDS-rich and DTAB-rich isotropic solutions (Kamenka et al., 1992). Changes in size of microstructures formed in systems of alkyltrimethylammonium halide (C_nTAX), NaCl and bile-salts were recorded by cryo-TEM (Swanson-Vethamuthu et al., 1996). With increasing concentration of the bile-salt, micellar size decreased, leading to transition from thread-like micelles to spheroidal micelles.

Mixtures of the nonionic surfactant $C_{10}E_5$ and perfluorooctyl sulfonates $C_8F_{17}SO_3X$ with two counterions ($X=N^+(C_2H_5)_4$, Li^+) form micellar solutions of either elongated or spheroidal micelles, respectively. At increasing concentration of the $N^+(C_2H_5)_4$ counterion, the solution becomes more viscous, and dense circular structures (micelles?) start to appear (Knoblich et al., 1995).

Cryo-TEM of the nonionic surfactant octadecylamide oligo(oxyethylene)ether showed spherical and elongated micelles at low concentrations, while elongated micelles were observed at higher concentrations (Khan et al., 1996). That study provided a good demonstration of the distribution of aggregates in the vitrified specimen. Because the edge of the vitrified liquid film over a hole is thicker than the film in the hole center area, one sees a projection of many overlapping micelles near the hole edge. In the thinner areas individual micelles can be seen. Examples of images taken from this system can be seen in Figure 1 where at 1.0 % (Fig. 1a) a mixture of spheroidal micelles (arrowheads) and thread-like micelles (arrows) can be seen. Some of the shorter threads were closed into rings (R). At 10 % (Fig. 1b), long thread-like micelles are visible. In the thicker areas (T), the projection of many overlapping micelles is seen, while at the thinner areas (t), individual micelles may be detected (arrows). The dots (arrowheads) are in all probability thread-like micelles, aligned closely parallel to the electron beam. At 20 % (Fig. 1c), in the thicker areas (T) it is difficult to make out individual micelles, while the thread-like nature of the micelles becomes quite clear as one goes to the thinner areas (M).

Long thread-like micelles render the solution viscoelastic. A combined cryo-TEM and rheology study was performed on aqueous solutions of CTAC, NaCl and different concentrations of sodium salicylate. At low salicylate concentrations, globular micelles, 5 nm in diameter, were seen, while the solution was Newtonian. At increasing counterion concentration the solution turned viscoelastic, and entangled thread-like micelles, 5 nm in diameter and up to several micrometers long, were recorded (Clausen et al., 1992). Another study which used the combination of cryo-TEM and rheometry focused on the transition from spherical to thread-like micelles in CTAB solutions upon the addition of methylsalicylic acid or hydroxybenzoic acid. The transition was found to occurs over a range of acid concentrations, where both types of micelles where shown to coexist. When the majority of the surfactant was in the thread-like structure, the solutions were viscoelastic (Lin et al, 1994a).

Viscoelastic samples that are not allowed to relax between blotting and vitrification may display alignment of the thread-like micelles. To avoid this effect and to reach a state closer to that of the unsheared bulk solution, samples should be allowed to relax sufficiently after blotting. From rheological data for CTAC/NaCl/NaSal, relaxation times at low strain are on the order of 1-10 seconds, about the same duration between blotting and plunging (Clausen et al., 1992).

Zana and Talmon (1993) investigated the microstructures formed in a novel class of surfactants called "dimeric" or "gemini" surfactants. These are made of two quaternary ammonium surfactant molecules, m carbon atoms long, linked at the head-group level by a hydrocarbon spacer of s ethylene groups, with bromide counterions. Cryo-TEM has shown how the spacer length provides an additional parameter affecting surfactant packing. For example, in the $m=12$ dimeric surfactant series, when the spacer length is increased from $s=2$ to $s=20$, the recorded structures change from thread-like micelles at short spacers ($s=2, 3$), to spheroidal micelles at intermediate length ($s=4$ to 12), to thread-like again ($s=14$), and to vesicles at $s=20$ (Danino et al., 1995a).

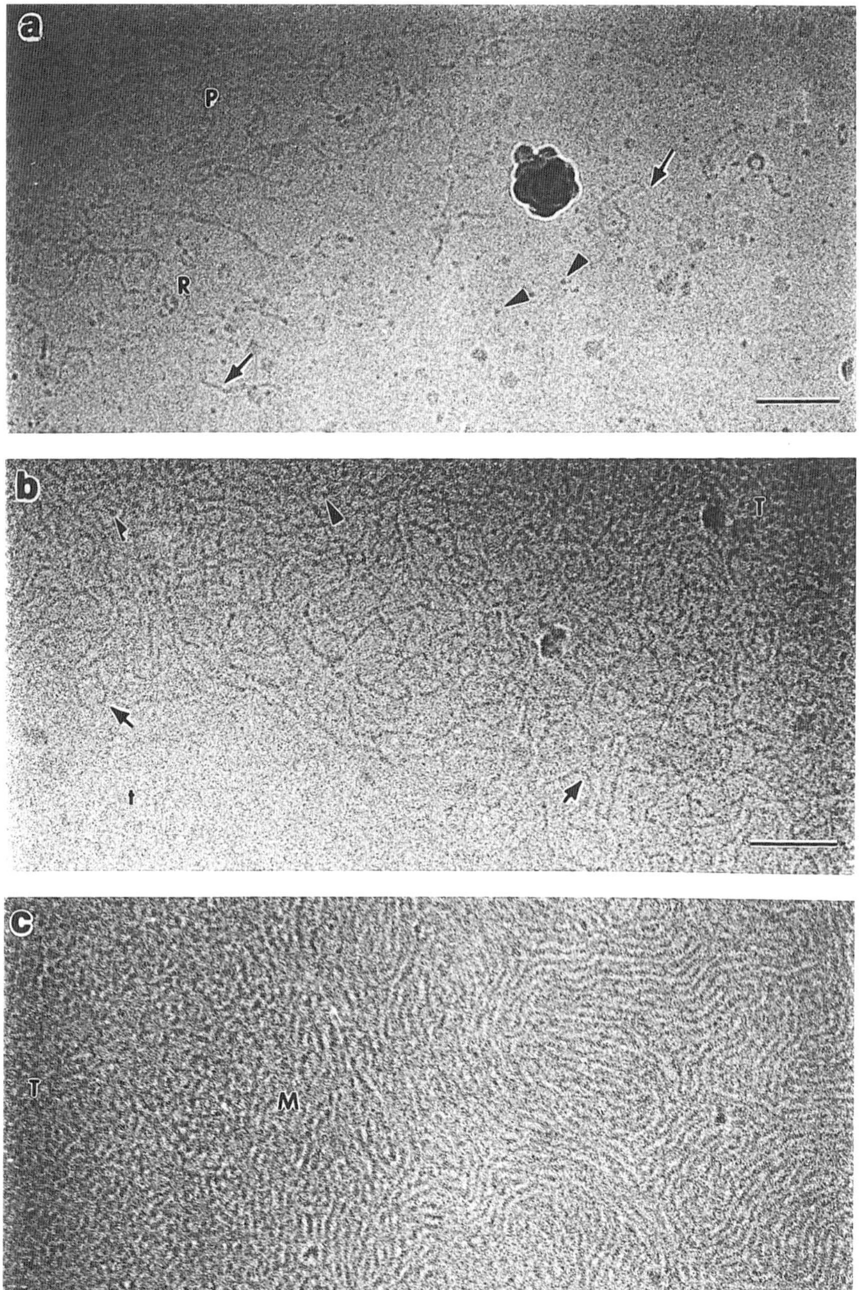

Figure 1. Cryo-TEM images of aqueous solutions of octadecylamide oligo(oxyethylene)ether at 1.0 % (a), 10 % (b) and 20 % (c). For details, see text. Bar = 100 nm (from Khan et al. (1996) with permission).

Some surfactants that form very long thread-like micelles exhibit peculiar rheological properties that could be explained by micellar branching (Candau et al., 1993). The first unequivocal observation of such branched thread-like micelles in a one-surfactant system was reported by Danino et al. (1995b) for the aqueous solution of the triquaternary ammonium (trimeric) surfactant 12-3-12-3-12 (using the same notation as for the dimeric surfactants). Branched thread-like micelles in earlier reports might have been in fact intertwined micelles (Harwigsson et al., 1994). Branched micelles have been observed also in mixtures such as those of dodecyldimethylamine oxide , sodium laureth sulfate and NaCl (Lin, 1996), and in various surfactant systems used for drag reduction, e.g., Lu et al. (1997). Clearly branched micelles have been also seen during the solubilization of lecithin vesicles with $C_{12}SO_4^-$ (Silvander et al., 1996), and in the mixture of C_{18}TAB, sodium salt of cholic acid (NaC), and the bile-salt sodium deoxycholate (Swanson-Vethamuthu et al., 1996).

2.2 Vesicles

Cryo-TEM of vitrified specimens had been used early on to examine vesicular and liposomal dispersions of the synthetic systems of sodium 4-(1'heptylnonyl) benzene sulfonate (SHBS) and octyldodecyldimethylammonium bromide (ODDAB), as well as in the biological colloidal system of egg yolk lecithin (L-α-phosphatidylcholine) (Talmon, 1986). A more comprehensive study on vesicles formed by SHBS was performed by combining video-enhanced light microscopy (VELM) and cryo-TEM to allow direct visualization of sizes spanning six orders of magnitude (Miller et al., 1987). By VELM one can follow in real time the undulation of individual layers within birefringent liposomes and the caged movement of smaller vesicles entrapped within larger vesicles. However, smaller structures are beyond the resolution limit of the light microscope. The same dispersions examined by cryo-TEM revealed the coexistence of vesicles within vesicles and coiled tubules within vesicles. Large liposomes were absent in the vitrified samples, and were probably located in the regions of the vitrified ice that were too thick for viewing, or were altogether excluded from the sample during preparation. The structural similarity of the VELM and cryo-TEM results suggested that the two methods visualized a continuity of self-similar structures.

Many double-tailed surfactant systems exhibit vesicle and liposome formation, that can be easily visualized by direct-imaging cryo-TEM. Recent examples are by Regev et al. (1994) and by Danino et al. (1994, 1995a). The latter work is on the gemini surfactants mentioned earlier. Cryo-TEM has also been useful in showing how vesicles of the 12-20-12 surfactant are solubilized either by the 12-10-12 surfactant or its "monomeric analog", DTAB. The solubilization process in these cases does not involve thread-like micelles as intermediate microstructure. The vesicles break up in the case of DTAB. The 12-10-12 surfactant causes the formation of a bimodal population of large liposomes and small vesicles. Final break-up ends in spheroidal micelles in both cases (Danino et al., 1996). Also silicone surfactants that are made of a methylated siloxane backbone coupled to one or more polar groups exhibit vesicles or vesicles coexisting with micelles, as shown by cryo-TEM (Lin et al., 1994b). Intermediate structures such as open vesicles and pieces of lamellae are observed at the vesicle-to-micelle transition.

Solutions of surfactant mixtures give rise to synergistic effects; mixtures of cationic and anionic surfactants appear to be particularly promising (Lucassen et al., 1981). This becomes evident as one probes their microstructure. As mentioned above, micelles were observed in the SDS/DTAB system either in the SDS- or the DTAB-rich regions. In between those two regions there is a solubility gap, were large vesicles were observed by cryo-TEM (Kamenka et al., 1992). Unilamellar vesicles and some multilamellar particles were observed also in the mixed system of the cationic surfactant cetyltrimethylammonium tosylate and the anionic surfactant sodium dodecylbenzenesulfonate (Chiruvolu et al., 1995). Another system of oppositely charged surfactants, CTAB and sodium octylsulfate (SOS), was explored by cryo-TEM. Spontaneously formed vesicles were observed over a wide range of compositions in either the CTAB-rich or the SOS-rich regions (Yatcilla et al., 1996). In the system of CTAB and dodecylbenzenesulfonic acid (HDBS), a variety of microstructures were seen by cryo-TEM: spherical unilamellar vesicles, multilamellar vesicles, and bilayered microtubules. In the presence of ferrous chloride the aggregates became larger but the bilayers were not destroyed (Yaacob and Bose, 1996).

The transition from large vesicles and liposomal aggregates formed by double-chained quaternary ammonium surfactants to small unilamellar vesicles and micelles is induced by the presence of mixed counterions. This was studied by time-resolved fluorescence quenching (TRFQ), VELM, and cryo-TEM in solutions of ditetradecyldimethylammonium with mixtures of bromide/acetate counterions. On reducing the ratio of bromide-to-acetate counterions, the solution forms smaller unilamellar vesicles, while large multilamellar structures disappear and are replaced by extended "worm-like" microtubules. When the acetate counterion reaches 74%, very small spherical aggregates in addition to small unilamellar vesicles were observed (Miller et al., 1988). This effect of counterions was later studied also by Regev and Khan (1994).

Some synthetic surfactants have been used as model membranes. For example, "rough vesicles" were imaged by cryo-TEM in dispersions of dihexadecyl phosphate (DHP), sonicated at 80°C. Sonicated dispersions cooled to room temperature were dominated by open and folded bilayer fragments (Hammarström et al., 1995).

Lipid liposomes can be loaded with drugs to become targeted drug carriers. Vesicles with core of precipitated drug doxorubicin were seen by cryo-TEM (Lasic et al., 1992). Vesicles may be used to encapsulate solid particles. Direct images of stable aqueous suspensions of clay particles within the Aerosol OT vesicles were shown by Li et al. (1996).

2.3 Lyotropic mesophases

At sufficiently high concentrations, amphiphiles in solution form lyotropic liquid crystalline phases. The Pm3n cubic phase of HCO-60 linoleic acid in water (see above) was investigated using the optimal combination of cryo-TEM to suggest the nature, size and packing symmetry of the structural building blocks, and SAXS to determine quantitatively the exact symmetry. These two complementary direct and indirect methods made it possible to establish the parameters of the unit-cell by adjusting the model to satisfy scattering peak intensities (Burns et al., 1990). Because the mesomorphic phase was too viscous to allow specimen preparation for direct imaging, "on-the-grid processing" was used. An additional electron microscopic technique, freeze-fracture-replication, was used in that study, to allow imaging of bulk specimens to back the direct imaging cryo-TEM results.

Aqueous solutions of the nonionic surfactant, octaoxyethyleneglycol monododecylether ($C_{12}E_8$) undergo phase transition: e.g., from micellar to cubic upon increasing surfactant concentration, or from hexagonal-to-micellar upon increasing temperature. This was examined using cryo-TEM, applying when necessary, "on-the-grid processing". The cubic phase was produced at 9 °C by taking advantage of the concentration gradient which exists in the thin liquid films spanning the holes of the polymer support film. The hexagonal phase was produced by slow controlled partial drying of the specimen prior to quenching. Images of the two phases show areas of ordered long parallel lines, the "lattice fringes" (to borrow the term from material science) of a cubic phase, or a hexagonal phase viewed perpendicularly to the long axis of the thread-like micelles. Image analysis of the digitized hexagonal phase micrographs involved Fourier transform, an application of an appropriate filter, and inverse Fourier transform to produce the filtered image (Fig. 2). It was shown that two hexagonal crystals grew in the liquid film before vitrification. These were separated by the image analysis process, and defects were identified (Danino et al., 1997a).

Similarly, liquid crystals of CTAB in the non-aqueous polar solvent glycerol were prepared by the on-the-grid processing, and the hexagonal liquid crystalline phases were imaged (Lin et al., 1997). Another variant of "on-the-grid processing" was applied to follow a dynamic process by vitrification: the emergence of a lyotropic lamellar phase in a contact or phase-penetration experiment, namely the addition of sodium hydroxide solution to bulk dodecylbenzenesulfonic acid (HDoBS). In the bulk HDoBS, long-range order was detected in localized areas. When water is placed on the grid and allowed to penetrate the acid, the order appears as fingerprint structures. Upon contact with NaOH solution, the surfactant NaDoBS is formed, and again fingerprints of constant periodicity are recorded (Sein et al., 1995). Repeating the experiment with increasing contact times to follow the development of the process amounts to "time-resolved cryo-TEM", described above with pH and temperature jumps.

Anomalous phases were observed in the aqueous mixture of two nonionic polyoxyethylene silicone surfactants. The transition from micellar solution through the anomalous L_3 phase to the surfactant-rich phase, L_2, was followed (Doumaux et al., submitted). The anomalous phase appeared macroscopically as a turbid dispersion, while microscopically a mesh structure was observed. In some cases the L_α phase was imaged. It was suggested that the L_3 phase could be converted to the L_α upon shearing. The L_2 phase was suggested to be made of sheet-like aggregates.

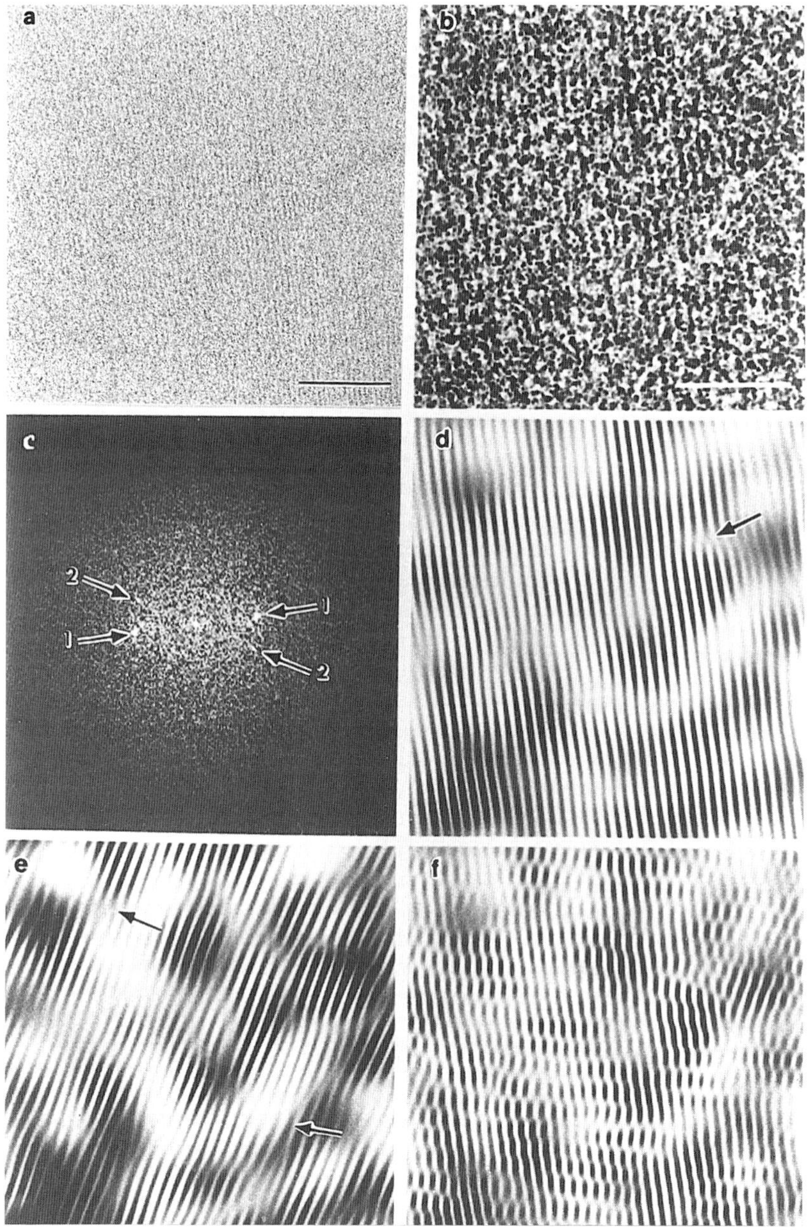

Figure 2. The hexagonal phase of $C_{12}E_8$ prepared on the grid: (a) cryo-TEM image showing ordered structure, bar = 100 nm; (b) a digitized portion of the micrograph in 2a, bar = 50 nm; (c) the Fourier transform of the image in 2b (denoted are two pairs of reflections "1" and "2"; (d) the inverse Fourier transform (IFT) of 2c using the pair of reflections "1"; (e) IFT of 2c using the pair of reflections "2"; (f) superposition of the two structures in 2d and 2e which corresponds to a filtered image of 2b. (From Danino et al. (1997a) with permission).

3. THE STUDY OF BIOLOGICAL COLLOIDAL SYSTEMS

Cryo-TEM has found much application in biological research, as in the study of viruses and cellular organelles. Here we bring just a few examples of the applications of the technique in the study of colloidal systems that are made of biologically originated components. In many cases these biological systems behave very much like their synthetic counterparts.

3.1 Phospholipid and phospholipid/surfactant systems

Phospholipids are the major component of cell membranes and one of the three major components of bile. The most abundant phospholipid in cell membranes is phosphatidylcholine (lecithin), a zwitterionic double-tailed amphiphile. Phospholipids can be dissolved in an organic solvent, dried as a film on the container wall, and vortexed in water to form a dispersion of multilamellar vesicles or liposomes. Sonication of the liposomes leads to small unilamellar vesicles. Large unilamellar vesicles can be prepared by solubilizing the phospholipid in a surfactant solution and then dialyzing or diluting the surfactant out (Yeagle, 1987). These liposomes have been used as model systems for the study of membrane solubilization and reconstitution.

Phospholipid vesicles have been imaged using cryo-TEM by many research groups (Talmon, 1986; Frederik et al., 1990; Vinson et al., 1991; Klösgen and Helferich, 1993; Winterhalter and Lasic, 1993; Chiruvolu et al., 1994a; Mui et al., 1995). Phospholipid vesicles may be used as drug-delivery carriers. An interesting example was given by Lasic et al. (1992), who directly imaged by cryo-TEM lipid liposomes loaded with the anti-cancer drug doxorubicin, which had been precipitated inside the liposomes. While most phospholipids form vesicles and liposomes when dispersed in water, recent cryo-TEM work on a phosphogluco-lipid from the membrane of the bacterium *Acholeplasma laidlawii* has shown long rigid thread-like micelles (Danino et al., 1997b).

Solubilization of phospholipid membranes by micelle-forming surfactants can be followed by reconstitution, which makes it possible to incorporate molecules of interest, such as proteins, in model membranes. Cryo-TEM has been most useful in elucidating the mechanism of membrane solubilization and reconstitution, and the intermediate microstructure involved in those processes. In fact, this is a good example were direct imaging proved a long-established model, based on indirect methods, wrong.

Vinson et al. (1989) studied the solubilization of egg lecithin vesicles upon the addition of octylglucoside (OG). The process was also followed by 90° light scattering and by spectrophotometry to assess turbidity changes. They found that the process involved some initial vesicle growth, followed by vesicles perforation and the appearance of membrane fragment. Upon additional surfactant concentration increase, the vesicles completely disintegrated, some becoming lacy structures, and thread-like micelles appeared, to become spheroidal mixed micelles towards the completion of the solubilization process.

Edwards et al. (1989) studied the solubilization of lecithin by Triton X-100; Edwards and Almgren (1991) studied the solubilization by the nonionic surfactant $C_{12}E_8$ with similar results.

The intermediate structures in the sodium cholate/lecithin system was directly studied by cryo-TEM and compared to changes in turbidity of the system (Walter et al., 1991). In that study it was shown that the cylindrical micelles grew from the edges of the bilayer sheets. Recently the reverse process of reconstitution was studied by Kaplun et al. (1997a) in the same system. Little hysteresis has been noted as compared to solubilization. The effect of ionic strength on the microstructures formed in aqueous solutions of phospholipids and sodium cholate was studied by means of static and dynamic light scattering as well as by cryo-TEM and VELM (Meyuhas et al., 1997). Thread-like micelles were observed in the system while the complementary methods of cryo-TEM and VELM were necessary due to the size range of microstructures in the system. In another study involving a lecithin/bile-salt system, cryo-TEM and static and dynamic light scattering were combined to give the quantitative determination of shape, size, polydispersity, and flexibility of micelles and vesicles in the solutions (Egelhaaf and Schurtenberger, 1994).

When the cationic surfactant CTAC was used to solubilize lecithin (Edwards et al., 1993), an unusual intermediate structure was found in the coexistence range in the form of perforated open membrane fragments and vesicles. Most recently, three different anionic alkyl sulfate surfactants were used to solubilize lecithin vesicles (Silvander et al., 1996). The length of the alkyl chain was found to have a profound effect on the solubilization process and the intermediate structures formed: thread-like micelles were observed only in the shorter alkyl chain surfactant (C_{10}).

Yet another route of solubilization has been described using cryo-TEM (Seras et al., 1996). Vesicles of diglycerol hexadecylether, cholesterol and dicetyl phosphate were solubilized by OG. At low surfactant concentrations, vesicles fusion and deformation occurred. As the surfactant concentration was increased, vesicles became smaller and formed clusters. At higher surfactant concentrations, mixed spheroidal micelles were observed.

Cryo-TEM and time-resolved cryo-TEM have been used to identify intermediate structures such as "inter-lamellar attachments" (ILAs) in membrane fusion. By on-the-grid pH jumps in the dioleylphosphatidylethanolamine (DOPE) Siegel et al. (1989) observed intermediates in the transition between phospholipid lamellar phase (L_α) and inverse hexagonal phase (H_{II}). ILAs were also the intermediate structures observed in DOPE liposomes that underwent fusion induced by a pH or Mg^{+2} jump. "On-the-grid-processing" was carried out by mixing a drop of the vesicular dispersion with a drop of an appropriate buffer (Talmon et al., 1990). Temperature-jump cryo-TEM using a strong light beam was applied in the study of lamellar-to-inverted hexagonal phase transition (Siegel et al., 1994) in a specially designed variant of the CEVS, where the beam from a flash-tube hits the sample just before it enters the cryogen. This followed an earlier temperature jump apparatus built by Chestnut et al. (1992). Membrane fusion products and fusion intermediates such as lipidic particles, vesicles with two concentric bilayers and attachments between vesicles were observed by Frederik et al. (1989) in systems of phosphatidylcholine derivatives and additives as cholesterol. Another approach to temperature-jump experiments, using an iodine-laser was presented by Groll et al. (1996) who studied the kinetics of the phase transition of dimyristoylphosphatidyl choline (DMPC) vesicles.

Dilute mixtures of dipalmitoylphosphatidylcholine (DPPC) and diheptanoylphosphatidylcholine (DHPC) show discoid structures at high temperatures, but below the chain melting or main transition temperature, large rippled bilayers coexist with the discoid structures. These ripple structures are identified as the $P_{\beta'}$ phase (Vinson et al., 1991).

A phase of highly entangled tubular vesicles was identified by light microscopy and cryo-TEM in system composed of DMPC and a reverse-micelle-producer, geraniol (Chiruvolu et al., 1994b). X-ray diffraction confirmed that the tubule walls are multilamellar. Another type of lipid investigated by cryo-TEM is the rhamnolipid. Self-aggregation of this biosurfactant is sensitive to pH and salt concentration. The increase in pH was found to reduce the size of the aggregates from lamellar bilayer sheets, vesicles and tubules to small spheroidal micelles (Champion et al., 1995).

3.2 Bile-models and native-bile

The process of gallstone formation may be better understood through cryo-TEM observations of native and model bile system, to capture the early stages of cholesterol crystallization. Unfortunately, most of the microstructural studies in this area were conducted using artifact-prone staining-and-drying electron microscopy techniques. The combination of cryo-TEM and light microscopy has been used to characterize the complex structural path leading to cholesterol crystallization. Kaplun et al. (1994, 1997b) visualized the microstructures formed in the early stages of the process in two model systems and in native human bile. Dilute isotropic solutions of bile models, composed of cholesterol, lecithin and sodium taurocholate, formed small spheroidal micelles. Subsequently, uni- oligo- and multilamellar vesicles appeared in coexistence with the micelles. The sizes of vesicles were compatible with structures seen at the same time by VELM. Thereafter, filaments, tubular and helical microstructures, as well as classical plate-like cholesterol monohydrate crystals were recorded by light microscopy. Eventually large plate-like crystals were observed by light microscopy, while cryo-TEM of the supernatant revealed only small spheroidal micelles. Very good correlation was noted between the native bile samples and the models used. A very interesting finding from that study is some evidence for initial precipitation of cholesterol between the lamellae of the liposomes formed in bile (Figure 3). More work is needed to substantiate this hypothesis.

In another study, a supersaturated bile-model prepared by mixing a dispersion of lecithin/cholesterol vesicles and sodium cholate micellar solution first yielded a clear mixed micellar solution. The micellar solution then "relaxed" by forming cholesterol-rich vesicles, which reorganized into cholesterol crystals of various shapes, and subsequently grew into mature plate-like cholesterol crystals (Fudim-Levin et al., 1995).

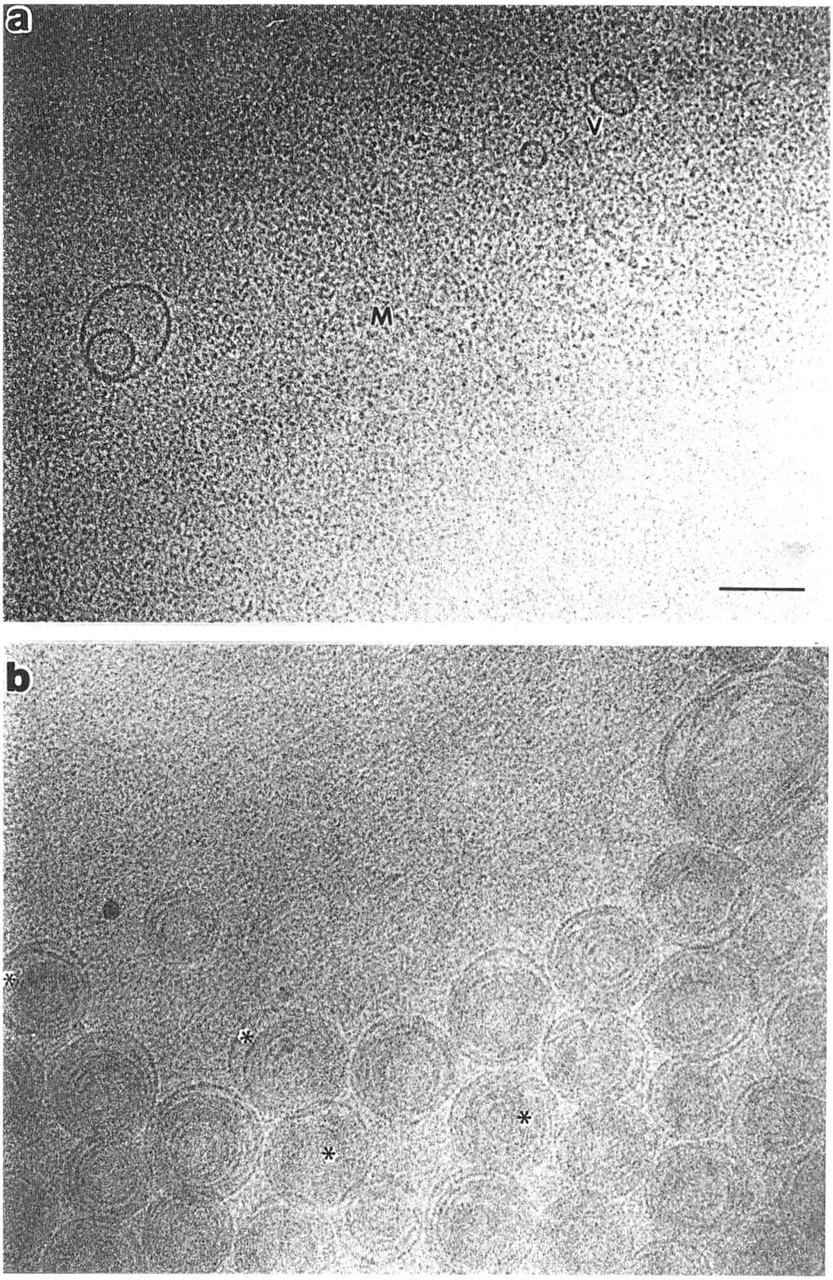

Figure 3. Cryo-TEM micrographs of typical microstructures in a pathophysiologically bile model: (a) spheroidal micelles (M) and unilamellar vesicles (V) in a specimen vitrified 1 hour after dilution, and (b) multilamellar vesicles 48 hours after dilution. Note dense material in some of the vesicles (*). Bar = 100 nm. (From Kaplun et al. (1997b) with permission).

3.3 DNA interactions

DNA supercoiling may be observed by direct imaging cryo-TEM, e.g., DNA-plasmids suspended in low concentration salt buffer studied by Dubochet et al. (1992). Due to growing interest and developments in gene therapy and genetic engineering, the use of liposomes as gene carriers in cell transfection has called for cryo-TEM imaging of the complexes formed. Thus the interaction between cationic liposomes of DOPE and DNA-plasmids was investigated by Gustafsson et al. (1995). In cryo-TEM images at low DNA/lipid ratios, DNA-plasmids entrapped between the lamellae in clusters of aggregated multilamellar structures were observed. At higher DNA/lipid ratios, free plasmids or protruding DNA strings appeared in the vicinity of the complexes.

4. POLYMER SOLUTIONS

The first polymer solution to be directly visualized by cryo-TEM was poly(γ-benzyl-L-glutamate) (PBLG) (Cohen et al., 1990). Micrographs of gelled films of PBLG in benzyl alcohol reveal interconnected microfibrils. Similar irregular isotropic network was seen in the same gelled system by freeze-fracture replication. A "foam-like" morphology with open cells was also seen in PBLG/benzene gels. The cryo-TEM direct images of the interconnected network of microfibrils in the PBLG/benzyl alcohol system as can be seen in Figure 4, were used to construct a physical model for the interpretation of SAXS and SANS data of the system. Thus complete qualitative characterization of the system was achieved (Cohen, 1996).

Attempts were made to visualize aqueous polymer solutions of sodium alginate and poly(ethyleneoxide) (Imae, 1993). A "super-network" structure was reported, but the images are most probably the result of electron-beam radiation damage. Aqueous polymer solutions of poly(methacrylic acid) (PMA) and its complex with poly(ethyleneoxide) were imaged more recently (Prevysh et al., 1997). Spherical aggregates and short rods, about 4 nm in diameter, were imaged in the case of the PMA solution, in agreement with SAXS, while bigger aggregates of about 20 nm were seen in the complex-containing solution.

Diblock copolymer micellar solutions of toluene were successfully studied by cryo-TEM (Esselink et al., 1995). Spheroidal micelles, about 30 and 45 nm in diameter of polystyrene/poly(2-vinylpyridine) with two different molecular weight of the poly(2-vinylpyridine) blocks were observed in the frozen toluene film. Thermal fixation was carried out in liquid nitrogen to avoid the dissolution of the sample in liquid ethane, as reported by the authors. From the apparent size of the copolymer micelles with different block lengths, it was concluded that only the poly(2-vinylpyridine) core was visible, whereas the polystyrene corona was invisible to the electron beam (Oostergetel et al., 1995). Nevertheless one should note the excellent contrast in those images.

Dilute aqueous solution of the polysaccharides κ-carrageenan and agarose were directly imaged by cryo-TEM. The helical state of these polysaccharides was seen as large elongated aggregates (Sugiyama et al., 1994). Long microfibrillar structures were seen in dilute solutions of helical κ-carrageenan in varied electrolyte solutions. It was postulated that those aggregates were made of more than one

Figure 4. Cryo-TEM image of thin film of a 1 % PBLG gel showing an interconnected network of microfibrils. Bar = 200 nm. (From Cohen (1996) with permission).

carageenan helix, contrary to Sugiyama et al. who interpreted the aggregates as individual helix (Borgström et al., 1996).

Some polymer-solvent systems exhibit an upper cloud point. Such a system, ethyl(hydroxyethyl)cellulose (EHEC)/water was recently studied by cryo-TEM (Goldraich, 1996). Below the cloud point, no specific microstructures were observed due to lack in contrast. However, when the temperature was raised and phase separation took place, dark polymer-rich domains and bright water-rich domains were clearly observed (Figure 5).

Pluronics, triblock copolymers of poly(ethylene oxide) and poly(propylene oxide) blocks, micellize and form lyotropic liquid crystalline phases not unlike those of low molecular weight amphiphiles. Aqueous solutions of the Pluronic F127, made of the blocks $(EO)_{99}(PO)_{65}(EO)_{99}$, were studied by cryo-TEM and SANS (Mortensen and Talmon, 1995). The cubic phase imaged by TEM was probably formed by micelles which were pushed together and closely packed as the specimen was thinned during preparation, similar to the formation of low molecular weight lyotropic phase on the grid (Danino et al., 1997a). The work on Pluronics has been extended to triblock copolymers of a more hydrophobic central block, namely polyisobutylene (PIB). The additional hydrophobicity has been shown to lead to thread-like micelles, as shown in Figure 6 (Mortensen et al., 1997).

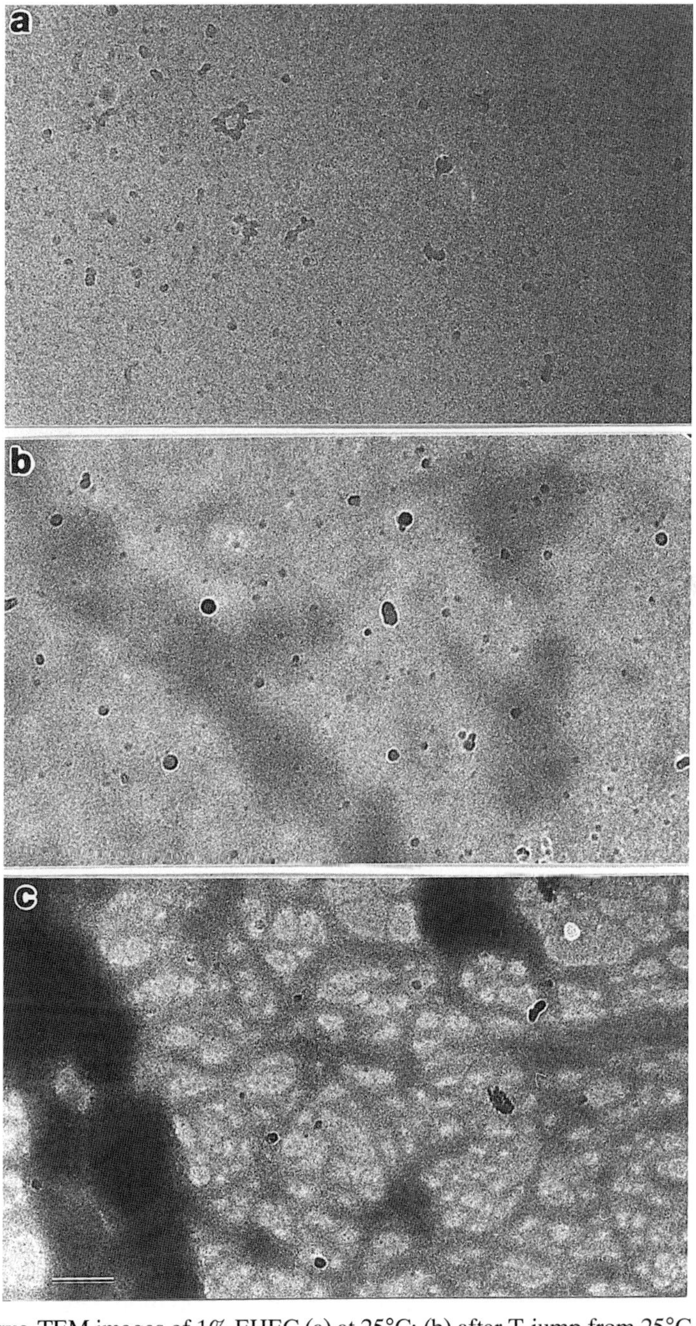

Figure 5. Cryo-TEM images of 1% EHEC (a) at 25°C; (b) after T-jump from 25°C to 35°C; (c) after T-jump from 25°C to 50°C. Bar = 100 nm. (From Goldraich (1996)).

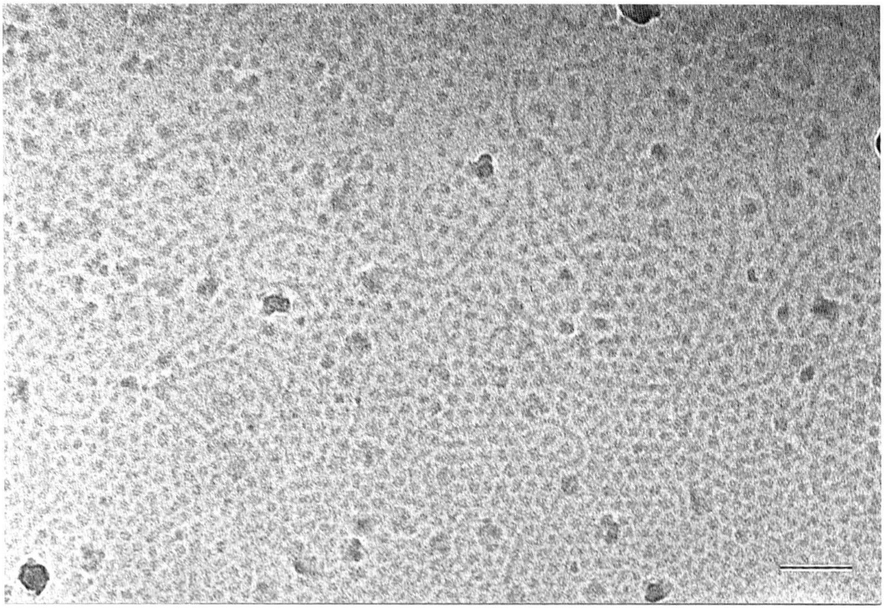

Figure 6. Cryo-TEM image of a 1% aqueous solution copolymer PEO-PIB-PEO. Bar=100 nm.

Polyamphiphiles self-aggregate to form enclosed hydrophobic microdomains similarly to those in micellization of conventional surfactants. The existence of those microdomains in aqueous solutions has been well established using fluorescence probing techniques (Binana-Limbelé and Zana, 1990), but their size and shape remained a matter of controversy. Cryo-TEM was instrumental in elucidating the microstructures formed in such solutions, and how the hydrophobic side-chains affect self-aggregation (Cochin et al., 1992).

5. POLYMER/SURFACTANT SYSTEMS

5.1 Uncharged polymers

Many indirect studies were reported regarding the interaction between uncharged polymers and charged surfactants, especially the case of poly(ethyleneoxide) (PEO) and SDS (e.g., Cabane and Duplessix, 1982, 1985, 1987). This system was studied also by cryo-TEM in combination with SAXS (Süss et al., 1995). In the homogeneous solution of PEO and SDS, only the SDS spheroidal micelles can be detected by cryo-TEM and the polymer in its dispersed state is "invisible" to the electron beam due to lack of contrast. The micelles do not change form in comparison to SDS micelles in water, however, SAXS results indicate reduction in micelle size in the presence of polymer. The addition of salt to SDS is known to modify their structure and form elongated micelles. In the presence of PEO, the bound micelles are expected to reduce their size to form spheroidal micelles (Nagarajan, 1985). This phenomenon was documented by cryo-TEM and quantified by SAXS (Süss et al., 1995). Spheroidal micelles of SDS were seen in the same study in the presence of photographic-grade gelatin below the CMC of pure SDS in water.

The interactions between the nonionic surfactants $C_{16}E_6$, $C_{12}E_5$, the ionic surfactant CTAB/sodium salicylate (NaSal) and the nonionic polymers poly(vinylmethylether) (PVME) and poly(propyleneoxide) (PPO) were studied by cryo-TEM and rheological measurements (Li et al., 1995). While the micellar solution of $C_{12}E_5$ does not show an appreciable change in microstructure upon the addition of the polymers, the $C_{16}E_6$ solution undergoes a transition from thread-like micelles to ribbon-shaped discoid micelles upon the addition of PPO. In the case of CTAB/NaSal, the expected worm-like-to-spheroidal micelles transition is observed upon the addition of the polymers.

5.2 Charged polymers

Although polymer and surfactant of like charges are not expected to interact, it has been established that when the polymer is sufficiently hydrophobic, that assumption is not true (Iliopoulos et al., 1991). When the anionic surfactant SDS is added to a solution of the anionic polysoap poly(disodium maleate-co-hexadecylvinylether), the result is loss of viscoelasticity of the polysoap solution (Zana et al., 1993). This is due to the breakdown of the giant thread-like micelles formed in the pure polysoap solution to smaller isolated chains, as was observed by cryo-TEM. A similar phenomenon was recorded by cryo-TEM when the nonionic surfactants $C_{10}E_6$ and $C_{12}E_6$ was added to a solution of the same polysoap (Kamenka et al., 1994). In both cases cryo-TEM produced images of single molecules of the polymer stabilized by the added surfactant. However, the efficiency of the nonionic surfactant to break-up the long thread-like aggregates seems much higher since the molar concentration ratio of surfactant-to-polysoap required to break the micelles is much lower.

The interaction between polymer and surfactant of opposite charges may be divided into three stages based on macroscopical observations: pre-precipitation, precipitation and resolubilization, as surfactant is added to the polymer solution. This was studied recently in the system of JR-400, quaternary ammonium-substituted hydroxyethylcellulose polymer, and SDS by cryo-TEM and light microscopy (Goldraich et al., 1997). In the pre-precipitation zone, where the solution is still homogeneous and light microscopy does not reveal any structures, disc-like structures, small vesicles and membrane fragments appear in the TEM images (Figure 7a). Just before precipitation, when the solution appears macroscopically turbid, light microscopy reveals large flocs, while cryo-TEM shows that the flocs are made of spheroidal aggregates; elongated structures, quite similar to thread-like micelles, are also observed. The sticky precipitate can be studied by light microscopy, but is made of aggregates too big to visualize by cryo-TEM. Upon resolubilization a turbid solution forms again, and flocs can be seen using light microscopy and cryo-TEM (Figure 7b). Addition of more SDS leads to a homogeneous solution again; light microscopy does not show any microstructures, while cryo-TEM reveals disc-like structures, small vesicles and thread-like micelles (Figure 7c). When SDS is in sufficiently high excess, spheroidal micelles are observed.

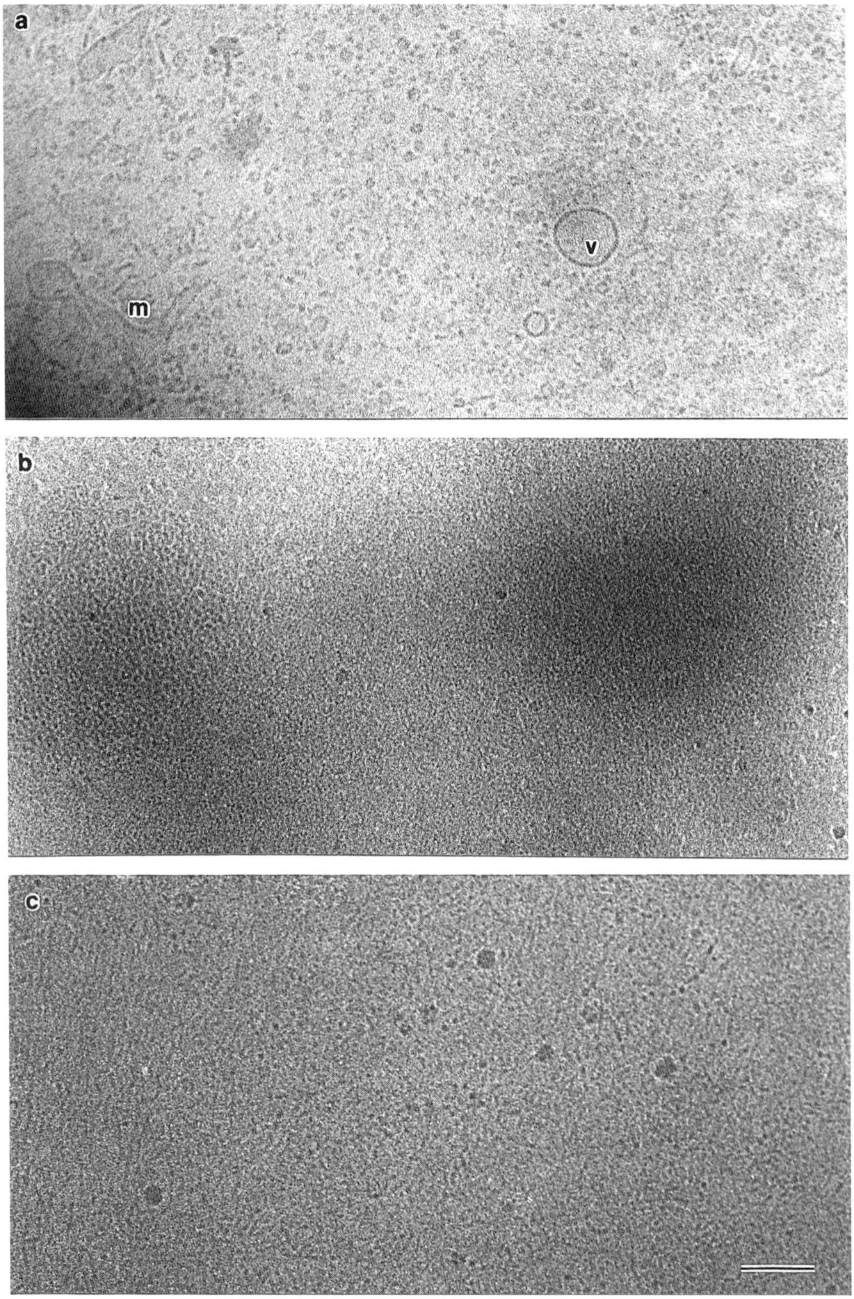

Figure 7. Cryo-TEM images of a 0.1% aqueous solution of JR-400 with added SDS at different charge ratios: (a) 0.5 SDS/JR-400 showing vesicles (v) and membranes (m); (b) 3 SDS/JR showing microphase-separation; (c) 5 SDS/JR-400 showing elongated structures. Bar=100 nm. (From Goldraich et al. (1997) with permission).

The interaction of another ionic polymer, poly(sodium 4-styrenesulfonate), with worm-like micelles of nonionic surfactants, e.g., $C_{16}E_6$, was studied by cryo-TEM by Li et al. (1995). In that case the addition of the polymer did not break up the micelles before precipitation of a polymer/surfactant complex.

5.3 Gelation processes

We have used cryo-TEM to study the microstructural changes that accompany the reversible thermal gelation of EHEC, an uncharged cellulose derivative with either CTAB or SDS. Our work soon to be written up follows the thorough study of the system by Lindman and coworkers (e.g., Carlsson et al., 1988; Nyström and Lindman, 1995; Nyström et al., 1995). In this study we took advantage of the ability of the FT-CEVS to perform temperature jumps, while keeping the specimen preparation chamber saturated with water vapor at all times. The homogeneous solution of EHEC and SDS at room temperature shows spheroidal micelles. As the temperature is raised, the solution becomes a clear gel, which eventually phase-separates to form a turbid gel. Images of the clear gel exhibit a network-like structure of the polymer network decorated by SDS micelles. In images of the turbid gel (Figure 8), one observes dark polymer-rich domains and brighter water-rich domains.

Similar gelation processes were observed in the system of EHEC, the cationic surfactant tetradecyl oxycarbonyl-N,N,N-trimethyl methanaminium chloride (TeBo), and an organic salt, timolol maleate, a system designed to facilitate the application of ophthalmic drugs (in this particular case timolol maleate). Indirect microstructural studies were conducted by SANS (Cabane et al., 1996). Cryo-TEM images were taken in three different surfactant-to-polymer ratios, at temperatures corresponding to different stages of the gelation process. Micro phase-separation was observed, again, through the appearance of dark polymer-rich domains and lighter water-rich one. The results are to be written up soon.

6. CONCLUDING REMARKS

Cryo-TEM is now almost a routine technique, following years of development and applications by a number of research groups working on various aspects of complex fluids. Although it does require special equipment and skilled users, the equipment is readily available form commercial suppliers, and there is sufficient body of literature that may be used s a starting point for new users. Based on work already done on self-aggregating systems, it is clear that the technique is ideal for the study of microstructured liquid system, including those based on polymers, amphiphilic copolymers, and polymer-surfactant mixtures. Ideally cryo-TEM, an excellent qualitative technique, is used with non-imaging techniques to achieve complete microstructural characterization of the systems under study. The extensions of the basic techniques, e.g., on-the-grid-processing and various time-resolved experiments, point to the versatility of the cryo-TEM, and how the creative imagination of a skilled researcher can extend its applicability.

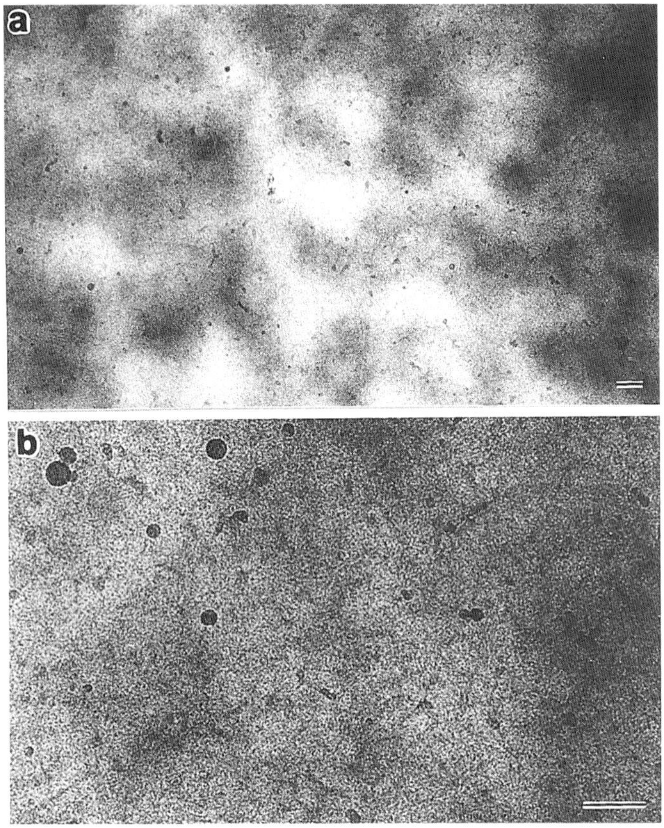

Figure 8. Micrographs of 1 % EHEC/4 mM SDS solution showing microphase-separated gel after T-jumping the sample from 25 to 60 °C (a), and showing finer details of the gel clusters (b). Bars = 100 nm. From Goldraich (1996).

ACKNOWLEDGEMENTS

We thank Dr. Dganit Danino and Ms. Judith Schmidt for their invaluable help in preparing this review. Cryo-TEM at the Technion has been supported in part by grants from the United States-Israel Binational Science Foundation (BSF), Jerusalem, and the Fund for the Promotion of Research at the Technion. YT thanks Prof. Björn Lindman and his colleagues in Physical Chemistry 1, Lund University, for their hospitality during several fruitful and enjoyable summer visits and many useful discussions on various aspects of colloid science and cryo-TEM.

REFERENCES

J.R. Bellare, H.T. Davis, L.E. Scriven, and Y. Talmon, Proceedings of the 11th Int'l Cong. on Electron Microscopy, Kyoto, vol. 1, 1986, p. 369.

J.R. Bellare, H.T. Davis, L.E. Scriven, and Y. Talmon, J. Electron Microsc. Tech., 10 (1988) 87.

W. Binana-Limbelé and R. Zana, Macromolecules, 23 (1990) 2731.

J. Borgström, L. Piculell, C. Viebke, and Y. Talmon, Int. J. Biological Macromolecules, 18 (1996) 223.

J.L. Burns, Y. Cohen, and Y. Talmon, J. Phys. Chem., 94 (1990) 5308.

B. Cabane and R. Duplessix, J. Physique, 43 (1982) 1529.

B. Cabane and R. Duplessix, J. Colloids Surfaces, 13 (1985) 19.

B. Cabane and R. Duplessix, J. Physique, 15 (1987) 651.

B. Cabane, K. Lindell, S. Engström, and B. Lindman, Macromolecules, 29 (1996) 3188.

S.J. Candau, A. Khatory, F. Lequeux, and F. Kern, J. Physique IV, 3 (1993) 197.

A. Carlsson, G. Karlström, B. Lindman, and O. Stenberg, Coll. Poly. Sci., 266 (1988) 1031.

J.T. Champion, J.C. Gilkey, H. Lamparsky, J. Retterer, and R.M. Miller, J. Colloid Interface Sci., 170 (1995) 569.

M.H Chestnut, D.P. Siegel, J.L. Burns, and Y. Talmon, Micros. Res. Tech., 20 (1992) 95.

S. Chiruvolu, S. Walker, J. Israelachvili, F.-J. Schmitt, D. Leckband, and J.A. Zasadzinski, Science, 264 (1994a) 1753.

S. Chiruvolu, H.E. Warriner, E. Naranjo, S.H.J. Idziak, O. Radler, R.J. Plano, J.A. Zasadzinski, and C.R. Safinya, Science, 266 (1994b) 1222.

S. Chiruvolu, J. Israelachvili, E. Naranjo, Z. Xu, J.A. Zasadzinski, E.W. Kaler, and K.L. Herrington, Langmuir, 11 (1995) 4256.

T.M. Clausen, P.K. Vinson, H.T. Minter, Y. Talmon, and W.G. Miller, J. Phys. Chem., 96 (1992) 474.

D. Cochin, F. Candau, R. Zana, and Y. Talmon, Macromolecules, 25 (1992) 4220.

Y. Cohen, Y. Talmon, and E.L. Thomas, in "Physical Networks, Polymers and Gels", W. Burchard and S.B. Ross-Murphy, (eds.) Elsevier Applied Science, London, 1990, p. 147.

Y. Cohen, J. Polym. Sci. B: Polym. Phys., 34 (1996) 57.

P.G. Cummins, J.B. Hayter, J. Penfold, and E. Staples, Chem. Phys. Lett., 138 (1987) 436.

D. Danino, A. Kaplun, Y. Talmon, and R. Zana, in "Structure and Flow in Surfactant Solutions", C.A. Herb and R.K. Prud'homme (Eds.), ACS Symposium Series 578, 1994, p. 105.

D. Danino, Y. Talmon, and R. Zana, Langmuir, 11 (1995a) 1448.

D. Danino, Y. Talmon, H. Levy, G. Beinert, and R. Zana, Science, 269 (1995b) 1420.

D. Danino, Y. Talmon, and R. Zana, J. Colloid Interface Sci., (1996) (to appear).

D. Danino, Y. Talmon, and R. Zana, J. Colloid Interface Sci., (1997a) (to appear).

D. Danino, A. Kaplun, G. Lindblom, L. Rilfors, G. Orädd, J.B. Hauksson, and Y. Talmon, Phys. Chem. Lipids, (1997b) (to appear).

H.A. Doumaux, H. Chang, H.T. Davis, D.F. Evans, and R.M. Hill, (1997) (to be submitted).

J. Dubochet, M. Adrian, I. Dustin, P. Furrer, and A. Stasiak, Methods in Enzymology, 211 (1992) 507.

K. Edwards and M. Almgren, J. Colloid Interface Sci., 147 (1991) 1.

K. Edwards, M. Almgren, J.R. Bellare, and W. Brown, Langmuir, 5 (1989) 473.

K. Edwards, J. Gustafsson, M. Almgren, and G. Karlsson, J. Colloid Interface Sci, 161 (1993) 299.

S.U. Egelhaaf and P. Schurtenberger, J. Phys. Chem., 98 (1994) 8560.

S.U. Egelhaaf, D. Sc. Thesis, ETH Zürich (1995).

F.J. Esselink, A.N. Semenov, G. ten Brinke, G. Hadziionnou, and G.T. Oostergetel, Macromolecules, 28 (1995) 3479.

Y. Fink and Y. Talmon, Proceedings ICEM I3 -Paris, Vol. 1 (1994) 37.

P.M. Frederik, M.C.A. Stuart, and A.J. Verkleij, Biochim. Biophys. Acta, 979 (1989) 275.

P.M. Frederik, M.C.A. Stuart, P.H. Bomans, W.M. Busing, and K.N. Burger, J. Microscopy, 161 (1990) 253.

E. Fudim-Levin, A. Bor, A Kaplun, Y. Talmon, and D. Lichtenberg, Biochim. Biophys. Acta, 1259 (1995) 23..

M. Goldraich, D. Sc. Thesis, Technion (1996).

M. Goldraich, J.R. Schwartz, J.L. Burns, and Y. Talmon, Colloids and Surfaces, (1997) (to appear).

R. Groll, A. Böttcher, J. Jäger, and J.F. Holzwarth, Biophys. Chem., 58 (1996) 53.

J. Gustafsson, G. Arvidson, G. Karlsson, and M. Almgren, Biochim. Biophys. Acta, 1235 (1995) 305.

L. Hammarström, I. Velikian, G. Karlsson, and K. Edwards, Langmuir, 11 (1995) 408.

I. Harwigsson, O. Söderman, and O. Regev, Langmuir, 10 (1994) 4731.

I. Iliopoulos, T.K. Wang, and R. Audebert, Langmuir, 7 (1991) 617.

T. Imae, Polymer J., 25 (1993) 201.

N. Kamenka, M. Chorro, Y. Talmon, and R. Zana, Colloids and Surfaces, 67 (1992) 213.

N. Kamenka, A. Kaplun, Y. Talmon, and R. Zana, Langmuir, 10 (1994) 2960.

A. Kaplun, Y. Talmon, F.M. Konikoff, M. Rubin, A. Eitan, M. Tadmor, and D. Lichtenberg, FEBS Letters, 340 (1994) 78.

A. Kaplun, Y. Talmon, and A. Walter, Biophys. J., (1997a) (to appear) .

A. Kaplun, F.M. Konikoff, A. Eitan, M. Rubin, A. Vilan, D. Lichtenberg, T. Gilat, and Y. Talmon, Microscopy Res. Tech., (1997b) (to appear).

A. Khan, A. Kaplun, Y. Talmon, and M. Hellsten, J. Colloid Interface Sci., 181 (1996) 191.

P.K. Kilpatrick, W.G. Miller, and Y. Talmon, J. Colloid Interface Sci., 107 (1985) 146.

B. Klösgen and W. Helferich, Eur. Biophys. J., 22 (1993) 329.

A. Knoblich, M. Matsumoto, K. Murata, and Y. Fujiyoshi, Langmuir, 11 (1995) 2361.

D.D. Lasic, P.M. Frederik, M.C.A. Stuart, Y. Barenholz, and T.J. McIntosh, FEBS, 312 (1992) 255 .

X. Li, Z. Lin, J.J. Cai, L.E. Scriven, and H.T. Davis, J. Phys. Chem., 99 (1995) 10865.

X. Li, Y. Guo, L.E. Scriven, and H.T. Davis, Colloids and Surfaces A, 106 (1996) 149.

Z. Lin, J.J. Cai, L.E. Scriven, and H.T. Davis, J. Phys. Chem., 98 (1994a) 5984.

Z. Lin, R.M. Hill, H.T. Davis, L.E. Scriven, and Y. Talmon, Langmuir, 10 (1994b) 1008.

Z. Lin, Langmuir, 12 (1996) 1729.

Z. Lin, L.E. Scriven, and H.T. Davis, Langmuir, 8 (1992) 2200.

Z. Lin, J.J. Cai, L.E. Scriven, and H.T. Davis, J. Phys. Chem., 98 (1994) 5984.

Z. Lin, H.T. Davis, and L.E. Scriven, Langmuir, (1997) (to appear).

B. Lu, X. Li, J.L. Zakin, and Y. Talmon, Non-Newtonian Fluid Mech., (1997) (to appear).

E.H. Lucassen-Reynders, J. Lucassen, and D. Giles, J. Colloid Interface Sci., 81 (1981) 150.

L.J. Magid, J.C. Gee, and Y. Talmon, Langmuir, 6 (1990) 1609.

D. Meyuhas, A. Bor, I. Pinchuk, A. Kaplun, Y. Talmon, M.M. Kozlov, and D. Lichtenberg, J. Colloid Interface Sci., (1997) (to appear).

D.D. Miller, J.R. Bellare, D.F. Evans, Y. Talmon, and B.W. Ninham, J. Phys. Chem., 91 (1987) 674.

D.D. Miller, J.R. Bellare, T. Kaneko, and D.F. Evans, Langmuir, 4 (1988) 1363.

K. Mortensen and Y. Talmon, Macromolecules, 28 (1995) 8829.

K. Mortensen, Y. Talmon, B. Gao, and J. Kops, Macromolecules, (1997) (to be submitted).

B.L.-S. Mui, H.-G. Dobereiner, T.D. Madden, and P.R. Cullis, Biophys. J., 69 (1995) 930.

R. Nagarajan, Colloids and Surfaces, 13 (1985) 1.

B. Nyström and B. Lindman, Macromolecules, 28 (1995) 967.

B. Nyström, H. Walderhaug, F.K. Hansen, and B. Lindman, Langmuir, 11 (1995) 750.

G.T. Oostergetel, F.J. Esselink, and G. Hadziioannou, Langmuir, 11 (1995) 3721.

V. Prevysh, M. Goldraich, and Y. Cohen, J. Polym. Sci. Part B: Polym. Phys., (1997) (to appear).

O. Regev and A. Khan, Progr. Colloid Polym. Sci., 97 (1994) 298.

O. Regev, C. Kang, and A. Khan, J. Phys. Chem., 98 (1994) 6619.

A. Sein, J.F.L. van Breemen, and J.B.F.N Engberts, Langmuir, 11 (1995) 3565.

M. Seras, K. Edwards, K. Almgren, M. Carlsson, M. Ollivon, and S. Leisieur, Langmuir, 12 (1996) 330.

D.P. Siegel, J.L. Burns, M.H. Chestnut, and Y. Talmon, Biophys. J., 56 (1989) 161.

D.P. Siegel, W.J. Green, and Y. Talmon, Biophys. J., 66 (1994) 402.

M. Silvander, G. Karlsson, and K. Edwards, J. Colloid Interface Sci., 179 (1996) 104 .

J. Sugiyama, C. Rochas, T. Turquois, F. Taraval, and H. Chanzy, Carbohydrate Polymers, 23 (1994) 261.

D. Süss, Y. Cohen, and Y. Talmon, Polymer, 36 (1995) 1809.

M. Swanson-Vethamuthu, M. Almgren, G. Karlsson, and P. Bahadur, Langmuir, 12 (1996) 2173.

Y. Talmon, J. Colloid Interface Sci., 93 (1983) 366.

Y. Talmon, Colloids and Surfaces, 19 (1986) 237.

Y. Talmon, Ber. Bunsenges. Phys. Chem., 100 (1996) 364.

Y. Talmon, J.L. Burns, M.H. Chestnut, and D.P. Siegel, J. Electron Microsc. Tech., 14 (1990) 6.

P.K. Vinson, J.R. Bellare, H.T. Davis, W.G. Miller, and L.E. Scriven, J. Colloid Interface Sci., 142 (1991) 74.

P.K. Vinson and Y. Talmon, J. Colloid Interface Sci., 133 (1989) 288.

P.K. Vinson, Y. Talmon, and A. Walter, Biophys. J., 56 (1989) 669.

A. Walter, P.K. Vinson, A. Kaplun, and Y. Talmon, Biophys. J., 60 (1991) 1315 .

M. Winterhalter and D.D. Lasic, Chemistry and Physics of Lipids, 64 (1993) 35.

I. Yaacob and A. Bose, J. Colloid Interface Sci., 178 (1996) 638.

M.T. Yatcilla, K.L. Herrington, L.L. Brasher, E.W. Kaler, S. Chiruvolu, and J.A. Zasadzinski, J. Phys. Chem., 100 (1996) 5874.

P. Yeagle, "The membranes of cells", Academic Press, Orlando, 1987.

R. Zana, A. Kaplun, and Y. Talmon, Langmuir, 9 (1993) 1948.

R. Zana and Y. Talmon, Nature, 362 (1993) 228.

Rheology Of Transient Networks Formed By The Association Of Hydrophobically Modified Water Soluble Polymers

Tom Annable*[,a] Richard Buscall[b] and Rammile Ettelaie[b]

[a] ZENECA Resins, The Heath, Runcorn, Cheshire, UK.
[b] ICI Technology Centre, Wilton, Middlesborough, Cleveland, UK.

1. INTRODUCTION

In the past fifteen or so years there has been considerable research interest in a particular class of block copolymers with commercial potential as rheology modifiers. These polymers are the hydrophobically modified water soluble polymers commonly known as associative thickeners (AT). Various types of AT are available and these are now widely used as thickeners for coatings, cosmetic formulations and in oil-recovery. For example in the coatings industry, the unique properties of these types of macromolecules have allowed new possibilities in the development of high performance water-borne latex coatings with not only improved rheological properties but also improved pigment dispersion and gloss. The use of associative thickeners in coating formulations has been studied by various workers [eg. 1-9]. These papers illustrate the broad range of interactions between associating polymers and the dispersed phases and surfactants contained in a formulated latex paint. It is not just their industrial application which makes such polymers interesting to the researcher, they also prove to be ideal systems for fundamental study. The combination of hydrophobic blocks and hydrophilic backbone causes the polymers systems to associate intermolecularly to form transient networks. The dynamics and structure of these networks have been studied by a broad range of techniques including rheology [10-16], NMR and light scattering [17-24], fluorescence techniques [25-31] and electron spin resonance [32,33].

The most widely studied class of associative polymer is the hydrophobically modified ethoxylated urethanes, HEUR [1-13,17-31]. These polymers consist of poly(ethylene glycol) chain extended by small hydrophobic urethanes and terminated by hydrophobic alcohols. Such polymers can be considered to be ABA block copolymers with hydrophobic A blocks and a water-soluble B chain. The molecular weight of these polymers is typically of the order of $1x10^4$ to $5x10^4$ Daltons and is beneath the entanglement molecular weight, M_e, at the concentrations of usual interest. Typical hydrophobic groups range from n-$C_{12}H_{25}$ to n-$C_{22}H_{45}$. Other telechelic polymers which have been studied include the poly(butylene oxide)-poly(ethylene oxide) ABA and BAB block copolymers [34] and fluorocarbon end-capped HEUR [35]. The greater hydrophobicity of fluorocarbon groups produces stronger associations than the corresponding hydrocarbons and produces dramatic effects on their viscosity behaviour. Transient networks formed in water in oil microemulsions containing PEO-polyisoprene-PEO block copolymers have also been reported [36].

Other hydrophobically modified polymers which have been studied include hydrophobically modified hydroxyethyl cellulose, HMHEC [14,32], hydrophobically modified polyacrylates [33] and hydrophobically modified polyacrylamide, HMPAM [37,38]. These polymers differ from the telechelic polymers in a number of important ways. First, the higher molecular weight, particularly for HMPAM, means that the chain dynamics are often strongly influenced by reptation as a consequence of chain entanglements. Secondly, the number of hydrophobic groups per chain is greater than two and these are typically distributed randomly down the polymer backbone. These differences between the two classes of thickener, the telechelic HEUR type and the random HMHEC/HMPAM systems, produces profound differences in the relaxation dynamics in such systems. In this chapter we discuss the rheological behaviour of associating polymers in solution and as mixtures with both surfactants and polymer latices. We start by describing transient network theory as developed to describe the rheology of telechelic associating polymers and use it to describe both the dynamics and structure of transient networks.

2. DYNAMICS OF TRANSIENT NETWORKS

The dynamic nature of the network formed by the association of AT has a significant influence on their rheological behaviour. To first consider a *static* network, the motion of each polymer chain is necessarily coupled to those of the neighbouring chains to which it is connected at both ends at all times. Thus, the stress in such a network is only released as a result of a co-operative motion of many chains, involving different parts of the network and taking place on a broad range of length scales. For such a system, the relaxation process is expected to be slow and to involve a wide distribution of relaxation times. However, the *transient* nature of an AT network allows for the hydrophobic end groups to detach from the micelles. If the activation energy for this process is ΔE_m, then the rate of detachment is given by

$$\beta_0 = \omega_0 \exp\left(-\frac{\Delta E_m}{kT}\right) \tag{1}$$

where ω_0 is a characteristic frequency of thermal motion, k the Boltzmann constant and T the temperature. Once detached at one end, a chain is free to move independent of the rest of the network, at least at AT concentrations of interest in here (ie. beneath the entanglement concentration) . For these concentrations the motion of the chain is governed by Rouse dynamics with a characteristic time τ_r. Hence, the typical time for a single chain to relax becomes the sum of the detachment time and the Rouse time:

$$\tau = \beta_0^{-1} + \tau_r \tag{2}$$

In many instances the above relaxation mechanism is likely to be much faster than the co-operative motion involving large number of chains. As such it will dominate the process of relaxation in these systems. For an ensemble of AT chains with a single end-cap length and for which the end-cap disengagement time is significantly larger than the Rouse time, it would be expected that the rheological behaviour of the system would be governed by a single relaxation time.

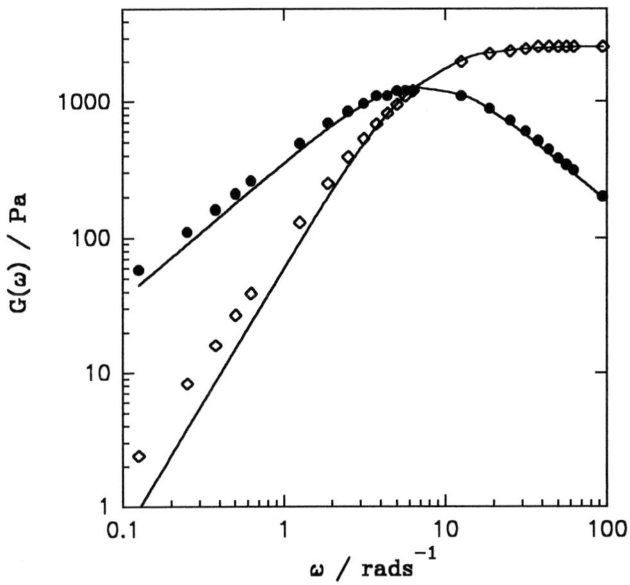

Figure 1. Viscoelastic spectrum for C16/35 at a concentration of 5%. Symbols = experimental data, lines = best fit to Maxwell model as given by equation 3.

The viscoelastic data for a single component HEUR-AT of molecular weight 35000, end-capped with hexadecanol is shown in Figure 1. (We shall notate the HEUR-AT as Cx/y where x is the end-cap length, in this case 16 carbon units, and y is the molecular weight/1000, in this case 35. Thus this polymer is C16/35). The data is fitted to a single viscoelastic fluid element, the so-called Maxwell model for which the viscoelastic moduli are given by

$$G'(\omega) = \frac{G_\infty \omega^2 \tau^2}{1+\omega^2\tau^2} \qquad G''(\omega) = \frac{G_\infty \omega \tau}{1+\omega^2\tau^2} \tag{3}$$

where G'(ω) and G''(ω) are the storage and loss components of the viscoelastic modulus repectively, ω is the angular frequency and τ is the stress relaxation time. This simple model provides a very good description of the data and thus to a first approximation, the pure AT solution possesses a single viscoelastic relaxation time and the picture of the timescale for detachment dominating over the Rouse chain dynamics seems to be a good one for these systems. There is evidence, however, of some minor deviation in the fitting at low frequencies with the implication that there might be some slower co-operative modes of relaxation taking place. In terms of the viscoelastic properties, the end-cap disengagement time certainly appears to be the dominant process governing stress relaxation, but slower modes of *diffusion* have been observed in nmr and dynamic light scattering studies [17,18]. These have generally been interpreted in terms of the theory of Ngai based on the diffusion of finite strongly interacting clusters [39]. Although this picture provides a good description of the spectroscopic data it is not easily reconciled with a dominant single viscoelastic relaxation time due to the polydispersity of cluster sizes. It is also not clear why two strongly interacting transient clusters in close proximity to each other, would not choose to join up into a single giant cluster. We have argued that the anomalous (or non-Fickian) diffusion processes as observed in nmr studies, might indicate that the network structure is fractal at short length scales. The long diffusion modes seen in light scattering measurements might arise from the decay of concentration fluctuations which would require the cooperative motion of many chains. However, it is, we believe, still unclear where the precise origins of these slower diffusion modes lie and indeed whether these relate to the minor deviations from Maxwellian behaviour observed in the viscoelastic data. For the purposes of describing the rheology of pure HEUR-AT solutions, the systems can be regarded as possessing a single viscoelastic relaxation time. The observed relaxation time is strongly dependent on the size of the hydrophobic group. For example at a concentration of 2wt%, the AT C14/35 has a relaxation time close to 0.01s, C16/35 ~ 0.1s and C18/35 ~ 1s.

In systems containing mixtures of HEUR-AT with differing hydrophobe lengths, mixed micellar aggregates containing a statistical mix of end-caps would be expected to form. It is to be expected, therefore, that each component in the blend would retain its own individual stress relaxation time since the activation energy for disengagement is related to the size of the hydrophobe exiting the micelle. Thus a blend of n components would be expected to show n relaxation times. The loss moduli for binary and tertiary blends of C12/35, C16/35 and C20/35 are shown in Figure 2. The data are fitted to a sum of Maxwell elements with the number of elements equal to the number of components in the blend

$$G''(\omega) = G_\infty \sum_{i=1}^{n} \frac{f_i \omega \tau_i}{1+\omega^2 \tau_i^2} \tag{4}$$

The data are well described by this fitting procedure with the relaxation spectra consisting of a series of sharp lines corresponding to the relaxation time of the individual AT components. It is intriguing that such systems could be used to design fluids with specific relaxation time distributions. Thus the viscoelastic response and the steady shear rheology could be controlled by the adjustment of the ratios of the components in the blend.

It is clear from the data presented above that the network dynamics of HEUR associating polymers can be described by a very simple picture in which the end-cap dissociation process dominates over the chain dynamics. Hence a single viscoelastic relaxation time can be observed in systems with a high degree of molecular weight polydispersity. This is the case for telechelic low molecular weight associating polymers. The situation becomes more complex when polymers with multiple hydrophobes such as comb polymers are considered. In this case, the disengagement of a hydrophobe from an association does not cause stress relaxation in all

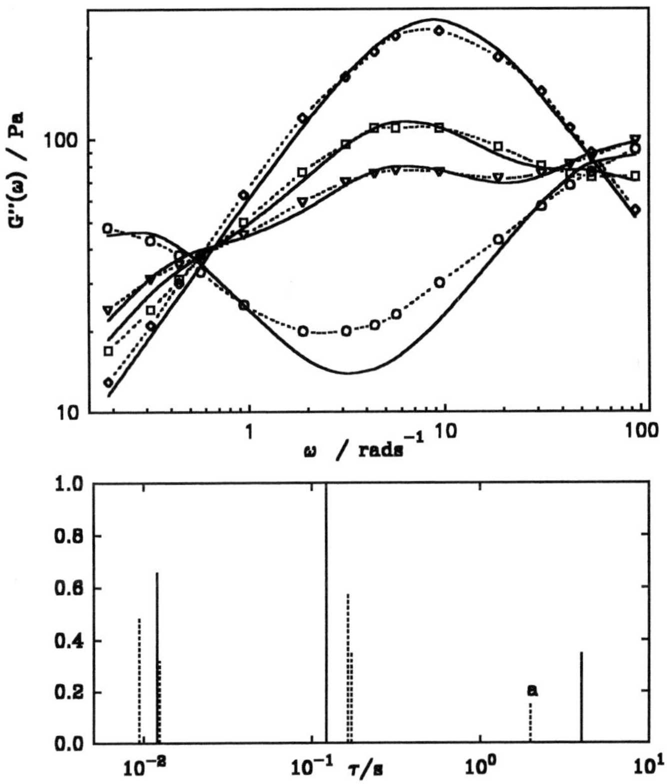

Figure 2. Top: Loss modulus for blends of C12/35, C16/35 and C20/35 at a total concentration of 3%. at blend ratios of 0:100:0 (diamonds), 25:50:25 (squares), 40:20:40 (triangles) and 50:0:50 (circles). Solid lines = best fit with number of Maxwell elements equal to the number of blended components. Bottom: Line relaxation spectra obtained from the above fits (at *a* two lines overlap).

other segments of the polymer chain as would be the case for the telechelic polymers. Thus for complete stress relaxation in a single polymer chain, several disengagement processes may need to occur. The dynamics and stress relaxation in such systems has been simulated by Groot and Agterhof [40] using off-lattice Monte-Carlo simulation techniques. Two stages of relaxation were observed in their simulations of systems capable of forming reversible binary associations. The relaxation at early-times was found to follow a $-^2/_3$ power law and was independent of association lifetime and concentration. However, despite the fact that the polymers in the simulation were not allowed to be restrained within a tube (since in the simulations, the chains were allowed to cross), the long-time behaviour was found to be consistent with the scaling predicted from reptation arguments. As mentioned above a single dissociative step produces only partial stress release in a chain connected to a network via multiple associations. For further stress release, another segment must be released from another binary association. This produces a picture analogous to the slow curvilinear diffusion of a polymer chain.

This behaviour can be described by considering the process of stress relaxation in a single chain having m segments. When the ith hydrophobe on the chain is released, the two neighbouring segments i and i+1 momentarily form an effective single segment of twice the original size. The effective segment is of course still attached to the network at two ends by the unreleased hydrophobes i-1 and i+1. If the Rouse time for this effective segment is small compared to the detachment/attachment time scale of the hydrophobes, the stress between the two parts will have enough time to equilibrate. That is to say that if the stresses on the segments i and i+1 were σ_i and σ_{i+1} prior to the detachment of the ith hydrophobe, afterward they will become $(\sigma_i+\sigma_{i+1})/2$, when the hydrophobe is released and subsequently rejoins a micelle. Thus, as pointed out by Groot and Agterof [40] and in contrast to the HEUR systems, the detachment of a hydrophobe does not release all the stress on a chain but is only a step towards doing so. The equation describing the time evolution of the stress on the ith segment can be expressed as follow

$$\sigma_i(t+\delta t) = \frac{\delta t}{\tau}\left(\frac{\sigma_i(t)+\sigma_{i+1}(t)}{2}\right) + \frac{\delta t}{\tau}\left(\frac{\sigma_i(t)+\sigma_{i-1}(t)}{2}\right) + (1-\frac{2\delta t}{\tau})\sigma_i(t) \tag{5}$$

where the rate of detachment of a hydrophobe from a micelle is $1/\tau$. The above equation is just the finite difference form of a 1D diffusion like equation

$$\frac{\partial\sigma}{\partial t} = \frac{1}{2\tau}\frac{\partial^2\sigma}{\partial i^2} \tag{6}$$

Solving the above equation with the boundary conditions σ=0 at i=0 and i=m, and assuming an initial uniform stress of $\sigma(0)$ in all segments, we get

$$\sigma(i,t) = \sigma(0) \sum_{p=odd} \frac{4}{p\pi} \sin\left(\frac{p\pi i}{m}\right) \exp\left(\frac{-p^2\pi^2 t}{2\tau m^2}\right) \tag{7}$$

and the average stress in a single chain is therefore found to be

$$\overline{\sigma(t)} = \frac{\sigma(0)}{m}\int_0^m \sigma(i,t)di \quad = \quad \sigma(0) \sum_{p=odd} \frac{8}{p^2\pi^2} \exp\left(\frac{-p^2\pi^2 t}{2\tau m^2}\right) \tag{8}$$

This is indeed the form that Groot and Agterof observe in their computer simulations [40] of comb associative polymers, in the long time limit. Note that the time scale in Equation 8 is given by

$$\tau_0 = \frac{2\tau m^2}{\pi^2} \tag{9}$$

which is proportional to the detachment time of the hydrophobes in a micelle. The result $\tau_0 \propto \tau$ has also been predicted from the computer simulation studies [40]. Note that in our calculations, we assume a fully developed network at moderate concentrations (below the entanglement concentration but above the concentration at which the network topology affects the relaxation spectrum).

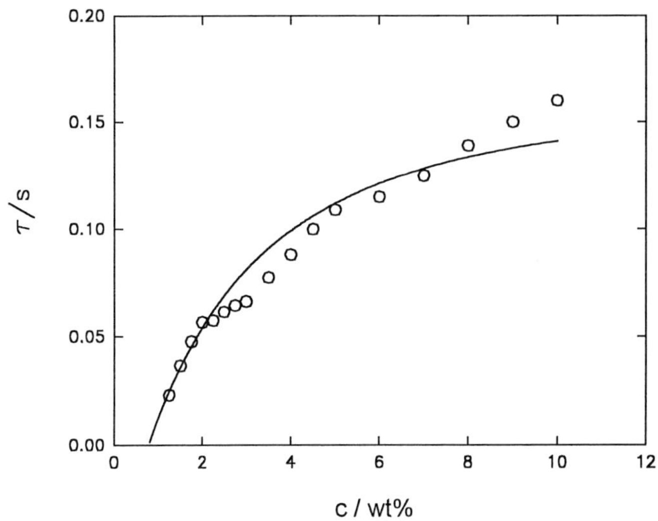

Figure 3. Effect of polymer concentration on the viscoelastic relaxation time for C16/35. Symbols = experimental data, lines = fit to equation 10.

Groot and Agterhof [40] found that at all but the highest concentrations studied τ_0 was found to be proportional to the polymer concentration. At the highest concentrations, the terminal relaxation time was found to approach the lifetime of the stickers, in accord with Equation 9. The experimental single viscoelastic relaxation time for HEUR-AT is also found to vary with concentration [10], see Figure 3. The simulation discussed above showed that the lifetime of an association was independent of polymer concentration. Thus, the most probable cause of the concentration dependence is the changing network structure. The network structure of AT will be discussed in more detail in the following section, but as the concentration decreases, the probability of the formation of "superbridges" (in which several active chains are linked together to form a single strand) increases, see Figure 6. If such states occur, then end-group dissociation by any one of the active chains would produce relaxation in all remaining chains within the superbridge. Thus a short relaxation time implies the formation of long superbridges. It has been shown [10] that the formation of superbridges in telechelic AT produces a concentration dependent relaxation time of the form:

$$\frac{\tau}{\tau_0} = \frac{1-\sum_{i=0}^{2} P(i)}{1-\sum_{i=0}^{1} P(i)} \tag{10}$$

where $P(i)$ is the probability for the formation of an association with cross-link functionality i (see following section) and τ_0 is the disengagement time. Such a prediction is superimposed on the relaxation time data for Figure 3. Reasonable correlation is observed although the experimental data does not plateau at high concentration in quite the same fashion. Similar reasoning for this concentration dependence is given by Groot and Agterhof [40]. The predictions of their simulations are also shown to be in good general agreement with the experimental findings.

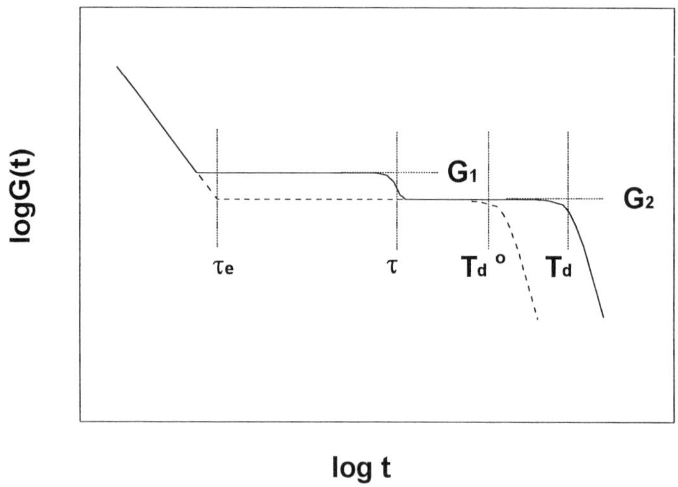

Figure 4. Schematic representation of the time dependent relaxation moduli for transient networks comprising linear chains with stickers (solid lines) and linear chains without stickers (dashed lines), from reference 41.

Leibler et.al. [41] have considered the dynamics of reversible networks built from linear flexible chains with S stickers per chain. They consider, theoretically, the case in which polymers are above their entanglement molecular weight. The presence of multiple association sites on the polymer backbone acts to hinder curvilinear diffusion along the tube length since diffusion can only occur if one or more sticker is disengaged thus releasing part of the polymer chain. Thus by a random sequence of such steps, reptative motion can occur. The timescale for diffusion is affected by both the fraction of free stickers $(1-p)$ and the number of stickers per chain, S, which for the high molecular weight limit is shown to be:

$$D_{self} \cong \frac{a^2}{2\tau S^2}(1 - \frac{9}{p} + \frac{12}{p^2}) \tag{11}$$

where a = tube diameter and τ = disengagement time. This result is similar to that obtained for unhindered reptation with disengagement time being the equivalent of the Rouse time of an entanglement strand, τ_e, and S being the equivalent of the number of entanglements per chain. Using such a model, Leibler et.al. predicted that there should be several important timescales for stress relaxation in such entangled transient networks. These are represented in Figure 4. At times $< \tau_e$, relaxation should be comparable to that of a polymer without stickers. On timescales $\tau_e<t<\tau$, a rubbery plateau is predited with the modulus being simply the product of the sum of the number of entanglements plus cross-links and the stored energy per strand, kT. At timescales in the vicinity of τ, the modulus falls to the modulus in the absence of stickers. This second plateau remains until the terminal relaxation time of the gel T_d which is significantly retarded from that for unmodified polymers T_d^0 due to the effect of the stickers in slowing down the reptation process.

The network dynamics of associating polymers can be seen to be highly dependent on various factors. These include number of hydrophobes, molecular weight, concentration, end-cap length and temperature (which will affect not only the disengagement time and the rate of segemnt relaxation but may also affect the fraction of free hydrophobes when $\Delta E_m \leq kT$). For simple telechelic, low molecular weight HEUR-AT, simple Maxwellian behaviour is now well known. An increase in the number of hydrophobes per chain *or* an increase in molecular weight will produce a broadening in the relaxation time distribution. Finally, one other pseudo-polymeric system that has been modelled is the rod-like micelles formed by surfactants such as CTAB [42]. These systems are akin to linear *entangled* polymers which may break at any point along their length. Such systems could conceivably be formed by telechelic polymers in which only binary associations are possible. Cates [42] showed that such extended chains are built up with a characteristic length $l=exp(E_m/kT)$. The resultant "superchain" is predicted to have a single stress relaxtion time $\tau_{eff} = [\tau_{break}\tau_{rep}(l)]^{\frac{1}{2}}$ provided that the the time for a single break along the chain, τ_{break} is less than the reptation time τ_{rep}. A single viscoelastic relaxation time has indeed been observed in such systems [43].

3. VISCOELASTIC MODULUS

A thorough mathematical analysis of the rheology of transient networks, building on that of Green and Tobolsky [44] has been presented by Tanaka and Edwards [45-47]. To calculate the viscoelastic modulus, it is necessary to calculate the number of elastically active polymer strands. The work of Tanaka and Edwards shows that the number density of the attached, and hence elastically effective, chains at any one time is given by

$$\nu \sim n\left(\frac{w(N)^2\exp(E_m/kT)}{1 + w(N)^2\exp(E_m/kT)}\right) \tag{12}$$

with the remaining n-ν chains having at least one end group outside a micelle. The quantity n is the number density of the AT chains and E_m the micellisation energy/head group (from now on taken to be ~ ΔE_m). w(N) is a weak function of N, the number of repeat unit per chain, and is of the order of unity for N>>1. For large micellisation energies E_m >>kT equation (12) predicts that ν ~ n. However, even for these cases a fraction of chains can still be elastically ineffective [10]. The reason for this is that a chain can have both of its hydrophobic end groups in a single micelle. A loop of this type clearly plays no part in supporting any stress. Its presence is therefore irrelevant to the elastic behaviour of the network. The existence of loops implies that the average crosslink functionality of a micelle, i.e. the number of bonds that connect it to the network, is on average less than its aggregation number. This gives rise to the possibility for more complex configurations such as "superbridges". Such superbridges are formed when a series of bi-functional micelles (ie. with cross-link functionality = 2) happen to be connected together to effectively form a long single chain, see Figure 6. These provide further means by which the chains can become elastically ineffective. It is well known from the theory of rubber elasticity [44] that the actual size of chains are not particularly significant in determining the elastic modulus of the network. It is rather their number density that is the quantity of importance, even for a polydisperse distribution of chain sizes. Thus the entire superbridge has to be taken into account as only a single effective chain, when calculating the elastic modulus of the network. Once the number of elastically effective chains in the network ν_{ef} is found the elastic modulus of the network is determined by the expression

$$G_\infty = \nu_{ef} kT \tag{13}$$

and the viscosity through the familiar relation

$$\eta = G_\infty \tau \tag{14}$$

The first step in calculating ν_{ef} involves the determination of the number of chains that are present as loop as opposed to link configurations. We [10] have shown that the probability of a chain, with one end already in a micelle, returning to have the other end in the same micelle is simply (for an ideal chain , the entropy of a chain backfolding is roughly the same as that for the formation of a single link)

$$F_2 = \frac{1}{n_m} \tag{15}$$

where n_m>1 denotes the average number of micelles in the neighbourhood of the chain (that is within a volume of size αR_g^3 , where R_g is the radius of gyration of the polymer and α a constant that turns out to be of the order unity). The probability of the same chain being a link is hence F_1=1- ($1/n_m$). Taking into account the fact that each link is connected to two micelles whereas a loop connects to a single micelle, the ratio of loops to links in the entire system is

$$\frac{F_1'}{F_2'} = \frac{F_1}{2F_2} \tag{16}$$

from which we easily obtain the result $F'_1 = F_1 / (2-F_1)$. In deriving the above results it was assumed that $\Delta E_m >> kT$, so that very few end groups are unattached at any one time.

Calculation of n_m requires a knowledge of the way that the micelles are distributed throughout the system, as specified by their radial distribution function g(r). Unfortunately this information is not available at present. In order to make further progress one needs to make certain assumptions regarding the spatial distribution of the micelles. The simplest of these is to take g(r)=1, *ie.* a random arrangement. AT solutions, at the concentrations of interest in this paper, are found to be optically clear. Thus, at least for length scales of the order of wavelength of light or larger, this assumption seems to be reasonable. It turns out that at higher polymer concentrations the calculations based on this approximation are in good agreement with our rheological data. However, at low concentrations, close to the point where the spanning network is first observed, the g(r)=1 assumption will break down. As mentioned above, we have argued that the structure of the network, at least up to some length scale ξ, might be fractal [48]. It is hoped that measurements of this kind may provide one with the structural information needed to make more precise calculations of the rheological properties in the near future. For present however we have adopted the following alternative approximation for estimating n_m. By analogy with a randomly closed pack set of hard spheres, we assume that at the overlap concentration each polymer chain has an average of 6 neighbouring chains. At this point the equivalent hard sphere volume fraction for the chains is 0.64. For any other concentrations then the number of neighbouring chains Z is taken to vary linearly with the effective volume fraction ϕ,

$$Z = \frac{6\phi}{0.64} \tag{17}$$

where ϕ is given by

$$\phi = \frac{4\pi R_g^3}{3} n = \frac{4 \times 10^4 \pi c N_A R_g^3}{3M} \tag{18}$$

for polymers of molecular weight M at a concentration c. Next, a balance on the number of end groups, micelles and Z gives

$$n_m = \frac{(2 - F_1)Z + 2}{N_{agg}} \tag{19}$$

where N_{agg} denotes the mean aggregation number of the micelles. Note that out of an average of $n_m N_{agg}$ head groups surrounding the chain two of them belong to the chain itself. Using Equations 15, 18 and 19 one can now calculate the values of n_m, F_1 and F_2 for any polymer concentration c. The mean micelle functionality $\langle\psi\rangle$ is then

$$< \psi > = N_{agg}F_1' = \frac{F_1 N_{agg}}{2 - F_1} \tag{20}$$

In the absence of any loops (i.e. F_1=1) the average functionality is of course the same as N_{agg}. This is the case at high polymer concentrations where each chain is surrounded by many micelles. As the polymer concentration is reduced an increasing number of the AT chains adopt loop configurations. Such loops as well as being elastically ineffective themselves, also decrease the average functionality of the micelles (equation 20), resulting in a higher number of dangling ends, superchains etc. Our picture then predicts that the value of (G_∞/nkT) should drop as n is reduced. We have found the following result for (G_∞ /nkT)

$$\frac{G_\infty}{nkT} = F_1' - \frac{2}{N_{agg}}[P(1) + P(2)] + \frac{[P(1)]^2}{N_{agg}\,[N_{agg}F_1' - 2P(2)]} \tag{21}$$

where P(ψ) denotes the probability of a micelle having a functionality ψ [10]. The micelle functionalities are assumed to be distributed in binomial fashion with the average value <ψ> given by Equation 20. As well as loops, Equation 21 takes into account the effect of superbridges and some simpler dangling ends in reducing the number of effective elastic chains. The importance of uni and bi-functional micelles in the formation of such configurations (see figure 6) is reflected through the appearance of their probabilities P(1) and P(2) in Equation 21.

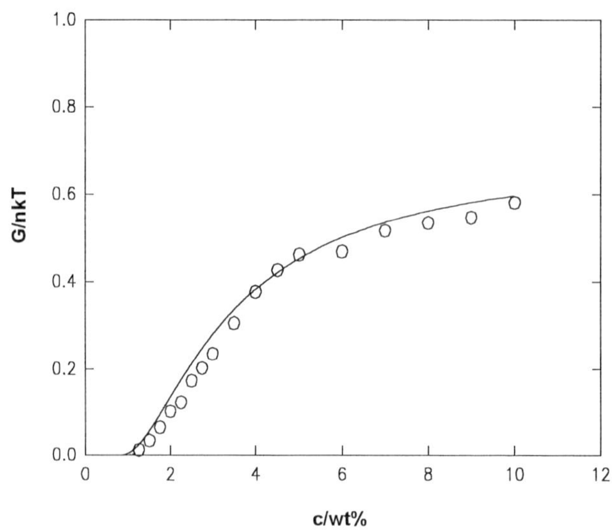

Figure 5. Reduced viscoelastic modulus for C16/35 as a function of polymer concentration. Symbols = experimental data, lines = best fit to Equation 21.

The model as described by Equation 21 is a simple picture of the concentration dependence of the viscoelastic modulus in solutions of telechelic HEUR-AT. This is compared to the experimental data for C16/35 in Figure 5 . A pre-factor is used which describes the efficiency of the end-capping reaction since a fraction of chains will have less than two end-caps. The theory and experiment are in good agreement. The picture is then that the concentration dependence arises from an entropic transition from a network comprising predominantly looped chains at low concentrations to one in which all chains are bridges at high concentrations. The model as described assumes that all chains are part of a single giant cluster and does not take account of any fragmentation of the network. Our simulation results [10] and those of Groot and Agterhof [40] indicate that there is a rapid transition at concentrations ~ c* to a fully developed network. At concentrations below c*, there is considerable spectroscopic evidence to support the view that initially isolated chains associate above a broad critical aggregation concentration into micelle like aggregates [26]. As the concentration increases, clusters of micellar aggregates form which, in the region of c*, join up to form a single cluster. In this region the viscoelastic properties are developed. An identical picture of the network development was independently given by Winnik et.al. [28]. The picture thus developed is shown in Figure 6. As well as describing the concentration of the viscoelastic modulus, the model is further supported by the fact that it provides a picture of the concentration dependence of the relaxation time through the formation of superloops (see above), provides the correct scaling for the molecular weight dependence of the relaxation time [10] and provides a good description of the influence of surfactants on the modulus [48].

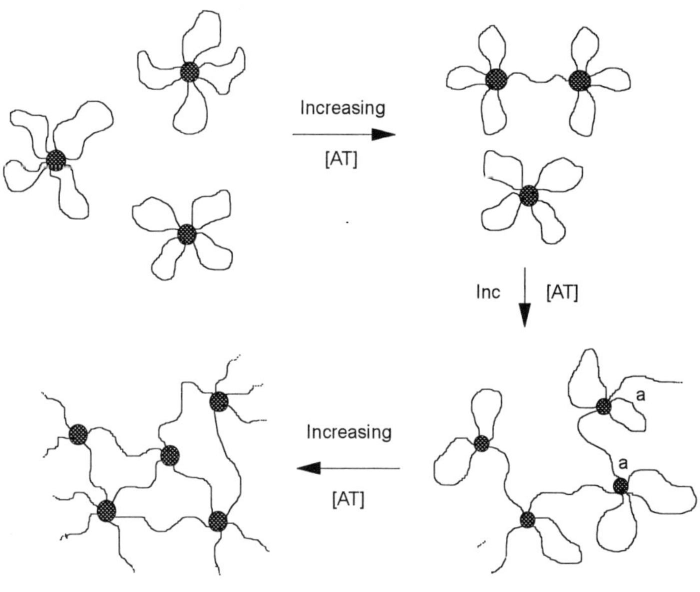

Figure 6. Schematic representation of the effect of concentration on the development of an AT network. The associations marked **a** show micelles of cross-link functionality 2 which give rise to the formation of superchains (see text).

As described above, Tanaka and Edwards [45-47] have calculated the fraction of free chain ends, Equation 12, to be vanishingly small for the micellisation energies found for HEUR systems ($\Delta E_m \sim 20kT$). This is further supported by the observation [26] that the critical aggregation concentration of these polymers occurs at low concentrations. (We have measured the c.a.c. for C16/35 using similar spectrosopic techniques and found it to be rather broad and occur in the concentration range of 0.01-0.1wt%). Thus, at the concentrations at which the viscoelastic properties begin to develop (>1wt% for C16/35), the fraction of free chain ends will be very small. It is justifiable then that chain states involving free chain ends can be considered to be statistically insignificant and can be discounted when developing models for the rheological behaviour of such telechelic systems.

Clearly, the modelling of sytems with greater than two hydrophobes per chain is considerably more complex since the number of different chain states is greatly increased. As described above, the theory of Leibler et.al., describing systems with multiple hydrophobes and of sufficient molecular weight such that the chains are entangled, predicts two plateau moduli providing the timescales of end-cap disengagement and the Rouse time for a polymer strand are sufficiently separated. These two plateau moduli are depicted in Figure 4. The moduli are given by

$$G_1 \cong cRT\left[\frac{p}{N_s} + \frac{1}{N_e}\right] \tag{22}$$

and

$$G_2 \cong cRT\left[\frac{1}{N_e}\right] \tag{23}$$

where c is the number concentration of monomers and N_e and N_s are the average number of monomers between entanglements and stickers respectively.

3. STEADY-SHEAR PROPERTIES

The viscosity shear-rate dependence of a solution of C16/35 is displayed in Figure 7. Also shown is the dyanmic viscosity obtained from oscillatory measurements. The data are typical of that observed for solutions of telechelic AT [10,12,49]. Three distinct regions are apparent. At low to moderate shear-rates, the viscosity is independent of shear-rate. In this region the dynamic viscosity is equal to the steady shear viscosity. The viscosity in this region is given by Equation 14. At high shear-rates, the response is shear-thinning. Under steady shear conditions, instabilities and gel fracture are frequently encountered in this region making the flow curve difficult to measure. Indeed, Winnik [50] has reported that a power-law index of -1 is often observed at higher shear-rates, this being indicative of such effects. It has been proposed that the action of shear can induce phase separation [51]. A thorough treatment of such behaviour is, however, lacking at present.

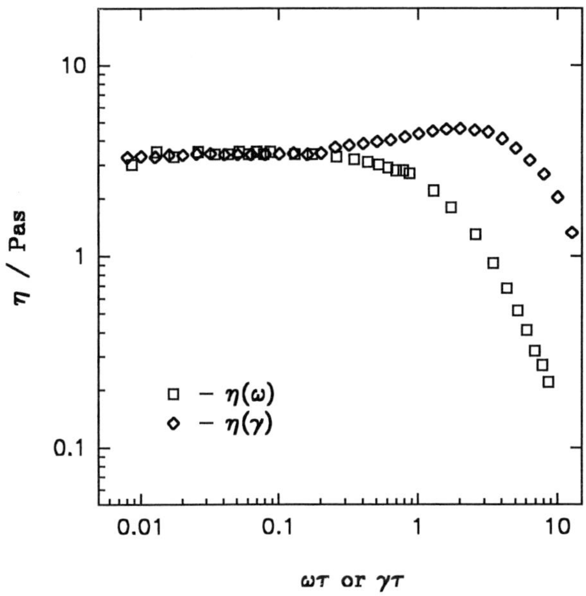

Figure 7. Comparison of the reduced shear-rate dependence of viscosity and the frequency dependence of the dynamic viscosity for C16/35 at 1.5%.

Despite these flow instabilities, there is no doubt, that shear-thining does occur. The primary cause is, we believe, the diminishing relaxation time. At high shear-rates (timescales in excess of the relaxation time), the chains become stretched [10,45,49]. The elastic energy in the chains produces an increase in the rate of detachment of the end-groups from associations and hence a drop in the relaxation time. This in itself produces shear-thinning, Equation 4. In addition to this, there is the potential for some network rearrangement. In spectroscopic experiments, in which the effect of shear on the network has been examined, it has been observed that there is no change in the number density of associations in the shear-thinning regime [25]. In other words, shear does not act to destroy associations at these shear-rates. Winnik et.al. [28] proposed that shear acts to increase the fraction of loop to link states. This, we have shown, can be related to the falling relaxation time since these are in dynamic equlibrium

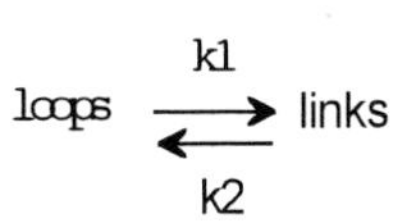

Clearly k_1 will be unaffected by shear-rate (since loops are unstressed) whereas k_2 will increase due to the elastic energy on the links. Thus there will be a shift in the equilibrium to

the left, ie. loops will form at the expense of links and the cross-link density will fall. In addition to the fall in τ, this will result in a decreasing viscosity as the shear-rate is increased. The relative importance of the diminshing relaxation time and the reducing cross-link density to the rate of shear thinning will depend on the AT concentration, as this determines the probabilities for loop and link formation. At high concentrations, it is likely that there is very little shift towards loop formation since a detached end-group will most probably form a new link state. Thus, any shear-thinning at such concentrations is almost certainly be the result of the change in τ. Similar arguments will hold for systems with multiple hydrophobes although the detachment rate of end-groups attached to looped strands may well be affected by shear rate since neighbouring strands within the polymer chain could have formed link states.

The final region that is evident in the flow curve, lying at shear-rates on the timescale of the viscoelastic relaxation time, is a weak shear-thickening regime prior to shear-thinning. Similar shear-thickening effects have been seen in poly(vinyl alcohol)/sodium borate [52] and t-butylstyrene/methacrylic acid copolymers [53]. In these latter systems, a strong case for a shift from intramolecular to intermolecular association as the chains are extended, is expounded by both research groups. This is an attractive interpretation for systems in which multiple association sites per chain are possible. However it is unlikely that this is the cause of shear-thickening effect in telechelic associating polymers since as just discussed, shear would be expected to shift the balance between loops and links in the opposite direction. Jenkins [49] has presented an alternative theoretical description of shear-thickening in which he generalises Yamamoto's network model [54,55] to include non-Gaussian chain behaviour and a conformation dependent junction breakage potential. Jenkins' picture is one in which the network junctions are strong enough to support stress and thus chain extension occurs under shear. The shear-thickening then results from the finite extensibility of the network chains. The model predicts that the magnitude of the viscosity enhancement decreases with decreasing molecular weight, in line with experimental results [49]. Marucci et.al. [56] present two further models based on similar non-Gaussian chain effects. A further view of shear-thickening has been taken by Wang [56,57] in transient networks in which a significant fraction of free chains are present.

4. EFFECT OF SURFACTANTS

The non-specificity of the association process results in a broad range of interactions of AT with colloidal phases such as polymer latices and surfactant micelles. Surfactants are known to profoundly affect the rheology of hydrophobically modified polymers such as HEUR [48,58,59] and hydrophobically modified cellulose derivatives [60,61]. The magnitude of these effects is strongly dependent on the surfactant concentration and on the level of AT. Unmodified water soluble polymers are known to interact with surfactants such as sodium dodecyl sulphate, SDS but here we are concerned with the stronger hydrophobic interaction of surfactants with the attached alkyl groups. It has been shown using fluorescence spectroscopy, that in HEUR systems this is the dominant interaction over SDS/PEG interactions [27]. The effect of SDS on the normalised elastic modulus of C16/20 is shown in Figure 8. Similar trends have been observed for the viscosity dependence in similar systems [58]. It can be seen from Figure 8 that the SDS acts to increase or decrease the measured modulus dependent on both the polymer concentration and SDS concentration. At low polymer concentrations, a maximum is observed which diminishes as the polymer concentration is increased. At high polymer concentrations, SDS acts only to destroy the AT network. There is also evidence of a spread of relaxation times in systems containing higher SDS levels [48].

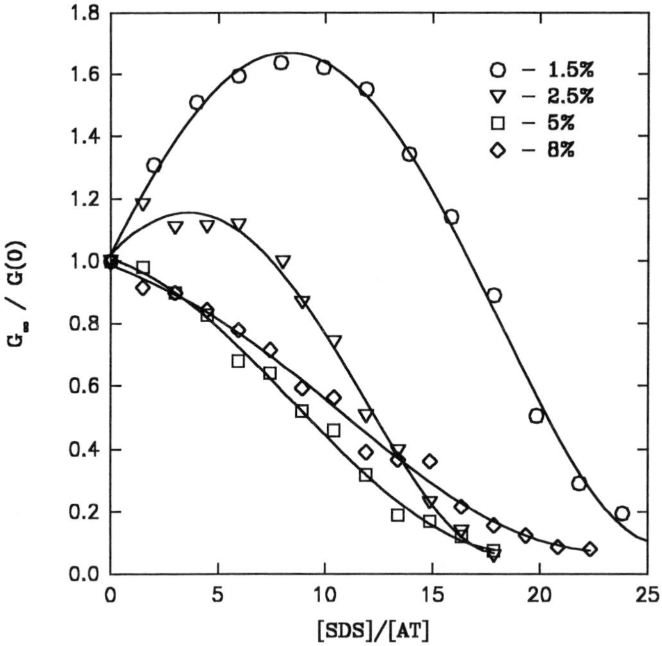

Figure 8. Effect of sodium dodecyl sulphate on the viscoelastic modulus of C16/35 at various polymer concentrations. The viscoelastic modulus is normalised by its value in the absence of surfactant.

We have modelled the modulus dependence using the loops-links model described above. In order to do this it is assumed that mixed micelles are freely formed thus the addition of surfactant acts to reduce the number of AT hydrophobes per micelle. In other words we assume that one SDS molecule joining a AT association will release one AT hydrophobe. This 1:1 equivalence is unlikely due to the differing aggregation numbers of HEUR and SDS but affects only the concentration range over which changes occur, rather than the magnitude of the predicted modulus. (ie. A Rescaling of the [SDS]/[AT] axis.) We can then describe the effect of SDS on the aggregation number of AT chains within a micelle N_{eff} by

$$N_{eff} = \frac{N_{agg}}{1+r} \tag{24}$$

where r=[SDS]/[AT] and N_{agg} is the aggregation number in the absence of surfactant. The theoretical predictions of such a simple model are shown in Figure 9. The magnitude of the modulus changes are in good qualitative agreement with the experimental data shown in Figure 8. The changes arise from the following simple picture. At low polymer concentrations, the fraction of chains present in looped configurations is high as the number concentrations of micelles is small, see Equation 15 and Figure 5. As surfactant is added, more micelles are formed. Some of these will be mixed micelles involving both AT hydrophobes and SDS. This increased micellar concentration will cause some of the chains in looped states to take on linked configurations. This is the mechanism for any modulus, and viscosity, increase. At high polymer concentrations, the chains are already predominantly link states since the number density of associations is high. Thus there is no mechanism for any modulus increase. At high surfactant concentrations, the average micelle cross-link functionality will fall and eventually all link states will attach to elastically ineffective low functionality micelles. Piculell et.al. [62] have calculated the extent of this region using similar mechanistic arguments. At very high surfactant concentrations, the modulus falls to zero as the number of micelles is in excess of the number of hydrophobes, hence all chain ends lie in separate uni-functional micelles and the modulus falls to zero.

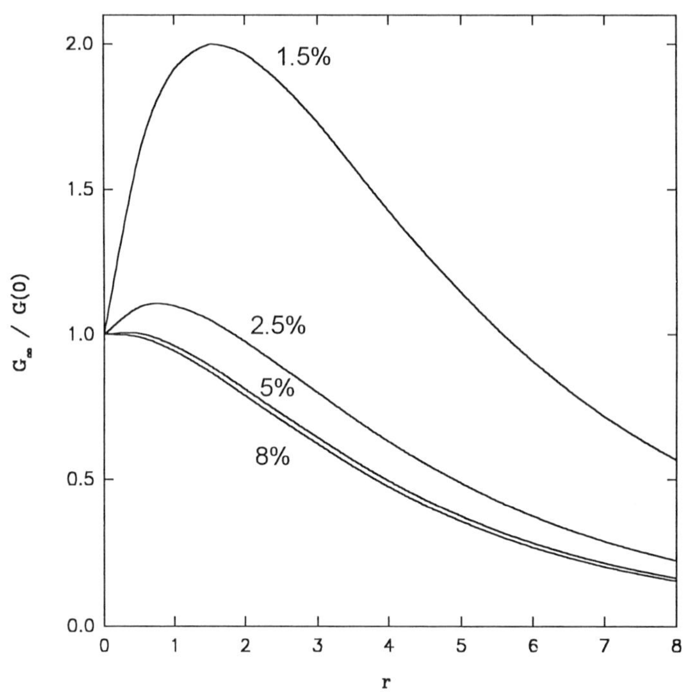

Figure 9. Theoretical prediction of the effect of surfactant on the reduced viscoelastic modulus of an AT of molecular weight 35000.

Similar effects have been reported for mixtures of AT with nonionic surfactants such as the nonylphenol ethoxylates $NP(EO)_nH$ [47,63]. It has been shown that the HLB of the surfactant can have a significant effect on the maximum viscosity or modulus of the mixtures [47,63]. In particular, low HLB surfactants produce enhanced effects relative to those with high HLB. We have attributed this phenomenon to the low aqueous phase solubility of the lower HLB surfactants and the consequent low occurrence probability of micelles with low cross-link functionality. At higher levels of surfactant, the surfactant phase itself may dominate the rheological properties. The effect on HEUR-ATs on the surfactant rheology in this region has been studied by Glass et.al. [58].

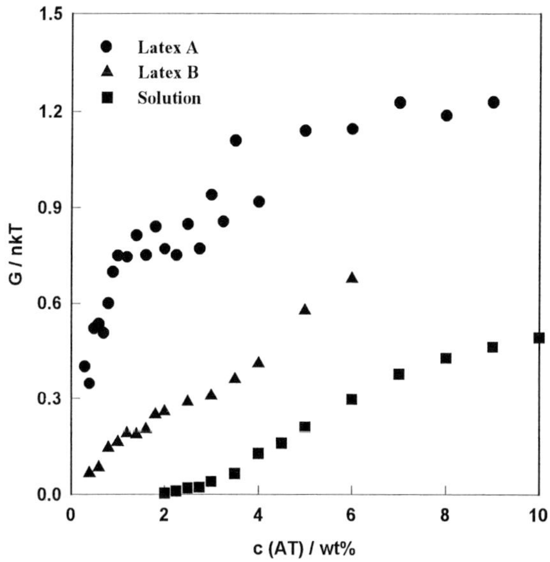

Figure 10. Reduced modulus as a function of AT concentration for C16/20 in solution and in latices A and B. Latex volume fraction = 0.3.

5. INTERACTION WITH POLYMER LATICES

Associating polymers are widely used as thickening agents in latex coatings with desirable rheological properties. Thus it is of considerable interest to develop an understanding of the structure-property relationships in such mixtures. The interaction of AT with polymer latices has been widely studied. In addition to rheological measurements [49.64], techniques such as fluorescence [65], adsorption equlibria [49,66], light scattering [67] and NMR [24] have been used to elucidate the nature of the interaction. It is generally thought that the hydrophobic groups of the AT adsorb on the relatively hydrophobic latex surface. This adsorption process has been shown to be described by a Langmuir type isotherm. The adsorption density and the strength of the interaction will clearly depend on numerous factors including hydrophobe length, surface coverage, surface energy and competitive adsorption with surfactants. An interaction between the hydrophilic PEG chain in HEUR-AT and the carboxylic acid stabilising groups, present in the vast majority of commercially important latices, has also been proposed. Mixtures of associating polymers in polymer latices generally exhibit a strong synergy between the AT network and the dispersed phase, giving rise to enhanced viscosities [49,64]. Often, colloid stability problems can be observed in poorly stabilised thickened dispersions. This has been studied by Santore et.al. [67], but here we restrict the discussion to stable mixtures.

The effect of two commercial latices on the normalised viscoelastic modulus of C16/20 is shown in Figure 10. The HEUR concentration is expressed as wt% on aqueous phase. Latex A is a surfactant stabilised 100nm styrene-acrylic latex whilst latex B is an all acrylic latex polymerised in the presence of polyethyleneglycol methacrylate and with a particle size of 100nm. In both systems a synergy is observed although the styrene-acrylic shows enhanced effects relative to the all acrylic latex B. The aqueous phase serum was isolated from both latices and AT used to prepare solutions of C16/20. These were found to have very similar rheology to that obtained for the same polymer in pure water at the same concentrations. Thus the free surfactant level was insufficient significantly affect the rheology of the HEUR network, and thus any rheological changes are likely to be due to latex/HEUR interactions. Thus the enhanced synergy in latex A is likely to be due to a stronger interaction with the AT network than for latex B. This is probably due to hydrophobe adsorption and thus the latex acts as a multifunctional network site and acts to reinforce the network. In addition, the latex acts as a non-deformable filler particle.

The viscoelastic response of C16/20 in latex A is shown in Figure 11. The measurements are made at a latex volume fraction of 0.4 and an AT concentration of 2wt%. The data can be seen to be poorly described by a single Maxwell element but good fitting is obtained using two elements. In contrast, for C16/35 in Latex B, the response is dominated by a single viscoelastic relaxation time and a good fit to the Maxwell model is observed at all AT concentrations. For latex A, the viscoelastic behaviour is not described by a single relaxation process, as observed in solution, and good description of the experimental results can be only be obtained when two relaxation processes are assumed. The strongly interacting latex A mixture appears to show two well separated relaxation times. It seems plausible that these may result from hydrophobe dissociation from an association and hydrophobe desorption from the particle surface respectively. The short relaxation time is comparable to that observed for a pure solution and compares with that observed for the latex B mixture. Thus the longer timescale process could reflect the desorption kinetics of a hydrophobe from the latex surface. Clearly, it seems that the more hydrophobic styrene containing latex appears to show a greater capacity to interact and reinforce the AT network than the relatively hydrophilic acrylic latex B. In this latter case, it may be that the latex acts only as a filler and does not interact at all. The behaviour at low associating polymer concentrations is hard to explain. The concentration at which viscoelastic effects are measurable is significantly less in both latices than in pure solution. At these concentrations, the polymer is well below its overlap concentration c* and thus it is difficult to see how connectivity could be achieved. Clearly the physical presence of latex particles could affect the conformation of AT chains and may, therefore, affect the loop to bridge ratio. Also, for an interacting latex, it is possible that either single chain or cluster bridging between particles could be seen at this relatively high latex volume fraction.

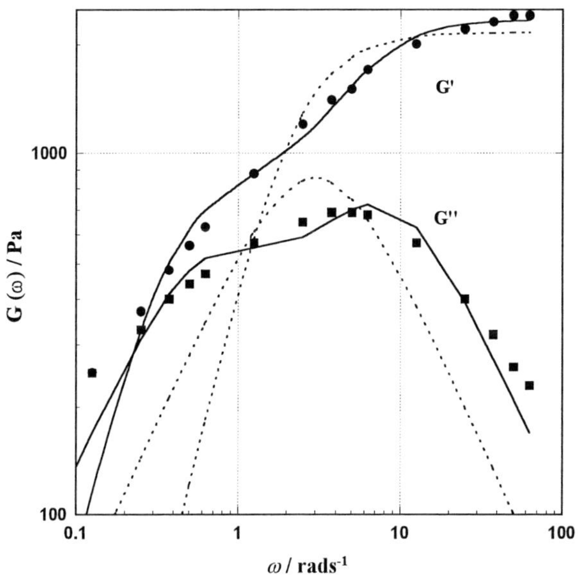

Figure 11. Viscoelastic spectrum for C16/35 in Latex A at a volume fraction of latex of 0.4 and an AT concentration of 2 wt%. Dashed lines = best fit to Maxwell model, solid lines = fit obtained with two Maxwell elements.

6. CONCLUSIONS

In this chapter we have given a brief review of the rheological properties of associating polymers in solution and as mixtures with polymer latices and surfactants. As mentioned in the introductory section, there are many different types of associating polymers, but here the emphasis has been on the telechelic HEUR polymers. Where appropriate, we have tried to indicate how the polymer architecture and molecular weight may affect the rheology. For the telechelic systems, simple Maxwellian viscoelastic response is observed, consistent with a picture in which stress relaxation occurs through end-group disengagement followed by much faster Rouse relaxation of the polymer chain. Such a mechanism is also consistent with the observation that the viscoelastic spectrum of HEUR-AT with different hydrophobe lengths possess the same number of relaxation times as there are components in the blend. As a consequence, it is possible, through control of the blend ratios of the components, to produce fluids with designed relaxation spectra. Polymers with more than one hydrophobe per chain possess relaxation spectra that are considerably more complex due to the partial stress relaxation occurring after hydrophobe disengagement.

Many of the rheological properties of telechelic ATs can be described by a simple model in which looped and bridged chains co-exist within an infinite network. Such a picture can be readily used to describe various rheological features including the concentration dependencies of the viscoelastic modulus, relaxation time and viscosity (through relations 21, 10 and 14), the correct scaling for the molecular weight dependence of the relaxation time, and the effect of surfactants on the elastic modulus. The viscoelastic properties of AT in the presence of dispersed phases such as polymer latices is an area of considerable interest as such polymers are widely used as thickeners. The effect of polymer latices on the AT rheology is profound and there is considerable evidence of an interaction, probably hydrophobe adsorption, between the two components. Clearly these systems are considerably more complex than pure AT solutions, particularly for latices of commercial interest in which excess surfactant and a significant level of acid comonomers are often used. At present, a qualitative picture of the network structure exists but a proper theoretical treatment of the rheology of such systems is currently lacking. This is an area worthy of further work.

Clearly, there has been significant advances in the understanding of the rheology of associating polymers in pure solution and as mixtures with formulation components in particular surfactants. However, this is still an area of considerable research interest and there is much still to be understood. It is likely that the rheology of transient networks will continue to be an active research field for the foreseeable future.

REFERENCES

(1) A. Karunasena and J.E. Glass, Prog.Org.Coatings, 17 (1989) 301.
(2) Z.Ma, J.P. Kaczmarski and J.E. Glass, Prog.Org.Coatings, 21 (1992) 69.
(3) J.E. Glass, R.H. Fernando, S.K. Egland-Jongewaand, R.G. Brown, J. Oil Colour Chemists Assoc., 67 (1984) 256.
(4) J.E. Hall, P. Hodgson, L. Krivanek and P. Malizia, J. Coat. Technol., 58 (1986) 65.
(5) F.G. Schwab, Adv.Chem.Ser., 213 (1986) 369.
(6) F.M. Anwari and F.G. Schwab, Adv.Chem.Ser., 223, (1989) 527.
(7) J.E. Glass, Adv.Chem.Ser., 213 (1986) 391.
(8) J.H. Bieleman, F.J.J. Riesthaus and P.M. Van der Velden, Polym. Paint Colour J., 176 (1986) 450.
(9) A. Karunasena, R.G. Brown and J.E. Glass, Adv.Chem.Ser., 223 (1989) 411.
(10) T.Annable, R Buscall, R. Ettelaie and D. Whittlestone, J. Rheol., 37 (1993) 695.
(11) D.J. Lundberg, J.E. Glass and R.R. Eley, J. Rheol., 35 (1991) 1255.
(12) G. Fonnum, J. Bakke and F.K. Hansen, Colloid Polym.Sci., 271 (1993) 380.
(13) R. May, J.P. Kaczmarski and J.E. Glass, Macromolecules, 29 (1996) 4745.

(14) J.W. Goodwin, R.W. Hughes, C.K. Lam, J.A. Miles and B.C.H. Warren, Adv.Chem.Ser., 223 (1989) 365.
(15) B. Nystrom, K. Thuresson and B. Lindman, Langmuir, 11 (1995) 1994.
(16) T. Aubry and M. Moan, J.Rheol. 38 (1994) 1681.
(17) B. Nystrom, H. Walderhaug and F.K. Hansen, J.Phys.Chem., 97 (1993) 7743.
(18) H. Walderhaug, F.K. Hansen, S. Abrahmsen, K. Persson and P. Stilbs, J.Phys.Chem., 97 (1993) 8336.
(19) S. Abrahmsen-Alami, P. Stilbs, J.Phys.Chem., 98 (1994) 6359.
(20) K. Persson, G. Wang and G. Olofsson, J.Chem.Soc.Faraday Trans., 90 (1994) 3555.
(21) S. Abrahmsen-Alami, K. Persson, P. Stilbs and E. Alami, J.Phys.Chem., 100 (1996) 4598.
(22) K.Persson, P.C. Griffiths and P. Stilbs, Polymer, 37 (1996) 253.
(23) B. Roa, Y. Uemura, L. Dyke and P.M. Macdonald, Macromolecules, 28 (1995) 531.
(24) Y. Uemura and P.M. Macdonald, Macromolecules, 29 (1996) 63.
(25) B. Richey, A.B. Kirk, E.K. Eisenhart, S. Fitzwater and J. Hook, J.Coat.Technol., 63 (1991) 31.
(26) Y.Wang and M.A. Winnik, Langmuir, 6 (1990) 1437.
(27) Y-Z. Hu, C-L. Zhao, M.A. Winnik and P.R. Sundarajan, Langmuir, 6 (1990) 880.
(28) A. Yekta, J. Duhamel, H. Adiwdjaja, P. Brochard and M.A. Winnik, Langmuir, 9 (1993) 881.
(29) A. Yekta, J. Duhamel, P. Brochard, H. Adiwidjaja and M.A.Winnik, Macromolecules, 26 (1993) 1829.
(30) A. Yekta, B. Xu, J. Duhamel, H. Adiwidjaja and M.A. Winnik, Macromolecules, 28 (1995) 956.
(31) E. Alami, M. Almgren and W. Brown, Macromolecules, 29 (1996) 5026.
(32) R. Tanaka, J. Meadows, G.O. Phillips and P.A. Williams, Carbohydr.Polym., 12 (1990) 443.
(33) C. Senan, J. Meadows, P.T. Shone and P.A. Williams, Langmuir, 10 (1994) 2471.
(34) Y-W. Yang, Z. Yang, Z-K. Zhou, D. Attwood and C. Booth, Macromolecules, 29 (1996) 670.
(35) E.J. Amis, N. Hu, T.A.P. Seary, T.E. Hogen-Esch, M. Yassini and F. Hwang, Adv.Chem.Ser., 248 (1996) 279.
(36) M. Odenwald, H-F. Eicke, Chr. Friedrich, Colloid Polym.Sci., 274 (1996) 568.
(37) C.L. McCormick and C. Brent Johnson, Adv.Chem.Ser. 223 (1989) 437.
(38) Y. Uemura, J. McNulty and P.M. Macdonald, Macromolecules, 28 (1995) 4150.
(39) K.L. Ngai in "Structure and Dynamics of Strongly Interacting Colloids and Supramolecular Aggregates in Solution", S.H. Chen, J.S. Hwang and P. Tartaglia (eds.), NATO ASI Ser C, Kluwer Academic Publishers, Dordrecht, The Netherlands, 369 (1992) 221.
(40) R.D. Groot and W.M. Agterhof, Macromolecules, 28 (1995) 6284.
(41) L. Leibler, M. Rubinstein and R.H. Colby, Macromolecules, 24 (1991) 4701.
(42) M.E. Cates, J.Phys.Chem., 94 (1990) 371.
(43) S.J. Candau, E. Hirsch, R. Zana and M. Adam, J. Colloid Interface Sci., 122 (1988) 430.
(44) M.S. Green and A.V. Tobolsky, J.Chem.Phys., 14 (1946) 80.
(45) F. Tanaka and S.F. Edwards, J.Non-Newtonian Fluid Mech. 43 (1992) 247.

(46) F. Tanaka and S.F. Edwards, J.Non-Newtonian Fluid Mech. 43 (1992) 273.
(47) F. Tanaka and S.F. Edwards, J.Non-Newtonian Fluid Mech. 43 (1992) 289.
(48) T. Annable, R. Buscall, R. Ettelaie, P. Shepherd and D. Whittlestone, Langmuir, 10 (1994) 1060.
(49) R.D. Jenkins, PhD dissertation, Lehigh University, 1990.
(50) M.A. Winnik, paper presented at "Associating Polymers '95" Conference, Loen, Norway, (1995).
(51) J.C. Thibeault, paper presented at "Associating Polymers '95" Conference, Loen, Norway, (1995).
(52) J.M. Maerker and S.W. Sinton, J.Rheol., 30 (1986) 77.
(53) J.M. Ballard, R. Buscall and F.A. Waite, Polymer, 29 (1988) 1287.
(54) M. Yamamoto, J.Phys.Soc.Japan, 11 (1956) 413.
(55) M. Yamamoto, J.Phys.Soc.Japan, 12 (1957) 1148.
(56) G. Marrucci, S. Bhargava and S.L. Cooper, Macromolecules, 26 (1993) 6483.
(57) S-Q. Wang, Macromolecules, 25 (1992) 7003.
(58) W. Binana-Limbele, F. Clouet and J. Francois, Colloid Polym.Sci., 271 (1993) 748.
(59) A.P. Mast, R.K. Prud'homme and J.E. Glass, Langmuir, 9 (1993) 708.
(60) R. Tanaka, J. Meadows, P.A. Williams and G.O. Phillips, Macromolecules, 25 (1992) 1304.
(61) F. Guillemet and L. Piculell, J.Phys.Chem., 99 (1995) 9201.
(62) R.D. Jenkins, B.R. Sinha and D.R. Bassett, Abstracts of Papers; 35 (1992) 202, The 4th Chemical Congress of N. America, NY 1991, American Chemical Society, Washington DC., 1991, PMSE 35.
(63) L. Piculell, K. Thuresson and O. Ericsson, Faraday Discuss., 101 (1995) to be published.
(64) P.R. Howard, E.L. Leasure and S.T. Rosier, J.Coat.Technol., 64 (1992) 87.
(65) K. Char, C.W. Frank and A.P. Gast, Langmuir, 5 (1989) 1335.
(65) R.D. Jenkins, M. Durali, C.E. Silebi and M.S. El-Aasser, J. Colloid Interface Sci., 154 (1992) 501.
(66) H.D. Ou-Yang and Z. Gao, J. Phys.II.Fr., 1 (1991) 1375.
(67) M.M. Santore, W.B. Russel and R,K, Prud'homme, Macromolecules, 22 (1989) 1317.

Applications of block copolymers

K. Holmberg

Institute for Surface Chemistry, P.O. Box 5607, S-114 86 Stockholm, Sweden

1. INTRODUCTION

Block copolymers have a long history as industrial surfactants. The major types of block copolymers, such as those made from ethylene oxide (EO) and propylene oxide (PO) or EO and styrene, are cheap and easy to tailor-make for specific applications. The water-soluble EO-PO block copolymers are stable over a wide pH range and compatible with all other types of surfactants. In addition, they are mild to the skin and hair, a property related to their high molecular weight.

In the manufacture of an amphiphilic block copolymer for a specific application there are several degrees of freedom as compared with the synthesis of conventional, low-molecular weight surfactants:
(1) the size of both the hydrophilic and the hydrophobic part can be varied at will,
(2) the molecular weight can be varied within wide ranges while maintaining constant hydrophilic-lipophilic balance, and
(3) the properties and function of a block copolymer at an interface, e.g. oil-water, can be governed by the molecular architecture. As an example, an EO-PO-EO triblock polymer is preferred as steric stabilizer of oil-in-water emulsions whereas water-in-oil emulsions may be better served with a copolymer of PO-EO-PO type.

In more recent times slow degradability in the environment of the major type of water-soluble block copolymers, the EO-PO based compounds, has become a major obstacle for both household and industrial use of the products. Improved biodegradability is probably the strongest driving force for the development of new surfactants today and products which do not meet the OECD guideline in rate of degradation into carbon dioxide and water are challenged by alternative products, even if these are more expensive or not as good in terms of technical performance. Environmental considerations have already limited the use of EO-PO block copolymers and are likely to do so even more in the future.

Even so, water-soluble block copolymers are still an important surfactant class. The various types of EO-PO based products, often referred to as Pluronics or Polaxamers, which are commercial trade names, constitute the vast majority but ethylene oxide-butylene oxide (EO-BO) copolymers are also on the market. The three smallest alkylene oxide monomers, EO, PO and BO, all give a linear polymer backbone consisting of a repeating C-C-O- segment (see Figure 1).

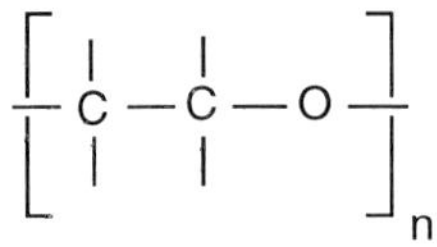
polyoxyethylene

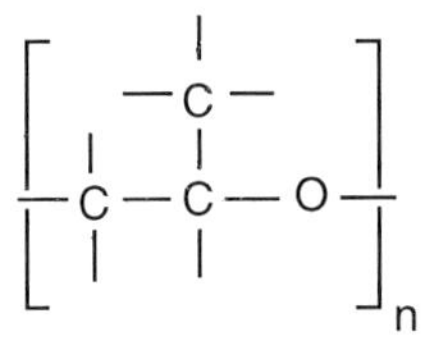
polyoxypropylene

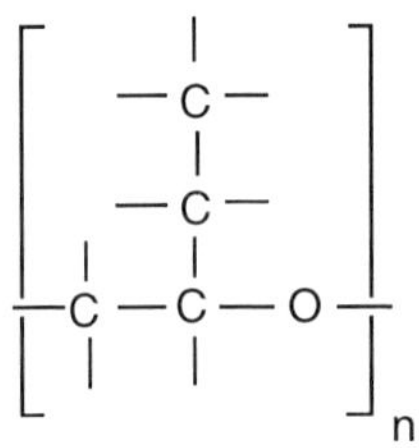
polyoxybutylene

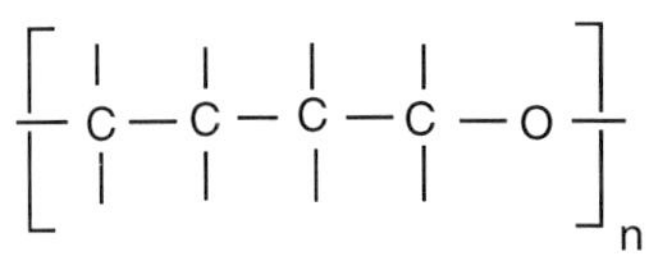
polytetrahydrofuran

Figure 1. Structure of three polyoxyethylenes and of polytetrahydrofuran.

They can all be prepared by anionic polymerization with the use of an alkoxide initiator which is a smooth and robust process compared with cationic or free radical polymerization. The sequential synthesis of alternating hydrophilic and hydrophobic blocks is a straight-forward process which can be carried out with a high degree of reproducibility. Tetrahydrofuran (THF) is a structural isomer of BO and the patent literature contains examples of EO-THF copolymers (1,2). Since copolymers based on THF as hydrophobic monomer do not contain alkyl substituents along the chain, it is likely that these will break down more rapidly in the environment than those based on BO or PO. Biodegradation data on water-soluble THF based copolymers seem not to have been reported in the open literature, however.

Sometimes it is advantageous to use a more hydrophobic segment than polyoxypropylene in the block copolymer. Well-known examples of block copolymers with an extremely hydrophobic sement are the silicone-EO-PO (and also silicone-EO) polymers used to stabilize non-aqueous foams. The silicone and the polyoxyalkylene parts of the molecule are combined by postreaction of the two prepolymer segments. A related type of amphiphilic block copolymer is that based on styrene or methyl methacrylate as the hydrophobic monomer and EO as the hydrophilic moiety. The main use of these products is in emulsion polymerization. EO-styrene block copolymers are prepared analogously to the EO-PO products, i.e. by anionic polymerization through sequential addition of the monomers.

This chapter presents the more important technical uses of the main types of amphiphilic block copolymers. Applications of EO-PO copolymers have been reviewed several times before from slightly different viewpoints (3-6). Also uses of the EO-styrene and EO-silicone type copolymers have been summarized before (7,8).

2. STRUCTURE-PROPERTY RELATIONSHIPS

There exist a variety of commercial EO-PO block copolymers. The products may be linear containing only polyoxyethylene and polyoxypropylene blocks or they may be based on a polyol or a polyamine, such as glycerol or ethylene diamine, in which case a branched block copolymer is obtained (Figure 2).

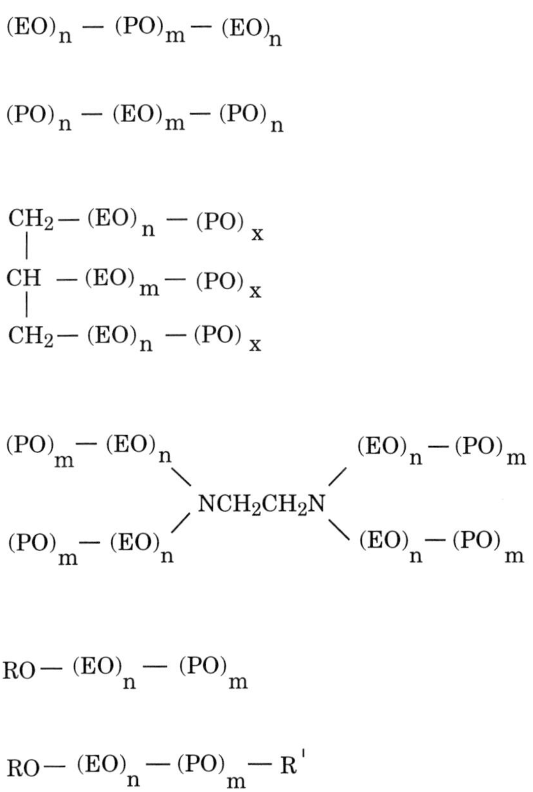

Figure 2. Schematic structures of copolymers based on ethylene oxide (EO) and propylene oxide (PO). R and R' mean a C8-C18 alkyl or alkyaryl and a C1-C4 alkyl group, respectively.

Ethylene diamine based products are often referred to by the BASF-Wyandotte trade name of Tetronics. Triblock products based on only EO and PO are by far the most common. EO-PO-EO producs are sometimes referred to as normal Pluronics and PO-EO-PO products are consequently called reverse Pluronics.

The properties of the triblock copolymers are mainly governed by the ratio of EO to PO. In the HLB scale of Davies, in which the hydrophilic-lipophilic balance (HLB) of a surfactant is calculated from group contributions, one EO has a value of 0.33 and one PO a value of -0.15 (9). Using the formula

$$HLB = 7 + \Sigma \text{ (group numbers)}$$

the HLB value of the copolymer can be readily obtained and the values can be compared with those of other nonionic surfactants such as alcohol ehoxylates (10). Commercial Pluronics exist within very broad range of HLB values which is indicative of the wide spread of applications of the products, with hydrophobic emulsifiers for non-aqueous dispersions on the low HLB end to pigment dispersants at the high HLB extreme.

The properties of an EO-PO block copolymer are also influenced by which segment, polyoxyethylene or polyoxypropylene, that is in the central position but the segment order seems to be of less importance for the physico-chemical properties than the EO to PO ratio. However, compounds having the reverse block structure generally have higher chemical stability since these polymers are terminated by secondary hydroxyl groups which are less reactive than the primary hydroxyls of polyoxyethylene ends.

A related type of EO-PO based copolymer is ethoxylated and propoxylated fatty alcohol or alkylphenol (see Figure 2). The alkyl or alkylaryl chain is more hydrophobic than the polyoxypropylene segment; thus, these products are unsymmetrical with regard to solubility characteristics of the segments. Such products have been found particularly useful for demulsifying and defoaming applications. It is likely that the more hydrophobic character of the alkyl or alkylaryl part makes them very effective in replacing other surface active compounds at oil-water and air-water interfaces.

3. APPLICATIONS

3.1. Emulsifiers

EO-PO block copolymers are used both as emulsifier and as emulsion stabilizer. The emulsifier application generally requires low molecular weight polymers, usually below 2000, while post-stabilization of a ready-made emulsion works best with higher molecular weight material. For O/W emulsions, normal Pluronics with relatively high EO content are usually employed both as emulsifier and emulsion stabilizer.

EO-PO block copolymers have been used in emulsion polymerization of a variety of monomers (11-18). Relatively hydrophilic compounds, such as copolymers with an EO to PO molar ratio of 2:1 to 4:1 give best results. More hydrophobic EO-PO copolymers have been used for reverse suspension polymerization of water-soluble monomers in hydrocarbon (19).

In oil-water systems the block copolymer is anchored at the interface by having the more hydrophobic polyoxypropylene segment dissolved in the oil phase and the polyoxyethylene segments oriented into the aqueous phase. As stabilizer of latex particles the block copolymers orient themselves analogously; a prerequisite of effective stabilization is that there is a strong interaction between the hydrophobic polymer segment and the particle surface. The thickness of the adsorbed polymer layer has been found to increase with increasing molecular weight of the polyoxyethylene segments (MW_{POE}) with the layer thickness being proportional to $(MW_{POE})^{0.55}$ (20).

In a systematic study using EO-PO-EO block copolymers with the same polyoxypropylene segment of 56 PO units and with different length of the polyoxyethylene segments the area per adsorbed molecule was determined for two systems, a hexadecane-in-water emulsion and a polystyrene latex (16). As expected, the area per molecule increased with increasing number of EO in the hydrophilic segments. The experimentally determined values were 6.6, 8-11 and 13-17.5 nm^2, whereas the calculated values were much larger, 8.9, 17 and 60 nm^2, respectively. Extrapolating the experimentally determined values to zero EO gave an area of 4.5 nm^2 for the 56 PO segment. These results are interesting since they tell that

(1) the polyoxyethylene segments, not the polyoxypropylene segment govern packing density;

(2) the polyoxyethylene chains are extremely tightly packed, in particular for the high EO polymers. This means that the chains must extend further from the surface than they do in unhindered conformation. This elongation of the polyoxyethylene chains should enhance steric stabilization of dispersed system. It could be an explanation of the good performance of EO-PO copolymers as emulsion stabilizers.

There is an interest to design block copolymers for emulsion polymerization such that the hydrophobic group matches the latex polymer. This approach should secure maximum compatibility between the surfactant and the latex polymer; thus good anchoring of the emulsifier to the growing particle surface can be expected (21-26).

In work along this line a series of well-defined EO-styrene block copolymers (Figure 3) were synthesized and evaluated as stabilizer in emulsion polymerization of styrene (25,26). The molecular weight of the polyoxyethylene and polystyrene blocks varied from 3000 to 9000 and from 1000 to 7000, respectively. Results from the emulsion polymerization indicated that a molecular weight of the polystyrene block of 1000, corresponding to 10 monomer units, is enough to provide proper anchoring. The experiments also showed that nothing is gained on having larger molecular weight of the polyoxyethylene chains than 3000, corresponding to 68 EO units. The EO to styrene molar ratio also turned out to be of importance.

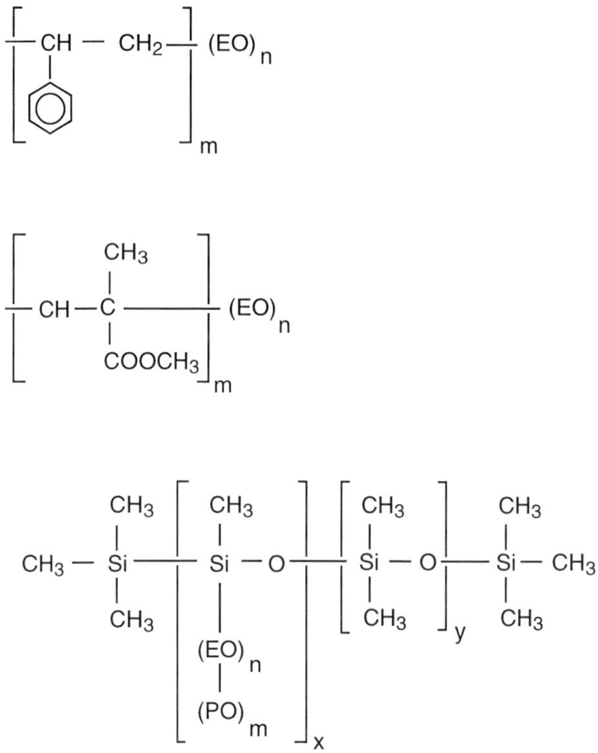

Figure 3. Schematic structures of styrene- EO copolymers (top), methyl methacrylate-EO copolymers (middle) and EO-PO copolymers grafted to a polysiloxane backbone (bottom).

In analogous experiments block copolymers of EO and methyl methacrylate (Figure 3) were used as stabilizer in emulsion polymerization of methyl methacrylate (8,27). Similar to the results from the polystyrene work, neither the anchoring poly(methyl methacrylate) segment, nor the stabilizing polyoxyethylene chain need be very long. Small particle size and good colloidal stability were obtained with molecular weights of the hydrophobic and hydrophilic segments around 800 and 2000, respectively. Optimum weight percentage of EO in the copolymer was in the range of 70 to 80. When the EO content was smaller, the copolymer partitioned too much into the monomer phase, leading to insuffïcent concentration of polyoxyethylene chains on the surface to provide proper steric stabilization. When the EO content was higher than 80 %, the solubility of the copolymer in water was too high, again leaving insufficient amounts on the particle surface.

EO-styrene block copolymers have also been used to formulate water-in-oil microemulsions. Such microemulsions have been used as minireactors for polymerization of acrylamide (28,29).

EO-PO block copolymers (and also EO-BO products) are widely used as emulsifier in agricultural formulations. Both insecticides, bactericides and fungicides have been emulsified (30-32). Usually, the block copolymer is used in conjunction with an anionic surfactant. In such formulations the role of the copolymer is manyfold: it aids in the emulsification process (the anionic surfactant is often the primary emulsifier), it provides emulsion stability, it acts as a defoamer at the application stage (the anionic surfactant is often a potent foamer), and it improves the biological effect by functioning as a wetting agent.

In the agrochemical field the block copolymer is frequently used together with an anionic surfactant in so called emulsifiable concentrates. These are formulations which when added to water give oil-in-water emulsions. Emulsifiable concentrates have gained a large market share in formulations of agrochemicals, mainly because of their ease of handling. The oil-in-water emulsion forms either spontaneously or by only gentle agitation. In order to obtain spontaneous emulsion formation, the choice of surfactant is critical. The anionic surfactant is usually the calcium or magnesium salt of an alkylbenzenesulfonate. A block copolymer with relatively high EO to PO ratio often constitutes the nonionic surfactant (31-33). A combination of cationic surfactant and an EO-PO block copolymer may also be of interest (34-37). The leaf surface is normally negatively charged and a cationic surfactant may therefore adsorb strongly. Since such an adsorption leads to a depletion of surfactant from the oil-water interface, the emulsion may undergo rapid coalescence on the leaf surface. This is the same type of controlled emulsion breaking that is commonly used in road surfacing where the cationic stabilizer of the bitumen emulsion preferentially adsorbs at the mineral filler surface on application. The balance in HLB between the emulsifiers used in agrichemical formulations is crucial; trial and error has lead to the combination of a more hydrophobic ionic and a more hydrophilic nonionic surfactant. Such formulations are made both with liquid active substances, in which case the surfactants are added directly to the liquid, and with solid (or high viscous) active substances, which are dissolved in a suitable oil prior to addition of the surfactant.

The excellent dermatological properties of EO-PO block copolymers are taken advantage of in the use of these products as emulsifier in the personal care sector (38). A large variety of creams, lotions, etc. contain high EO content block copolymers as stabilizing agent. For these applications another aspect of this relatively high molecular weight surfactant class is taken advantage of: the possibility to adjust formulation rheology. EO-PO block copolymers are well suited to give a comfortable, creamy feel to a concentrated emulsion.

In recent years silicone-EO-PO (Figure 3) and silicone-EO copolymers have become popular as surfactant both in skin care and hair care emulsions (39-44). Silicone-containing copolymers are claimed to be superior to standard EO-PO block copolymers in giving smooth touch, nontackiness, wetness and brilliance. The silicone segments posses a very strong driving force for the interface; thus, these copolymers give emulsions with extreme stability. (The main drawbacks of the silicones, high price and poor biodegradability, are not as important in personal care formulations as in many other applications.)

Other industrial applications of EO-PO block copolymers as emulsifier include heavy crude-in-water emulsions (45,46) and lubricants (47). The copolymers have also found use as emulsifier in pharmaceutical formulations.

This application area is treated separately by Malmsten in another chapter of this book.

3.2. Demulsifiers

EO-PO block copolymers have an established position as demulsifier in oil production. In all types of oil recovery water is coproduced along with the crude. Because natural oil contains considerable amounts of polar components which can serve as emulsion stabilizers, oil and water often appear in the form of an emulsion stable enough to withstand the shear forces involved in the pumping operation. The emulsion needs to be separated before transport in pipeline or by tanker.

True EO-PO block copolymers have found use as crude oil demulsifiers (48-52). The EO to PO ratio is crucial for the performance and the optimum ratio depends on the characteristics of the oil. In a typical crude oil demulsification process an EO to PO molar ratio of 2:1 gave best result (53). The copolymer molecular weight is typically between 3000 and 4000, i.e. slightly higher than the products used for emulsification. Incorporation of ethylene diamine or higher ethylene amines into the chain, which leads to branching of the polymer backbone (se Figure 2), seems to improve performance, possibly due to favourable interaction between the tertiary amine groups and negatively charged components, e.g. naphthenic acids, in the oil. Also important from a commercial point of view are EO-PO copolymers linked to a strongly hydrophobic entity such as biphenol A-based epoxy resin or a novolac, i.e. a low molecular weight alkylphenol-formaldehyde resin (53). Such products offer many degrees of freedom: polymer molecular weight, size of the resin compared to the size of the EO-PO block copolymer, EO to PO ratio, etc. A large variety of products exist on the market and structure-function relationships are usually far from clear.

3.3. Defoamers and low-foaming surfactants

Block copolymers based on EO and PO are efficient defoamers. Foam control agents in general should be hydrophobic; the best anti-foam agents have very limited water solubility. Since the HLB of EO-PO block copolymers is very temperature dependant, so is their ability to depress foams. The block copolymers, therefore, work best at temperatures just above their cloud point.

EO-PO block copolymers are used as foam control agents in a variety of industrial applications, the most prominent being mashine dishwashing compounds and sugar beet processing. However, in these, as well as other applications, use of PO containing copolymers is diminishing as a result of the growing concern about biodegradation rate of surfactants (see Section 4).

Both normal and reverse block copolymers are used as antifoaming agents. For both types of compounds optimal antifoam properties are obtained at EO to PO molar ratios of between 1:4 and 1:9 and molecular weights of above 2000 (3). Also branched block copolymers, based on either ethylene diamine or polyols (see Figure 2) are used as foam control agents. The branched compounds typically have EO to PO molar ratio in the same range as their linear counterparts.

As is discussed below, the hydrophobic segment, i.e. polyoxypropylene, is the cause of the poor biodegradability characteristics of EO-PO block copolymers. In order to improve biodegradation rate the polyoxypropylene block may partly be replaced by a hydrocarbon chain. Consequently, alkoxylated fatty alcohols of the generic formula $RO\text{-}(EO)_n\text{-}(PO)_m$ were launched as environmentally friendly defoamers in mashine dishwashing and other applications (Figure 2). In order to achieve satisfactory biodegradation rate the two polyoxyalkylene segments were kept short; both n and m typically being below 10 (54).

A further development towards better biodegradation involves capping of the terminal secondary hydroxyl group by a lower alkyl, thus giving products of the general formula RO-$(EO)_n$-$(PO)_m$-R', where R' is methyl, ethyl, propyl or butyl (55,56). The improved biodegradability stems from the fact that excellent foam depressing effect is obtained with these products already at very short polyoxypropylene segments, m typically being smaller or equal to 3. In addition, these copolymers are particularly well suited for use in the highly alkaline mashine dishwashing formulations since the alkyl ether formed by the capping reaction is much more inert to high pH than a hydroxyl group.

The main foaming component in mashine dishwashing is believed to be degradation products of proteinaceous soil. Also polar lipids and synthetic surfactants added to the food as emulsifier and dispersant may contribute to the generation of stable foams. The foam depressing surfactant is usually added to the dishwashing compound in an amount of 1-3 %. Nonionic surfactants with antifoaming properties are also used in rinse aid formulation. In mashine dishwashing there is a particular problem in the selection of the antifoam since the surfactant should also be an effective wetting agent (54). It is not straight-forward to design a surfactant formulation that meets these two requirements since the HLB demand is different for the two types of functions. As mentioned above, a nonionic antifoam works best above the cloud point. A wetting agent, on the other hand, should be reasonably water-soluble. In the compromise that has to be made, priority is usually given to the antifoam property of the surfactant.

In the sugar beet industry foaming is a major problem. (In cane sugar production, however, severe foams are seldom encountered.) Antifoams are used in the beet process and washing, as well as in the sugar extraction process, which is usually referred to as the diffusion process. The main foam-generating substances are saponins and other surface active compounds present under the skin of the sugar beat root and released during the processing. Also surface active vegetable components derived from the soil may contribute to foam generation (54). The antifoam formulations are usually complex blends of several components with different functions (57). EO-PO block copolymers are widely used as the antifoam ingredient. In this application the traditional copolymer based on only polyoxyethylene and polyoxypropylene segments still dominates. The very low oral toxicity is the main reason why this copolymer class is preferred to the RO-$(EO)_n$-$(PO)_m$ type (see Section 4).

3.4 Foam stabilizers

A special type of hydrophobic copolymer, silicone-EO-PO or silicone-EO, is used for creating and stabilizing non-aqueous foams, particularly polyurethane foams. Silicone-EO-PO polymers can be seen as EO-PO block copolymers grafted to a silicone backbone (Figure 3). The silicone type surfactant is much more effective than conventional surfactants in stabilizing such foams and a variety of products, usually based on polydimethylsiloxane as the silicone block, are commercially available. The polyoxyalkylene part of the molecule is attached to the polydimethylsiloxane either by a Si-C bond or by a Si-O-C bond. Whereas the former is stable over the whole pH range, the latter undergoes hydrolysis both at acidic and basic conditions (58). It is believed that the silicone surfactants stabilize non-aqueous foams by both setting up a surface tension gradient and increasing the surface viscosity (59). It has been reported that copolymers based on branched polysiloxanes are more effective foam stabilizers than their linear counterparts (60,61). Use of silicone-polyoxyalkylene copolymers as stabilizer of non-aqueous foams has recently been reviewed (59).

3.5. Miscellaneous applications

Use of EO-PO block copolymers in laundry detergents and for hard surface cleaning is described in the patent literature (62-67). In many cases the copolymer is used in relatively small amounts, as a cosurfactant in combination with an anionic surfactant such as sodium alkylbenzenesulfonate or sodium dodecylsulfate. Such combinations are claimed to be particularly effective in removing both oily and particulate soil (68). It has also been claimed that with optimized combinations of anionic surfactant and EO-PO block copolymer excellent detergency can be obtained also with phosphate-free detergent formulations (69). The actual use of the copolymers in cleaning formulations today is probably very limited, however, as a result of the increasing environmental concern.

Various types of alkyl terminated EO-PO block copolymers are also of interest as detergent surfactants. Examples include RO-$(EO)_n$-$(PO)_m$, which are low foaming (70), RO-$(PO)_m$-$(EO)_n$, in which the hydrophobic part is a combination of a medium chain alcohol such as octanol and a few PO units (71) or the more complex structure RO-$(EO)_n$-$(PO)_m$-$(EO)_x$ (70).

Concentrated dispersions have been stabilized by EO-PO block copolymers. With hydrophobic particles the copolymer acts as dispersant by anchoring the polyoxypropylene segment to the solid surface and allowing the polyoxyethylene chains to extend into solution (72,73). With silica the situation is different. Both the polyoxyethylene and the polyoxypropylene segments adsorb, giving a thin, relatively uniform layer of polymer on the surface (20). The literature contains examples of the use of EO-PO block copolymers as dispersants in the cosmetics area (38) and for pigments (74) and dyes (75,76). Taken together, the area of pigment or particle dispersant is not a commercially important one for EO-PO block copolymers, however. This may partly be due to the poor biodegradability of the products but also to the fact that polyelectrolyte dispersants often give better stabilization of these dispersed systems.

EO-PO block copolymers are used in dental care products, both in tooth pastes, dentrifrices and mouthwash compositions (77-82). Rheological properties are crucial in such formulations and the thickening and gelling properties of properly selected copolymers have been taken advantage of (78-82). The products have been used to enhance the prophylactic action of antiplaque formulations containing a bactericidal agent (83-86) and also, in the form of a phosphorylated derivative, to provide a nonfouling tooth surface (87,88).

4. ENVIRONMENTAL CONSIDERATIONS

Compared with other nonionic surfactants EO-PO block copolymers are characterized by having
(1) very low aquatic toxicity,
(2) slow biodegradation both under aerobic and anaerobic conditions.

Typical straight EO-PO block copolymers have values of acute aquatic toxicity, determined as the LC_{50} value after 96 h exposure, greater than 100 mg/l (89). (LC_{50} means the concentration of active substance lethal to 50 % of the fish species used.) The alkyl-EO-PO compounds, developed as a more rapidly biodegradable class of antifoam agents (see Section 3.3), are considerably more toxic with LC_{50} values in the range 0.25 to 4.4 mg/l (54). The latter toxicity interval is approximately the same as for normal fatty alcohol ethoxylates (54).

Homopolymers of PO, i.e. poly(propylene glycol)s, are known to break down slowly in the environment (90). Conversely, poly(ethylene glycol)s biodegrade at a reasonable rate provided that the molecular weight is not too high (91,92). Both aerobic and anaerobic biodegradation have been reported (93,94). Several polyoxyethylene degrading bacteria have been isolated from the environment (54,91). The poor biodegradability of the polyoxypropylene chains is probably due to the pendant methyl groups which cause steric hindrance to enzymatic attack but also to the fact that this segment is hydrophobic and thus less available in the aqueous domains where most hydrolytic enzymes operate.

As expected, the rate of biodegradation of the EO-PO block copolymers depends on the EO to PO ratio. However, also the high EO copolymers fail to pass the standard OECD test for ready biodegradability, the main requirement of which is that 60 % of the theoretical amount of carbon dioxide is to be liberated within 28 days (Modified Sturm Test) (54).

Also for block copolymers of the general type alkyl-EO-PO, as well as alkyl-PO-EO, biodegradability is governed by the polyoxypropylene segment. The alcohol has only a marginal effect unless it is highly branched (89,95). Although somewhat better than the straight EO-PO block copolymers, the RO-$(EO)_n$-$(PO)_m$ compounds usually also fail the test for ready biodegradability (54).

REFERENCES

1. R. Becker, M. Huellmann, C. Palm and W. Franzischka, Eur. Pat. Appl. EP 636,638 A1 950201 to BASF A.G. (1995).
2. L. Wei, PCT Int. Appl., WO 9613477 A1 960509 to Mobil Oil Corp. (1996).
3. I.R. Schmolka, in Nonionic Surfactants (M.J. Schick, ed.) Surfactant Science Series 1, Marcel Dekker, New York, 1966, p. 300.
4. I.G. Lundsted and I.R. Schmolka, in Block and Graft Copolymerization (R.J. Ceresa, ed.), Vol. 2, Wiley, New York, 1976.
5. I.R. Schmolka, J. Am. Oil Chem. Soc., 54 (1977) 110.
6. M.W. Edens, in Nonionic Surfactants. Polyoxyalkylene Block Copolymers (V.M. Nace, ed.) Surfactant Science Series 60, Marcel Dekker, New York, 1996, p. 185.
7. P. Bahadur and G. Riess, Tenside 28 (1991) 3.

8. I. Piirma, Polymeric Surfactants, Surfactant Science Series 42, Marcel Dekker, New York, 1992.
9. J.T. Davies, Proc. Int. Congr. Surface Activity, 2nd Ed., Vol. I, Butterworth, London, 1959, p. 426.
10. J.T. Davies and E.K. Rideal, Interfacial Phenomena, Academic Press, New York, 1961.
11. D.M. French, J. Polymer Sci., 32 (1958) 395.
12. B.B. Kine and G.H. Redlich, in Surfactants in Chemical/Process Engineering (D.T. Wasan, M.E. Ginn and D.O. Shah, eds.) Surfactant Science Series 28, Marcel Dekker, New York, 1988, p. 263.
13. W.D. Hergeth, P. Bloss, F. Biedenweg, P. Abendroth, K. Schmutzler and S. Wartewig, Makromol. Chem., 191 (1990) 2949.
14. Y.C. Chen, V. Dimonie and M.S. El-Aasser, J. Appl. Polym. Sci., 45 (1992) 487.
15. Y.C. Chen, V. Dimonie and M.S. El-Aasser, J. Appl. Polym. Sci., 46 (1992) 691.
16. I.T. Kim and P.F. Luckham, Colloids Surfaces, 68 (1992) 243.
17. M.A. Faers and P.F. Lukham, Colloids Surfaces A, 86 (1994) 317.
18. M. Ara and A. Katoh, U.S. Patent 5,378,291 to Nihon Parkerizing Co. (1995).
19. M. Niessner, S. Wickel, W. Schneider, J. Beck, H. Hartmann and T. Meyer, U.S. Patent 5,258,473 to BASF A.G. (1993).
20. E. Killmann, H. Maier and J.A. Baker, Colloids Surfaces, 31 (1988) 51.
21. D. Rogez, S. Marti, J. Nervo and G. Riess, Makromol. Chem., 176 (1975) 1393.
22. G. Riess, J. Nervo and D. Rogez, Polym. Eng. Sci., 8 (1977) 634.
23. G. Riess and D. Rogez, ACS Polymer Preprints, 23 (1982) 19.
24. T.N. Khan, R.H. Mobbs, C. Price, J.R. Quintana and R.B. Stubbersfield, Eur. Polym. J., 23 (1987) 191.
25. I. Piirma, Makromol. Chem., Macromol. Symp., 35/36 (1990) 467.
26. G.L. Jialanella, E.M. Firer and I. Piirma, J. Polym. Sci. A: Polym. Chem., 30 (1992) 1925.
27. W.L. Wu, and I. Piirma, Polym. Bull., 31 (1993) 531.
28. Y.S. Leong, G. Riess and F. Candau, J. Chim. Phys., 78 (1981) 279.
29. F. Candau and F. Ballet, in Microemulsions (I.B. Robb, ed.) Plenum Press, New York, 1981, p. 49.
30. G. Frisch and K. Albrecht, U.S. Patent 5,074,905 to Hoechst A.G. (1991).
31. T. Tadros. Surfactants in Agrochemicals. Surfactant Science Series 54, Marcel Dekker, New York, 1994.
32. Y. Ishimoto, Y. Manabe, S. Zen, K. Takahashi and M. Imai, Japan. Patent 7,285,803 A2 951031 to Sumitomo Chemical Co. (1995).
33. E. Fischer, Farm. Chem., 126 (1963) 18.
34. G.W.J. Lee and T.F. Tadros, Colloids Surfaces, 5 (1982) 105.
35. G.W.J. Lee and T.F. Tadros, Colloids Surfaces, 5 (1982) 117.
36. G.W.J. Lee and T.F. Tadros, Colloids Surfaces, 5 (1982) 129.
37. L.M. Boize, G.W.J. Lee and T.F. Tadros, Pest. Sci., 14 (1983) 427.
38. I.R. Schmolka, Cosmetics and Toiletries, 97 (1982) 61.
39. S. Tsubaki and I. Noda, U.S. Patent 5,472,686 A 951205 to Nippon Unicar Co. (1995).
40. S. Aoyanagi and I. Kobayashi, Japan. Patent 7,277,923 A2 951024 to Kanebo Ltd. (1995).

41. H. Yamashita and M. Oosawa, Japan. Patent 7,309,714 A2 951128 to Kanebo Ltd. (1995).
42. A.J. O'Lenick, Jr., U.S. Patent 5,475,125 A 951212 to Siltech Inc. (1995).
43. C. Dupuis, Europ. Patent 684,041 A1 951129 to Oreal S.A. (1995).
44. K. Sakuta, Japan. Patent 7,304,627 A2 951121 to Shinetsu Chem. Ind. Co. (1995).
45. N. Moriyama, T. Ogura and A. Hirake, U.S. Patent 5,024,676 to Kao Corp. (1991).
46. S.P. Current, U.S. Patent 5,110,487 to Chevron Corp. (1992).
47. F.E. Bailey, Jr, and J.V. Koleske, Alkylene Oxide and Their Polymers, Surfactant Science Series 35, Marcel Dekker, New York, 1991, p. 212.
48. H.-F. Fink, G. Körner and G. Rossmy, U.S. Patent 4,029,596 to Th. Goldschmidt A.G. (1977).
49. P.K. Das and S. Hartland, Chem. Eng. Commun., 92 (1990) 169.
50. F. Staiss, R. Böhm and R. Kupfer, Fossil Fuels, Derivatives, and Related Products, 6 (1991) 334.
51. S.E. Taylor, Chem. Ind., 20 (1992) 770.
52. R.A. Mohammed, A.I. Bailey, P.F. Luckham and S.E. Taylor, Colloids Surfaces A, 83 (1994) 261.
53. M. Amaravathi and B.P. Pandey, Res. Ind., 36 (1991) 169.
54. T.G. Blease, J.G. Evans, L. Hughes and P.Loll, in Defoaming (P. Garrett, ed.) Surfactant Science Series 45, Marcel Dekker, New York, 1993, p. 299.,
55. K.H. Schmidt, H. Pruhs, T. Attenschöpfer and R. Piorr, Europ. Patent 197,434 A1 861015 to Henkel K.G.A.A. (1986).
56. L. Person and B. Mourrut, Europ. Patent 276,050 A1 880727 to ICI France S.A. (1988).
57. I. Lichtman and T. Gammon, in Kirk-Othmer Encyclopedia of Chemical Technology, 3rd Ed., Vol. 7, 1979, p. 430.
58. E.G. Schwarz in Silicone Technology, Vol. 14 (P.F. Bruins, ed.) Interscience, New York, 1970, p. 71.
59. D.L. Schmidt in Foams (R.K. Prudhomme and S.A. Khan, eds.) Surfactant Science Series 57, Marcel Dekker, New York, 1996, p.287.
60. M.J. Owen, T.C. Kendrick, B.M. Kingston and N.C. Lloyd, J. Colloid Interface Sci., 24 (1967) 141.
61. M.J. Owen and T.C. Kendrick, J. Colloid Interface Sci., 27 (1968) 46.
62. K.M. Hellsten and S.O. Santamaki, U.S. Patent 3,850,831 to Mo och Domsjö AB (1974).
63. L.A. Gilbert, U.S. Patent 4,284,524 to The Procter & Gamble Co. (1981).
64. W.R. Camp, C.J. Bunczk and P.A. Burke, U.S. Patent 5,043,090 to Kiwi Brands Inc. (1991).
65. C.J. Bunczk and P.A. Burke, U.S. Patent 5,049,299 to Kiwi Brands Inc. (1991).
66. P.D. Deck, J.B. Rivera and W.L. Harpel, U.S. Patent 5,114,607 to Betz Laboratories Inc. (1992).
67. M. Ikeda and J. Kataoka, U.S. Patent 5,269,957 to Taiho Industries Co. (1993).
68. R.J. Holland, U.S. Patent 5,152,933 to BASF Corp. (1992).

69. I.I. Secemski and J.L. Lynn, U.S. Patent 5,049,303 to Lever Brothers Co. (1991).
70. C.G. Naylor, J. Amer. Oil Chem. Soc., 63 (1986) 1201.
71. F.H. Stahel, U.S. Patent 4,299,994 to Sandoz Inc. (1981).
72. F. Miano, A.I. Bailey, P.F. Luckham and T.F. Tadros, Colloids Surfaces, 62 (1992) 111.
73. F. Miano, A.I. Bailey, P.F. Luckham and T.F. Tadros, Colloids Surfaces, 68 (1992) 9.
74. R.J. Holland, M.J. Anchor and C.G. Utz, U.S. Patent 5,169,894, unassigned (1992).
75. R. Bruttel and P. Kvita, U.S. Patent 4,961,755 to Ciba-Geigy Corp. (1990).
76. T.B. Gunnell, J.B. Hines, Jr. and C.N. Barry, U.S. Patent 5,331,097 to Milliken Research Corp. (1994).
77. M.R. Burke, M. Prencipe and J.M. Buchanan, U.S. Patent 5,292,502 to Colgate-Palmolive Co. (1994).
78. S, Mori and C. Makino, U.S. Patent 5,035,881 to Sunstar Kabushiki Kaisha (1991).
79. S. Mori and C. Tomita, U.S. Patent 5,292,528 to Sunstar Kabushiki Kaisha (1994).
80. A.L. Bianchi, J.T. Freiberg and K.D. Konopa, U.S. Patent 5,028,413 to Bausch & Lomb Inc. (1991).
81. R.L. Mitchell and G.A. Durga, U.S. Patent 5,096,698 to Colgate-Palmolive Co. (1992).
82. N. Nabi, A. Gaffar and J. Afflitto, U.S. Patent 5,275,805, unassigned (1994).
83. C.J. Kleber and M.S. Putt, U.S. Patent 4,976,954 to Purdue Research Foundation (1990).
84. E.J. Carlin, A.K. Talwar, L.T. Principe and S.S. Dills, U.S. Patent 5,100,650 to Warner-Lambert Co. (1992).
85. K.D. Konopa, U.S. Patent 5,292,527 to Bausch & Lomb Inc. (1994).
86. A. Gaffar, N. Nabi, J. Afflitto and O. Stringer, U.S. Patent 5,294,431 to Colgate-Palmolive (1994).
87. J. Olsson, A. Carlén and K. Holmberg, J. Dent. Res., 69 (1990) 1586.
88. J. Olsson, M. Hellsten and K. Holmberg, Colloid Polym. Sci., 269 (1991) 1295.
89. K. Boch, L. Huber and P. Schoberl, Tenside, 25 (1988) 86.
90. A. Hettche and E. Klahr, Tenside, 19 (1982) 127.
91. D.P. Cox, Ann. Rev. Microbiol., 23 (1978) 173.
92. R. Swisher, Surfactant Biodegradation, Marcel Dekker, New York, 1987.
93. L.D.L. Jenkins, J. Appl. Bact., 47 (1979) 75.
94. R. Dwyer, Appl. Env. Microbiol., 46 (1983) 185.
95. C. Naylor, F. Castalds and B. Hayes, J. Am. Oil Chem. Soc., 65 (1988) 160.

Block copolymers in pharmaceutics

M. Malmsten

Institute for Surface Chemistry, P.O. Box 5607, S-114 86 Stockholm, Sweden

1. Introduction

Over the last few decades block copolymers have been extensively developed industrially. With the occurence of a wide range of new block copolymers also the usefullness of these in practical applications has increased. As with other industrial applications many uses of these copolymers in the development of pharmaceutical formulations are largely related in one way or another to the amphiphilic nature of these substances. This amphiphilicity results in block copolymer self-assembly, e.g., micellization and liquid crystalline phase formation. The self-assembled structures formed are attractive for pharmaceutical applications for the reason, e.g., of enhanced drug solubility or improved drug chemical stability in the formulation, achievement of a sustained and controlled release of the drug after administration, controlled rheological behaviour etc.. Furthermore, the amphiphilic nature of block copolymers makes them ideal as stabilizers of pharmaceutical colloidal dispersions, e.g., emulsions, liposomes, or (nano)particles, since they typically contain one block which experiences poor solvency and thus is capable of achieving a firm anchoring of the copolymer at the drug carrier surface, and one or several blocks experiencing good solvency, resulting in an efficient steric repulsion, and thus in a good colloidal stability (1,2). Apart from controlling the colloidal stability of pharmaceutical dispersions, adsorbed block copolymers may also affect the biological response to these dispersions. For example, a prolonged circulation in the blood stream after intraveneous administration of colloidal drug carriers, and a reduced uptake in, e.g., liver, spleen, and marrow, may be achieved by adsorption or other incorporation of block copolymers at the carrier surface (3-8).

Since most pharmaceutical applications of block copolymers in one way or another involve aqueous solutions, water-soluble block copolymers containing one or several hydrophobic and one or several hydrophilic blocks are of special interest. Of these, block copolymers consisting of poly(ethylene oxide) (PEO) and poly(propylene oxide) (PPO) have been subject of a particularly strong interest, which is due to both relatively unique physical properties (e.g., particularly efficient serum protein rejection when adsorbed at intraveneously administered colloidal drug carriers and reversed temperature dependent self-assembly and gelation; see below), the relatively low toxicity of this type of copolymers, and also due to this type of copolymers being one of the first to be industrially developed in a broad way (9). Since a vast majority of the studies of pharmaceutical applications of block copolymers concerns PEO-PPO-PEO block copolymers many of the examples given in this chapter will involve this type of copolymers. In the present discussion we focus on pharmaceutical applications of various physicochemical properties displayed by block copolymers, including solubilization, self-assembly, adsorption, and stabilization. However, "concentrated systems" such as copolymer films or (nano)particles will not be treated since the pharmaceutical performance of these regarding, e.g., degradation or sustained release of incorporated drugs, is more strongly related to general material properties than to the amphiphilic nature of block copolymers, which is the main theme of the present discussion. Furthermore, since this aspect of pharmaceutical applications of block copolymers is discussed separately elsewhere in this book, targeting of pharmaceutical formulations and drugs through the use of biospecific entities, e.g., ligand/receptor, substrate/enzyme, and antigen/antibody pairs, will not be extensively discussed.

2. Micellization

One of the most notable manifestations of the amphiphilic nature of block copolymers is their ability to self-assemble and to form a variety of associated structures. Just as for low molecular weight surfactants block copolymer assemblies display a rich diversity, including micelles and liquid crystalline phases (e.g., lamellar, hexagonal, and cubic phases), as well as the reversed counterparts of these structures and a range of microemulsion structures (10-18). These structures have been described in detail elsewhere in this book. The main reason for the use of block copolymer self-assemblies, and block copolymer micelles in particular, is in one way or another related to their solubilization capacity for sparingly soluble drugs. The solubilization in itself may be desirable for a number of reasons, including achievement of a controlled release rate, an increased chemical stability of the drug, e.g., regarding hydrolysis or proteolysis, or an improved bioavailability. The solubilization of hydrophobic drugs in block copolymer micelles and the release rate of solubilized drugs from micellar solutions have been investigated by numerous authors. In general it is found that while the solubilization capacity of block copolymer solutions below the critical micellization concentration (cmc) is poor, once micelles are formed, the solubilization increases (15,19-26). In general, the solubilization of hydrophobic substances in PEO-PPO-PEO block copolymer micelles is higher for aromatic than for aliphatic substances (26), which could be expected from the presence of the PO ether oxygen. Furthermore, as could be expected from mixing entropy considerations, the solubilization capacity is larger the smaller the solubilized molecule (26). Naturally, the larger the number of solubilizing micelles, i.e., the higher the concentration, the larger the solubilization capacity of the block copolymer micellar solutions (21,22,27). Furthermore, for PEO-PPO-PEO block copolymers the higher the temperature and the relative PO content, the higher the micelle aggregation number and the smaller the core water content, which results in an increased solubilization capacity due to a partition enhancement to the micellar core (19,22).

Once solubilized into the hydrophobic micellar core the drug is largely protected from the aqueous surroundings, which may stabilize it, e.g., against hydrolysis. For example, Lin et al. studied the hydrolysis of indomethacin, a non-steroidal anti-inflammatory agent, when solubilized into a series of PEO-PPO-PEO block copolymer micelles, and found that the solubilization efficiently reduced the hydrolysis rate. In particular, the hydrolysis rate was more reduced at higher copolymer concentration and molecular weight (constant EO/PO ratio). The protective capacity of the block copolymers thus follows their solubilization capacity (19,23). This behaviour is illustrated in Figure 1.

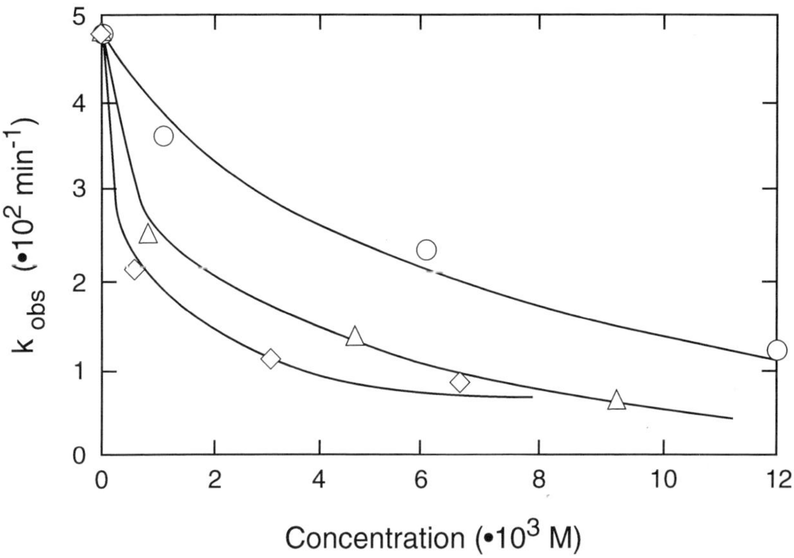

Figure 1. Degradation rate constant (k_{obs}) of indomethacin as a function of polymer concentration for Pluronic F68 (circles), F88 (triangles), and F108 (diamonds) in alkaline aqueous solution at 37°C. From ref.(23).

An interesting aspect of the solubilized drug is its release into the surrounding solution, which could be expected to depend on primarily the drug partition between the micellar hydrophobic core and the aqueous solution, as well as on the micellar dynamics. Whereas charged or hydrophilic substances, which are not extensively solubilized in the micellar interior, are released quite fast, the release rate is much slower for uncharged and hydrophobic substances. For example, La et al. studied the solubilization of indomethacin in PEO-poly(β-benzyl L-aspartate) (PBLA) block copolymer micelles in an attempt to reduce side-effects of this drug, e.g., irritation of the gastrointestinal mucosa from direct exposure and central nervous system toxicity due to high plasma levels following conventional treatment (28). It was found that while the release of indomethacin from the PEO-PBLA micelles at low pH is quite slow, the release rate increases strongly on increasing pH above pK_a of the drug ($pK_a \approx 4.5$). Thus, at low pH the hydrophobic and uncharged indomethacin is solubilized into the hydrophobic micellar core, and the release is slow, whereas upon charging of the indomethacin molecules the partition is shifted towards the aqueous solution, resulting in an increased release rate (Figure 2).

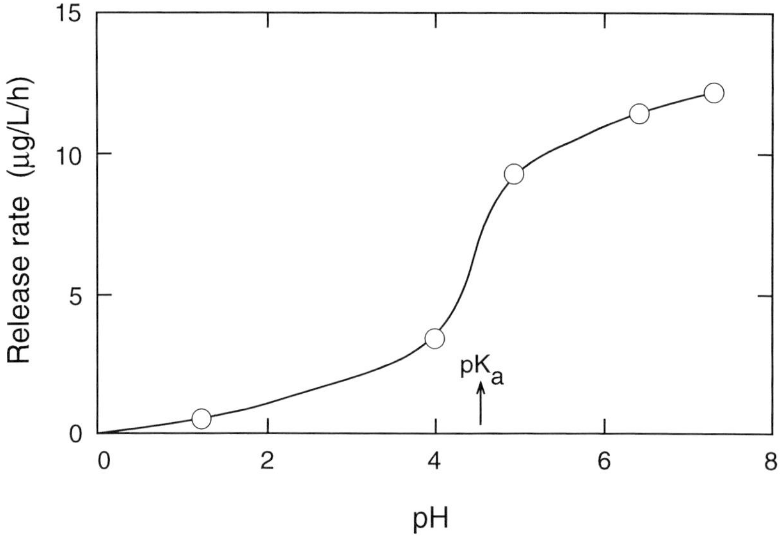

Figure 2. Release rate of indomethacin from PEO-PBLA micelles in aqueous solution at 37°C and different pH. From ref.(28).

A notable feature of block copolymers is the generally very long residence time of the individual polymer molecules in the micelles. In particular, it has been observed (29) that block copolymer micelles remain stable for very long times (~ hours) on reducing the concentration below cmc. This means that the micelles are largely intact over the corresponding time span also after administration to the body and thus in contact with excess water. Together with the low cmc's displayed in many of these systems this is a key requirement for many pharmaceutical applications of block copolymer micelles, and particularly so for parenteral uses of such systems.

Although the vast majority of pharmaceutical applications of block copolymer micelles has involved PEO-PPO-PEO copolymers, also other copolymers are of interest in these applications. For example, Rolland et al. prepared micelle-forming PEO-poly(isoprene)-PEO block copolymers for parenteral applications (30). At the same time as the micelles formed by these polymers could be expected to have a good solubilizing capacity of hydrophobic drugs, the small size of the micelles (≈100 nm) and the protein rejecting capacity of the PEO palisade layer result in a long circulation time in the blood stream and in a decreased uptake in RES-related tissues, which is beneficial from both a bioavailability and local toxicity /side-effect occurence point of view. A similar behaviour is displayed also by PEO-PPO-PEO and a range of other PEO-containing block copolymers (cf. discussion on sterically stabilized colloids in parenteral administration below).

Due to the occurence of long-term storage diseases of certain non-biodegradable polymers of high molecular weight (e.g., poly(vinyl pyrrolidone)), also (partly) biodegradable micelle forming block copolymers have been developed (27). Apart from PEO-PBLA mentioned above, examples of these include PEO-poly(γ-benzyl L-glutamate) (PEO-PBLG), PEO-poly(aspartatic acid) (PEO-PAsp), and PEO-poly(L-lysine) (PEO-PLys) (27,31). As with other biodegradable polymer systems, the degradation of these copolymers may be controlled in a number of ways (32).

A particularly well characterized and thoroughly investigated system is PEO-PAsp conjugates with adriamycin (ADR), a potent anti-cancer drug. By appropriate design of the copolymer conjugate, small micelles (15-60 nm) are formed, in which ADR is solubilized/anchored to the micelle interior. Furthermore, as for many other block copolymer systems the release of unimers from the micelles is exteremely slow (~ days) (33). Thus, the release rate of ADR from the block copolymer conjugate micelles is expected to proceed largely through the biodegradation of the Asp segments (34). By the use of the conjugate micelles a very long circulation time in the bloodstream and a slow uptake in liver, spleen, and marrow was observed. As can be seen in Figure 3, the circulation time of the PEO-P(Asp(ADR)) conjugate is much longer than that of both free ADR and P(Asp(ADR)) (31,34,35). In this context it is interesting to note that longer PEO chains are more efficient in increasing the circulation time in the bloodstream than short PEO chains (27). This is in accordance both with the behaviour of PEO-stabilized colloidal drug carriers after parenteral administration and with the protein rejecting properties of PEO-modified surfaces in a wide range of biomedical applications (see below).

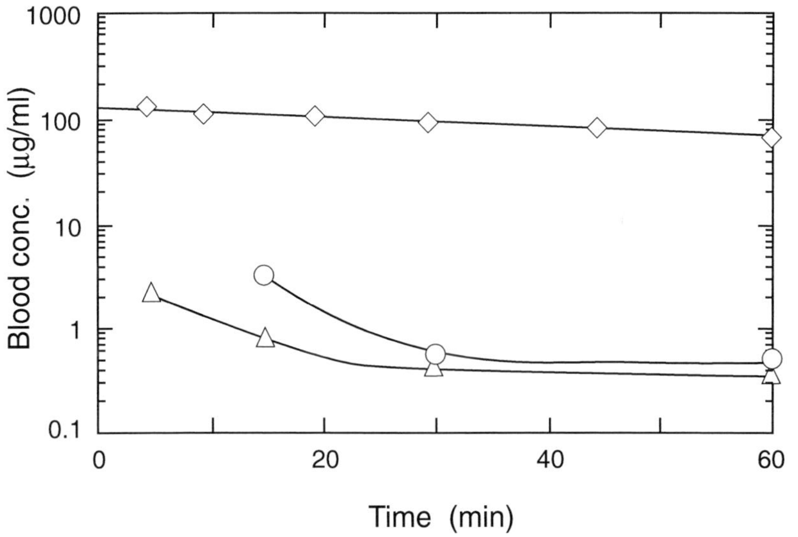

Figure 3. Blood concentration of ADR (10 mg/kg, circles), P(Asp(ADR)) (25 mg/kg, triangles), and PEO-P(Asp(ADR)) (20 mg/kg, diamonds) after intraveneous injection in tumor bearing mice. From ref.(31).

Although ADR possesses a considerable anti-cancer activity in itself, it has been found that at concentrations higher than about 10 mg/kg the mean survival time is reduced, which is due to the occurence of serious side-effects. On the other hand, the side-effects occur at 1-2 orders of magnitude higher ADR concentration in the case of the conjugate micelles. Thus, the use of these micelles increases the maximum ADR concentration possible to use in therapy without causing toxic side-effects. Due to this and to the longer circulation time in the bloodstream the cytotoxicity of the block copolymer conjugate micelles is superior to that of ADR in itself, as can be seen in Figure 4 (31).

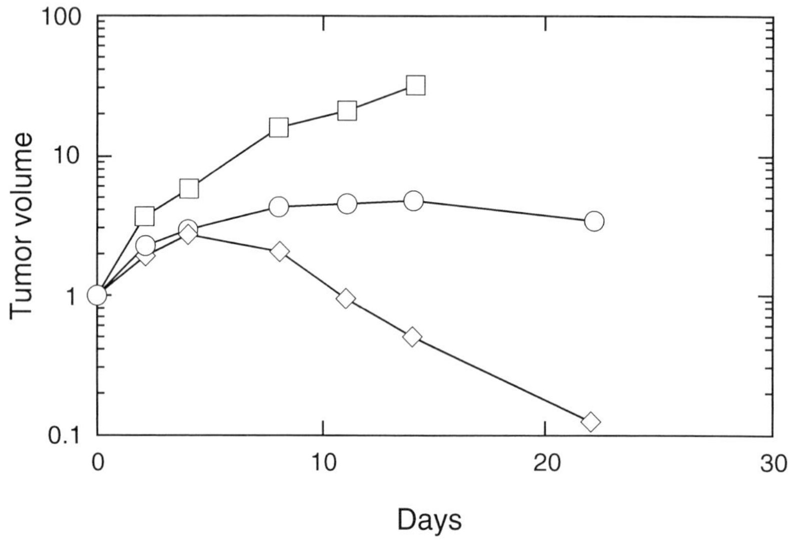

Figure 4. *In vivo* antitumor activity against murine C26. Squares, circles, and diamonds refer to control, ADR (10 mg/kg), and PEO-P(Asp(ADR)) (200 mg/kg), respectively. From ref.(31).

Block copolymer conjugates have also been used successfully for selective administration of drugs to certain tissues or cells (so-called targeting). For example, Kabanov et al. observed specific targeting of fluorescein isothiocyanate solubilized in PEO-PPO-PEO block copolymers to the brain when the copolymer was conjugated with antibodies to the antigen of brain glial cells (α_2-glycoprotein). Furthermore, incorporation of haloperidol into such micelles was found to result in a drastically increased therapeutic effect (24). However, since these applications of block copolymer micelles are discussed in detail elsewhere in this volume, they will not be discussed further here.

The solubilizing capacity of block copolymer micelles also have a wide range of other uses in pharmaceutical applications. For example PEO-PPO-PEO block copolymers have been investigated together with low molecular weight surfactants regarding their ability to dissolve gallstones following oral administration, and found to be able to support such a dissolution (36,37). Furthermore, the resistance of *Mycobacterium avium* complex (MAC) to antibiotics is thought to be enhanced by its outer glycolipid layer. Therefore, PEO-PPO-PEO block copolymers were used to solubilize these glycolipids, or alternatively to be incorporated into the glycolipid layer, thereby changing its spontaneous curvature, in any case resulting in a disruption of the glycolipid layer. It was found that particularly "hydrophobic" PEO-PPO-PEO block copolymers were quite potent in increasing the efficiency of antibiotic therapy of MAC infections (38).

Solubilization, or disruption of the *Stratum corneum* lipid structures, is a frequently used approach in transdermal drug delivery in order to improve the topical bioavailability (39-41). Due to their solubilizing capacity, it is not surprising that also PEO-PPO-PEO block copolymes have been investigated in this respect. For example, Shen et al. investigated the percutaneous absorption of salicylic acid and sodium salicylate and found that PEO-PPO-PEO block copolymers increased the percutaneous absorption of both substances. Again, the effect was found to be more pronounced for relatively hydrophobic block copolymers (42). However, since the situation is complex and the mechanisms for penetration enhancement in topical administration uncertain, and since other studies indicate little or no penetration enhancement due to these copolymers, their effect is still an open issue (43).

3. Gelation and Liquid Crystalline Phase Formation

Just as low molecular weight surfactants, block copolymers have been found to form a range of liquid crystalline phases, such as lamellar, hexagonal, and cubic phases, as well as their reversed counterparts. Which of these systems are found for a given system depends on, e.g., the copolymer composition, molecular weight, concentration, as well as on temperature, presence of low molecular weight cosolutes such as inorganic salts, (ionic) surfactants, and "oil", etc (10,12-18). Due to the often poorly organized liquid crystalline phases formed particularly in high molecular weight block copolymer systems, but also due to the poor knowledge among pharmaceutical researchers about self-assembly and phase formation in block copolymer systems, liquid crystalline phases formed by these polymers are frequently referred to as "gels" in the pharmaceutical literature. Despite the fundamental differences from gels formed by network forming polymers, e.g., polysaccharides (32), regarding the origin of the rheological properties these different types of "gels" are frequently compared regarding pharmaceutical performance. Considering this, but also that information on both the type of the liquid crystalline phase formed and its structure are essentially invariably lacking in reports on pharmaceutical studies of these systems, the term "gel" will be taken to include liquid crystalline phases also in the present discussion.

The most extensively studied class of gels in the pharmaceutical literature is that formed by PEO-PPO-PEO block copolymers. The reason for this is that these gels have a number of advantages from the point of view of pharmaceutical applications. First, the fairly low toxicity displayed by most of these copolymers (see below) makes them interesting not only for dermal administration, but also for a number of other administration routes, including nasal, occular, oral, buccal, rectal, and vaginal administration. Furthermore, the presence of hydrophobic domains allows a much higher solubilization capacity of hydrophobic drugs than that achievable with conventional gels, e.g., formed by polysaccharides. However, the single most interesting property of the PEO-PPO-PEO block copolymer gels is their reversed thermal behaviour. Thus, these systems are typically quite low viscous at low temperature (e.g., in refridgerator or at room temperature). However, on contact with any body surface the temperature of the formulation is increased and the gelation initiated (at a temperature which can be predetermined). Due to the cooperativity of the self-assembly, this gelation typically occurs over a narrow temperature range (≈5°C) and the elastic modulus of the gels formed is typically quite high (≈10^4 Pa) (44,45).

It is important to note that just as for other self-assembled structures used in pharmaceutical applications, the presence of low molecular weight cosolutes, e.g., drugs, will generally influence the structure and stability of the self-assemblies. This is the case also for block copolymer micelles and gels. For PEO-PPO-PEO block copolymer gels it has been found that ionic amphiphilic molecules result in an increase in both the phase separation temperature ("cloud point"; CP) and in the gelation temperature (46). The reason for this is that the ionic amphiphilic molecules (surfactants) bind to the polymer, thus effectively introducing charges on the polymer backbone and improving its solvency, thus precluding self-assembly and requiering poorer solvency for the PPO moiety in order to drive the self-assembly (47).

The effects of electrolytes and non-amphiphilic ionic species on the self-assembly of PEO-PPO-PEO block copolymers are somewhat more complex and depend on the identity of the solute. For certain electrolytes, e.g., NaCl, a decrease in both CP and gelation temperature is observed, whereas the reverse is found for other electrolytes, e.g., NaSCN (29). The origin of these effects is still somewhat uncertain, but most likely it is related to the distribution of the electrolytes within the system. Thus, while NaCl is depleted close to interfaces, presumably due to an image charge repulsion originating from the interfacial dielectric constant discontinuity, NaSCN is accumulated at these interfaces. This results in an increase and a decrease in the effective surface tension, respectively, causing an increased and decreased tendency for self-assembly, respectively (48-50).

Of particular interest in most pharmaceutical applications of these gels are the effects of hydrophobic solutes on the gel formation and stability. In general, it is found that hydrophobic substances causes a reduction in the gelation temperature as well as an increase in the gel strength. For example, Gilbert et al. investigated the effects of benzoic acid and p-hydroxybenzoate esters on the sol-gel transition of a PEO-PPO-PEO block copolymer (Pluronic F127) and found that these solutes caused a decrease in the gelation temperature displayed by this system, and that the more hydrophobic the solute the larger the decrease in the gelation temperature (51). This decrease is due to the presence of the cosolute effectively enhancing self-assembly, which is analogous to the behaviour of many low molecular weight surfactants (52).

Just as for block copolymer micelles one of the reasons for the use of solubilization in block copolymer gels is the achievement of a sustained and controlled release of the solubilized drug. For PEO-PPO-PEO block copolymers, the release of solubilized hydrophobic drugs has been found to decrease with an increasing polymer concentration (53-56), which could be expected from the effectively reduced diffusion rate at a higher polymer concentration. Note, however, that despite the very large effects of polymer concentration on viscocity, the polymer concentration dependence of the diffusion rate of low molecular weight solutes in polymer systems is fairly weak (57,58). As could be expected, the release rate increases with an increasing drug concentration (53,55,56). Furthermore, the release rate depends on the partition of the drug within the system and thus also on the properties of the drug, and in particular its charge and hydrophobicity. As can be seen in Figure 5, the more hydrophobic the substance, i.e., the larger its partition to the hydrophobic domains of the block copolymer gels, the slower the release rate (54) .

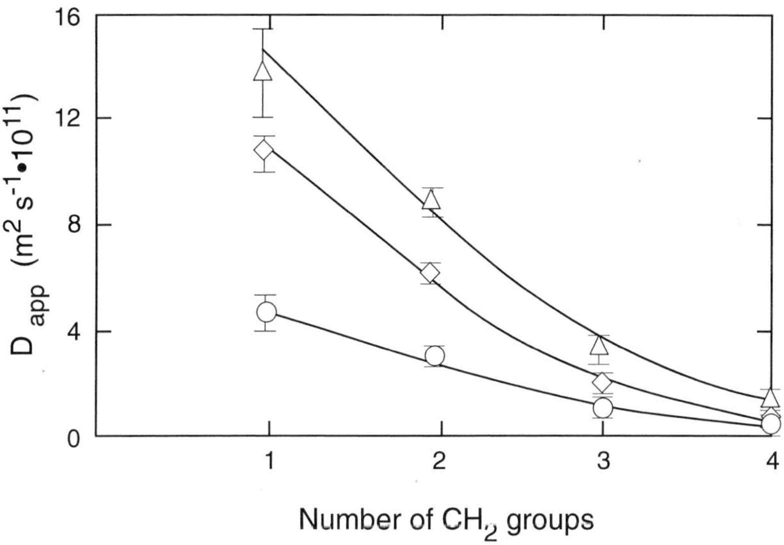

Figure 5. Apparent diffusion coefficient of p-hydroxybenzoate esters from 25% Pluronic F127 gels as a function of ester chain length at 30°C (circles), 37°C (diamonds), and 50 °C (triangles). From ref.(54).

Moreover, the release rate is found to increase with an increasing temperature in these systems (53-56). This is expected from the point of view that the temperature increase most frequently results in an increase in the diffusion rate. However, the structure of the self-assemblies formed depends on temperature in these systems, and in particular an increased temperature results in a micellar growth and in an increase in the micellar aggregation number, as well as in a more compact micellar core (15,44,59,60). Since the release rate depends on the partition of the drug between the hydrophobic domains and the aqueous continuous phase, and since the partition to the hydrophobic domains could be expected to be promoted by this, a temperature increase could also, in principle, result in a release rate decrease. Obviously, the effect on the diffusion rate is larger than this latter effect, which can nevertheless be observed by worsening the solvency of the polymer and promoting its self-assembly at constant temperature. Doing so, Chen-Chow et al. observed that increasing the NaCl concentration at constant polymer and drug concentration resulted in a decreased release rate of lidocaine for the reasons given above (55).

Due to the somewhat limited solubilization capacity of PEO-PPO-PEO copolymer micelles for very hydrophobic substances (26) it is interesting to try to identify polymers with similar properties regarding, e.g., the reversed thermoreversible gelation, but which contain domains more hydrophobic than those formed by PPO. A particularly interesting class of block copolymers in this respect is PEO-PBO-PEO copolymers, where PBO is poly(butylene oxide). By a comparison of the self-assembly of PEO-PPO, PEO-PBO, and PEO-alkyl copolymers, Yang et al. concluded that the hydrophobicity of the PBO block is much higher than that of the PPO moiety (61). Booth et al. also investigated the gelation of several of these copolymers, and found a similar gelation behaviour as for the PEO-PPO-PEO block copolymers (62-64). Moreover, the release of salicylic acid from one of these gels was investigated and found to be quite slow and of zero-order kinetics under the conditions investigated (63).

Apart from an effectively enhanced solubility of sparingly soluble drugs and the resulting sustained and controlled release of the drug after administration, a reason for seeking solubilizing hydrophobic drugs into block copolymer gels is to improve their chemical stability (cf micellar solubilization). For example, Tomida et al. investigated PEO-PPO-PEO block copolymer gels based on Pluronic F127 containing effectively water-insoluble indomethacin regarding their suitability as topical drug delivery systems, and found that the hydrolysis rate of this model drug was significantly reduced in the block copolymer gels compared to that in aqueous solution (65). The protective properties of these gels make them interesting, e.g., for oral administration of substances sensitive to acid-catalyzed hydrolysis by *in situ* gelling block copolymer systems. Another potentially interesting area of the protective properties of these gels is within oral administration of protein and peptide drugs, which without such protection have only a very limited bioavailability, primarily due their proteolytic degradation (66). However, whenever simultaneously employing proteins/peptides and surface active agents there is a risk for activity loss of the former due to conformational changes caused by surfactant binding (47). However, nonionic low molecular weight surfactants generally display little tendency to associate and form complexes with proteins (47). Due to the similarity in many respects between these low molecular weight surfactants and PEO-PPO-PEO block copolymers the latter are expected to be fairly gentle regarding denaturation of protein and peptide drugs. Indeed, Fults et al. investigated the incorporation of urease in a Pluronic F127 gel and found that incorporation of this enzyme in the gel only resulted in a limited (≈10%) activity loss even at high polymer concentrations (67). Similarly, Jonston et al. investigated the solubilization of interleukin-2 in Pluronic F127 gels and found that this formulation resulted in no activity loss or denaturation on storage at room temperature (68). Thus, it appears that PEO-PPO-PEO block copolymer gels may be used also for protein and peptide drug delivery. Note, however, that these copolymers in some cases may have negative effects on the oral absorption of drugs. Thus, Florence et al. investigated the effects of PEO-PPO-PEO coatings of polytyrene nanoparticles on the uptake of the latter after oral administration, and concluded that the adsorbed copolymers blocked uptake in the small intestinal region of the gut (69). Although this finding concerns particles stabilized with PEO-

PPO-PEO block copolymers rather than micelles or other types of self-assemblies formed by these polymers, similar effects may occur also for the latter type of systems.

Nasal administration offers several advantages in drug delivery, including fast absorption, avoidance of first-pass effects, and a way within protein and peptide drug delivery to avoid their proteolytic degradation in the gastrointestinal tract. With this in mind, Abd El-Bary et al. investigated the bioavailability of glibenclamide when administred nasally in a PEO-PPO-PEO block copolymer gel (Pluronic F127). Results *in vitro* showed that the release rate was reduced with an increasing polymer concentration and at a lower temperature, whereas results *in vivo* on the relation between pharmacokinetic properties and dose response showed that the drug in internasal gel form was more efficient in reducing the glucose level in alloxinated rabbits than a tablet formulation (56).

Controlled delivery of drugs to the eye is limited by the efficient protective mechanisms which exist in the percorneal area. In order to achieve a high absorption of drugs into the eye a prolonged contact time with the corneal tissue is therefore required apart from a good corneal penetration of the drug. For this purpose viscous polymer gels have been investigated regarding occular drug delivery. The gelation on heating for PEO-PPO-PEO block copolymer solutions and their capacity to solubilize both hydrophilic and hydrophobic drugs make them interesting for these applications. For example, Miller and Donovan examined Pluronic F127 for the delivery of pilocarpine and concluded that the gel formulation enhanced the activity of this substance compared to the aqueous solution (70,71). However, other investigators have found that PEO-PPO-PEO block copolymer gels seem not to offer any large advantage over, e.g., other polymer gel systems in terms of a sustained effect (72). Furthermore, the typically rather high polymer concentration required for gelation (~20 wt%), a finite toxicity, and the surface active and solubilizating properties of these polymers make their use in occular drug delivery still an open issue (72,73).

A more promising area of application of PEO-PPO-PEO block copolymer gels is within rectal administration. As an example of this, Miazaki et al. investigated the rectal administration of indomethacin, the usefullness of which is severely reduced by gastrointestinal side-effects. For this reason, indomethacin PEO-PPO-PEO gel preparations based on Pluronic F127 were administred rectally to rabbits and the drug plasma levels were compared with those after rectal administration of commercial suppositories (74). From the integration of the plasma concentration over time it was concluded that the bioavailability of the two formulations were virtually identical. However, for the commercial suppositories there was a pronounced peak in the plasma concentration after less than one hour (Figure 6). Since it has been shown (75,76) that the frequency and severity of indomethacin side-effects, e.g., on the nervous system, is well correlated with peak plasma concentrations above about 5 μg/ml and since the PEO-PPO-PEO block copolymer gel formulation displayed no such peak plasma concentration it was suggested that these side-effects should also be reduced (74). Furthermore, histopathological investigation showed that damage to the mucosal membrane by the copolymer was rarely fould in the rectum. Moreover, individual differences following rectal administration of the gel formulation were small compared to those after administration of the other formulation (77). Finally, as discussed in more detail above solubilization of indomethacin in PEO-PPO-PEO block copolymer gels reduces the hydrolysis rate of this substance, which is expected to contribute to the good bioavailability observed.

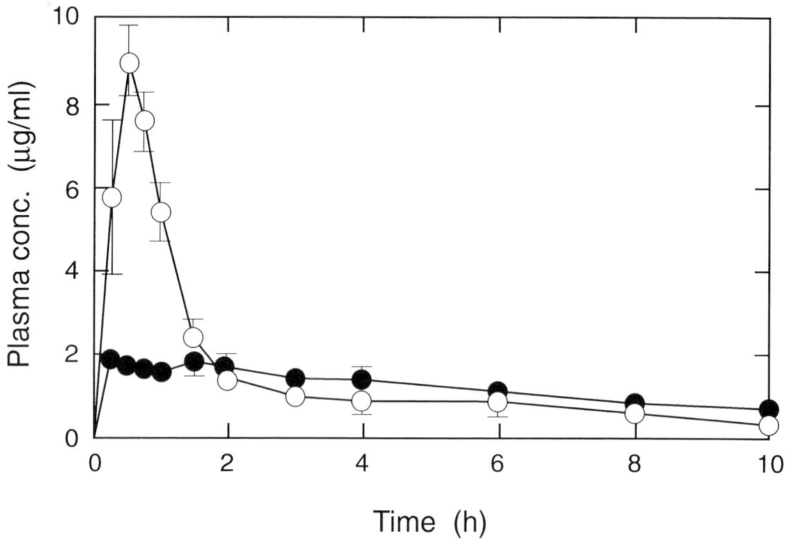

Figure 6. Plasma concentration of indomethacin after rectal administration to rabbits from commercial suppositories (open circles) and Pluronic F127 gel formulations (filled circles). From ref.(74).

Injectable *in situ* gelling formulations have also been investigated in a range of different contexts. For example, Wang and Johnston investigated sustained release of interleukin-2 (IL-2) following intramuscular injection in rat (78). The backgroud to the study is that anti-tumor treatment using IL-2 has shown positive results for several cancers in both experimental animal models and in humans, the primary limitation of the use of high-dose IL-2 therapy being the toxicity associated with it. Since it has been found that the antitumor effect of IL-2 is correlated with the time IL-2 remains in the serum rather than with the peak serum IL-2 concentration a sustained release formulation of IL-2 would be expected to improve the therapeutic efficiency at the same time as toxic side-effects are reduced. Considering this, the observed reduced peak serum IL-2 concentration and the longer circulation of IL-2 following intramuscular administration for a PEO-PPO-PEO block copolymer gel formulation (Pluronic F127) compared to the aqueous IL-2 solution (Figure 7), is promising for IL-2 intramuscular therapy (78).

In order to seek an improved efficiency against Sarcoma-180 ascites tumor in mice, as well as reduced toxic side-effects, Pluronic F127 gels were evaluated regarding peritoneal administration of mitomycin C (MMC) (79). A prolongation of the life-span of the tumor-bearing mice was noted for the block copolymer formulation compared to the free drug. In particular, a high chemotherapeutic efficiency of MMC was found at high doses, which would be toxic in the case of the drug alone. It was concluded that the block copolymer formulation was superior against intraperitoneally inoculated tumor as compared with MMC aqueous solution (79).

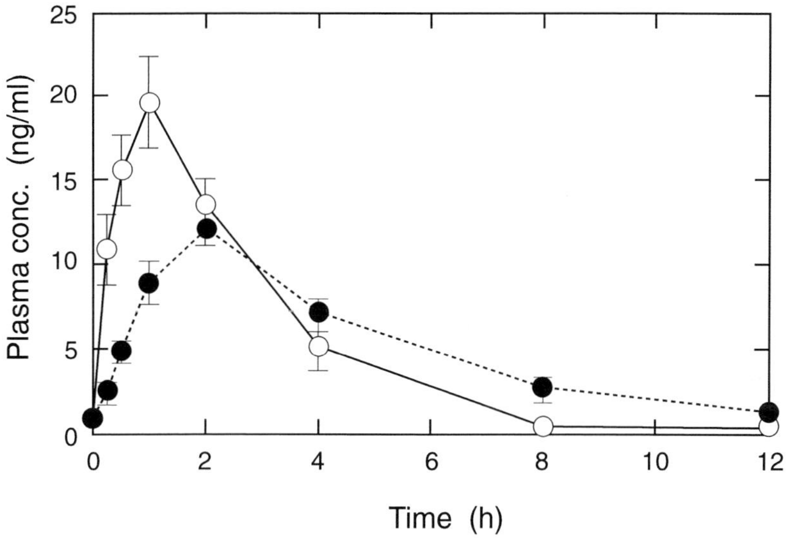

Figure 7. Plasma IL-2 concentration following intramuscular injection in rats of an aqueous IL-2 solution (open circles) and an IL-2/Pluronic F127 gel formulation (filled circles). From ref.(78).

Another application where PEO-PPO-PEO block copolymer gels have shown promise is as wound dressings in the treatment of thermal burns. Application of such a dressing should be uncomplicated, whereafter the dressing should adhere to the uninjured skin surrounding the wound sufficiently strong to resist mechanical damage such as lifting and slipping, but also come off easily when removed. Furthermore, the adherence should be uniform since small areas of non-adherence leads to fluid-filled pockets where bacteria may proliferate. Also, there should be no ingrowth of tissue into the dressing or other strong attachments which would preclude easy removal. An important aspect of the treatment of burn wounds is the fluid balance, since large amounts of fluid are lost through evaporation and exudation, resulting in a fall in body temperature as well as in metabolic disturbances. For this reason, but also since this assists epithelization, dressings should absorb fluid and maintain a high humidity at the wound. Equally important, wound dressings should provide a bacterial barrier, which could be achieved either by the dressing itself or by the inclusion of antibacterial agents, the release of which should preferably be sustained (80).

In the treatment of burn wounds a number of different types of dressings have been investigated, including preformed polymer films, spray-on films, gels, foams, and composites. The advantage of polymer gels is that they follow the contours of the wound surface, that they display a low wound adherence, and that antibacterial agents may be readily incorporated. Dissadvantages are mainly related to the short durability of these systems (80). Due to their reversed thermoreversible gelation also PEO-PPO-PEO block copolymer gels have been investigated in burn wound therapy. For example, Nalbandian et al. found that Pluronic F127 is an efficient base for bacteriocidal silver nitrate and silver lactate following full thickness thermal burns in rats (81). No inhibition of skin growth and repair was noted and the dressings were equally efficient against *Pseudomonas aeruginosa* and *Proteus mirabilis*. The dressings showed promise as artificial skin against electrolyte imbalances, heat loss, and bacterial invation (81). The effects of the gel formulations on the mortality rate following thermal burns are shown in Figure 8.

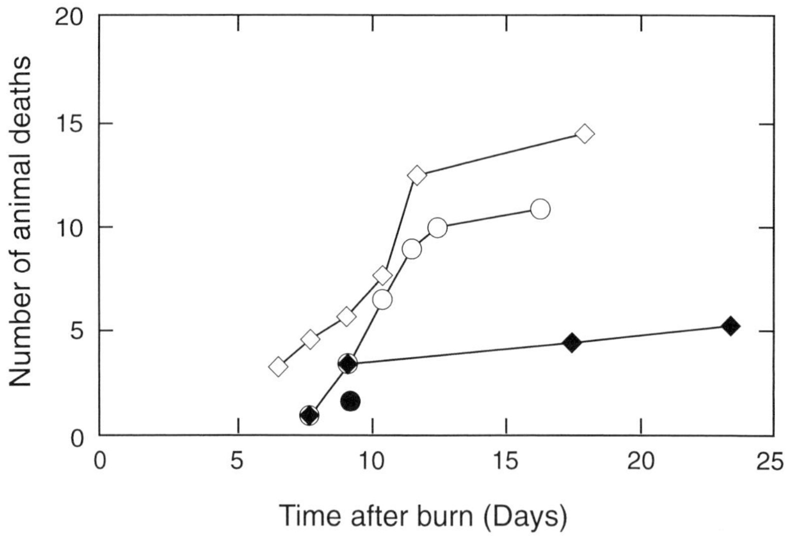

Figure 8. Mortality after 18-22% full thickness skin thermal burns in rats for Pluronic F127+silver lactate (filled circle), Pluronic F127+silver nitrate (filled diamonds), Pluronic F127 (open circles), and control (open diamonds). From. ref.(81).

In parenthesis, it is interesting to note that PEO-PPO-PEO block copolymers have also been successfully administered intraveneously in early burn wound treatment. Thus, Paustian et al. found a dramatic improvement in full thickness burn wounds in rats treated intraveneously with Pluronic F127 30 min after burn wound inflection (82). Histologically, skin biopsies showed less of the microscopic damage usually associated with full thickness burns in the polymer treated animals than found for the saline control. Also, the block copolymer therapy had a positive effect on the inflammatory process in the wound area. The mechanisms by which the intraveneously administered block copolymers achieve their positive effects on wound healing is somewhat uncertain at present. However, it has been observed that intraveneously administered PEO-PPO-PEO block copolymers effectively reduce the blood viscocity (82-84). In patients with extensive full skin thickness burns an increase in the blood viscocity occurs, causing sludging of red blood cells in small vessels and capillaries in the wound area. Quite possibly, the positive effects of intraveneously administered PEO-PPO-PEO block copolymers on thermal wound healing are related to the avoidance or reduction of such sludging (82). [Although the mechanisms behind the blood viscocity decrease due to PEO-PPO-PEO block copolymers is somewhat unclear at present, it is interesting to note that Carr et al. investigated the effects of Poloxamer 188 on the assembly, structure, and dissolution of fibrin clots, and found that fibrin self assembly was accelerated by the presence of the copolymer, presumably as a result of bridging flocculation (2,83).]

An interesting related area of intraveneously administered PEO-PPO-PEO block copolymers is that of reduction of the development of tumor metastatis. The reason for this is that the ability of a circulating tumor cell to develop into a metastatis appears to be related to its adhesion to the endothelium and microclot formation, since studies of tumor metastatis following trauma indicate cellular aggregation or sludging at the trauma site (85). Since PEO-PPO-PEO block copolymers have been found to reduce blood viscocity and such sludging (see above), this type of polymer is interesting for decreasing the incidence of tumor metastatis. Realizing this, Silk et al. investigated the effects of Pluronic F68 on the development of metastatis and found that treatment with this copolymer after intraveneous administration of Walker 256 tumor cells resulted in a decreased incidence of pulmonary metastatis in rats (85).

Although PEO-PPO-PEO block copolymers by far are the ones most frequently studied in connection with pharmaceutical applications, also other amphiphilic gel forming block copolymers are of interest in these applications. For example, Sawhney et al. investigated biodegradable hydrogels based on photopolymerized PEO-co-poly(α-hydroxy acid di acrylate) macromers (86). By a suitable choice of monomers the degradation displayed by these hydrogels could be tailored from less than one day to 4 months. Also the cross-linking density was found to affect the degradation rate, and the more tightly cross-linked the polymer, the slower the degradation. Other gels investigated in relation to drug delivery applications include block copolymers of PEO and poly(lactide acid), poly(ε-caprolactone), poly(ethylene terephtalate), poly(ester carbonate), poly(ethylene succinate), pol(γ-benzyl L-glutamate), and poly(hydroxyalkyl L-glutamine) (87). These systems have been investigated for their mechanical properties, degradation, tissue reaction, and morphology. Drug release properties, on the other hand, have not been extensively investigated. However, Zhu et al. and Rosati et al. investigated the release of norethisterone from PLA-PEO-PLA (PLA being poly(lactide)) and the release of theophylline from PEO-poly(glycolide-co-trimethylene carbonate), respectively (88,89). Furthermore, the release of hydrocortisone from PEO-poly(hydroxyalkyl L-glutamine) (PHAG) gels was investigated and it was found that the release of this substance, partitioning mainly towards the PEO domains, was influenced by the ratio of the two compounds, the microstructure morphology, and the drug solubility. In particular, a combination of low PEO content, low drug solubility, and longer alkyl chains resulted in long-term constant release from these gels. Largely the effects observed could be explained by a barrier effect of the continuous PHAG phase (90).

4. Adsorption and Stabilization

One of the most useful consequences of the amphiphilic nature of block copolymers is their tendency to accumulate at different surfaces. In general, the adsorption tendency depends on the molecular weight of the copolymers in a similar manner as for homopolymers. Furthermore, up to a limit, the larger the solvophobic domain(s) and the smaller the solvophilic domain(s), the higher the adsorbed amount. Not surprisingly, block copolymers frequently, but not always, adsorb with the solvophobic part toward the surface, whereas the solvophilic domains protrude toward the bulk solution, forming either "mushrooms", "brushes", or tail-loop-train distributions, depending on the surface coverage and on whether the solvophilic block has an adsorption tendency or not (1). Furthermore, interfacial micelle formation has been found for certain systems, e.g., Pluronic F127 at silica (91). These and other aspects of block copolymer interfacial behaviour are discussed in detail elswhere in this book.

In particular, this surface activity has consequences for the interaction between colloidal entities coated with block copolymers. Of special interest in pharmaceutical applications are block copolymers containing one or several hydrophobic blocks and one or several hydrophilic blocks. Thus, while the hydrophobic domains act as anchors and assure an extensive and strong adsorption of the block copolymer at various surfaces, e.g., liposomes, emulsion droplets, and polymer (nano)particles, the hydrophilic blocks experience good solvency conditions which results in a repulsive interaction between the entities coated with the block copolymer, and thus in a system colloidal stability (1,2). The mechanism behind this so-called steric stabilization is that when the polymer-coated colloids approach each other through thermal motion, the polymer layers overlap at an interparticle distance depending on the adsorbed layer thickness (1,2,92). When the adsorbed layers interpenetrate the polymer concentration in the overlap region increases, which is unfavourable when the stabilizing moieties experience good solvency conditions and favourable at poor solvency conditions for the stabilizing chains. This results in a repulsive effective interparticle interaction at the typically relatively good solvency conditions employed in pharmaceutical applications (e.g., PEO-containing block copolymers in aqueous solution at moderate temperatures) (2).

Somewhat analogous to the stabilization of colloidal systems is the generally observed phenomena of protein rejection due to adsorbed or immobilized (co)polymers at good solvency conditions for the stabilizing moieties, and particularly for PEO-derivatives in aqueous solution. Thus, it has been found in numerous systems that PEO-modified surfaces obtained, e.g., through the adsorption of PEO-containing block copolymers, are very efficient in reducing the adsorption of a large number of proteins (93-98). Although the mechanisms behind the protein rejection particularly of PEO-chains have been extensively discussed it now seems clear that this in many ways is analogous to the steric stabilization of colloidal dispersions (4,99,100). Thus, as a protein approaches the interface through thermal motion the polymer layer is forced to become more concentrated in order for the protein to be able to adsorb, provided that the protein is unable to adsorb on top of the polymer layer, provided that the effective mesh size of the layer is smaller than the protein dimensions, and provided that protein-induced polymer desorption does not occur. Since PEO layers do not contain charges (101) and since their water content is typically quite high (93), electrostatic attractive interactions between the protein and the polymer layer are absent, and the van der Waals interactions expected to be quite weak (99). Thus, provided that the PEO-layer is thick enough to mask the interaction between the protein and the underlying surface, there is no driving force for protein adsorption. Therefore, the combination of the effectively repulsive protein-polymer interaction and the absence of a strong adsorption driving force results in a low protein adsorption.

Although PEO-containing copolymers, and PEO-PPO-PEO block copolymers in particular, are used extensively in order to stabilize colloidal systems in a host of pharmaceutical applications, a particularly interesting application area is within parenteral administration of colloidal drug carriers. Examples of such colloidal systems stabilized with PEO-containing amphiphiles include emulsion droplets (8,102), liposomes (103,104), polymer particles (6,7), dispersed cubic phase (105,106), solid lipid nanoparticles (107), solid drug particles (108,109), etc. On intraveneous administration of colloidal drug carriers these are frequently cleared from circulation in the blood stream quite efficiently and accumulated in certain tissues, e.g., liver, spleen, and marrow (110-113). From a drug delivery point of view, this has some problematic consequences. First, since the contact between the drug carrier and tissues not related to the reticuloendothelial system (RES) is limited, the drug bioavailability is typically low (unless, of course, the drug is aimed for any of these tissues). Furthermore, since effectively the whole dose is accumulated in RES-related tissues, the local toxicity in these tissues may be quite high, causing toxic side-effect. With this in mind, a key concern in parenteral administration of drugs aimed for other than RES-related tissues via colloidal drug carriers is that of increasing the circulation time of the drug (and of the drug carrier).

The rate and extent of RES-mediated uptake of colloidal drug carriers have been found to depend on the size of the drug carrier (the larger the carrier the faster the uptake), as well as on its surface properties, e.g., its hydrophobicity and charge, as well as the presence of specific functional groups. There is growing evidence that RES-uptake and endocytosis is initiated by adsorption of certain serum proteins (so-called opsonins) at the drug carrier surface. For example, complement proteins (e.g., C3 and C1q) may adsorb at the surface which may result in activation of the complement cascade. Other examples of opsonins include IgG, e.g., interacting with the complement system (C1q), "adhesion" proteins such as fibrinogen and fibronectin, as well as other types of proteins, e.g., the Hageman factor and CRP (110-113). Naturally, the interplay between the opsonins is extremely rich and little understanding is available of how to design a surface specifically binding particular serum proteins. However, it has been found in several studies (114,115) that there is an inverse correlation between the total amount of serum proteins adsorbed at a drug carrier surface on one hand and the circulation time on the other. More specifically, the lower the overall serum protein adsorption at the carrier surface the longer the circulation in the blood stream.

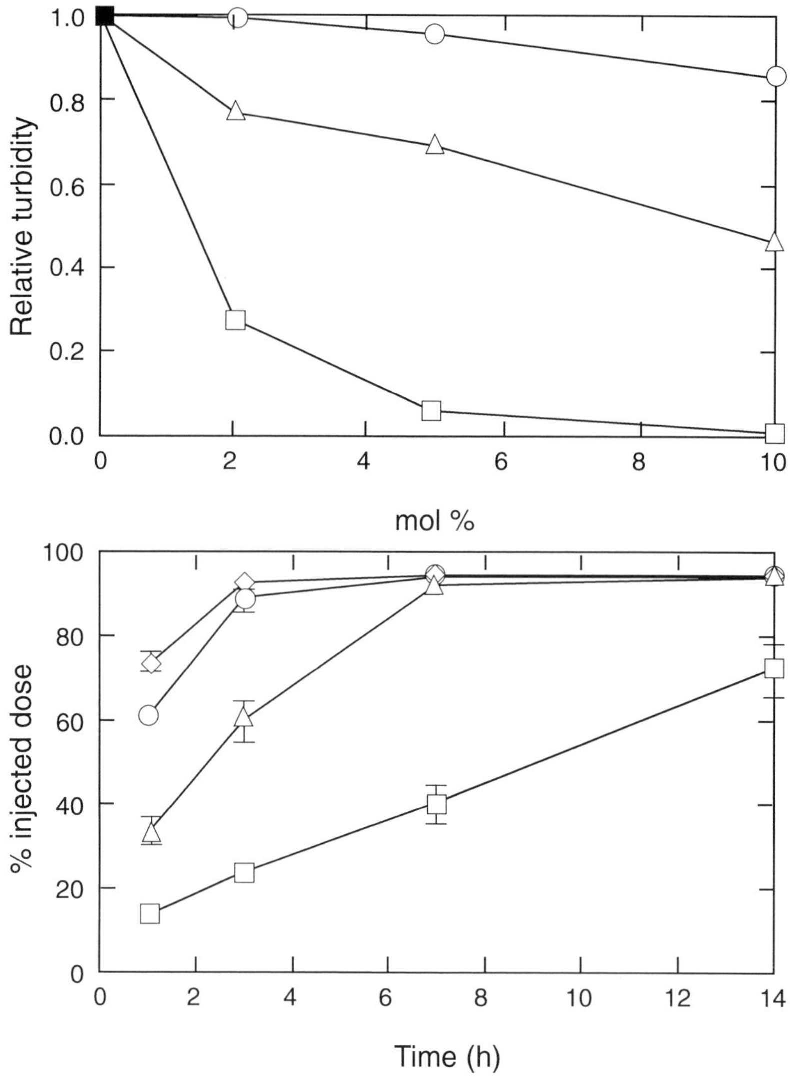

Figure 9. (a) Effects (normalized to that for bare phosphatidylcholine/ cholesterol liposomes) of increasing concentrations of PEO_{750}-PE (circles), PEO_{2000}-PE (triangles), and PEO_{5000}-PE (squares) on steptavidin-induced agglutination of liposomes containing boitin-cap-PE. (b) Uptake in liver and spleen of intraveneously injected ^{111}In-labeled liposomes in male Balb/c mice. Legends as in (a). Diamonds show results obtained for the uncoated liposomes. From ref (117).

Since phospholipids or related polar lipids are frequently used to stabilize pharmaceutical dispersions (e.g., emulsion droplets or lipoprotein mimics), considerable work has been carried out in order to investigate how their properties may be controlled in order to obtain a required biological response. From this work it has been found that PEO-modified phospholipids are particularly efficient in reducing the RES uptake and prolonging the circulation time of colloidal drug carriers (103,104,116-119). Note, however, that in order to achieve this the steric layer needs to be sufficiently dense and thick for reasons given above. This is shown also in Figure 9 for phosphatidylcholine (PC)/cholesterol liposomes containing varying amounts of PEO-modified phosphatidylethanolamine (PE) of different length (117). Analogous effects were also obtained, e.g., for polystytene-PEO particles by Dunn et al., who found that uptake of the particles by non-parenchymal liver cells *in vitro* decreased with an increasing PEO surface density of the particles, at the same time as *in vivo* studies in rats showed a prolonged circulation time for particles with a dense PEO coating (120). Analogous effects have been found also for poly(D,L-lactide)-PEO particles by Verrecchia et al. (121). Furthermore, covalent PEO-modification of genetically produced proteins to be administered intraveneously, which would otherwise cause an immunicity response and a consequtive rapid clearance of the protein from blood stream circulation, is an efficient way, although depending on the PEO chain density, to avoid problems associated with the parenteral administration of these proteins (122). Although these results concern grafted PEO chains, similar results are expected for adsorbed PEO-containing block copolymers, e.g., of the PEO-PPO-PEO type (1,2), and the results therefore illustrate that the polymer layer has to be sufficiently dense and thick in order to be able to exert a strong protein rejecting capacity on a wide range of proteins.

From an industrial point of view, an interesting alternative to use PEO-modified phospholipids is to use emulsifiers (e.g., phosphatidylcholine) chosen due to their low cost and emulsification efficiency, and once the colloidal system of interest has been formed, postadsorb, e.g., PEO-PPO-PEO block copolymers. Using adsorbed rather than grafted PEO layers, several investigators have observed very efficient prolongation of blood circulation time and control of the biodistribution *in vivo*. For example, in a series of studies, Davis and Illum have investigted effects of the adsorption of linear as well as branched PEO-PPO block copolymers on the circulation time and tissue distribution of polystyrene model particles, and found a prolonged circulation time in blood and a reduced accumulation in, e.g., liver and spleen (3,6,7). Similar results have been found for these systems by Tan et al. (5). Furthermore, Illum et al. investigated the effects of Poloxamine 908 on the circulation time of emulsion droplets administered intraveneously to rabbits and again found the general features outlined above also for this system (Figure 10) (8). Analogous results have been found also by, e.g., Davis and Hansrani (102).

Naturally, also liposomes to be administered intraveneously may be modified by either covalently bound or physically adsorbed PEO chains in order to increase their circulation time and control their tissue distribution (103,104). A major concern for these systems, however, is that of liposome leakage induced by post-adsorbed copolymers. For example, Woodle et al. observed that PEO-PPO-PEO block copolymers induce leakage of liposomes, including those prepared by high temperature phase transition lipids (123). Similar results were reported also by

Jamshaid et al. (124). This is analogous to liposome leakage induced by some EO-containing low molecular weight surfactants but not by others (125,126). Whether or not block copolymer-induced leakage is a proplem for liposome formulations most likely depends on both the phospholipid and the copolymer used, and this aspect of liposome administation therefore requires further investigation.

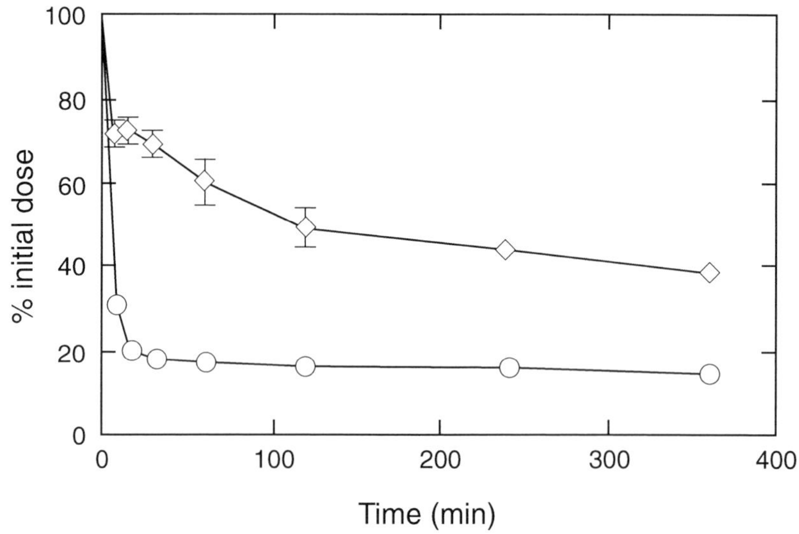

Figure 10. Blood clearance of ^{123}I-labelled emulsions stabilized by egg lecithin (circles) and Poloxamine 908 (diamonds) intraveneously administered to rabbit. From ref.(8).

A practical concern of copolymers in general, and particularly their use in adsorption applications such as stabilization of emulsion droplets, polymer particles, or liposomes, is that of the finite polydispersity and heterogeneity of these substances, and the resulting batch-to-batch variation. Porter et al. investigated this aspect of parenteral administration of copolymer-stabilized colloidal particles for Poloxamer 407 and found a marked variation in tissue distribution for different samples, despite the average composition and molecular weight, as well as the adsorbed layer thickness and resulting particle zeta potential displaying rather limited variations (127). These findings show that in practical formulation work involving PEO-PPO-PEO block copolymer adsorption care has to be taken in order to minimize these effects.

A somewhat special pharmaceutical application of adsorbed and stabilizing PEO-PPO-PEO block copolymers is that of artificial blood, where these copolymers are used to stabilize perfluorochemical (PFC) emulsions (128-131). PFC emulsions used as artificial blood typically consist of droplets about 100-200 nm in diameter, and in many respects these emulsions resemble lipid emulsions used for parenteral nutrition (132,133). PFC's are not metabolized but excreted unchanged by the lungs following temporary storage by the monocyte-macrophage system. Injected

PFC emulsions increase the oxygen-carrying capacity by dissolution of oxygen rather than by oxygen binding, and they therefore bear little resemblence to hemoglobin. As a consequence of this, virtually all dissolved oxygen is extracted by the tissues before hemoglobin is off-loaded. Although the choice of which PFC should be used in artifical blood is crucial from both a toxicity and an efficiency point of view, systems have been found which efficiently and safely can act as oxygen carrier (130). Several different emulsifiers have been employed in artificial blood, but PEO-PPO-PEO block copolymers were the first to be used commercially, and are still frequently used both in practical systems and in model investigations (130). However, there are observations that the PFC emulsions stabilized with PEO-PPO-PEO block copolymers activate complement. For example, Vercellotti et al. found that Fluosol, a PFC emulsion stabilized by Pluronic F68, lead to complement activation as observed by C3 conversion, decrement of CH_{50}, and generation of C5a-related PMN aggregating activity. Furthermore, indications were found to imply activation via the alternative pathway. These effects were found also for Pluronic F68 in itself (129). Similar results for the Fluosol emulsion were obtained by Ingram et al. (131). Contrary to Vercellotti et al., however, the latter investigators found that none of the individual components of the Fluosol emulsion, including Pluronic F68, activate complement by themselves, whereas a reduction of CH_{50} levels similar to that observed for Fluosol was obtained for Pluronic F68 in combination with either of the PFC's present in Fluosol. This lead the investigators to infer that the complement activation was due to the micelle (sic) structure. However, the mechanisms behind such a micelle-induced complement activation is unclear at present.

5. Adjuvants

In the development of vaccines it is frequently found that purified immunogens do not provoke a strong immune response by themselves, but rather require the simultaneous presence of immune response promoters, or so-called adjuvants. It is well known that adjuvants influence both the duration and the intensity of the immune response, as well as the actual type of immune response (134-137). Of particular interest to the present discussion is that certain surface active agents act as adjuvants. It is believed that the adjuvant activity of these substances is related to their surface activity, since reducing their amphiphilicity in a number of ways leads to a reduced or eliminated adjuvant activity. On the other hand, not all surface active substances act as adjuvants. Among the surface active substances displaying adjuvant activity are PEO-PPO block copolymers when administered with an oil phase. Although the mechanism of the adjuvant activity of these block copolymers is largely unknown, it has been inferred that the polymers used allow proteins to bind to the surface of the oil droplets, and in particular complement and other host mediator systems are activated by the contact with the copolymer surface.

There have been several studies of the effects of the size and composition of the PEO-PPO block copolymers on their adjuvant activity, and it has been found that the longer the the PPO moiety, the shorter the PEO moiety, and the more hydrophobic the copolymer the stronger the adjuvant activity (134-137).

Considering that the adjuvant activity is thought to be at least partly related to protein adsorption at the droplets stabilized by these copolymers (e.g., C3 adsorption and activation in relation to surface hydroxyls due to the PEO ends) (134-138), this behaviour is expected, and analogous to the general behaviour observed for protein adsorption at polymer-modified surfaces (see above). On the other hand, decreasing the molecular weight keeping the EO/PO ratio constant results in a decreased adjuvant activity (137), which is opposite from the behaviour expected from a protein rejection point of view (see above). Furthermore, PEO-PPO-PEO block copolymers are generally more efficient adjuvants than the corresponding PPO-PEO-PPO block copolymers (137), which again is opposite from the trends expected from a protein rejection point of view, since loops are poorer in exerting a steric repulsion than tails of the corresponding size for entropic reasons (1,2). From these latter observations it is clear that the mechanisms behind the adjuvant activity of PEO-PPO block copolymers are more complex than just involving protein adsorption and complement activation, and that further studies are required in order to clarify the mechanisms behind this adjuvant activity.

Meanwhile, however, the adjuvant activity of PEO-PPO block copolymers is frequently employed in the development of new vaccine therapies. Just to mention one example, de Souza and Playfair employed Pluronic L121 as adjuvant in a new formulation for blood-stage malaria treatment, and found that this formulation induces potent protective immune response against an otherwise lethal malaria infection in mice (Figure 11) (139).

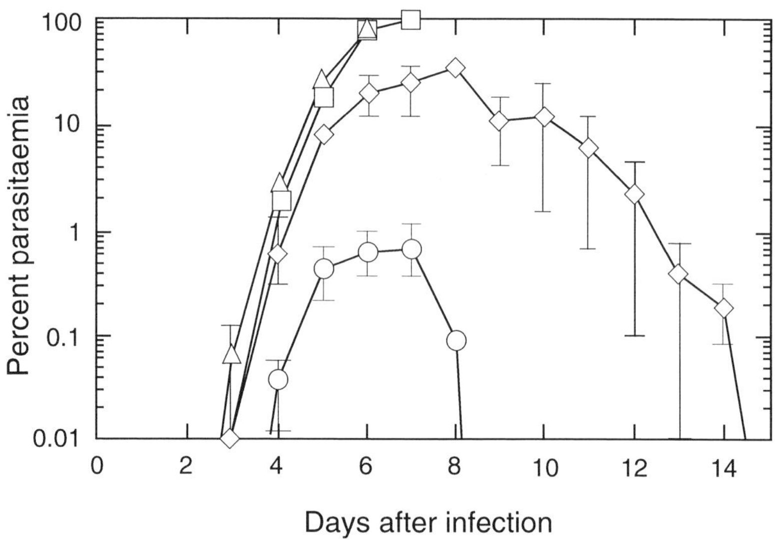

Figure 11. Course of infection in mice vaccinated against blood-stage malaria with (circles) and without (diamonds) adjuvant. Shown also are data obtained after injection of adjuvant alone (squares), as well as data for the control group (triangles). From ref.(139).

6. Toxicity Aspects

An important aspect of any substance to be used in a pharmaceutical formulation is that of the toxicity of the substance and of the resulting formulation. Considering the wide range of pharmaceutical applications of PEO-PPO-PEO block copolymers it is not surprising that considerable toxicity data have been reported for this class of copolymers, e.g., concerning chronic oral toxicity, skin and eye irritation, acute intraveneous toxicity, and reproduction consequences (9). From these studies it largely follows that the toxicity of these substances is fairly low, and decreasing with an increasing molecular weight and PEO content. Much of the toxicity studies have been performed by the polymer manufacturers, but also numerous independent investigators confirm the general picture outlined. For example, Siebenbrodt and Keipert investigated the opthalmic application of microemulsions containing Poloxamer L64 and found an acceptable tolerance to this formulation (140). Furthermore, Johnston and Miller investigated the muscle toxicity caused by injection of various PEO-PPO-PEO copolymers into rabbits via gross morphological examination of the muscle tissue and by monitoring creatine phosphokinase (CPK) levels after injection (141). It was found that the toxicity of the copolymers was correlated to their hydrophobicity, and the more hydrophobic the copolymer, the more severe the lesion produced following injection and the greater the increase in the plasma CPK concentration. Moreover, Kier et al. investigated, e.g., the acute toxicity after peroral and intraveneous administration of EO/PO statistical copoymers and found that this decreased with an increasing copolymer molecular weight (142). Also, Port et al. evaluated the toxicity following intraveneous administration of Pluronic F38, and although this copolymer was found to cause dose dependent morphology changes related to epithelial cells and Type II macrophages in the lung, the paucity of additional changes led these investigators to conclude that although Pluronic F38 may be readily phagocytized it is well tolerated even when administered intraveneously in large doses (143).

On the other hand, the toxicity of Pluronic F68 was studied in rat when administered intraveneously for a month in daily doses by Magnusson et al. (144). Although this PEO-PPO-PEO block copolymer induced no detectable morphological abnormalities in either lung or kidney up to a daily dose of 50 mg/kg, heigher doses resulted in the presence of foam cells in the lungs and focal cortical degenerative changes in the kidneys. It was inferred by these investigators that Pluronic F68 is able to induce phospholipidosis in rat, which quite possibly is related to an inhibition of phospholipase activities, possibly originating from enzyme conformational changes induced by complex formation between the enzyme(s) and this copolymer (47). Furthermore, Davidorf et al. investigated the occular toxicity of vitreal Pluronic F127, one of the most high molecular weight and hydrophilic of the PEO-PPO-PEO block copolymers, thus a substance for which a relatively low toxicity would be expected, and found that eyes containing Pluronic F127 showed a marked destruction of the retina 2 weeks after surgery (73). Thus it is clear that the toxicity of these block copolymers is a complex issue which has to be assessed in each application.

7. Acknowledgement

Professor Krister Holmberg is gratefully acknowledged for comments on the manuscript. This work was financed by the Foundation for Surface Chemistry, Sweden.

8. References

1) G.J. Fleer, M.A. Cohen Stuart, J.M.H.M. Scheutjens, T. Cosgrove, B. Vincent, "Polymers at Interfaces", Chapman & Hall, London, 1993.
2) D.H. Napper, Polymeric Stabilization of Colloidal Dispersions, Academic Press, London, 1983.
3) S. Stolnik, L. Illum, and S.S. Davis, Adv. Drug Del. Rev. 16 (1995) 195.
4) V.P. Torchilin and V.S. Trubetskoi, Polym. Sci. 36 (1994) 1585.
5) J.S. Tan, D.E. Butterfield, C.L. Voycheck, K.D. Caldwell, and J.T. Li, Biomaterials 14 (1993) 823.
6) S.S. Davis and L. Illum, Biomaterials 9 (1988) 111.
7) L. Illum, S.S. Davis, R.H. Müller, E. Mak, and P. West, Life Sci. 40 (1987) 367.
8) L. Illum, P. West, C. Washington, and S.S. Davis, Int. J. Pharm. 54 (1989) 41.
9) I.R. Schmolka in P.J. Tarcha (ed.) Polymers for Controlled Drug Delivery, CRC Press, Boca Raton, 1991.
10) G. Wanka, H. Hoffmann, and W. Ulbricht, Macromolecules 27 (1994) 4145.
11) P. Linse, Macromolecules 26 (1993) 4437.
12) K. Mortensen and J.S. Pedersen, Macromolecules 26 (1993) 805.
13) P. Alexandridis, U. Olsson, and B. Lindman, Macromolecules 28 (1995) 7700.
14) P. Alexandridis, U. Olsson, and B. Lindman, J. Phys. Chem. 100 (1996) 280.
15) P. Alexandridis and T. A. Hatton, Colloids Surf. A 96 (1995) 1.
16) K.-W. Zhang and A. Khan, Macromolecules 28 (1995) 3708.
17) P. Linse, J. Phys. Chem. 97 (1993) 13896.
18) I. Goodman (ed.), Developments in Block Copolymers - 1, Applied Publishers, Ltd., London, 1982.
19) P.N. Hurter and T.A. Hatton, Langmuir 8 (1992) 1291.
20) B.D. Tarr and S.H. Yalkowsky, J. Parenteral Sci. Technol. 41 (1987) 31.
21) J.H. Collett and E.A. Tobin, J. Pharm. Pharmacol. 31 (1979) 174.
22) Y. Saito, Y. Kondo, M. Abe, and T. Sato, Chem. Pharm. Bull. 42 (1994) 1348.
23) S.-Y. Lin and Y. Kawashima, Pharm. Acta Helv. 60 (1985) 345.
24) A.V. Kabanov, E.V. Batrakova, N.S. Melik-Nubarov, N.A. Fedoseev, T.Y. Dorodnich, V.Y. Alakhov, V.P. Chekhonin, I.R. Nazarova, and V.A. Kabanov, J. Controlled Release 22 (1992) 141.
25) A.V. Kabanov, V.P. Chekhonin, V.Y. Alakhov, E.V. Batrakova, A.S. Lebedev, N.S. Melik-Nubarov, S.A. Arzhakov, A.V. Levashov, G.V. Morozov, E.S. Severin, and V.A. Kabanov, FEBS Letts. 258 (1989) 343.
26) R. Nagarajan, M. Barry, and E. Ruckenstein, Langmuir 2 (1986) 210.
27) G.S. Kwon and K. Kataoka, Adv. Drug Del. Rev. 16 (1995) 295.

28) S.B. La, T. Okano, and K. Kataoka, J. Pharm. Sci. 85 (1996) 85.
29) M. Malmsten and B. Lindman, Macromolecules 25 (1992) 5440.
30) A. Rolland, J. O'Mullane, P. Goddard, L. Brookman, and K. Petrak, J. Appl. Polym. Sci. 44 (1992) 1195.
31) M. Yokoyama, Crit. Rev. Ther. Drug Carrier Syst. 9 (1992) 213.
32) K. Park, W.S.W. Shalaby, and H. Park, Biodegradable Hydrogels for Drug Delivery, Technomic Publishing Co, Inc., Lancaster, 1993.
33) K. Kataoka, Pure Appl. Chem. 31 (1994) 1759.
34) M. Yokoyama, M. Miyauchi, N. Yamada, T. Okano, Y. Sakurai, K. Kataoka, and S. Inoue, J. Controlled Release 11 (1990) 269.
35) A. Mazzoni, R.A. Gambetta, F. Trave, and F. Zunino, Cancer Drug. Del. 3 (1986) 163.
36) I.A.D. Bouchier, Gut 29 (1988) 137.
37) U. Wosiewitz, F. Sabinski, C. Haus, S. Güldütuna, and U. Leuschner, J. Hepatol. 14 (1992) 7.
38) R.L. Hunter, C. Jagannath, A. Tinkley, C.A. Behling, and F. Nolte, Antimicrob. Agents Chemother. 39 (1995) 435.
39) S.E. Friberg, I. Kayali, W. Beckerman, L.D. Rhein, and A. Simion, J. Invest. Dermatol. 94 (1990) 377.
40) S.E. Friberg, J. Soc. Cosmet. Chem. 41 (1990) 155.
41) J. Engblom and S. Engström, Int. J. Pharm. 98 (1993) 173.
42) W.-W. Shen, A.G. Danti, and F.N. Bruscato, J. Pharm. Sci. 65 (1976) 1780.
43) M.J. Cappel and J. Kreuter, Int. J. Pharm. 69 (1991) 155.
44) W. Brown, K. Schillén, M. Almgren, S. Hvidt, and P. Bahadur, J. Phys. Chem. 95 (1991) 1850.
45) M. Malmsten and B. Lindman, Macromolecules 26 (1993) 1282.
46) M. Scherlund and M. Malmsten, to be submitted.
47) E.D. Goddard and K.P. Ananthapadmanabhan (eds.), Interactions of Surfactants with Polymers and Proteins, CRC Press, Boca Raton, 1993.
48) R. Aveyard and S.M. Saleem, J. Chem. Soc. Faraday Trans. 1 72 (1976) 1609.
49) L. Piculell and S. Nilsson, Prog. Colloid Polym. Sci. 82 (1990) 198.
50) S. Nilsson, L. Piculell, and M. Malmsten, J. Phys. Chem. 94 (1990) 5149.
51) J.C. Gilbert, J.L. Richardson, M.C. Davies, K.J. Palin, and J. Hadgraft, J. Controlled Release 5 (1987) 113.
52) B. Lindman and H. Wennerström in Topics in Current Chemistry, Vol.
53) S. Miyazaki, S. Takeuchi, C. Yokouchi, and M. Takada, Chem. Pharm. Bull. 32 (1984) 4205.
54) J.C. Gilbert, J. Hadgraft, A. Bye, and L.G. Brookes, Int. J. Pharm. 32 (1986) 223.
55) P.-C. Chen-Chow and S.G. Frank, Int. J. Pharm. 8 (1981) 89.
56) A. Abd El-Bary, N. Foda, and S. Tayel, Pharm. Ind. 53 (1991) 1151.
57) M. Malmsten and B. Lindman, Macromolecules 25 (1992) 5446.
58) A.H. Muhr and J.M. Blanshard, Polymer 23 (1982) 1012.
59) P. Linse and M. Malmsten, Macromolecules 25 (1992) 5434.
60) M. Malmsten, P. Linse, and K.-W. Zhang, Macromolecules 26 (1993) 2905.
61) Y.-W. Yang, N.-J. Deng, G.-E. Yu, Z.-K. Zhou, D. Attwood, and C. Booth, Langmuir 11 (1995) 4703.
62) Y.Z. Luo, C.V. Nicholas, D. Attwood, J.H. Collett, C. Price, and C. Booth, Colloid Polym. Sci. 270 (1992) 1094.

63) Y.-Z. Luo, C.V. Nicholas, D. Attwood, J.H. Collett, C. Price, C. Booth, B. Chu, and Z.-K. Zhou, J. Chem. Soc. Faraday Trans 89 (1993) 539.
64) Z. Yang, S. Pickard, N.-J. Deng, R.J. Barlow, D. Attwood, and C. Booth, Macromolecules 27 (1994) 2371.
65) H. Tomida, N. Kuwada, and S. Kiryu, Acta Pharm. Suec. 25 (1988) 87.
66) V.H.L. Lee, Pharm. Int. 8 (1986) 208.
67) K.A. Fults and T.P. Johnston, J. Parenteral Sci. Technol. 44 (1990) 58.
68) T.P. Johnston, M.A. Punjabi, and C.J. Froelich, Pharm. Res. 9 (1992) 425.
69) A.T. Florence, A.M. Hillery, N. Hussain, and P.U. Jani, J. Controlled Release 36 (1995) 39.
70) S.C. Miller and M.D. Donovan, Int. J. Pharm. 12 (1982) 147.
71) J.R. Robinson and G.M. Mlynek, Adv. Drug Del. Rev. 16 (1995) 45.
72) C.A. Le Bourlais, L. Treupel-Acar, C.T. Rhodes, P.A. Sado, and R. Leverge, Drug. Dev. Ind. Pharm. 21 (1995) 19.
73) F.H. Davidorf, R.B. Chambers, O.W. Kwon, W. Doyle, P. Gresak, and S.G. Frank, Retina 10 (1990) 297.
74) S. Miyzaki, C. Yokouchi, T. Nakamura, N. Hashiguchi, W.-M. Hou, and M. Takada, Chem. Pharm. Bull. 34 (1986) 1801.
75) H. Bechgaard, R.R. Brodie, L.F. Chasseaud, P. Houmoller, J.O. Hunter, P. Siklos, and T. Taylor, Eur. J. Clin. Pharmacol. 21 (1982) 511.
76) G. Alvan, M. Orme, L. Bertelsson, R. Ekstraud, and L. Palmer, Clin. Pharmacol. Ther. 18 (1975) 364.
77) S. Miyazaki, T. Nakamura, C. Yokouchi, M. Takada, Chem. Pharm. Bull. 35 (1987) 1243.
78) P.-L. Wang and T.P. Johnston, Int. J. Pharm. 113 (1995) 73.
79) S. Miyazaki, Y. Ohkawa, M. Takada, and D. Attwood, Chem. Pharm. Bull. 40 (1992) 2224.
80) K.J. Quinn, J.M. Courtney, J.H. Evans, J.D.S. Gaylor, and W.H. Reid, Biomaterials 6 (1985) 369.
81) R.M. Nalbandian, R.L. Henry, and H.S. Wilks, J. Biomed. Mater. Res. 6 (1972) 583.
82) P.W. Paustian, J.C. McPherson, R.R. Haase, R.R. Runner, K.M. Plowman, D.F. Ward, T.H. Nguyen, and J.C. McPherson, Burns 19 (1993) 187.
83) M.E. Carr, P.L. Powers, and M.R. Jones, Thromb. Haemostas. 66 (1991) 565.
84) R.L. Hunter, C. Papadea, C.J. Gallagher, D.C. Finlayson, and I.J. Check, Thromb. Haemostas. 63 (1990) 6.
85) M. Silk and E. Sigman, Cancer, 29 (1972) 171.
86) A.S. Sawhney, C.P. Pathak, and J.A. Hubbell, Macromolecules 26 (1993) 581.
87) Y.H. Bae and S.W. Kim, Adv. Drug Del. Rev. 11 (1993) 109.
88) K.J. Zhu, L. Xiangzhou, and Y. Shilin, J. Appl. Polym. Sci. 39 (1990) 1.
89) L. Rosati and D.J. Casey, Polym. Mater. Sci. Eng. 59 (1988) 516.
90) C.S. Cho, Y.H. Bae, and S.W. Kim, ACS Symp. Ser. 520, American Chemical Society, Washington, DC, 1993.
91) M. Malmsten, P. Linse, and T. Cosgrove, Macromolecules 25 (1992) 2474.
92) M. Malmsten and F. Tiberg, Langmuir 9 (1993) 1098.
93) M. Malmsten and J.M. Van Alstine, J. Colloid Interface Sci. 177 (1996) 502.

94) M. Malmsten, B. Lassen, J.M. Van Alstine, and U.R. Nilsson, J. Colloid Interface Sci. 178 (1996) 123.
95) M. Malmsten, B. Lassen, K. Holmberg, V. Thomas, and G. Quash, J. Colloid Interface Sci. 177 (1996) 70.
96) J.M. Harris (ed.), Poly(Ethylene Glycol Chemistry) - Biotechnical and Biomedical Applications, Plenum, New York, 1992.
97) S.W. Kim and J. Feijen, Crit. Rev. Biocompat. 1 (1986) 229.
98) K.L. Prime and G.M. Whitesides, J. Am. Chem. Soc. 115 (1993) 10714.
99) S.I. Jeon, J.H. Lee, J.D. Andrade, and P.G. de Gennes, J. Colloid Interface Sci. 142 (1991) 149.
100) S.I. Jeon and J.D. Andrade, J. Colloid Interface Sci. 142 (1991) 159.
101) J.M. Van Alstine, N.L. Burns, J.A. Riggs, K. Holmberg, and, J.M. Harris, Colloids Surf. A 77 (1993) 149.
102) S.S. Davis and P. Hansrani, Int. J. Pharm. 23 (1985) 69.
103) M.N. Jones, Adv. Colloid Interface Sci. 54 (1995) 93.
87, Springer Verlag, Berlin, 1980.
104) D. Lasic and F. Martin (eds.) "Stealth Liposomes", CRC Press, Boca Raton, 1995.
105) S. Engström, Lipid Technol. 2 (1990) 42.
106) S. Engström, Polym. Prepr. 31 (1990) 157.
107) R.H. Müller, W. Mehnert, J.-S. Lucks, C. Schwartz, A. zur Mühlen, H. Weyhers, C. Freitas, and D. Rühl, Eur. J. Pharm. Biopharm. 41 (1995) 62.
108) R. Gust, G. Bernhardt, T. Spruss, R. Krauser, M. Koch, H. Schönberger, K.-H. Bauer, S. Schertl, and Z. Lu, Arch. Pharm. 328 (1995) 645.
109) G.G. Liversidge and P. Conzentino, Int. J. Pharm. 125 (1995) 309.
110) H.M. Patel, Crit. Rev. Ther. Drug Carrier Syst. 9 (1992) 39.
111) Y. Tabata and Y. Ikada, Adv. Polym. Sci. 94 (1990) 107.
112) T. Eldem and P. Speiser, Acta Pharm. Technol. 35 (1989) 109.
113) J.H. Senior, Crit. Rev. Ther. Drug Carrier Syst. 3 (1987) 123.
114) A. Chonn, S.C. Semple, and P.R. Cullis, J. Biol. Chem. 267 (1992) 18759.
115) A. Chonn and P.R. Cullis, J. Liposome Res. 2 (1992) 397.
116) G. Blume and G. Cevc, Biochim. Biophys. Acta 1029 (1990) 91.
117) A. Mori, A.L. Klibanov, V.P. Torchilin, and L. Huang, FEBS Letts. 284 (1991) 263.
118) A.L. Klibanov, K. Maruyama, V.P. Torchilin, and L. Huang, FEBS Letts. 268 (1990) 235.
119) T.M. Allen, C. Hansen, F. Martin, C. Redemann, and A. Yau-Young, Biochim. Biophys. Acta 1066 (1991) 29.
120) S.E. Dunn, A. Brindley, S.S. Davis, M.C. Davies, L. Illum, Pharm. Res. 11 (1994) 1016.
121) T. Verrecchia, G. Spenlehauer, D.V. Bazile, A. Murry-Breiler, Y. Archimbaud, and M. Veillard, J. Controlled Release 36 (1995) 49.
122) C. Delgado, G.E. Francis, and D. Fisher, Crit. Rev. Ther. Drug Carrier Syst. 9 (1992) 249.
123) M.C. Woodle, M.S. Newman, and F.J. Martin, Int. J. Pharm. 88 (1992) 327.
124) M. Jamshaid, S.J. Farr, P. Kearney, and I.W. Kellaway, Int. J. Pharm. 48 (1988) 125.
125) M. Malmsten, B. Bergenståhl, L. Nyberg, and G. Odham, J. Am. Oil Chem. Soc. 71 (1994) 1021.

126) B. Kronberg, A. Dahlman, J. Carlfors, J. Karlsson, and P. Artursson, J. Pharm. Sci. 79 (1990) 667.
127) C.J.H. Porter, S.M. Moghimi, M.C. Davies, S.S. Davis, and L. Illum, Int. J. Pharm. 83 (1992) 273.
128) P. Schneider, Transfus. Sci. 13 (1992) 357.
129) G.M. Vercellotti, D.E. Hammerschmidt, P.R. Craddock, and H.S. Jacob, Blood 59 (1982) 1299.
130) T .F. Zuck and J.G. Riess, Crit. Rev. Clin. Lab. Sci. 31 (1994) 295.
131) D.A. Ingram, M.B. Forman, and J.J. Murray, J. Cardiovasc. Pharm. 22 (1993) 456.
132) I.D.A. Johnston (ed.), Current Perspectives in the Use of Lipid Emulsions, Marcel Dekker, New York, 1983.
133) P. Becher (ed.), Encyclopedia of Emulsion Technology: Applications, Vol. 2, Dekker, New York, 1983.
134) R.L. Hunter, J. McNicholl, and A.A. Lal, Aids Res. Human Retroviruses 10 (1994) S95.
135) R. Hunter, M. Olsen, and S. Buynitzky, Vaccine 9 (1991) 250.
136) R. Hunter, F. Strickland, and F. Kézdy, J. Immunol. 12 (1981) 1244.
137) R.L. Hunter and B. Bennett, J. Immunol. 133 (1984) 3167.
138) R.B. Sim (ed.), Activators and Inhibitors of Complement, Kluwer Academic, Dordrecht, The Netherlands, 1993.
139) J.B. de Souza and J.H.L. Playfair, Vaccine 13 (1995) 1316.
140) I. Siebenbrodt and S. Keipert, Eur. J. Pharm. Biopharm. 39 (1993) 25.
141) T.P. Johnston and S.C. Miller, J. Parenteral Sci. Technol. 39 (1985) 83.
142) L.D. Kier, L.M. Wagner, T.V. Wilson, A.P. Li, R.D. Short, and G.L. Kennedy, Drug Chem. Toxicol. 18 (1995) 29.
143) C.D. Port, P.J. Garvin, and C.E. Ganote, Toxicol. Appl. Pharmacol. 44 (1978) 401.
144) G. Magnusson, T. Olsson, and J.-A. Nyberg, Toxicol. Letts. 30 (1986) 203.

Micelles of amphiphilic block copolymers as vehicles for drug delivery

Alexander V. Kabanov[a] and Valery Yu. Alakhov[b]

[a]Department of Pharmaceutical Sciences, College of Pharmacy, University of Nebraska Medical Center, 600 South 42nd Street, Omaha, NE 68198-6025

[b]Supratek Pharma Inc., c/o Institute Armand-Frappier, Immunology Research Center, 531 boul. des Prairies, building 18, Laval, Quebec, H7B 1B7, Canada

This chapter describes the use of block copolymer micelles in drug delivery. Three major systems are considered : (1) micelle-forming conjugates of drugs and block copolymer; (2) drugs non-covalently incorporated into the block copolymer micelles ("micellar microcontainers"), and (3) polyelectrolyte complexes formed between polynucleotides and cationic block copolymers ("block ionomer complexes"). The structure-function relationships in these systems and the data on their application with various drugs are discussed.

1. NEW APPEARANCE OF AN OLD FIELD: "*SMART*" COMPLEXES

A chemist exposed to pharmaceutical sciences should recognize that modern pharmaceutics and, particularly, drug delivery rely heavily on colloidal and polymer chemistry. Indeed, many of the current drug formulations represent colloids, such as suspensions, emulsions, powders, aerosols, etc. In many cases the pharmaceutical dispersions include polymers and surfactants to control wetting, flocculation, stability and bioavailability of the drug dosage form [1]. Current pharmaceuticals are primarily *lyophobic colloids*, i.e. they are thermodynamically unstable and require additional energy to be applied for their dispersion. During the last few years, however, many novel drug delivery systems have appeared. These systems represent *lyophilic colloids*, i.e. they spontaneously form ("*self-assemble*") from macroscopic phases and are thermodynamically stable [2]. The shift from lyophobic to lyophilic colloids is in our opinion a thermodynamic interpretation of a major trend in modern pharmaceutics. This trend is explained by the increasing need for the multifunctional systems capable of fulfilling several tasks, including carrying the drug to the target site in the body, transport of the drug into the cell and subsequently directing the drug into the desired intracellular compartment. The complexity of the function implies more complex structural organization that mimics natural biopolymer assemblies. These multifunctional species can only be designed using self-assembly principles. Further, to enable easier tissue and cell penetration the dimensions of the drug delivery species are shifting towards the range of 10 to 100 nm. Due to the high rate of particle collision and high net surface energy the lyophobic colloids usually do not form in this size range [2]. As a result the lyophilic colloidal systems are gaining increasing attention as vehicles for designing sophisticated drug delivery and drug targeting systems.

The idea of drug targeting, was formulated by Paul Ehrlich about a century ago as a need for a "magic bullet" to improve drug performance in a body [3]. It took a genius to foresee a concept that was far ahead of actual technological development. The systematic attempts to materialize this concept were started in the mid 60's. Since then a better understanding of mechanisms of drug action, disposition in the body and interactions with cells have left us less "magic". The focus now is on the design of appropriate "bullets", which is still a big challenge.

It has been recognized that many useful drugs produce side effects by affecting non-target cells and systems of man [4]. As a result, the drug delivery systems are designed to provide site-specific drug targeting and hinder premature drug release in non-target sites. Drugs often are rapidly eliminated from the body thereby decreasing their efficacy [4]. Therefore it is needed to increase drug circulation time, so that the therapeutic effect is achieved with lower drug doses. Drug metabolism is another common reason for decrease of activity and increase of toxicity [5]. The drug delivery systems are designed to decrease drug metabolism. Many drugs are unable to penetrate histohematic barriers that separate the target cells from the perfusing blood [6], or they are poorly transported into target tissues, such as solid tumors [7]. In these cases the drug delivery systems must help the drug overcome the barriers for their transport in the body.

Most efforts in drug delivery during last few decades has focused on delivery of the drug to the target organ through the body. The problem of drug delivery into a cell and at the subcellular level is currently receiving increasing attention [8, 9]. Target cells often develop mechanisms to reduce cell sensitivity to the administered agents, frequently by hindering drug transport into a cell [10]. Overcoming these drug resistance mechanisms is a serious problem in drug delivery. Furthermore, there is an issue of drug transport within the target cell, which is at least as significant as delivery to the target organ. Current developments in cell biology suggest that intracellular vesicular trafficking involves targeting of vesicles to their correct destination which is controlled by unique compartment-specific proteins [11, 12]. The precise transport of naturally occurring molecules between intracellular organelles is crucial for normal cell functioning and viability. Most pharmaceutical drugs are "alien" to the cell and they lack proper recognition characteristics necessary to direct their transport inside the cell. With many drugs entering cells through vesicular transport (e.g. endocytosis) one common problem is to how increase drug release from the transport vesicles [8]. Further, after leaving the endocytic vesicles the drug must be delivered into the specific intracellular compartments [11]. Indeed to a drug molecule a cell is an enormous place in which it can easily "get lost" and sequestered in the wrong places within the cell instead of going to its intracellular target. Therefore, it is not surprising that many pharmaceutical drugs that are introduced into a cell reveal decreased efficacy, or even adverse effects because they are heading to the wrong places. One example that will be discussed in this chapter is anthracycline antibiotics that accumulate in intracellular vesicles in multiple drug resistant tumor cells, and do not reach their anticipated target, the nucleus [13]. This problem cannot be easily resolved by "bombarding" the cell with increasing drug doses, since drug toxicity will often limit this course of action. New delivery systems must be developed that will deliver drugs inside the cell with high efficacy and precision characteristic of natural trafficking mechanisms.

Overall, the diversity of circumstances under which the drug delivery systems are needed and the complexity of the problems to be resolved are understood much better now [14]. However, the keynote in this field remains the same as was seen by Ehrlich. There commonly is a narrow gap between a drug efficacy and toxicity, called the "therapeutic index" [15]. The major focus remains to widen this gap so that drug formulation become more effective and/or less toxic.

The self-assembling drug delivery systems are being developed that will help to resolve these problems. The major requirements to these systems have been formulated as follows [16]:

(i) The drug delivery particle should self-assemble from drug molecules, components providing for target recognition, and other components necessary for its functioning.

(ii) The size of the drug delivery particles should not exceed the size of viruses to enable its effective penetration in tissues and cells.

(iii) The drug delivery particles should be stable and biologically inert before reaching its target (non-toxic, non-immunogenic, not degradable, avoiding sequestration by the reticulo-endothelial system, not affecting non-target cells, etc.).

(iv) If needed the drug delivery particles should penetrate into cell and provide for drug transport to the desired intracellular compartment.

(v) Drug release should proceed only as a result of interactions of the drug delivery particle with target cells.

(vi) After interaction with a target, the components of drug delivery system should easily be removed from the organism.

Since many of the above enumerated properties are characteristics of viruses, some researches initially termed self-assembling drug delivery systems the "*artificial viruses*" [16, 17]. However, this is often confused with the viral vectors for gene therapy [18]. Furthermore, unlike natural viruses the drug delivery systems are incapable of self-reproduction. Therefore, we now prefer to call these systems "*SMART*", which stands for ***S***elf-assembling ***M***embrane ***A***ctive ***R***egulated ***T***ransporters. The search for the novel technologies in this area continues to be a very active area of research.

2. BLOCK COPOLYMER MICELLES AS DRUG CARRIERS

From a chemistry standpoint one simple way to design self-assembling drug delivery systems is to use micelles. The spontaneous assembly in supramolecular complexes with strictly controlled composition and structure is characteristic of most micelle systems. In such systems aggregates with various packing of surfactant molecules are formed, the parameters of which (dimensions in particular) may be purposefully varied with wide ranges [19]. The mixed micelles can be assembled from several different micelle-forming components, providing for the use of "building blocks" with various structure and functionality. Further, the micelles are able to solubilize nonpolar compounds providing for spontaneous incorporation of a variety of drugs. It is exactly these properties of micelles which make it possible to use them as universal drug carriers. Several micelle-based drug delivery systems have been investigated [20-23]. Among these approaches methods using *block copolymer micelles* as a basic element of a delivery system are experiencing rapid development (Figure 1).

The use of amphiphilic block copolymers in experimental medicine and pharmaceutical sciences has a long history. For example, intensive studies have been performed on gels and emulsions of poly(ethylene oxide)-*b*-poly(propylene oxide)-*b*-poly(ethylene oxide) (Pluronic®) as components of "artificial blood" formulations [24, 25], drug release systems [26], immuno-adjuvant [27-29], anti-tumor [29, 30], and anti-inflammatory agents [31]. The *in vivo* studies demonstrated that Pluronic copolymers posses reduced toxicity compared to the low molecular mass surfactants which permit their administration in man [32]. There were numerous reports suggesting that hydrophilic copolymers such as Pluronic F68 reduce vascular resistance and blood viscosity [33-35]. This work justified the use of hydrophilic Pluronic copolymers as therapeutic adjuvants to increase microvascular transport of physiologically active compounds, which is often reduced in normal and neoplastic tissues [36]. Indeed, an enhancement of vascular and tumor imaging with radiographic contrast agents was observed in the presence of Pluronic F68 [37]. Further, this copolymer was shown to decrease the incidence of pulmonary tumor metastases in rats [38], and used as a vehicle for injectable diazepam to reduce the thrombotic and inflammatory effects in rabbits [39]. Pluronic copolymers were also reported to significantly enhance the availability of various antibacterial and antifungal drugs [40]. The similar results and ideas were later described in the patent literature by different authors [41-44].

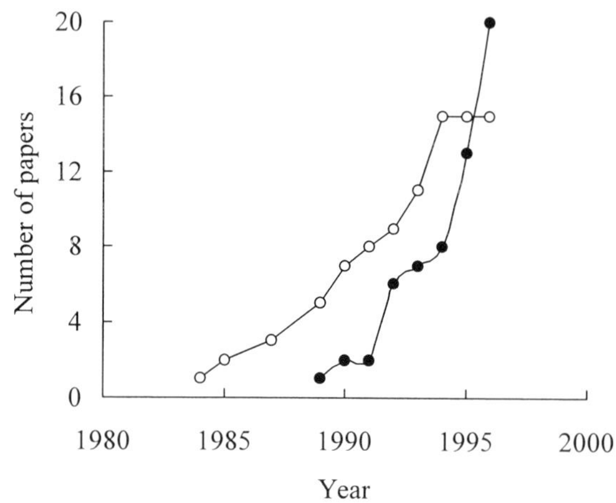

Figure 1 Development of work on (○) block copolymer and drug conjugates and (●) block-copolymer micelles as carriers for non-covalently incorporated drugs.

Various surfactants, including amphiphilic block copolymers have been used to increase solubility and bioavailability of water-insoluble drugs. For example, non-micellar solution of Pluronic F68 was investigated as a vehicle for administration of a novel mitomycin derivative [45]. In mid 80's Lin and Kawashima used Pluronic micelles to solubilize indomethacin [46, 47] and diazepam [48]. The focus of this work did not go beyond increasing solubility, stability and bioavailability of drugs with the use of surfactants. At approximately the same time several authors initiated the use of block copolymer micelles as drug carriers in the sense described in the previous Section [49-51]. At present, three different types of drug delivery systems based on block copolymer micelles have been investigated. These include (i) micelle-forming block copolymer conjugates, (ii) noncovalent complexes of drug and block copolymer micelles ("micellar microcontainers"), and (iii) block ionomer complexes.

The first work on micelle-forming block copolymer conjugates were reported in the early 80's by the group of H. Ringsdorf in Mainz [49, 52-54]. The concept advanced by Ringsdorf was based on the prior work investigating conjugates of drugs and homopolymers [55, 56]. The conjugation of hydrophobic drugs with water soluble polymers resulted in a decreased solubility of the polymer-drug conjugate [54]. This imposed a limit on the amount of drug molecules to be linked to a polymer carrier without precipitating the conjugate. Ringsdorf suggested an elegant solution to this problem by using the block copolymers in which only one segment is conjugated with the drug and another, poly(ethylene oxide), remained unmodified and water soluble. The resulting "missile" drugs were found to form micelles with a core comprised of drug-modified segments and a shell comprised of poly(ethylene oxide). This idea was further used by Yokoyama et al. [51; 57-67] to develop a novel anti-cancer drug representing a conjugate of a poly(ethylene oxide)-*b*-poly(aspartic acid) block copolymer and doxorubicin. These authors were also the first to report the anti-tumor activity of their micelle-forming block copolymer conjugates *in vivo* [51]. In another development Trubetskoy et al. recently proposed the use iodine-modified block copolymers as radiographic contrasting agents [68].

Another approach consisting of non-covalent incorporation of drugs in block copolymer micelles was developed in the late 80's [50]. This work used Pluronic® micelles as drug carriers for site-specific drug delivery. Drug molecules spontaneously incorporate into micelles during solubilization of the drug in Pluronic micelle solutions. The antibody or peptide "vectors" were attached to the surface of the micelles to provide for the targeting of the solubilized drugs. High efficacy of this system in delivering a neuroleptic drug to the brain by crossing the blood brain barrier (BBB) was demonstrated in animals [50]. This was also the first *in vivo* work on the drug delivery with block copolymer micelles. Despite initial concerns that non-covalent complexes between the drug and the micelles might be insufficiently stable in physiological conditions [62, 69], this approach soon attracted significant attention. Numerous *in vitro* and *in vivo* studies have demonstrated the improvement of the transport, safety and efficacy of drugs through their non-covalent incorporation into the block copolymer micelles [16, 50, 70-86]. At present the micelles formed by various amphiphilic block copolymers are being developed as vehicles for delivery of anti-cancer, anti-inflammatory, anti-viral, anti-bacterial and other drugs. Furthermore, enhancing effects of Pluronic micelles on DNA transport into cells have been recently reported [87].

The most recent approach using block copolymers implies that the drug and copolymer are linked through a cooperative system of electrostatic bonds. This approach was initially proposed for the delivery of polynucleotides [88]. In this case the block ionomers comprised of poly(ethylene oxide) and polycation segments are used as DNA carriers. The polycation segments of the block ionomers electrostatically bind to the DNA molecules, which results in the mutual neutralization of the charges of the reacting chains. Despite the charge neutralization the resulting particle (termed "*block ionomer complex*") remains in solution due to the presence of water soluble poly(ethylene oxide) segments. Further, these complexes reveal the ability to form micelles with the hydrophobic core formed by neutralized DNA and polycation segments and poly(ethylene oxide) shell. The block ionomer complexes represent a special class of chemical compounds that reveal combined properties of regular amphiphilic block copolymers and polyelectrolyte complexes. Several groups are actively working in this direction [88, 89].

Overall, the block copolymer based drug delivery systems have been successfully used to: (1) target drugs into cells and modify drug intracellular distribution to increase drug specific activity [86]; (2) modify drug pharmacokinetics and biodistribution by increasing drug longevity and site-specific drug targeting [16, 64, 82]; (3) target drugs to the brain across the BBB [16, 50]; (4) develop new anticancer drug conjugates with lowered side effects [60, 61]; (5) develop anticancer drug compositions effective against multiple drug resistant and metastatic tumors [81, 85, 86]; (6) increase stability of and deliver DNA into cells [87, 88]. These and other examples are discussed below in more detail.

3. CONJUGATES OF DRUGS AND BLOCK COPOLYMERS

3.1. Synthesis and solution behavior of micelle-forming drugs

The general scheme of a micelle-forming drug is based on the use of the AB di-block copolymers containing one segment that exhibits functional groups in the repeating units and one non-reacting water soluble segment, usually poly(ethylene oxide) [54]. A hydrophobic drug is covalently linked via the functional groups to one segment of the copolymer. The drug-modified segment becomes hydrophobic and the copolymer forms micelles in aqueous solutions. The hydrophobic core of such micelles consists of the drug-modified segments. The hydrophilic shell is formed by the unmodified segment. Ringsdorf's group have synthesized the conjugates of poly(ethylene oxide)-*b*-poly(L-lysine) copolymers with the antineoplastic drug - cyclophosphamide [49, 52, 53]. Similar methodology was further applied by Yokoyama et al. to develop a conjugate of poly(ethylene oxide)-*b*-poly(aspartic acid) copolymer with a differed antineoplastic agent - doxorubicin [51, 57-67]. In both cases the drug molecules were linked to the repeating units of the polypeptide segment of the copolymers. The resulting conjugates formed micelles having the diameter in the range of 30 to 100 nm [51]. The stability of the micelles in aqueous solution strongly depended on the ratio of the lengths of the poly(ethylene oxide) and polypeptide segments, and on the degree of substitution of the polypeptide units with the drug [62].

The possibility of introduction of antibody molecules into the block copolymer-drug conjugates has been investigated [58, 59]. In this work doxorubicin was first conjugated to the COOH groups of the polypeptide segment of the poly(ethylene oxide)-*b*-poly(aspartic acid) copolymer. Then the free end NH_2 groups of the polypeptide segment were used to link the antibody molecules. However, it has not been clear, whether the antibody molecules remained active after this modification, or whether they were able to target the conjugates in the body. Indeed, linking of the antibody molecules to the ends of a polypeptide segment, which becomes hydrophobic after modification with the drug should affect the micelle formation and may sterically mask the antibody active centers upon aggregation of the copolymer. It appears that a more rational approach to introduction of targeting molecules is to use water soluble poly(ethylene oxide) segments. In this case, the antibody molecules localize on the surface of the micelles where they are available for binding with the target [16].

Ringsdorf's group has also introduced micelle-forming drug conjugates that are based on water-soluble homopolymers [54, 91]. In this case part of the repeating units of the polymer (e.g., polyimine) are substituted with the hydrophobic drug molecules. As a result of hydrophobic modification of the polymer chain, the conjugates formed micelle-like structures. It was recently demonstrated that these structures can be either multi-molecular or uni-molecular depending on the length of spacer group separating the drug and the polymer [92]. It is known that the modification of soluble polymers with water insoluble substituents often leads to the formation of products in which the modified units are segregated along the polymer chain into segments [93]. Therefore it is possible that some of the drug-polymer conjugates reported also had a block-copolymer structure ABAB..., with the drug-containing segments joining the segments of unmodified repeating units.

3.2 Drug stability and release *in vitro* and *in vivo*

One important consequence of drug conjugation to the block copolymers is drug stabilization. Being incorporated into the core of the micelle the conjugated drug is protected from interactions with various agents that cause its destruction in aqueous solution. Therefore stabilization of the conjugated drug is usually observed *in vitro* [60]. It is also believed that during *in vivo* administration the hydrophilic shell of the block copolymer micelles masks the drug molecules from interactions with plasma proteins, cells and tissues that frequently decrease the amount of the drug in the blood [69]. In this respect the block copolymer micelles may be similar to Stealth® liposomes [94, 95], that are sterically stabilized by poly(ethylene oxide) chains grafted to a liposome surface. In Stealth® liposomes the poly(ethylene oxide) chains prevent opsonization (i.e., identification of liposomes by the immunoglobulines as a foreign entity) and uptake of the liposomes by the mononuclear phagocyte system [94]. As a result the elimination of the liposomes from circulation in blood is inhibited. An increased circulation time in the blood after parenteral administration was observed for the conjugates of doxorubicin with poly(ethylene oxide)-*b*-poly(aspartic acid) copolymer compared to non conjugated drug [64]. Furthermore, the relatively low uptake of these conjugates by the liver and spleen, the representative organs of reticuloendothelial system, suggests that the block copolymer micelles reveal the "stealth" properties characteristic of other particles with poly-(ethylene oxide)-modified surface [64]. In this sense the block copolymer conjugates provide significant advantages over hydrophobic drugs conjugated with water-soluble homopolymers, that are sequestered by organs of the reticuloendothelial system such as the liver [96].

Previous studies demonstrated that low molecular mass drugs, such as doxorubicin, linked to the polymer chains are inactive, since the conjugated drug is masked from interaction with its target by the spacious polymer chain [56]. The restoration of activity of the drug requires the cleavage of the chemical bond between the drug and the polymer and release of the drug in the free form. The problem associated with release of the drug from the micelle-forming conjugates was first investigated by Ringsdorf's group [54]. A set of conjugates with different hydrophobic spacers, that link the drug to the polyimine chain via the S-S bond was studied. The increase in the hydrophobicity of the spacer, that supposedly was accompanied by more effective insertion of the drug into the hydrophobic environment, also resulted in significant deceleration in the release of the drug in the external medium. It appears that, in the case of the micelle-forming drug conjugates, the increase in drug stability is gained at the expense of a decrease in the drug activity. Some attempts to overcome this problem with conventional drug-polymer conjugates have been made [97, 98]. Drug molecules were conjugated to the polymer carriers via the peptide links that are specifically digested by lysosomal enzymes. In these systems the drug molecules were not released in the blood serum but were rapidly cleaved from the carrier by a lysosome extract. Therefore this approach may permit design of long lasting missile drugs that are specifically released in the intracellular compartments during interaction with the target cell.

3.3 Anti-tumor activity of drug and block copolymer conjugates

One major disadvantage of drug-copolymer conjugates is that the covalent attachment of the drug to the block copolymer decreases interaction of the drug moiety with the target molecules and results in a substantial loss of activity. Indeed, the *in vitro* evaluation of the DNA cross-linking activity of the cyclophosphamide conjugates with poly(ethylene oxide)-*b*-poly(L-lysine) copolymers on L1210 leukemia cells revealed that the conjugated drug is less active compared to the free drug [54]. Similar results were obtained during *in vitro* evaluation of the cytotoxicity of doxorubicin conjugated with poly(ethylene oxide)-*b*-poly(aspartic acid) copolymer on P388 leukemia cells [60]. These studies suggested that the effects of the conjugated drugs on cells are significantly decelerated and the active doses are increased which appears to be due to the slow release of the drug from the polymeric carrier.

The *in vivo* performance of the doxorubicin conjugates with poly(ethylene oxide)-*b*-poly(aspartic acid) copolymer have been investigated in significant detail using various tumor models [60, 61]. As seen in the case of *in vitro* experiments, the therapeutically effective doses of the drug *in vivo* were increased after conjugation by 10 to 20 fold. However, the side effects of doxorubicin, that are usually observed during administration of the free drug, were also decreased significantly after the conjugation. In particular, the conjugate caused significantly less loss of animal body weight compared to the free drug, which is indicative of decreased toxicity [60, 61]. The hematotoxicity, was also decreased after conjugation of the drug with the block copolymer [61]. Finally, liver and kidney toxicity of the conjugate were observed at much higher doses compared to the free drug doses [61]. Since toxicity was decreased, the administration of higher doses of doxorubicin conjugate was possible, which resulted in the same or superior therapeutic effects compared to the free drug effects. In other words, parallel to the decrease in the drugs anticancer activity the toxicity was also decreased after conjugation, so that the therapeutic index was improved at least with some tumors.

While the Ringsdorf's conjugation approach [49] was historically the first in using block copolymer micelles as drug carriers it obviously has serious limitations for drug delivery. One of the limitations is the problem with drug release, already discussed. The recent studies on doxorubicin conjugated with poly(ethylene oxide)-*b*-poly(aspartic acid) copolymer suggested that these conjugates contained admixtures of non-conjugated drug [65]. It is possible that the anticancer activity reported with these conjugates was due to the free doxorubicin that was non-covalently incorporated into the block copolymer micelles. Further, compared to the micellar microcontainer approach the conjugation technique lacks universality since new conjugation chemistry has to be developed for each new drug molecule. This is probably one reason why the work on conjugates has leveled off (Figure 1). By contrast, after lag period in early 90's, the work on the micellar microcontainers has been developing very rapidly (Figure 1).

4. MICELLAR MICROCONTAINERS

4.1 Physico-chemical characterization of micellar microcontainers

Block copolymers. Several amphiphilic block copolymers have been used to noncovalently incorporate drug molecules. Some examples include poly(ethylene oxide)-*b*-

poly(propylene oxide)-*b*-poly-(ethylene oxide) tri-block copolymer (also named Pluronic, poloxamer, or Synperonic) [16]

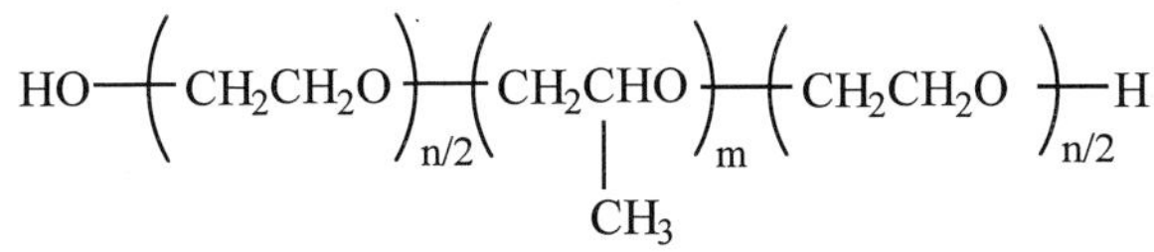

(I)

poly(ethylene oxide)-*b*-poly(isoprene)-*b*-poly(ethylene oxide) tri-block copolymer [70]

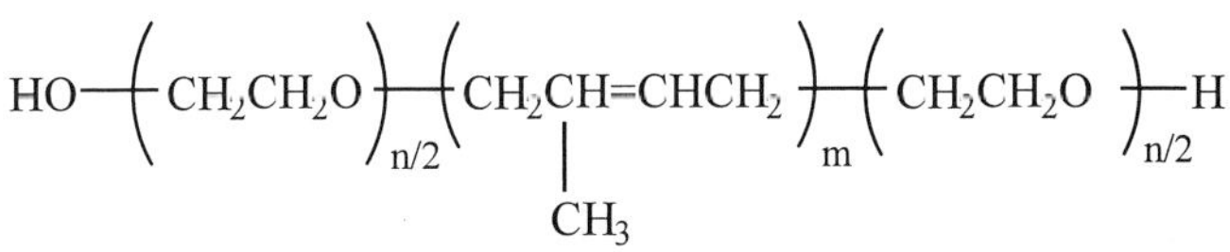

(II)

poly(ethylene oxide)-*b*-poly(styrene) block copolymer [75]

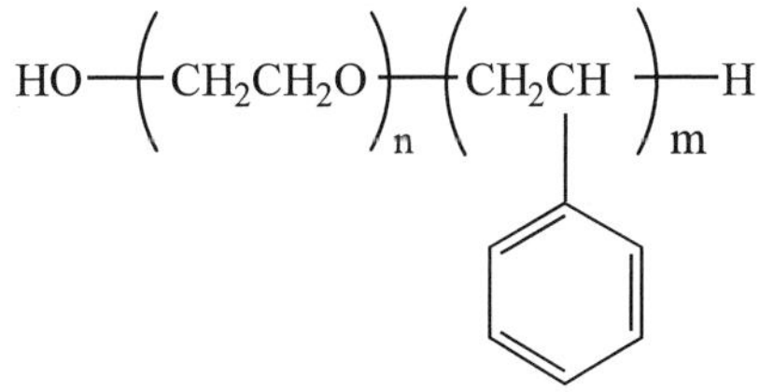

(III)

poly(ethylene oxide)-*b*-poly(D,L-lactide) di-block copolymer [82]

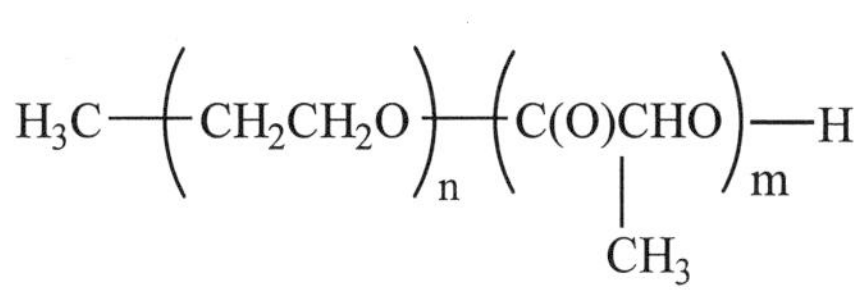

(IV)

and poly(ethylene oxide)-*b*-poly(β-benzyl L-aspartate) di-block copolymer [74]

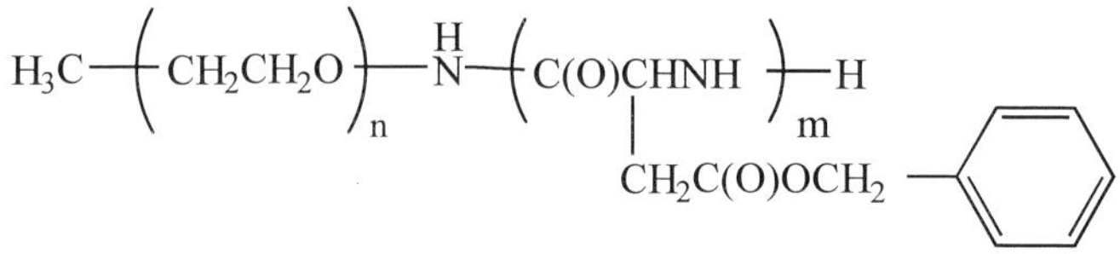

(V)

Copolymers **I-III** are usually synthesized by anionic polymerization with sequential addition of two monomers [99]. Copolymer **IV** is synthesized by the ring opening polymerization of lactide using α-methyl-poly(ethylene oxide) as initiator [82]. Copolymer **V** is synthesized by polymerization of β-benzyl L-aspartate N-carboxy-anhydride using α-methyl-ω-amino-poly(ethylene oxide) as initiator [74]. These reactions provide control over the chain length of the hydrophilic and hydrophobic segments. As a result the copolymers with different molecular masses and hydrophobic-lipophilic balance (HLB) are produced. For example, a variety of Pluronic copolymers are commercially available from BASF Co. (Parisppany, NJ, USA) [100]. The molecular masses of these copolymers vary from about a thousand to over a ten thousand. They range from hydrophobic compounds (HLB from 1 to 7) to hydrophilic compounds (HLB from 24 to 27) depending on the ratio of the lengths of the segments with different lipophilicity.

Micelle dimensions and aggregation numbers. In aqueous solutions the hydrophobic segments of the block copolymers self-assemble into a micelle core with hydrophilic chains providing a water soluble shell. The size of the block copolymer micelles usually ranges from 10 to 100 nm. The surfactant aggregation numbers commonly vary from several dozen to several hundred. Both the micelle size and aggregation number strongly depend on the chain lengths and the ratio of the hydrophobic and hydrophilic segments. Some theoretical considerations of these dependencies for di- and tri-block copolymers can be found in Refs. [101-104] and other chapters of this book. From the drug delivery standpoint it is important that the block copolymer micelles are small enough to penetrate small capillaries in the body tissues [16] and can be accommodated in the endocytic vesicles (100 nm or less) to enter target cells via endocytosis [72].

Critical micelle concentration (CMC). The CMC of the block copolymers is also strongly dependent on the lengths of the segments [105-106]. Usually, an increase in the length of the hydrophobic segment results in a CMC decrease [107]; the theoretical considerations of this dependency can be found elsewhere [108]. In certain cases, e.g., Pluronic copolymers, the CMC increases with an increase in the hydrophilic segment length [105, 106]; this dependency appears to be less significant for the copolymers characterized with larger differences in solubility parameters between the segments [107]. Some block copolymers, such as Pluronic, reveal very strong temperature dependencies of the micelle formation due to differential effects of the temperature on the hydration of the chain blocks. For example, hydrophilic Pluronic copolymers (HLB ≥ 14), form spherical micelles in aqueous solution in the temperature range of about 20°C to about 60°C. The increase in the temperature within this range results in dehydration of the propylene oxide segments leading to a significant decrease in CMC [105]. Further temperature increase results in dehydration of ethylene oxide segments accompanied by sharp structural rearrangements in the copolymer aggregates (sphere to rod transition [109]) and ultimately in the copolymer precipitation. Typically, the copolymers used for drug delivery at physiological temperature (37°C) have a CMC ranging from 0.1 to 10 μM [76]. The CMC is an important parameter from a drug delivery standpoint since it characterizes the *thermodynamic stability of the micelles* during dilution that may occur in biological fluids [76]. The CMC values for amphiphilic block copolymers are significantly lower than those of the low molecular mass surfactants, suggesting that block copolymer-based systems are more stable against dilution compared to regular micelles.

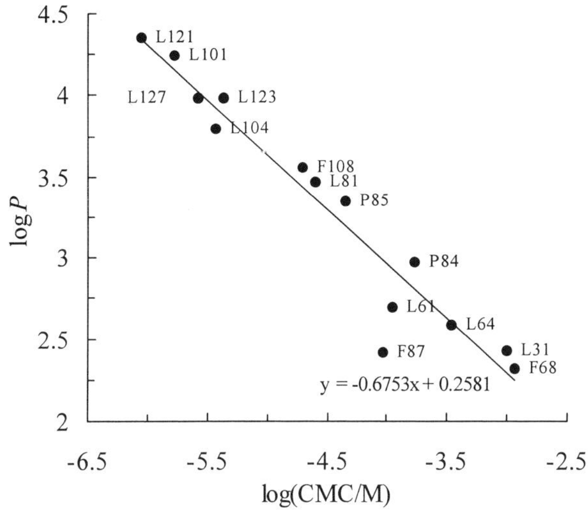

Figure 2 Relationship between partitioning coefficient of pyrene and CMC for various Pluronic block copolymers (paper in preparation).

<u>Partitioning of drugs.</u> Another parameter that is important from a drug delivery standpoint is the partitioning coefficient, P [76, 81]. It is defined as a thermodynamic constant that characterizes the drug distribution between the micelles and aqueous phase under equilibrium conditions (i.e. equivalence of the rates of dissociation and association of the drug-micelle complex):

$$P = [\mathrm{Drug}]_m / [\mathrm{Drug}]_w, \tag{1}$$

where $[\mathrm{Drug}]_m$ and $[\mathrm{Drug}]_w$ are the local concentrations of the drug in the micelle microphase and aqueous phase, respectively. The portion of the drug incorporated into the micelles under conditions of equilibrium (α) is given by the equation [76, 81] :

$$\alpha = \frac{P([\mathrm{Copolymer}] - \mathrm{CMC})}{100v^{-1} + (P-1)([\mathrm{Copolymer}] - \mathrm{CMC})}, \tag{2}$$

where v is the partial specific volume of the copolymer, and [Copolymer] is the copolymer concentration (% w/v). Equation (2) characterizes the *thermodynamic stability of the micelle and drug complex* during dilution. Many drug molecules are fairly hydrophobic and are retained in the micelle hydrophobic core over a broad range of block copolymer concentrations.

Like the CMC the partitioning coefficient in the block copolymer micelles depends on the lengths of the copolymer segments [102, 103]. Using an example of Pluronic copolymers a simple empirical relationship between P and CMC was established [110] :

$$\log P = -a \log \mathrm{CMC} + b\,, \qquad (3)$$

This dependence is illustrated for Pluronic copolymers by Figure 2. It suggests that by varying the molecular parameters within the set of homologous block copolymers one can adjust the efficacy of drug incorporation in the micellar microcontainer.

Relaxation phenomena and kinetic stability. In addition to elevated thermodynamic stability characterized by CMC and P, block copolymer micelles may exhibit high *kinetic stability* due to the long lasting relaxation processes. For example, micelles of poly(ethylene oxide)-*b*-poly(β-benzyl L-aspartate) conjugated with doxorubicin are preserved beneath CMC for several days [63]. Pluronic systems are usually equilibrated much faster, however, there is evidence that certain copolymers exist in micelle form for several hours below the CMC [111]. In most cases the partitioning of the low molecular mass compounds in the aqueous micelle systems rapidly reaches equilibrium [112]. However, there were also reports suggesting that the equilibration may last for several hours. For, example an extremely slow release of the admixture of non-covalently bound doxorubicin from the micelles of doxorubicin and poly(ethylene oxide)-*b*-poly(aspartic acid)conjugate was described [63, 65]. Slow drug release into the aqueous phase was also observed with doxorubicin in poly(ethylene oxide)-*b*-poly(β-benzyl L-aspartate) micelles [77]. It is not clear whether the release rate in these cases is limited by the rate of dissociation of the drug-micelle complex or rate of disintegration of the block copolymer micelles, which is also slow for these systems. The interference of the two processes is possible. The problem of equilibration is therefore of utmost importance for the design of micelle-based drug forms, and understanding the mechanisms of action of these systems both *in vitro* and *in vivo* is crucial for their success.

Some recent work has been focused on increasing the *kinetic stability* of the drug-micelle complexes to maximize the circulation time of the drug-containing vesicles [113]. It is important however to recognize that an increase in circulation time through stabilization of the drug-micelle complexes will not necessarily improve the therapeutic index of the drug form. Indeed, too strong of an attachment of drug to the micelle carrier should result in a decrease in drug release and activity. There must be some optimum ratio between poor drug retention in the microcontainer and the absence of drug release, at which the therapeutic index is maximal. Of course this optimum may vary depending on the route of administration, choice of target and drug properties. The strength of the block copolymer approach is that one can adjust the molecular parameters within one homologous block copolymer set to maximize the performance of the drug for a given drug delivery situation.

4.2 Effects of block copolymers on drug transport and activity in cells

The effects of block copolymers on drug transport in eukaryotic cells have been studied most extensively for the Pluronic systems [71, 72, 81, 86, 87]. These studies suggest the Pluronic copolymers exhibit multiple activities on cells providing for increase in uptake and changes in intracellular localization of drugs. Some important examples are described in this section.

Transport of negatively charged compounds. The uptake of negatively charged drugs into cells is usually hindered by the inability of these drugs to the cross cell plasma membrane [114-116]. The possibility to enhancing the uptake of nucleotides using the mixture of a Pluronic copolymer and cationic surfactant was recently demonstrated [71]. This work evaluated the effects of Pluronic P85 on the phosphorylation of intracellular proteins with $\gamma[^{32}P]$ATP. Since ATP molecules alone do not readily cross cell membranes [117] only the external membrane proteins incorporated radioactive phosphate during the treatment of cells with $\gamma[^{32}P]$ATP [71]. By contrast, in the presence of Pluronic P85 both the external and intracellular proteins were phosphorylated. The radioactive label uptake was significantly enhanced when a small quantity of cationic amphiphile, dodecylamine, was added to Pluronic P85 system [71]. No increase in the plasma membrane permeability was observed under these conditions since Pluronic systems did not release the intracellular ATP from the cells into the external medium [71]. Therefore, Pluronic systems provided for a *directed transport* of ATP from the external medium into the cell. The inhibition of ATP efflux from the cells was also observed with these systems, suggesting effects of the copolymer on the specific membrane transporters of ATP. However, it is unlikely that the enhanced ATP uptake was explained merely by inhibition of ATP efflux systems. First, there was no difference in inhibition of ATP efflux with Pluronic P85 and Pluronic P85-dodecylamine mixture, while the difference of ATP uptake between these systems was very pronounced [71]. Second, inhibition of ATP efflux was observed at a longer time of cell incubation with block copolymer when compared to the observation time for ATP uptake [71]. Since both Pluronic P85 and dodecylamine are able to transport into cells through what appears to be a vesicular transport their possible role in releasing ATP from the membrane vesicles into cells was suggested to explain the ATP uptake phenomenon [71].

The enhancement of transport into cells was recently observed with plasmid DNA formulated with poly(N-ethyl-4-vinylpyridinium) cation and Pluronic P85 [87]. In this case, the uptake of the plasmid into the cell and gene expression were greatly increased in the presence of the copolymer. As in the case of ATP transport, the presence of the cationic molecule was essential since Pluronic P85 alone did not influence DNA transport and transfection [87]. One limiting stage for cell transfection is the release of the plasmid from endosomic vesicles into the cytoplasm [122]. It is possible that the combination of amphiphilic block copolymers with cationic molecules exhibit fusogenic activity in the vesicles, which provides for the release of the negatively charged molecules into the cytoplasm. The identity of this mechanism has yet to be determined in further studies, however, the similarity between the ATP and DNA transport phenomena in the presence of the Pluronic copolymers is noteworthy.

Reversion of multiple drug resistance. One remarkable example of the effects of Pluronic copolymers on the transport and activity of drugs was recently described for MDR tumor cells [81, 86]. MDR is often found in many types of human tumors that have relapsed after initial positive response to chemotherapy [119, 120]. One major mechanism for the appearance of MDR is the overexpression of glycoprotein P (P-gp), a member of the ATP-binding cassette superfamily of transporters [121]. The P-gp is an energy-dependent membrane pump with a broad specificity for chemically unrelated hydrophobic compounds, especially cationic. In MDR tumors P-gp decreases net accumulation of antineoplastic drugs such as anthracycline antibiotics, thus contributing to tumor resistance to chemotherapy. It has been recently demonstrated that formulation of anthracyclines with hydrophobic Pluronic copolymers (e.g. P85, L61) results in a dramatic increase in the cytotoxic effect of these drugs with respect to MDR cells [81]. The resistance reversion indexes[1] observed with certain Pluronic systems were as high as 1000 [81, 86]. By contrast, the Pluronic copolymers only marginally affect the drug activity with respect to the parental sensitive cells [81, 86]. As a result in the presence of copolymer the MDR cells become much more sensitive with respect to the drug compared to the parental cell lines. This phenomenon, named "*hypersensitization*" of MDR cells [81] was observed with a wide variety of P-gp expressing resistant cell lines including human ovarian carcinoma (SKVLB), human breast carcinoma (MCF-7/ADR), human adenocarcinoma (KBv), and others [81, 86, 122, 123]. These studies also suggested that sensitization of resistant cells with Pluronic copolymers is a general phenomenon observed for a broad range of MDR type drugs including daunorubicin, epirubicin, doxorubicin, vinblastine and methotrexate [81].

The effects of Pluronic copolymers on the transport of anthracyclines in resistant and sensitive cells were reported [81, 86]. The influx of these drugs in MDR cells was significantly increased in the presence of the copolymers. By contrast, in the sensitive cells the influx was not altered or was increased less significantly. Furthermore, addition of Pluronic to the external medium of the resistant cells that were preincubated with the drug significantly decelerated the efflux of the drug from the cells [86]. These results suggest, that Pluronic copolymers inhibit the P-gp efflux system in the MDR cells, thus contributing to the increase in the net drug accumulation in these cells [81]. However, the hypersensitization of MDR cells by Pluronic copolymers cannot be explained merely by inhibition of P-gp efflux. Most currently known inhibitors of P-gp render MDR cells sensitive to MDR type drugs, so that the IC_{50} values approach those of sensitive parental cells.

One property of MDR cells is that anthracycline drugs are sequestered in the cytoplasmic vesicles [124-126], which apparently diminishes the amount of the drug transported into the nucleus [13]. Uptake studies demonstrated that Pluronic copolymers enhanced accumulation of anthracyclines in the nucleus of MDR cells [81]. Fluorescent microscopy studies revealed that in the absence of the copolymer doxorubicin accumulated in the cytoplasmic vesicles, with very low levels of fluorescence observed in the nucleus [86]. In the presence of the P-gp inhibitor verapamil, the fluorescence in the nucleus was increased, whereas the vesiculated form of the drug remained unaffected [86]. By contrast, when the cells were incubated with doxorubicin and Pluronic L61 composition, the drug fluorescence was mainly associated with the nucleus, with no fluorescence observed in the cytoplasm [86]. Therefore, by contrast to regular P-gp inhibitors the Pluronic copolymers caused a two-fold effect in the resistant cells. First, they inhibit P-gp, which increases the net accumulation of the drug in the cells. Second, they release the drug sequestered in the vesicles, thus inducing the redistribution of the drug into the nucleus. Sequestration of the positively charged compounds in the acidic cytoplasmic

[1] Resistance reversion index is determined as the ratio of the IC_{50} of the drug alone to the IC_{50} of the drug formulated with the block copolymer [81].

vesicles is attributed to the high pH gradients between the vesicles and cytosol in MDR cells [126]. The Pluronic copolymers were recently shown to eliminate these pH gradients in MDR cells, which possibly explain the effects of these copolymers on cationic drug release from the vesicles (in preparation).

The similarity and differences between the MDR reversion effects of Pluronic copolymers and other nonionic surfactants [127-132], such as Cremophor EL, Solutol HS-15, Triton X-100 and polyoxyethylated fatty acids are noteworthy. These surfactants, like Pluronic copolymers, increase the net accumulation of the MDR-type drugs in the resistant cells apparently by inhibiting the P-gp efflux system. In contrast to Pluronic copolymers, these compounds do not hypersensitize the MDR cells since the maximal resistance reversion indexes reported for them did not exceed 50 (compare with over a 1000 for Pluronic L61 [86]). Therefore the hypersensitizing effect of Pluronic copolymers is unique for the nonionic surfactants. These block copolymers are possibly among the most effective chemosensitizers of MDR cells that are currently known.

The role of self-assembly of Pluronic copolymers in their effects on MDR cells has been investigated. Figure 3 shows the dependency of resistance reversion index for doxorubicin in MCF7/ADR cells on the concentration of Pluronic L61 in extracellular media. The analogy between this curve and curves describing dependency of a property of surfactant solution such as surface tension, is remarkable. The drastic changes in the resistance reversion index are observed below the CMC. This dependency has a break at CMC followed by the leveling off by the copolymer effect. Similar relationships between the CMC and increase in the drug influx in MDR cells were observed for several Pluronic copolymers (data not shown in Figure 3). This data suggests that the reversion of drug resistance and inhibition of P-gp are caused by the Pluronic unimers [81]. The unimer concentration increases up to CMC and then remains constant resulting in the effect saturation.

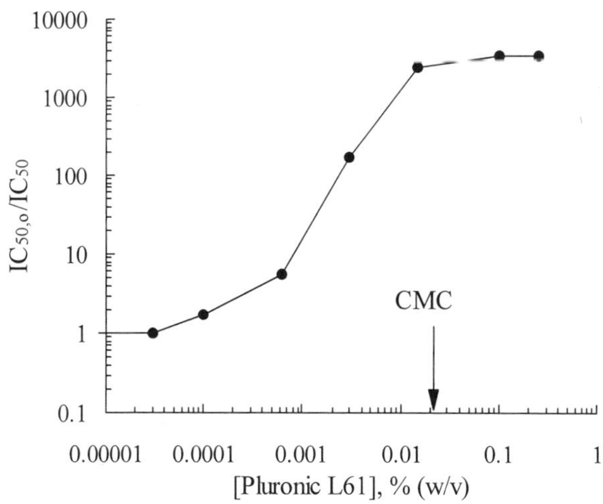

Figure 3 Dependency of resistance reversion index ($IC_{50,o}/IC_{50}$) for doxorubicin in MCF7/ADR cells on the concentration of Pluronic L61. The CMC is shown by the vertical arrow (paper in preparation).

Recent studies also revealed the dependency of the Pluronic effect in MDR cells on the lengths of the ethylene oxide and propylene oxide segments [122, 123]. The chemosensitizing effect of the copolymers and inhibition of the P-gp efflux system in MDR cells increase with (1) an increase in the copolymer net hydrophobicity and (2) an increase in the length of propylene oxide segment. There is a very low activity observed in MDR cells with hydrophilic copolymers such as Pluronic F68 and F108. By contrast, the hydrophobic copolymers, such as Pluronic L61, L81, L101 and L121 are the most active. A detailed study on the relationship between reversion of drug resistance and Pluronic molecular parameters will be published elsewhere.

Multiple mechanisms of block copolymer-mediated transport of drugs. The importance of P-gp in regulating the accumulation of selected drugs in MDR cancer cells has been well established [121]. However, P-gp is also expressed in "normal cells" such as the brain microvessel endothelial cells, intestinal epithelial cells, and hepatocytes [133-135]. The role of P-gp in controlling permeability in these normal cells, is currently receiving more attention [136, 137]. From a drug delivery standpoint, blocking P-gp function in brain microvessel endothelial cells or intestinal epithelial cells could have a significant impact [137-139]. There is a growing body of evidence that P-gp inhibitors, such as cyclosporin A (CSA) can enhance drug permeability across the BBB [138, 139]. Our initial *in vivo* work on Pluronic P85 micelle delivery to the brain targeted the transcytosis transport mechanism [16, 50]. However, the discovery of the P-gp inhibiting activity with the same copolymer [81] suggested that the enhanced delivery of drugs to the brain with Pluronic may occur through either pathway. Using primary cultured bovine brain microvessel endothelial cells (BBMEC) as an *in vitro* model of the BBB, Miller et al. [140] demonstrated that Pluronic copolymers can affect the uptake of drugs in brain microvessel endothelial cells through (1) inhibition of P-gp and (2) vesicular transport.

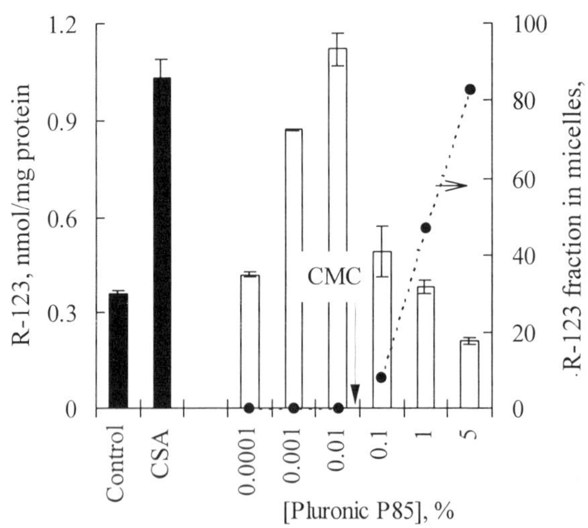

Figure 4. Accumulation of R-123 in BBMEC in presence of various concentrations of Pluronic P85 (white bars). Accumulation of free R-123 ("control") and R-123 in the presence

of CSA (2 μg/ml) are shown by shaded bars. The filled symbols (●) show fraction of R-123 solubilized into the micelles. The CMC is shown by an arrow. From ref. [140].

The relationship between drug accumulation in brain microvessel endothelial cells and Pluronic P85 concentration is illustrated in Figure 4. In this case the transport of a P-gp substrate rhodamine 123 (R-123) was examined. Below the CMC the increase in the uptake of R-123 was observed accounting for the P-gp inhibition by the Pluronic P85 unimers [140, 141]. Above the CMC both the free and micelle-incorporated fractions of the drug were present. While the free R-123 transported across the cell membrane through diffusion, the micelle-incorporated R-123 entered the cells through vesicular transport [140, 141]. The uptake of the R-123 containing micelles has been inhibited by the metabolic inhibitor, that did not affect the P-gp activity but inhibited the fluid-phase endocytosis in BBMEC [140]. The fraction of the micelle-incorporated drug (α) elevated and the fraction of the free drug ($1 - \alpha$) reduced with the increase in the copolymer concentration in accordance with the equation 2. This lead to the lowering of R-123 accumulation through the P-gp related mechanism and increase in R-123 accumulation through vesicular transport. As a result a dramatic changes in the mechanisms and rates of the drug accumulation into the cells occurs when one crosses the CMC [140]. This result suggests possibility of increasing drug transport in BBB by either inhibiting efflux or directing vesicular transport with Pluronic copolymers [140].

Receptor-mediated drug delivery with block copolymer micelles. The receptor-mediated transport of drugs into cells using Pluronic micelles as drug carriers has been reported [72]. These studies revealed that the Pluronic micelles, containing a fluorescent dye are taken up into a cell via an endocytosis mechanism. The uptake was dramatically increased when a superantigen molecule, capable of binding with a cell receptor, was covalently linked to a micelle surface. The uptake was competitively inhibited by the free superantigen suggesting that it was mediated by the specific interaction of the micelle-incorporated toxin with the cell receptor.

In vitro studies suggest that block copolymers exhibit multiple activities with respect to drug delivery in cells including (1) interaction with drug efflux mechanisms, (2) transporting drugs through vesicular mechanism, particularly, receptor-mediated endocytosis, and (3) releasing drugs from intracellular vesicles. These activities are strongly dependent on the molecular parameters of the copolymers and their self-assembly. Optimization of the micelle system for a given drug (or drug class) and delivery situation can be achieved by varying the HLB, molecular mass and concentration of the copolymer. Investigation of the block copolymer interactions in cells is essential for understanding the mechanisms of their effects on drug transport and activity, as well as for evaluating their potential use in drug delivery into cells and at the subcellular level.

4.3 *In vivo* evaluation of drugs in block copolymer micelles

Pharmacokinetics and biodistribution of the block copolymer systems. There are few reports regarding the pharmacokinetics and tissue distribution of block copolymer systems *in vivo* after parenteral administration. The tissue distribution of fluorescein solubilized in Pluronic micelles was studied in mice [16]. It has been demonstrated that the pattern of tissue distribution is highly dependent on the ratio of the lengths of the ethylene oxide and propylene oxide segments. Similar conclusions were recently made for the micelles of poly(ethylene oxide)-*b*-poly(D,L-lactide) copolymer containing indium-111 oxine [82]. It was found that the blood clearance and tissue distribution of indium-111 oxine after parenteral administration of the block copolymer systems in rats were dependent on the molecular masses of the ethylene oxide and lactide segments [82]. Further, tissue distribution with these systems were strongly dependent on copolymer concentration. Particularly, a significantly longer life time of the indium-111 oxine in the blood was observed with a high concentration of the copolymer present, when compared to that of a low copolymer concentration. It was suggested that in the case of the lower concentration of the copolymer a rapid dilution of the micelles occurred in the body fluids reverting the system to the copolymer unimers [82]. By contrast, in the case of the higher concentration of the copolymer, the micelles were preserved in the blood for the longer period of time, and were probably carrying the drug during circulation.

It is expected that the block copolymers can exhibit multiple activities *in vivo* affecting drug bioavailability. Particularly, interactions of the copolymer systems with the drug transport systems in various tissues is very possible. The evaluation of these effects is the subject of future studies. The data reported already [16, 82] suggest that the block copolymer micelles represent a versatile drug delivery system in which the bioavailability of drugs can be modified by adjusting the molecular parameters and the concentration of the block copolymers. It is interesting to note that the pharmacokinetic and tissue distribution properties of the block copolymers strongly depend on their molecular parameters [142]. Table 1 presents the results of the tissue distribution studies on of $[H^3]$-labeled Pluronic copolymers. It shows that the tissue distribution coefficients, $P_{org/blood}$, increase with a Pluronic order F68 < F108 < P85 < L61, suggesting that the hydrophobic copolymers and copolymers with longer propylene oxide segment are retained in the organs longer.

The unusual pharmacokinetics observed with the Pluronic copolymers after parenteral administration are noteworthy [142]. These curves revealed recirculation of the copolymers in the blood and organs observed during dozens of hours which is a unique phenomenon for polymer pharmaceuticals.

Table 1
Pharmacokinetic and tissue distribution parameters of Pluronic copolymers after i.v. administration in mice.[a] From Ref. [142].

Tissue	Pluronic F68		Pluronic F108		Pluronic P85		Pluronic L61	
	$AUC\|_0^{200}$	$P_{org/blood}$	$AUC\|_0^{200}$	$P_{org/blood}$	$AUC\|_0^{200}$	$P_{org/blood}$	$AUC\|_0^{200}$	$P_{org/blood}$
Blood	0.111	-	0.026	-	2.180	-	0.044	-
Liver	0.225	2.03	0.079	3.03	7.365	3.38	0.215	4.93
Spleen	0.066	0.59	0.032	1.23	4.240	1.94	0.118	2.68
Lungs					1.787	0.82		
Kidneys					1.748	0.80		
Heart					1.277	0.58		

[a] $AUC|_0^{200}$ is the "area under the curve" from 0 to 200 hours. $P_{org/blood}$ is the tissue distribution coefficient defined as the ratio of the $AUC|_0^{200}$ in organ to $AUC|_0^{200}$ in blood.

Use of Pluronic copolymers to enhance brain delivery. The potential use of Pluronic block copolymers for targeted brain delivery of drugs has been examined [16]. In these studies, the distribution of a florescent dye incorporated in Pluronic P85 micelles was examined in mice. The distribution of the dye was highest in the lung with a minimal amount appearing within the brain [16]. However, following conjugation to either insulin or antibodies to the brain-specific antigen, alpha-2 glycoprotein, the distribution pattern of the dye was shifted away from the lungs with significantly greater amounts of fluorescence accumulating in brain tissue. Studies examining the delivery of the neuroleptic agent haloperidol alone, and in the presence of Pluronic P85 further support the ability of Pluronic block copolymers to increase the delivery of drugs to the brain [16, 50]. In these studies, the neurological effects of haloperidol were increased in mice treated with Pluronic 85. Further enhancement of neurological activity of haloperidol was observed (greater than 100-fold) in mice treated with Pluronic 85 conjugated to either insulin or the alpha-2 glycoprotein antibody [50]. Based on the studies with haloperidol and fluorescently labeled Pluronic 85, it appeared that Pluronic could enhance drug delivery to the brain.

One mechanism proposed for the effects of vectorized Pluronic P85 micelles on the delivery of drugs to the brain is through an endocytosis pathway within the brain microvessel endothelial cells [140]. While endocytic activity is reduced in brain endothelial cells compared to peripheral endothelial cells, such a mechanism may also account for the effects of Pluronic copolymers on drug delivery to the brain. The transport of macromolecules through vesicular (endocytic) routes in brain microvessel endothelial cells has been postulated previously as a mechanism for the transport of peptides and proteins across the BBB [143, 144]. Studies in brain microvessel endothelial cells indicate that adsorptive and receptor-mediated endocytosis allows for an increased transport of peptides and proteins across the BBB [145, 146]. The reason for this appears to be due in part to an increase in the transcytosis, i.e. endocytosis with subsequent exocytosis to the opposite side of the cell [145, 146]. It is interesting to note that studies by Miller et al. [147] support insulin transport through the brain microvessel endothelial cells through a transcytosis mechanism. These studies showed that a significant portion of the internalized insulin was effluxed to the basolateral side compared to that observed with the fluid-phase endocytosis marker, which recirculated to the apical side [147].

Anti-tumor drugs in Pluronic micelles. Anti-tumor activity of anthracycline antibiotics formulated with Pluronic micelles has been evaluated [85, 122, 123]. This work compared the anti-tumor activity of epirubicin and doxorubicin solubilized in the micelles of Pluronic L61, P85 and F108 using murine leukemia and myeloma grown subcutaneously. The study revealed that the lifespan of the animals and inhibition of tumor growth were considerably increased in mice treated with drug/copolymer compositions compared with animals treated with free drugs. The anti-tumor activity of the drug/copolymer compositions depended on the hydrophobicity of the block copolymer [85, 122, 123]. The data suggested that higher activity was associated with more hydrophobic copolymers. In particular, a significant increase in the lifespan was observed with doxorubicin/Pluronic L61 composition [85]. The effective dose of this composition caused inhibition of the tumor growth and complete disappearance of tumor in 50 % of animals. The effective doses of doxorubicin in Pluronic compositions were the same or lower than those for the free drugs [85]. This is an important difference compared to the doxorubicin and block copolymer conjugates described in Section 3.3, for which an increase in the effective doses was observed. It is noteworthy that the anti-tumor activity of doxorubicin/Pluronic L61 composition depended on the concentration of the copolymer. A decrease in activity was observed when the Pluronic L61 concentration was decreased [85]. This effect was explained [85] by the lowering of the fraction of the micelle-incorporated drug, and reversion of the system to the copolymer unimers as a result of dilution. More concentrated systems revealed higher stability against dilution in the body fluids which presumably resulted in longer circulation of the micelle-incorporated drug [85].

The study on the pharmacokinetics of doxorubicin incorporated into Pluronic L61 micelles suggested that the clearance of the drug in the copolymer composition was significantly decreased compared to the clearance of the free doxorubicin [123]. Further, the metabolism of micelle-incorporated doxorubicin was also decreased compared to the free drug metabolism [123]. This is a very important advantage over the free doxorubicin, since accumulation of its C13-metabolite, doxorubicinol, causes cardiotoxicity, a major side effect during chemotherapy with doxorubicin. Indeed, recent studies suggested that the cardiotoxicity with doxorubicin in Pluronic L61 composition is decreased [123]. Furthermore, this composition also revealed a decreased hematotoxicity, which is another serious complication during chemotherapy [123]. The increased stability and decreased clearance of the drug are obviously contributing to the improvement of the therapeutic index observed with the doxorubicin/Pluronic L61 composition.

Since the Pluronic copolymers revealed hypersensitizing effects with respect to MDR tumors *in vitro* the activity of the doxorubicin in Pluronic L61 composition against human breast carcinoma MCF7/ADR has been evaluated using a nude mice model [122, 123]. This study revealed that the micelle-incorporated doxorubicin exhibited high activity against this MDR cell line while free doxorubicin was only moderately active. A similar result was obtained with the metastatic Lewis lung carcinoma 3LL-H59 [123]. The free doxorubicin exhibited low activity in treating these metastasis. By contrast, a dramatic decrease in the number and size of the metastatic sites was observed with the doxorubicin in Pluronic L61 composition. A detailed report of these studies will be published elsewhere.

In conclusion, the *in vivo* studies on micellar microcontainers has demonstrated high efficacy of this approach for drug delivery to the brain and cancer tumors. The systems based on amphiphilic copolymers will provide versatile tools for purposeful modification of the pharmacokinetic and biodistribution properties of drugs, as well as for increasing drug stability, decreasing drug side effects and site-specific targeting of drugs to select tissues.

5. BLOCK IONOMER COMPLEXES FOR DELIVERY OF DNA

Interpolyelectrolyte complexes formed between nucleic acids and polycations have been used for the delivery of polynucleotides [148]. A general disadvantage of these systems is their low solubility and disproportioning accompanied by precipitation of the neutralized (stoichiometric) complexes [148]. To overcome this problem, a cationic poly(ethylene oxide)-*b*-polyspermine diblok copolymer has been developed [88, 149]:

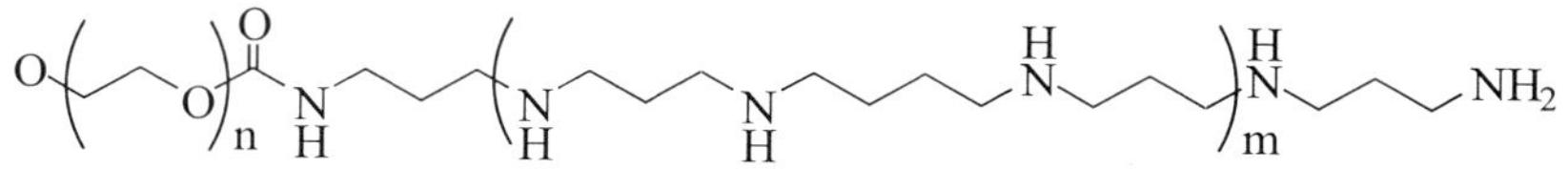

(VI)

This block copolymer spontaneously form complexes with DNA in aqueous solutions [88]. The protonated spermine segment binds to the DNA phosphate groups, which results in charge neutralization. The ethylene oxide segment provides for the aqueous solubility of the complex. The resulting complexes are similar to amphiphilic block copolymers. They form micelle-like aggregates in aqueous solution, which contain a hydrophobic core comprised of neutralized polyions, and a hydrophilic shell comprised of hydrated PEO chains [88]. Since the microphase in these complexes is formed by the polyion chains they have been named "*block ionomer complexes*" [150].

The detailed study of the solution behavior of the block ionomer complexes has begun recently by Kataoka et al. [151], as well as Eisenberg, Kabanov and Kabanov groups [150]. The initial work was performed on synthetic block polyelectrolyte systems : (1) poly(ethylene oxide)-*b*-poly(L-lysine) and poly(ethylene oxide)-*b*-poly(α,β-aspartic acid) block copolymers [151]; and (2) poly(ethylene oxide)-*b*-poly(methacrylic acid) block copolymer and poly(N-ethyl-4-vinylpyridinium bromide) homopolymer [150]. The data obtained suggest that these systems form water-soluble stoichiometric complexes, a major difference compared to stoichiometric complexes of homopolymers that precipitate [150, 151]. Further, block ionomer complexes are stable in a much broader pH range compared to the interpolyelectrolyte complexes prepared from corresponding homopolymer polyions [150]. This suggests cooperative stabilization of the polyion complexes by the ethylene oxide segments, with the stabilization effect of up to 10 kJ$\cdot$M^{-1} [150]. The block ionomer complexes self-assemble to form micelle-like aggregates as determined by light scattering [151] and fluorescent probe technique [150]. The stability of these aggregates in aqueous solutions depends on the length of the ethylene oxide segments. The structure of the aggregates depends on the lengths of the interacting polyion segments (Figure 5) :

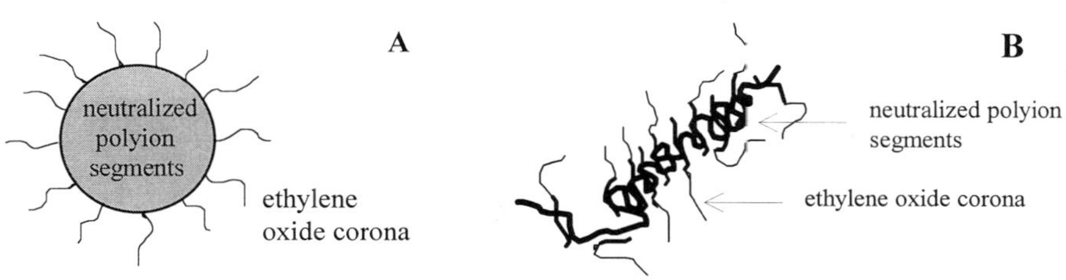

Figure 5. Schematic representation of the block ionomer complexes formed between poly(ethylene oxide)-*b*-poly(sodium methacrylate) and poly(N-ethyl-4-vinylpyridinium bromide): (A) The contour length of the N-ethyl-4-vinylpyridinium bromide chain is shorter or equals that of the sodium methacrylate segment; (B) the N-ethyl-4-vinylpyridinium bromide chain is longer than the sodium methacrylate segment. Based on data reported in Ref. [150].

The block ionomer complexes are salt-sensitive since they fall apart as the salt concentration increases beyond a critical value [150]. Furthermore, they can participate in the cooperative polyion substitution reactions [150]. Therefore the block ionomer complexes represent a special class of hybrid materials that combine properties of amphiphilic block copolymers and cooperative interpolyelectrolyte complexes.

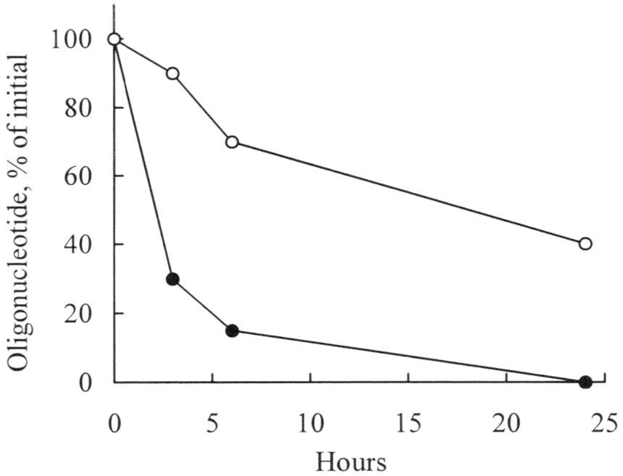

Figure 6. Degradation of free 19-mer oligonucleotide (●) and of its complex with poly(ethylene oxide)-*b*-polyspermine block copolymer (○) in mice blood serum at 37°C.

The polynucleotide chain incorporated into block ionomer complexes becomes more stable against enzyme digestion [152]. As a result the degradation of polynucleotides in the biological fluids is decelerated (Figure 6).

The first activity study reported for block ionomer complexes of DNA investigated the effect of poly(ethylene oxide)-*b*-polyspermine copolymer on the antisense oligonucleotide directed against splice junction sites of IE mRNAs 4 and 5 of Herpes simplex virus, type 1 [88]. A significant increase in the sequence-specific inhibitory effect of this oligonucleotide on virus reproduction was observed when the oligonucleotide was administered in the form of the block ionomer complex. This effect was attributed to the increased stability and transport of the oligonucleotide into the cells [153].

The block ionomer complexes of oligonucleotides were recently used to explore the function of a synaptic vesicle-related protein, amphiphysin in primary neurons [154]. These studies revealed that the complex between poly(ethylene oxide)-*b*-polyspermine and 20-mer phosphodiester oligonucleotide directed against amphiphysin effectively inhibited the expression of amphiphysin and neurite outgrowth in the cultured primary hypothalamic neurons [154]. The oligonucleotide effect was sequence-specific and was observed at the concentration of oligonucleotide as low as 250 nM, which appears to be the highest activity reported to-date for the unmodified phosphodiester in cells [154].

Finally, the studies on the *in vitro* transfection activity of the block ionomer complexes of plasmid DNA were recently reported suggesting that these systems significantly surpass other transfection systems such as lipofectin [153].

The investigation of block ionomer complexes for gene and oligonucleotide delivery is actively being pursued, with several groups involved in these studies worldwide. Some, new cationic block copolymers that are currently being examined with polynucleotides include poly(ethylene oxide)-*b*-poly(L-lysine) [89, 90] :

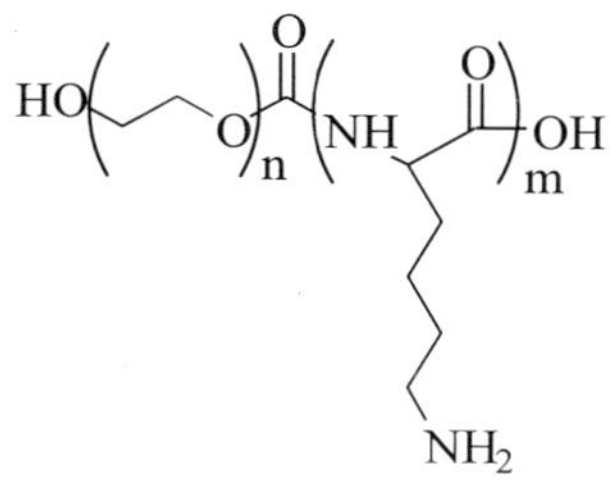

(VII)

and poly(ethylene oxide)-*b*-poly(1,3-butanediol-1,4-diaminobutylphosphonate) [152] :

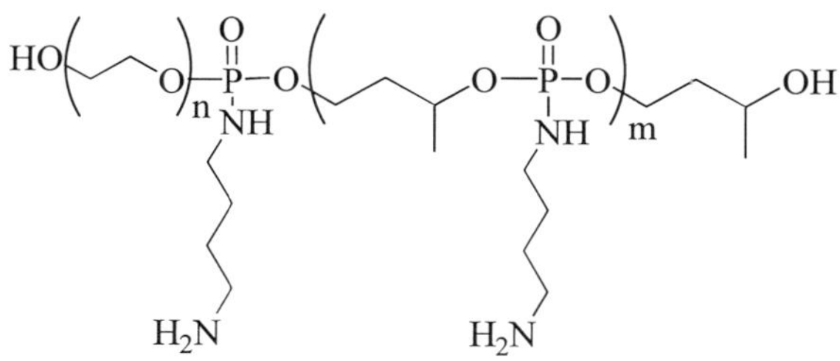

(IX)

and poly(ethylene oxide)-*b*-poly(L-lysine/L-glycine) [152] :

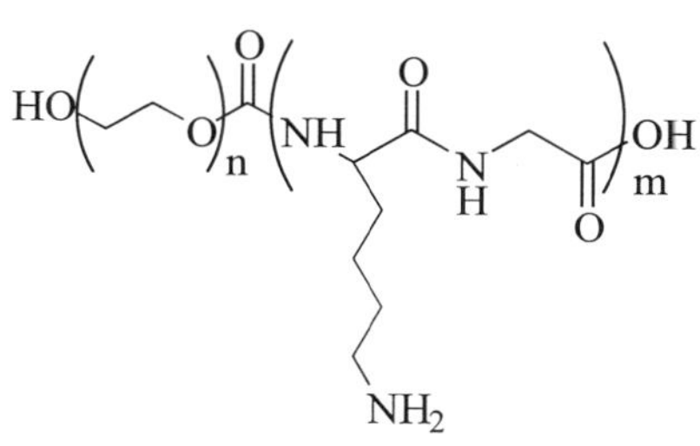

(VIII)

It is expected that the use block ionomer complexes of polynucleotides will provide for new and novel pharmaceutical forms for gene therapy. The solubility and thermodynamic stability appears to be a decisive advantage of these systems when compared to most other non-viral vectors for gene delivery that are water insoluble.

5. CONCLUSION

The decade of studies on *block copolymer micelles* as drug carriers has revealed the great potential of this approach and its importance for targeted drug delivery. It is expected that studies in this area will intensify in the near future, and new and important results will occur. Some more recent developments include the use of poly(ethylene oxide)-b-poly(aspartic acid) block copolymer for incorporation of cisplatin [155]. These systems self-assemble into micelle-like aggregates as a result of formation of electrostatic bonds between the cisplatin and repeating units of the aspartic acid segment of the block copolymer. The resulting block copolymer form of cisplatin exhibited activity *in vitro* against B16 melanoma. This data suggests that block polyelectrolytes can be used successfully to incorporate low molecular mass drugs through electrostatic interactions [156]. In another development the conjugates of Pluronic copolymer and low molecular ligands were developed that alter morphology through specific interactions with their soluble receptor [156]. The transitions induced by the receptors in these systems were competitively inhibited by the free ligand molecules. This novel class of materials was named the "*biospecific surfactants*" [156]. The biospecific surfactants formed rod-like aggregates ($\cong$ 50 x 250 nm) in the presence of their receptors. The aggregates were carrying drug molecules but were unable to release them into cells. By contrast, in the presence of free ligands the aggregates produced spherical micelles (25 nm), which delivered drug molecules into cells. Aggregated biospecific surfactants is a prototype for a *SMART* delivery system which is inert with respect to non-target cells. Upon meeting cells secreting the specific "signal" molecules (free ligands), the large aggregates produce micellar microcontainers which interact with the target cells. A similar approach was described recently for toxins masked by antibodies ("respecrins") that selectively activate during interaction with the target cell antigens and are inactive with respect to the non-target cells [8].

Acknowledgments : We would like to thank Dr. Donald Miller (UNMC), with whom we have collaborated for several years on mechanisms of the block copolymer transport in cells, for valuable comments and support. Paul Haney (UNMC) read this manuscript and made a number of valuable suggestion for which we are very grateful.

REFERENCES

1. A. Martin, P. Bustamante and A.H.C. Chun, Physical Pharmacy (4th edition), Lea & Febiger, Philadelphia, London, 1993.
2. R.J. Hunter, Introduction to Modern Colloid Science, Oxford University Press, Oxford, New York, 1993.
3. V. Raso, Semin. Cancer Biol., 1 (1990) 227.
4. A. Goodman Gilman, L.S. Goodman, T.W. Rall and F. Murad (eds). The Pharmacological Basis of Therapeutics, Macmillan Publishing Company, New York, Toronto, London, 1985.
5. R.D. Olson and P.S. Mushlin, FASEB J., 4 (1990) 3078.

6. E.A. Neuwelt (ed.) Implications of the Blood-brain Barrier and its Manipulation, Vols. I, II. New York, London, Plenum Medical Book Company, 1989.
7. R.E. Durand, Cancer Chemoth. Pharmacol., 26 (1990) 198.
8. A.V. Kabanov and V.Yu. Alakhov, J. Contr. Release 2 (1994) 15.
9. S.F. Hamm-Alvarez, Pharm. Res. 13 (1996) 489.
10. C.F. Higgins and M.M. Gottesman, TIBS, 17 (1992) 18.
11. T. Sollner and J.E. Rothman, Trends Neurosci., 17 (1994) 344.
12. N.A. Bradbury and R.J. Bridjes, Am. J. Physiol., 267 (1994) C1.
13. G.J. Shuurhuis, H. Broxterman, and A. Cervantes, J. Natl. Cancer. Inst., 81 (1989) 1887.
14. J.R. Robinson and V.H.L. Lee (eds.), Controlled Drug Delivery, Marcel Dekker, New York, Basel, 1987.
15. M. Gibaldi and D. Perrier, Pharmacokinetics, Marcel Dekker, New York, Basel, 1982.
16. A.V. Kabanov, E.V. Batrakova, N.S. Melik-Nubarov, N.A. Fedoseev, T.Yu. Dorodnich, V.Yu. Alakhov, V.P. Chekhonin, I.R. Nazarova and V.A. Kabanov, J. Contr. Release, 22 (1992) 141.
17. K. Kataoka, In Proc. 2nd Taniguchi Conference on Polymer Research, Y. Imanishi (ed.), Kagalen-Dojin Press, Kyoto, 1991.
18. C.P. Hodgson, Bio/Technology, 13 (1995) 222.
19. K.L. Mittal and B. Lindman (eds.) Surfactants in solution, vol. 1-3, Plenum Press, New York.
20. D.D. Lasic, Nature (London), 355 (1992) 279.
21. T.T. Kararli and V.W. Gupta, J. Pharm. Sci., 81 (1992) 483.
22. J. Swabrick and J.R. Siverly, Pharm. Res. 9 (1992) 1546.
23. K. Yamamoto, A.C. Shach, T. Nishihata, J. Pharm. Pharmacol. 46 (1994) 608.
24. C. Chubb and P. Draper, Proc. Soc. Exp. Biol. Med., 184 (1987) 489.
25. Y. Nogata, C.E. Mondon and A.D. Cooper, Metabolism, 39 (1990) 682.
26. K. Morikawa, F. Okada, M. Hosokawa and H. Kobayashi, Cancer Res., 47 (1987) 37.
27. K.C. Lowe and F.H. Armstrong, Adv. Exp. Med. Biol., 277 (1990) 267.
28. A.C. Allison and N.E. Byars, Mol. Immunol., 28 (1991) 279.
29. I.N. Topchieva, V.N. Erokhin, S.V. Osipova, M.M. Khrutskaya, T.A. Kupriyanova and S.N. Bukovskaya, Biomed. Sci., 2 (1991) 38.
30. D.J. Chaplin, M.R. Horsman and D.S. Aoki, Br. J. Cancer, 63 (1991) 109.
31. P.C. Chen-Chow and S.G. Frank, Int. J. Pharm., 8 (1981) 89.
32. T. Kodama, K. Arima, H. Sakamoto, M. Suga and A. Yachi, Gan To Kagaku Ryoho, 15 (1988) 2544.
33. H.H. Weetal and J. Bozichevich, Nature (London) 215 (1967) 1479.
34. A.R. Moore, B.C. Paton and B. Eiseman, J. Surg. Res., 8 (1968) 563.
35. F.L. Grover, R.S. Kahn, M.W. Heron and B.C. Paton, Arch. Surg., 106 (1973) 307.
36. L.E. Gerlowski and R.K. Jain, Microvas. Res., 31 (1986) 288.
37. D.M. Cassel, S.W. Young, W.R. Brody, H.H. Muller and A.L. Hall, J. Comput. Assist. Tomogr., 6 (1982) 141.
38. M. Silk and E. Sigman, Cancer, 25 (1972) 171.
39. A.V. Prancan, B. Ecanow, R.J. Bernardoni and M.S. Sadove, J. Pharm. Sci., 60 (1980) 970.
40. W. Saski and S.G. Shah, J. Pharm. Sci., 54 (1965) 277.
41. R.L. Hunter, A. Duncan, US Patent 4,873,083 (1989).

42. R.L. Hunter, US Patent 5,030,448 (1991).
43. R.L. Hunter, PCT WO 88/06038 (1988).
44. R.L. Hunter, PCT WO 95/15182 (1995).
45. W.T. Brander, W.C. Rose, J.E. Schuring, A.P. Florczyk, J.B. Huftalen and J.J. Catino, Cancer Res., 45 (1985) 6475.
46. S.Y. Lin and Y. Kawashima, Pharm. Acta. Helv., 60 (1985) 339.
47. S.Y. Lin and Y. Kawashima, Pharm. Acta. Helv., 60 (1985) 345.
48. S.Y. Lin and Y. Kawashima, J. Parenter. Sci. Technol., 41 (1987) 83.
49. H. Bader, H. Ringsdorf and B. Shmidt, Die Angewandte Makromolekulare Chemie, 123/124 (1984) 457.
50. A.V. Kabanov, V.P. Chekhonin, V.Yu. Alakhov, E.V. Batrakova, A.S. Lebedev, N.S. Melik-Nubarov, S.A. Arzhakov, A.V. Levashov, G.V. Morozov, E.S. Severin and V.A. Kabanov, FEBS Lett., 258 (1989) 343.
51. M. Yokoyama, M. Miyauchi, N. Yamada, T. Okano, Y. Sakurai, K. Kataoka and S. Inoue, J. Contr. Rel., 11 (1990) 269.
52. G. Hoerpel, Dissertation, Mainz, West Germany, 1983.
53. M.K. Pratten, J.B. Lioyd, G. Horpel and H. Ringsdorf, Makromol. Chem., 186 (1985) 725.
54. K. Dorn, G. Hoerpel and H. Ringsdorf, In Bioactive Polymeric Systems. An Overview., C.G. Gebelein and C.E. Carraher, Jr. (eds.), p. 533. Plenum Press, New York, London, 1985.
55. H. Maeda, L.W. Seymour and Y. Myamoto, Bioconjugate Chem. 3 (1992) 351.
56. R. Duncan and F. Spreafico, Clin. Pharmacokinet. 27 (1994) 290.
57. M. Yokoyama, S. Inoue, K. Kataoka, N. Yui and Y. Sakurai, Makromol. Chem., Rapid Commun., 8 (1987) 431.
58. M. Yokoyama, S. Inoue, K. Kataoka, N. Yui, T. Okano and Y. Sakurai, Makromol. Chem., 190 (1989) 2041.
59. M. Yokoyama, T. Okano, Y. Sakurai, K. Kataoka and S. Inoue, Biochem. Biophys. Res. Commun., 51 (1989) 1234.
60. M. Yokoyama, M. Miyauchi, N. Yamada, T. Okano, Y. Sakurai, K. Kataoka and S. Inoue, Cancer Res., 50 (1990) 1693.
61. M. Yokoyama, M. Miyauchi, N. Yamada, T. Okano, Y. Sakurai, K. Kataoka and S. Inoue, Cancer Res., 51 (1991) 3229.
62. M. Yokoyama, G.S. Kwon, T. Okano, Y. Sakurai, T. Seto and K. Kataoka, Bioconjugate Chem., 3 (1992) 295.
63. M. Yokoyama, T. Sugiyama, T. Okano, Y. Sakurai, M. Naito and K. Kataoka, Pharm. Res., 10 (1993) 895.
64. G.S. Kwon, M. Yokoyama, T. Okano, Y. Sakurai and K. Kataoka, Pharm. Res., 10 (1993) 970.
65. G.S. Kwon, S. Suwa, M. Yokoyama, T. Okano, Y. Sakurai and K. Kataoka, J. Contr. Release, 29 (1994) 17.
66. M. Yokoyama, G.S, Kwon, T. Okano, Y. Sakurai, and K. Kataoka, J. Contr. Rel., 28 (1994) 59.
67. M. Yokoyama, T. Okano, Y. Sakurai, T. Seto and K. Kataoka, J. Contr. Rel., 32 (1994) 269.

68. V.S. Trubetskoy, V.P. Torchilin, G.S. Gazelle and G.L. Wolf, Proc. Intl. Symp. Contr. Rel. Bioact. Mater., 21 (1994) 676.
69. M. Yokoyama, Crit. Rev. Ther. Drug Carrier Syst., 9 (1992) 213.
70. P.L. Morgan, S.E. Harding and K. Petrak, Biochem. Soc. Trans. 18 (1990) 1021.
71. Slepnev, L.E. Kuznetsova, A.N. Gubin, E.V. Batrakova, V.Yu. Alakhov and A.V. Kabanov, Biochem. Internat., 26 (1992) 587.
72. A.V. Kabanov, V.I. Slepnev, L.E. Kuznetsova, E.V. Batrakova, V.Yu. Alakhov, N.S. Melik-Nubarov, P.G. Sveshnikov and V.A. Kabanov, Biochem. Internat., 26 (1992) 1035.
73. A. Rolland, J. O'Mullane, P. Goddard, L. Brookman and K. Petrak. J. Appl. Polym. Sci. 44 (1992).
74. G.S. Kwon, M. Naito, M. Yokoyama, T. Okano, Y. Sakurai, and K. Kataoka, Langmuir 12 (1993) 945.
75. S.E. Dunn, A. Brindley, S.S. Davis,M.C. Davies and L. Illum, Pharm. Res. 11 (1994) 1016.
76. A.V. Kabanov, I.R. Nazarova, I.V. Astafieva, E.V. Batrakova, V.Yu. Alakhov, A.A. Yaroslavov and V.A. Kabanov, Macromolecules, 28 (1995) 2303.
77. G.S. Kwon, M. Naito, M. Yokoyama, T. Okano, Y. Sakurai, and K. Kataoka, Pharm. Res., 12 (1995) 192.
78. A. Rolland, F. Paul, K.M. Scott and C.J. Olliff, Pharm. Res., 12 (1995) 1435.
79. M. Daraper, M.Savage, J.H. Collett, D. Atwood, C. Price, C. Booth and Q.-G. Wang, Pharm. Res., 12 (1995) 1231.
80. Piskin, X. Kaitian, E.B. Denkbas and Z. Kucukyavuz, J. Biomater. Sci. Polym. Ed. 7 (1995) 359.
81. V.Yu. Alakhov, E.Yu. Moskaleva, E.V., Batrakova and A.V. Kabanov, Bioconjugate Chem., 7 (1996) 209.
82. S.A. Hagan, A.G.A. Coombes, M.C. Garnett, S.E. Dunn, M.C. Davies, L. Illum, S.S. Davis, S.E. Harding, S. Purkiss and P.R. Gellert, Langmuir 12 (1996) 2153.
83. S.B. La, T. Okano, K. Kataoka, J. Pharm. Sci. 85 (1996) 85.
84. X. Zhang, J.K. Jackson and H.M. Burt, Int. J. Pharm. 132 (1996) 195.
85. E.V. Batrakova, T.Yu. Dorodnych, E.Yu. Klinskii, E.N. Kliushnenkova, O.B. Shemchukova, S.A. Arjakov, V.Yu. Alakhov and A.V. Kabanov, Br. J. Cancer, in press (1996).
86. A. Venne, S. Li, R. Mandeville, A.V. Kabanov and V.Yu. Alakhov, Cancer Research, 56 (1996) 3626.
87. I. Astafieva, I. Maksimova, E. Lukanidin, V. Alakhov and A. Kabanov, FEBS Lett. 389 (1996) 278.
88. A.V. Kabanov, S.V. Vinogradov, Yu.G. Suzdaltseva and V.Yu. Alakhov, Bioconjugate Chem. 6 (1995) 639.
89. M.A. Wolfert, E.H. Schaht, V. Tonceva, K. Ulbrich, O. Nazarova and L.W. Seymour, Human Gene Ther. 7 (1996) 2123.
90. S. Katayose and K. Kataoka, In Advanced Biomaterials in Biomedical Engineering and Drug Delivery Systems, N. Ogata, S.W. Kim, J. Feijen and T. Okano (eds.), p. 319, Springer, Tokyo, 1996.
91. W. Klesse, Dissertation, Mainz, West Germany, 1982.
92. M. Nukui, K. Hoes, H. van den Berg and J. Feijen, Makromol. Chem., 192 (1991) 2925.
93. S.G. Starodubtzev, Yu.E. Kirsh and V.A. Kabanov, Eur. Polym. J., 10 (1987) 739.

94. T.M. Allen and C. Hansen, Biochim. Biophys. Acta, 1068 (1991) 133.
95. D. Needham, T.J. McIntosh and D.D. Lasic, Biochim. Biophys. Acta, 1108 (1992) 40.
96. L.E. Ambler, L. Brookman, J. Brown, P. Goddard and K. Petrak. J. Bioact. Compat. Polym., 7 (1992) 223.
97. H. Ringsdorf, B. Schmidt and K. Ulbrich, Makromol. Chem., 187 (1986) 257.
98. J. Kopecek and R. Duncan, J. Contr. Release, 6 (1987) 315.
99. I.R. Schmolka, J. Am. Oil Chem. Soc., 54 (1977) 110.
100. BASF Performance Chemicals, BASF Co., Parisppany, NJ, 1991.
101. R. Nagarajan and K. Ganesh, Macromolecules, 22 (1989) 4312.
102. P.N. Hurter, J.M.H.M. Scheutjens and T.A.Hatton, Macromolecules, 26 (1993) 5592.
103. P.N. Hurter, J.M.H.M. Scheutjens and T.A.Hatton, Macromolecules, 26 (1993) 5030.
104. P. Linse, Macromolecules, 26 (1993) 4437.
105. P. Alexandridis, V. Athanassiou, S. Fukuda and T.A. Hatton, Langmuir, 10 (1994) 2604.
106. P. Alexandridis, J.F. Holzwarth and T.A. Hatton, Macromolecules, 10 (1994) 2604.
107. I. Astafieva, X.F. Zhong and A. Eisenberg, Macromolecules, 26 (1993) 7339.
108. Z. Gao and A. Eisenberg, Macromolecules, 26 (1993) 7353.
109. K. Schlillen, W. Brown and R.M. Johnson, Macromolecules, 27 (1994) 4825.
110. E. Batrakova and A. Kabanov, in preparation.
111. M. Malmsten and B. Lindman, Macromolecules, 25 (1992) 5440.
112. M. Wilhelm, C.-L. Zhao, Y. Wang, R. Xu, M. Winik, J.-L. Mura, G. Riesse, M.D. Croucher, Macromolecules, 28 (1991) 2303.
113. G.S. Kwon and K. Kataoka, Adv. Drug. Delivery Rev., 16 (1995) 295.
114. D.W. Horne and K.A. Reed, Arch. Biochem. Biophys. 298 (1992) 121.
115. J.F. Poschet, S.M. Hammond and P.D. Fairclough, Biochem. Biophys. Acta. 1278 (1996) 223.
116. J.F. Cascalheira and A.M. Sebastiao, Eur. J. Pharmacol. 222 (1992) 49.
117. N. Brutovetsky, A. Backer, M. Klingenberg and E. Bamberg, Proc. Natl. Acad. Sci. USA, 93 (1996) 664.
118. O. Boussif, F. Lezoualc'h, M.A. Zanta, M.D. Mergny, D. Scherman, B. Demeneix and J.-P. Behr, Proc. Natl. Acad. Sci. USA, 92 (1995) 7297.
119. L.J. Goldstein, M.M. Gottesman and I. Pastan, Cancer Treat. Res., 57 (1991) 101.
120. L.J. Goldstein, M.M. Gottesman and I. Pastan, Crit. Rev. Oncol. Hematol., 12 (1992) 243.
121. P. Lepage and P. Gross, Curr. Opin. Nephrol. Hypertens., 2 (1993) 735.
122. V.Yu. Alakhov, E.V. Batrakova, T. Dorodnich, S. Li, A. Venne and A.V. Kabanov, In Proceedings of 1st International Symposium on Polymer Therapeutics, p.72, London, England, 1996.
123. V. Alakhov, E. Batrakova, S. Li, A. Venne and A. Kabanov, Pharm. Res. 13 (1996) S. 214.
124. G.J. Shuurhuis, H.J. Broxterman, J.H.M. de Lange, H.M. Pinedo, T.H.M. van Heijningen, C.M. Kuiper, G.L. Scheffer, R.J. Scheper, C.K. van Kalken, J.P.A. Baak and J. Lankelma, Br. J. Cancer, 64 (1991) 857.
125. M.A. Sognier, Y. Zhang, R.L. Eberle, K.M. Sweet, G.A. Altenberg and J.A. Belli, Biochem Pharmacol. 48 (1994) 391.
126. M. Schindler, G. Sharon, E. Hoff and S.M. Simon, Biochemistry, 35 (1996) 2811.

127. D.M. Woodcock, S. Jefferson, M.E. Linsenmeyer, P.J. Crowther, G.M. Chonjowski, B. Williams and I. Bertoncello, Cancer Res., 50 (1990) 4199.
128. J.S. Coon, W. Knudson, K. Clodfelter, B. Lu and R.S Weinstein, Cancer Res., 51 (1991) 897.
129. D.M. Woodcock, M.E. Linsenmeyer, G. Chojnowski, A.B. Krieger, V. Nink, L.K. Webster and W.H. Sawyer, Br. J. Cancer, 66 (1992) 62.
130. T. Zordan-Nudo, V. Ling, Z. Liu and E. Georges, Cancer Res., 53 (1993) 5994.
131. P.K. Dudeja, K.M. Anderson, J.S. Harris, L. Buckingham and J.S. Coon, Arch. Biochem. Biophys., 319 (1995) 309.
132. L.E. Buckingham, M. Balasubramanian, A.R. Safa, H. Shah, P. Komarov, R.M. Emanuelle and J.S. Coon, Int. J. Cancer, 65 (1996) 74.
133. F. Thiebault, T. Tsuruo, H. Hamada, M.M. Gottesman, I. Pastan and M.C. Willingham, Proc. Natl. Acad. Sci. USA, 84 (1987) 7735.
134. I. Sugawara, I. Kataoka, Y. Morishita, H. Hamada, T. Tsuruo, S. Itoyama and S. Mori, Can. Res., 48 (1988) 1926.
135. O. Fardel, D. Ratanasavanh, P. Loyer, B. Ketterer and A. Guillonzo, Eur. J. Biochem., 205 (1991) 847.
136. A.H. Schinkel, J.J.M. Smit, O. van Tellingen, J.H. Beijnen, E. Wagenaar, L. van Deemter, C.A.A.M. Mol, M.A. van der Valk, E.C. Robanus-Maandag, H.P.J. te Riele, A.J.M. Berns and P. Borst, Cell 77 (1994) 491.
137. M.M. Nerurkar, P.S. Burton and R.T. Borchardt, Pharm. Res., 13 (1996) 528.
138. Q. Wang, H. Yang, D.W. Miller and W.F. Elmquist, Biochem. Biophys. Res. Commun., 211 (1995) 719.
139. M. Fontane, W.F. Elmquist, and D.W. Miller, Life. Sci., 59 (1996) 1521.
140. D.W. Miller, E.V. Batrakova, T.O. Waltner, V.Yu. Alakhov and A.V. Kabanov, Pharm. Res., submitted.
141. E. Batrakova, M. Fontane, V. Alakhov, D. Miller and A. Kabanov, Pharm. Res., 13 (1996) S-214.
142. A.V. Kabanov and V.Yu. Alakhov, PCT WO 96/16541 (1996).
143. K.L. Audus, P.J. Chikhale, D.W. Miller, S.E. Thompson and R.T. Borchardt, Adv. Drug Res., 23 (1992) 3.
144. U. Bickel, T. Yoshikawa and W.M. Pardridge, W.M. Adv. Drug Del. Rev., 10 (1993) 205.
145. K.R. Smith and R.T. Borchardt, Pharm. Res. 6 (1989) 466.
146. U. Bickel, W.M.Y. Lee, J.Q. Trojanowski and W.M. Pardridge, W.M. Bioconjugate Chem., 5 (1994) 119.
147. D.W. Miller, B.T. Keller and R.T. Borchardt, J. Cell. Physiol. 161 (1994) 333.
148. A.V. Kabanov, and V.A. Kabanov, Bioconjugate Chem., 6 (1995) 7.
149. A.V. Kabanov, S.V. Vinogradov and V.Yu. Alakhov, In 7th International Symposium on Recent Advances in Drug Delivery Systems, Salt Lake City, p. 189, 1995.
150. A.V. Kabanov, T.K. Bronich, V.A. Kabanov, K. Yu and A. Eisenberg, Macromolecules, 29 (1996) 6797.
151. A. Harada and K. Kataoka, Macromolecules, 28 (1995) 5294.
152. A. Kabanov, S. Vinogradov, Yu. Suzdaltseva and V. Alakhov, Pharm. Res., 13 (1996) S-214.

153. A.V. Kabanov, S.V. Vinogradov, Yu. G. Suzdaltseva, E.Yu. Klinsky and V.Yu. Alakhov, In Proceedings of 1st International Symposium on Polymer Therapeutics, p.30, London, England, 1996.
154. O. Mundigl, G.-C. Ochoa, C. David, A.V. Kabanov and P. De Camilli, In 36th American Society for Cell Biology Annual Meeting, San Francisco, 1996.
155. M. Yokoyama, In 7th International Symposium on Recent Advances in Drug Delivery Systems, Salt Lake City, p. 99, 1995.
156. A.V. Kabanov, I.N. Nazarova, N.S. Melik-Nubarov, H.W. Meyer and A.A. Yaroslavov, *Langmuir*, in preparation.

Applications of Amphiphilic Copolymers in Separations.

Mårten Svensson[a], Hans-Olof Johansson[b] and Folke Tjerneld[b]

[a]Department of Physical Chemistry 1,
Center for Chemistry and Chemical Engineering,
Lund University, P. O. Box 124,
S-221 00 Lund, Sweden.

[b]Department of Biochemistry,
Center for Chemistry and Chemical Engineering,
Lund University, P. O. Box 124,
S-221 00 Lund, Sweden.

1. INTRODUCTION

The use of amphiphilic copolymers in separation systems is attracting increasing interest. One reason for this is the recent advances in the physical-chemical characterisation of the solution behaviour of amphiphilic copolymers. The knowledge of how these polymers behave opens the way to applications. For block copolymers of the type PEO-PPO-PEO (PEO = polyethylene oxide, PPO = polypropylene oxide) the interest has been focused mainly on the extraction of small hydrophobic solutes. The micelle formation exhibited by these polymers offers the possibility to solubilize hydrophobic compounds in the hydrophobic interior of the micelles. This means that amphiphilic copolymers in aqueous solutions are potential alternatives to organic solvents for extractions of organic molecules. Water solutions of copolymers are environmentally friendly solvents, and we will describe research on extractions using amphiphilic copolymer systems.

In biotechnology there is a strong interest in use of "smart polymers" in separation systems. The need is here for polymers which can react on external influence, such as temperature or pH change. With such polymers it is possible from the outside to affect the properties of a separation system. The interest has been directed towards amphiphilic copolymers and poly-ampholytes. As we show in this chapter the amphiphilic copolymers show drastic changes in solubility properties, such as self-association and phase separation, at e.g. temperature increase. With poly-ampholytes the solubility can be affected by changing the solution pH. These characteristic properties of copolymers in solution can be utilised for external control of separations.

An important separation technique in biotechnology is aqueous two-phase extraction. The two-phase systems are based on the phase separation in water solution of two polymers with different structure, and these systems are used for extraction of sensitive biomolecules or particles, e.g. proteins or membranes. In this chapter we show studies where amphiphilic block copolymers have been used as phase forming polymers in two-phase systems. The possibility is hereby created for micelle formation in one of the phases. The micelle formation depends on the temperature and the partitioning of substances between the phases can thus be regulated by e.g. the temperature.

The random (statistical) copolymers of EO (ethylene oxide) and PO (propylene oxide) units have been used in aqueous two-phase separations. The main interest in this application has been to utilise temperatures above the copolymer cloud point for separation of the copolymer from the purified biomolecule and to utilise these effects for copolymer recycling. The phase formation by temperature increase offers the possibility to create separation systems with only one polymer in water. We describe studies of these novel systems, where the partitioning of molecules between a water phase and a polymer phase formed above the copolymer cloud point is utilised for separation. Similar principles are applied in systems with thermoprecipitating copolymers, e.g. vinyl-acrylamide based copolymers. For these polymers the temperature or pH is used to trigger the precipitation. If the target biomolecule can be bound to the copolymer e.g. by affinity interaction it is then possible to precipitate the biomolecule-copolymer complex by temperature or pH change, which leads to effective separation of biomolecule from contaminants.

The rapid expansion of modern biotechnology creates new demands on effective separations. The amphiphilic copolymer systems are very attractive because of the multitude of possibilities for design of polymers with capability for capture and release of the target biomolecule and which also can be reused in the process. Furthermore these polymers are water-soluble, mild and not denaturing towards biomolecules. They can be derivatised e.g. with charged groups for introduction of pH sensitivity or with affinity ligands for the creation of specific binding of biomolecule to the copolymer. As we show in this chapter the amphiphilic copolymers have a range of properties which can be utilised in polymer-based separation systems.

2. SOLUBILIZATION IN AQUEOUS BLOCK COPOLYMER SOLUTIONS

2.1. Extraction of hydrophobic solutes

The use of micelle-forming block copolymers as solubilizing agents of hydrophobic compounds in aqueous solutions has great potential. A solute that is sparingly soluble in water (e. g. organic pollutants in waste water) can by the amphiphilic properties of a block copolymer be solubilized and removed/ extracted. This has been investigated in several works (see Table 1 below). The solubilization capacity of hydrophobic compounds in aqueous block copolymer solutions depends on several factors, the physical nature of the solubilized solute and the different blocks of the polymer, block copolymer concentration as well as the molecular weight. Also the composition ratio of the blocks and the temperature influences the uptake of hydrophobic substances in block copolymer solutions.

Some studies on solubilization of hydrophobic (mostly fairly small organic molecules) substances in amphiphilic copolymer solutions are listed in Table 1. The large extent of the studies demonstrates the application potential for block copolymers in this area. Review articles on solubilization in amphiphilic copolymer solutions, comprising both theoretical treatments and experimental results, have also been published [1, 2]. We have in this work refrained from giving a description of different model approaches in this area.

Table 1.
List of some solubilization studies in amphiphilic copolymer solutions

Solute	Polymer	Reference
o-Xylene	Pluronic L64	Tontisakis et al. [3]
Naphthalene	Pluronic P103, P104, P105	McFann et al. [4]
Naphthalene, Phenanthrene, pyrene	Pluronic P103, P014, P105, P108, P123 and Tetronics T1304, T1504, T1107, T1307	Hurter and Hatton [5]
2-Undecanone	Poly(sodium styrenesulfonate-co-2-vinylnaphthalene	Nowakowska et al. [6]

Benzene, toluene, o-xylene, ethylbenzene, n-hexane, n-heptane, n-octane, n-decane, cyclohexane	PEO-PPO (70:30) Mol. Weight: 12500, and poly(N-vinylpyrrolidone-styrene) (40:60)	Nagarajan et al. [7]
Pyrene, 1,6-diphenyl-1,3,5-hexatriene (DPH)	Pluronic F68, P85, F108	Kabanov et al. [8]
Indomethacin	Pluronic F68, P88, F108	Lin and Kawashima [9]
Octadecylrhodamine B	Pluronic F68, L64	Nakashima et al. [10]
Diazepam	Pluronic L64	Pandya et al. [11]
Benzene	Polystyrene-block-poly(methacrylic acid)	Kiserow et al. [12]
Hexane, heptane, octane, nonane, cyclohexane, cycloheptane, benzene, toluene, ethyl benzene, p-xylene, chlorobenzene, butyl chloride, chlorohexane, chloroform, CCl_4, propyl acetate, butyl acetate	Polystyrene-block-poly(methacrylic acid)	Tian et al. [13]
Toluene, naphthalene, phenanthrene	Poly(N-vinylpyrrolidone-styrene)	Haulbrook et al. [14]
Toluene	Poly(tert-butylene-b-styrenesulfonate)	Valint and Bock [15]
n-Hexane	Pluronic F68	Al-Saden et al. [16]
Styrene	Poly(sodium styrenesulfonate-co-2-vinylnaphthalene)	Nowakowska and Guillet [17]
Naphtalene	Pluronic 88, 87, 123 Tetronic 908,1307	Calvert et al. [18]

Toluene, benzene, chlorobenzene, p-xylene	pluronic 61, 62, 64, 65, 84, 103, 104, 123, R172, R174, R252	Gadelle et al. [19]

Solubilization experiments with aliphatic and aromatic solutes in Pluronic (PEO-PPO-PEO) and poly(N-vinylpyrrolidone-styrene) solutions (concentrations above cmc) have revealed unusual selectivity for aromatic solutes [7]. In this work (by Nagarajan et al.) it was shown that while aliphatic hydrocarbons were almost negligibly solubilized, the aromatic analogues were solubilized to a large extent. For example, the difference in solubilized amounts between benzene and n-hexane was found to be 20-50-fold. It was also shown that the block copolymers selectively solubilized aromatic hydrocarbons in aromatic/aliphatic-mixtures. This difference can partly be attributed to the fact that the aromatic and aliphatic hydrocarbons are good and poor solvents, respectively, for the non polar block in the polymers examined. The solubilization of polycyclic aromatic substances in aqueous solution of Poly(N-vinylpyrrolidone-styrene), a random copolymer with high molecular weight (3.4 million g/mol), was investigated by Haulbrook et al [14]. The solubilized amount increased linearly with polymer concentration for all three solutes (toluene, naphthalene, phenanthrene). Log K_{PW}, the concentration-based partition coefficient between polymer and water, was highest for phenanthrene (4.69) and lowest for toluene (3.39). A relationship between K_{PW} and K_{OW}, octanol-water partition coefficient (a standard measure of the hydrophobicity of the solute), was theoretically derived for this polymer system. This relationship may be used to predict solubilization capacity for other amphiphilic copolymer-solute systems, more accurately if the solutes are within the same structural class. A similar work was performed by Hurter and Hatton [5]. Three polycyclic aromatic hydrocarbons were partitioned between water and PEO-PPO-PEO micelles of varying structure and composition. A membrane-based separation process was also proposed in this work. A comparison between the solubility behaviour for the three polycyclic aromatic hydrocarbons (naphthalene, phenanthrene, pyrene) at different Pluronic P103 ($(EO)_{17}(PO)_{60}(EO)_{17}$) concentrations is shown in Figure 1a. It is clearly seen that pyrene (with four fused benzene rings) reach highest solubility enhancement. The solubilization capacity decreased with decreasing number of benzene rings (see Figure 1a). A plot of K_{mw}, partition coefficient between micelle (P103) and water, versus K_{ow} (see above) showed good linear relationship (see Figure 1b).

The solubilization of benzene in diblock copolymer, polystyrene-poly(methacrylic acid), was performed by Kiserow et al. [12]. The solubilization was found to be influenced by pH, due to the polyelectrolytic nature of the methacrylic acid shell, as well as copolymer concentration. The methacrylic acid shell is known to undergo hyper coiling at low pH. In Figure 2 the solubility capacity, weight ratio of benzene/polystyrene, is plotted against pH, and it is obvious that solubilization in this system is very dependent of pH. At low pH the copolymer/benzene solution is not stable and the copolymer precipitates, while at higher pH (>~5) the copolymer can solubilize benzene approx. 5 times its own weight.

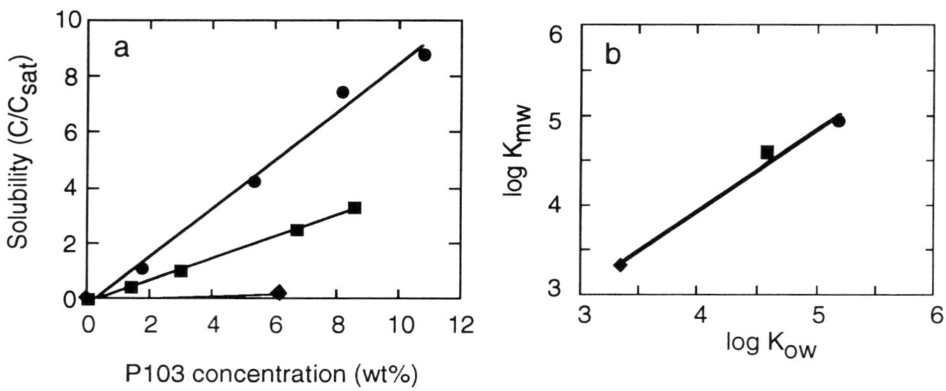

Figure 1. Solubility of pyrene (●), phenanthrene (■), and naphthalene (◆) in Pluronic P103 solution. Data extracted from ref [5]. Some data points excluded for simplicity.

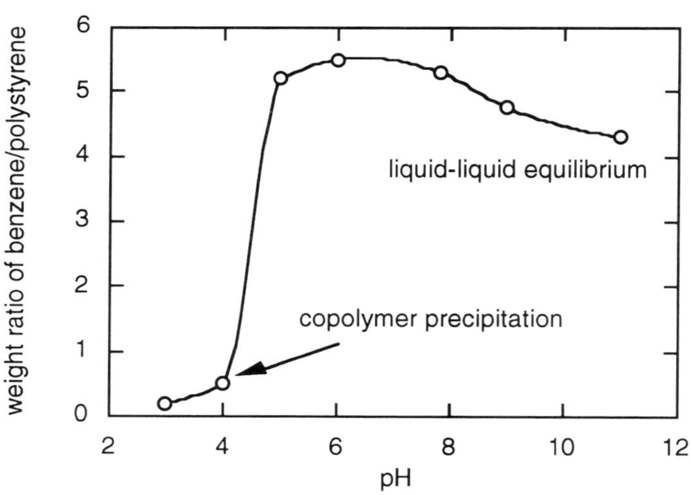

Figure 2. Solubilization limits of benzene in solutions of polystyrene-block-poly(methacrylic acid) expressed in weight ratio benzene/polystyrene versus pH. Data extracted from ref [12].

2.2. Effect of temperature and polymer architecture

Nowakowska et al. recently presented a solubilization study of 2-undecanone in aqueous solutions of poly(sodium styrenesulfonate-co-2-vinylnaphthalene (PSSS-VN) [6]. This copolymer can form hydrophobic domains (a pseudomicellar conformation), which can enhance the water solubility of otherwise sparingly soluble solutes. The solubilization of 2-undecanone was found to increase with increasing copolymer concentration and decreased with increasing temperature. By assuming a two-phase model (polymer core and water) distribution coefficients (K) were determined in two different ways; on mole and weight fractions as reference fractions. Both K-values decreased with an increase in temperature which may be explained by a decrease in the dielectric constant of water, which reduces the hydrophilic-hydrophobic interaction. Still, the distribution coefficients were very high (>1000). Moreover, a solubilization maximum was found with respect to pH, which had a strong correlation with the effective hydrodynamic volume of the polymer coil (see Figure 3). The solubility of 2-undecanone seems to be higher when the copolymer is more expanded, which may be explained by the ability of the solute to penetrate the copolymer coil.

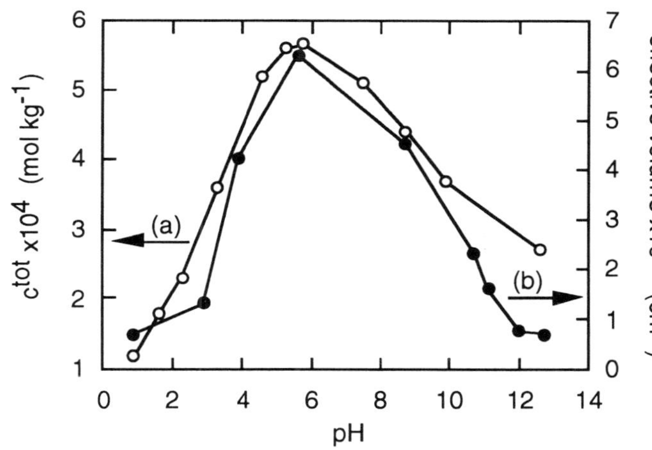

Figure 3. Plot of total molal concentration of solubilized 2-undecanone in poly(sodium styrenesulfonate-co-2-vinylnaphthalene (a) and effective hydrodynamic volume of the copolymer (b) versus pH. Concentration of copolymer = 1 x 10^{-5} M. Data extracted from ref [6].

Opposite temperature effects were found in solubilization experiments with micelle-forming Pluronic solutions. Kabanov et al [8], studied the micellization of some Pluronic block copolymers (F68, P85, and F108) by surface tension and fluorescent probes (pyrene, 1,6-diphenyl-1,3,5-hexatriene (DPH)).

They found that the partitioning coefficient of probe molecule between micelle and water (K=$[probe]_{micelle}/[probe]_{water}$) increased when the temperature was elevated from 20 to 37 °C, due to the increase in hydrophobicity of the Pluronic molecule. Moreover, it was concluded that the probes, both pyrene and DPH, was incorporated in the micelles. The same trend was obtained by Pandya et al. [11], which studied solubilization of diazepam in a 5 wt% solution of L64 ($(EO)_{13}(PO)_{30}(EO)_{13}$). The concentration of the solute in the polymer solution increased approx. 6 times when the temperature was increased from 30 to 55 °C, and was explained by an increase in aggregation number and a reduction of cmc of L64 upon temperature increase. Similar temperature effect was also found by Al-Saden et al. [16] on solubilization of n-hexane in Pluronic F68 solutions.

The solubility of indomethacin in aqueous solutions of Pluronic F68, F88, and F108 was studied by Lin and Kawashima [9]. Again, the solubility of the solute increased by an increase in temperature. Distinct transition points were found in plots of concentration of indomethacine versus concentration of Pluronic polymer. These transition points shifted to lower Pluronics concentrations when the temperature was elevated, demonstrating the balance between cmc and cmt (critical micelle temperature). The three Pluronic examined had the same ratio between the hydrophilic and hydrophobic groups, but differed in molecular weight, (F68<F88<F108). The partition coefficients of indomethacin between micellar phase and water phase increased with temperature and with types of Pluronic, F68<F88<F108, indicating that the size of the polymer (or micelle) plays a role in solubilization enhancement.

The role of block copolymer structure on solubilization of polycyclic hydrocarbons was investigated by Hurter and Hatton [5]. The composition of the blocks in linear Pluronic and branched star-like Tetronics was varied. Not surprisingly, a linear relationship between the wt% PPO in the copolymer and the partitioning coefficient between micelle and water (K_{mw}) for naphthalene (shown in Figure 4) was found. An increase of the hydrophobic part of the Pluronic polymer offers a more favourable environment for the hydrophobic solute. Still, when the K_{mw}-values were normalised with the PPO content (denoted K'_{mw}) of the polymers, an increase of the K'_{mw}-values with the molecular weight of the block copolymer was obtained. This implies that the molecular weight of the polymers affects the structure of the micelle which in turn influences the uptake of organic solutes in this type of systems. Moreover, it was found that the solubilization capacity of branched Tetronics were lower than for linear Pluronics [5]. A possible explanation of this phenomenon is that linear block copolymers can form tighter micelles than the branched which has more configurational constraints.

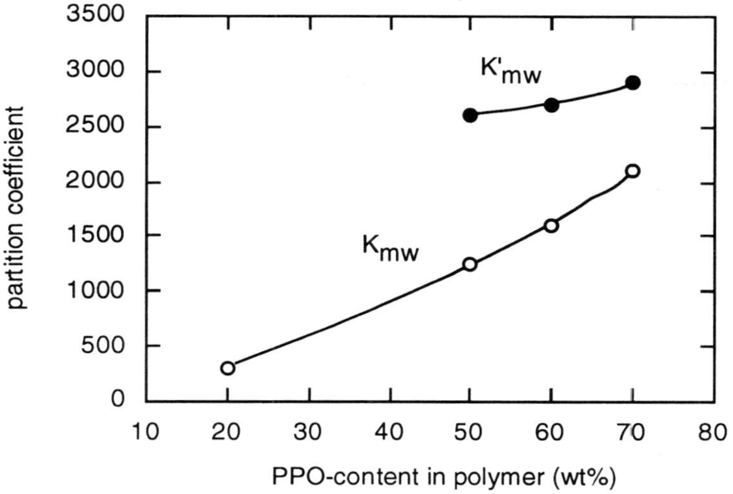

Figure 4. Relationship between PPO-content in Pluronic polymers and K_{mw} (partition coefficient between micelle and water) and K'_{mw} (PPO normalised partition coefficient). Data extracted from ref [5].

An investigation of the solubilizing behaviour when the solute concentration was varied in block copolymer solutions has been presented by Gadelle et al. [19]. They determined solubilization isotherms of toluene, benzene, chlorobenzene, and p-xylene in Pluronic copolymer solutions by head-space gas chromatography. They found that the partition coefficient (K) of the solutes between the polymer and the aqueous solution was very dependent on the free solute concentration, especially at low solute concentrations. Moreover, K increased for low molecular weight polymers and decreased for high molecular weight polymer with increasing solute concentration. This is likely to reflect the cmc:s of the different copolymers. Adding solutes to a low molecular weight block copolymer below cmc promotes aggregation which increases the partition coefficient. For a high molecular weight copolymer already above cmc an increase of the solute concentration decreases the partition coefficient. The authors hypothesis is that this is due to the increasing difficulty to replace the water in the micellar core at increasing solute concentration. More experimental data are needed to determine the amount of water in the PPO core of micelles.

3. PARTITIONING IN AQUEOUS TWO-PHASE SYSTEMS.

3.1. Random EO-PO copolymers.

Thermoseparating polymers separate from water solutions above a certain temperature, the cloud point temperature or the lower critical solution temperature (LCST) [20]. Such polymers have been used in two polymer aqueous two phase systems for protein purification. Depending on the type, the thermoseparated polymer may separate from the solution as a solid precipitate, a gel or a liquid phase. The polymers which separate as a liquid phase are used preferentially, since in these cases the phases are better defined and more easily handled. A recent review which lists a large number of thermoseparating polymers is given by Galaev et al. [21]. Examples of polymers with this property are ethylene oxide-propylene oxide random copolymers. The higher the propylene oxide (PO) content is the lower is the cloud point [22-23]. PEG which contains 100% ethylene oxide (EO) has a cloud point above 100°C which is a too high temperature for many biomolecules. EO50PO50, EO30PO70 and EO20PO80 have the cloud points of 50°C, 32°C and 18°C, respectively. The numbers indicate the weight percent of ethylene oxide and propylene oxide, respectively. Ucon 50 HB-5100 (=Ucon) which has the composition EO50PO50 and molecular weight of 4000, has been much studied in applications of aqueous two phase systems [24]. The cloud point is also affected by the molecular weight, for larger polymer the cloud point is lower [20]. The two phase formation starts at the cloud point. However, two macroscopic phases are formed quickly (within one hour) only if the temperature of the system is at least a few degrees above the cloud point. For a two component polymer-water system the bottom phase is usually polymer enriched, where the polymer concentration is often 40-60%. The top phase usually contains almost 100% water (see the principle for thermoseparation in Fig. 5a and the phase diagram in Fig. 5b).

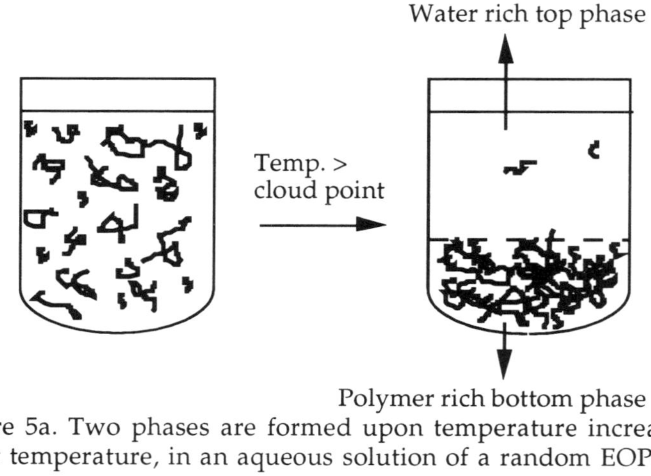

Figure 5a. Two phases are formed upon temperature increase above the cloud point temperature, in an aqueous solution of a random EOPO copolymer.

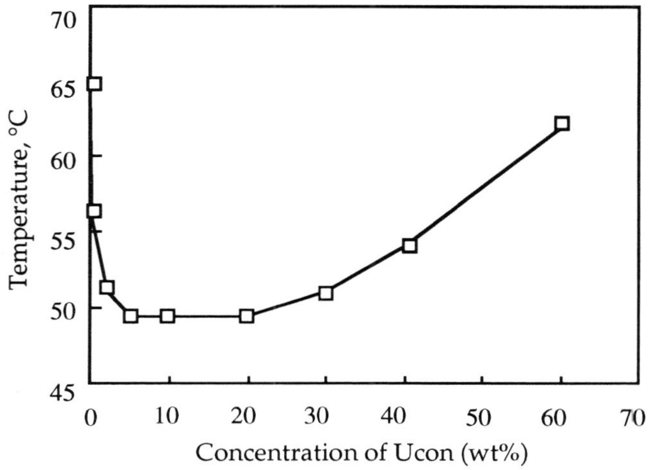

Figure 5b. Cloud point temperatures for Ucon 50 HB-5100 in water. Two phase systems are formed above the binodal line (Data from ref. 24).

The addition of a third component to a Ucon-water system usually results in one of the following phenomena:

1) The cloud point is decreased, and the additive is strongly partitioned to the water phase. These effects are observed if the additive is very hydrophilic, e.g. salt [25], glycine [26] and sugars [27].

2) The cloud point is decreased, and the additive is strongly partitioned to the polymer rich phase. These effects are observed if the additive is hydrophobic, e.g. phenol and butyric acid [24, 26, 28].

3) The cloud point is almost unchanged. The additive has an almost even partitioning between the phases. In this case the additive has an intermediate chemical character in a hydrophilic- hydrophobic scale, e. g. acetic acid 24] and ethanol [28].

4) The cloud point is increased or the system may loose the thermoseparating property. This effect can be obtained if the additive is a good solvent for the polymer and is highly concentrated in the system (e.g. acetic acid and butyric acid) [24, 28]. It can also be obtained if the additive is strongly amphiphilic, e.g. sodium valerate [26] and SDS [29].

Thus by adding different cosolutes it is possible to manipulate several important properties of the aqueous polymer system: cloud point, concentration of polymer in the polymer rich phase and hydrophobicity of the polymer rich phase (by for instance adding a hydrophobic organic solvent which partitions to the polymer rich phase) [30].

3.2. Temperature-induced separations of proteins and peptides.

The random EO-PO copolymers have been used in aqueous two phase systems for bioseparation. The first type of system studied was a two polymer aqueous two phase system, e.g. dextran/EOPO and hydroxypropyl starch/EOPO [30-34]. In these type of systems the EOPO copolymer is enriched in the top phase and the polysaccharide is enriched in the bottom phase. The top phase is more hydrophobic than the bottom phase. Hydrophobic target proteins can be strongly partitioned to the top phase. The separation of the target protein from the EOPO-copolymer is accomplished by heating the top phase above the cloud point. This will induce a new two phase system where the new top phase consists mainly of water and almost no polymer and the new bottom phase is strongly enriched with polymer. The target protein is completely excluded from the polymer rich phase. Thus the thermoseparation property in this system is used to separate polymer from protein and to recover the polymer.

The second type of studied system is a one-polymer aqueous two phase system. The system is composed of a random EOPO copolymer with a concentration of 10-20 weight % in water. By heating the system above the cloud point (typically ten degrees Celsius above), a two phase system is quickly formed, where the bottom phase is polymer enriched and the top phase polymer depleted as described above. The difference in hydrophobicity between the phases is large and hydrophobic cosolutes are easily separated from hydrophilic cosolutes. The system has been used to separate amino acids, oligopeptides and polypeptides [35-36]. The partition coefficient of amino acids in Ucon/water two phase systems has been studied [35]. The partition coefficients of the different amino acids reflect the hydrophobicities of the side groups, where tryptophan is the most hydrophobic amino acid. Hydrophilic amino acids such as lysine and glycine are strongly excluded from the polymer rich phase. The Ucon/water two phase system can therefore be used to purify tryptophan from other amino acids or other hydrophilic cosolutes. Oligopeptides have also been separated according to differences in hydrophobicity. For homopeptides the preference for the water phase or the polymer rich phase becomes more pronounced with increasing degree of polymerisation of the peptide (Fig. 6).

The mixing entropy in the polymer rich phase is lower than the mixing entropy in the water rich phase. This difference in mixing entropy between the phases corresponds to a driving force which tends to drive the cosolutes to partition to the water phase. One could say that the polymer rich phase exerts an "entropic repulsion" against all cosolutes. If the partitioning is performed at a higher temperature the polymer rich phase becomes more concentrated with polymer which leads to an increased entropic repulsion. This effect can be exemplified by tryptophan, which has a K value of 0.64 at 60 °C and 3.3 at 100 °C in a thermoseparated Ucon/water two phase system [36]. However, if the cosolute is sufficiently hydrophobic, the partitioning of the cosolute to the polymer rich phase may increase upon temperature increase. This effect has been observed for polytryptophan in the Ucon/water two phase system [36].

The partitioning of a charged cosolute to the EOPO-rich phase is facilitated by counter ions which have relatively low hydrophobicity. These effects are exemplified in Fig. 7 where the partitioning of positively charged tryptophan (pH 2) to the polymer rich phase was enhanced when its counter ion was exchanged from SO_4^{2-} to ClO_4^- [35].

Large monovalent ions such as $(C_2H_5)_3NH^+$, I^-, SCN^- and ClO_4^- have relatively low hydrophilicity and have been used as counter ions for target proteins in order to enhance the partitioning of the target protein to the EOPO copolymer rich top phase in two-polymer aqueous two phase systems [33-34]. The same strategy has been used in the one-polymer thermoseparated two phase system, where the partitioning of positively and negatively charged tryptophan was strongly enhanced to the polymer rich phase with the addition of ionic surfactants such as dodecyl sulphate and cetyl-trimethylammonium, respectively [35]. A positively charged amphiphilic polypeptide composed of randomly distributed lysinyl and tryptophanyl residues could be quantitatively transferred to the polymer rich or the water phase with the addition of $NaClO_4$ or Na_2SO_4, respectively [36].

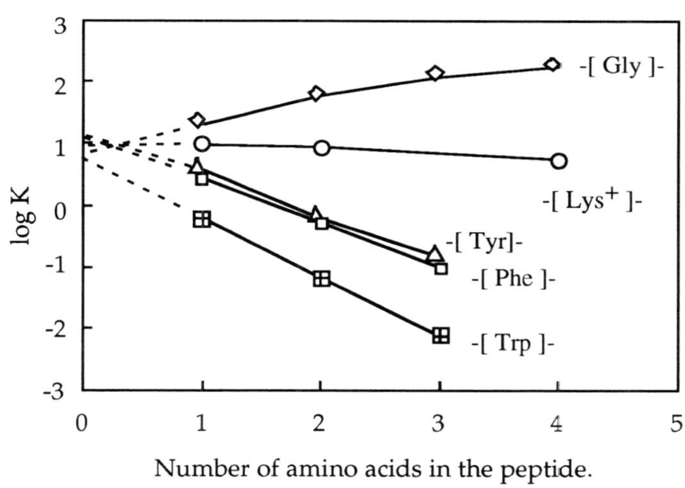

Fig. 6 Partitioning of oligopeptides in Ucon/water two phase systems. System composition: 20% Polymer 80% water (wt-%), 100 mM $NaClO_4$, concentration of: amino acids 1-10 mM, oligopeptides 1 mg/ml. 10 mM Na-phosphate buffer, pH 6.0. Temperature: 60°C. (Data from ref. 36).

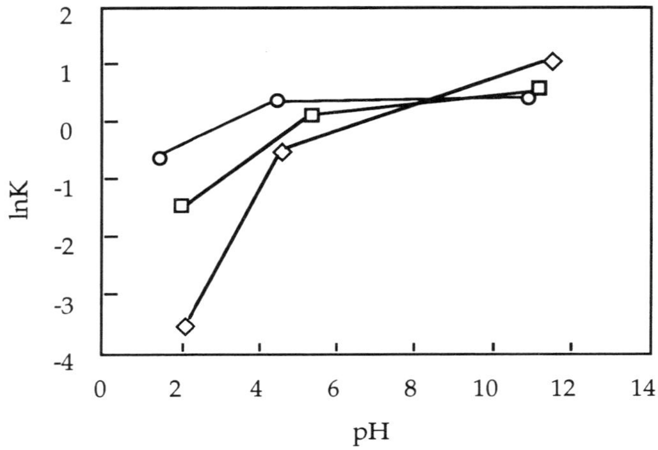

Figure 7. Partitioning of tryptophan in a Ucon/water two phase system. Effect of pH and different anions on the partitioning. Salt type: Na_2SO_4 (○), NaCl (□), $NaClO_4$ (◇). System composition: 20 wt% polymer, 80 wt% water, 100 mM salt (50 mM in the case of Na_2SO_4) and 2 mM Trp. Temperature 60°C. (Data from ref. 35).

3.3. Affinity ligands coupled to copolymers

In order to improve the partitioning of a target protein to one of the phases in an aqueous two phase system, affinity ligands to the target protein have been covalently attached to one of the polymer. Affinity partitioning in aqueous two phase systems is extensively discussed in ref [37]. Most studies have been done in EO-based polymers and dextran systems. Nguyuen et al studied affinity partitioning in copolymer/dextran or copolymer/pullulan systems [38]. They synthesised a copolymer from N-isopropylacrylamide (NIPAAM) and glycidyl acrylate. To this polymer a strong trypsin inhibitor was coupled, para-benzamidine. Trypsin was strongly partitioned to the phase containing the ligand-polymer. Recovery of the polymer was achieved by dissociating the trypsin ligand-polymer complex at low pH and adding ammonium sulphate, which precipitated the polymer but not the trypsin. A new way to recover the ligand-polymer has been examined by Alred et al. [39] by coupling the ligand to a thermoseparating EOPO random copolymer. The ligand was a triazine dye, procion Yellow HE-3G [see Figure 8].

The ligand has two reactive chlorine atoms where the end groups of the random copolymer Ucon 50 HB-5100 (=Ucon) were added. Before the reaction the hydroxyl end group of Ucon was modified to an amino group in order to improve the reaction. The partition coefficient of glucose-6-phosphate dehydrogenase in a Ucon/dextran system, increased from 0.065 to 12 when

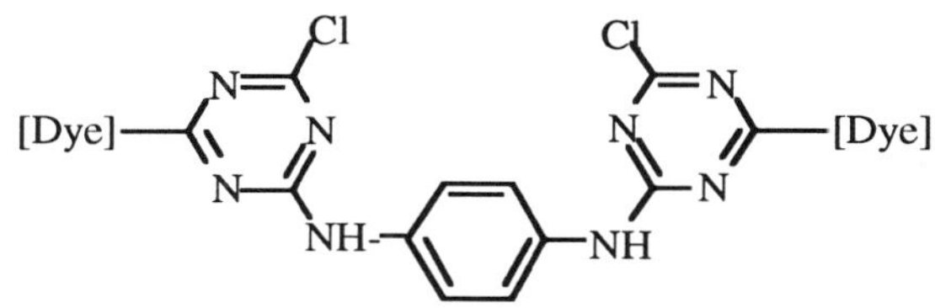

Figure 8. Procion Yellow HE-3G

ligand polymer was added (0.4% of total system). This system was exploited for purification of the enzyme from yeast homogenate and with a subsequent separation of enzyme from ligand-polymer and recovery of the ligand polymer (see Figure 9.). In the first step the enzyme was separated from the bulk proteins in a Ucon/dextran two phase system. In the second step 0.2 M of Na_2SO_4 and 0.2 M NaCl were added in order to decrease the cloud point of the polymer and to dissociate the enzyme from the Ucon-ligand. The temperature was raised to 40°C. In this step a new two phase system which contained a Ucon rich bottom phase and a water-salt enriched top phase was found. No protein was found in the Ucon rich bottom phase. The recovery of the enzyme in the top phase and Ucon-ligand in the bottom phase was 79% and 85%, respectively [39].

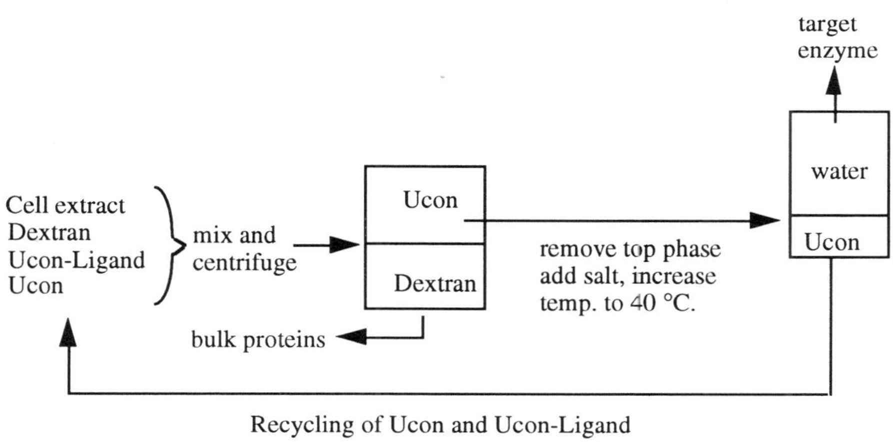

Figure 9. A purification process with affinity partitioning and polymer and ligand recovery by thermoseparation.

A similar type of purification process was studied by Garg et al [40]. A triazine dye, Cibacron Blue 3GA was coupled to a thermoseparating non-ionic surfactant, Triton X-114. A three phase system was formed with Triton, Reppal (a hydroxypropyl starch) and PEG. The top phase contained PEG, the middle phase Triton and the bottom phase contained Reppal. In this system lactate dehydrogenase (LDH) was strongly partitioned to the Triton rich phase which also contained the Triton-ligand conjugate. Separation of LDH was performed by thermoseparation of the Triton rich phase at 45 °C. No salt or cofactor such as NADH was needed for separation of enzyme from ligand-Triton.

3.4. Block copolymers in two-phase systems

By replacing one of the phase forming components in an aqueous two-phase system with an associating component, e. g. a block copolymer or a non-ionic surfactant, an additional pseudophase is introduced, namely the interior of the aggregate/micelle. Due to its nature, a micelle should have a favourable effect on the partitioning of hydrophobic biomolecules. This has been verified experimentally on partition of membrane proteins in two-phase systems with thermoseparating non-ionic surfactants [41, 42].

The partitioning of biomaterials in an aqueous two-phase system with Pluronic P105/hydroxypropyl cellulose (Klucel L, molecular weight 100 000) was recently investigated by Skuse et al. [43]. Pluronic P105 is a block copolymer with the composition $(EO)_{37}(PO)_{56}(EO)_{37}$ (subscripts denote no. of units). Physical characterisations (phase diagrams, interfacial tension, phase viscosities, and phase separation times) as well as partitioning of cells (salmonella enteriditis 3b and human erythrocytes) and proteins (glucose-6-phosphate dehydrogenase (G6PD)) were performed. The binodal was determined and three systems were chosen for further investigation (see Figure 10). The three systems had the same Pluronic P105 concentration (10 wt%) but different Klucel L concentrations (2, 4, and 6 wt%). The results from interfacial

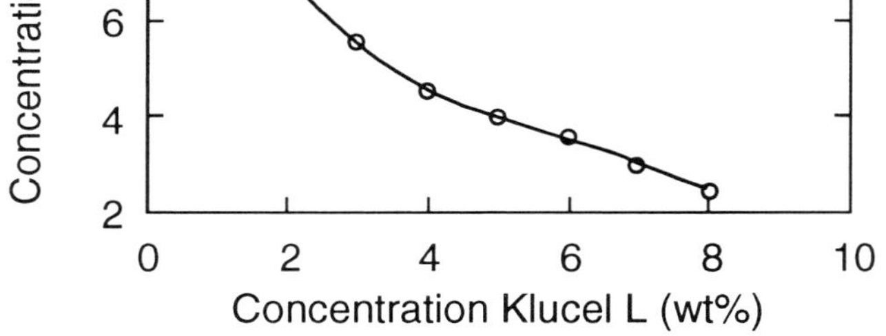

Figure 10. Phase diagram for the system Pluronic P105/Klucel L-system at 21 °C. System chosen for partitioning studies are denoted 1, 2 and 3. Data extracted from ref [43].

tension (IFT) measurements revealed that systems 2 and 3 were unsuitable for cell partitioning, due to the high values of IFT, 2.23 and 3.04 x 10^{-2} mNm^{-1}, respectively, whereas the IFT for system 1 (2.9 x 10^{-3} mNm^{-1}) was in the range typical for cell separations. Approximately 57 % of the salmonella enteriditis 3b cells were found in the P105 rich phase and the rest in the Klucel L rich phase. In the case of human erythrocytes, ca 100% of the cells partitioned to the top phase. Both these results differ from similar experiments performed in PEO/dextran-system [44]. The partitioning coefficients (K) of G6PD in systems 1-3 were 1.6, 1.52, and 1.94, respectively. The high K value of system 3 was likely to reflect the higher polymer concentration difference between the phases compared to systems 2 and 3. Similar results have been obtained for partitioning of G6PD in PEO/dextran-system [45,46] .

Binodals and protein partitioning in the system Pluronic F68/potassium phosphate/water were studied by Kitahara et al. [47]. They noted that phase separation, at 25 °C, occurred at lower polymer concentrations for F68-system than for corresponding PEO-system with same molecular weight as the F68, due to the more hydrophobic character of the Pluronic polymer. Moreover, when the temperature was lowered the one-phase area increased. The opposite effect was obtained when sodium chloride, a second salt, was added. Figure 11

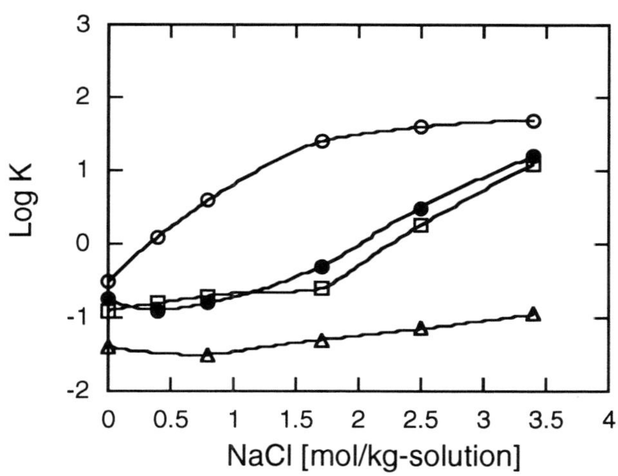

Figure 11. Effect of adding a second salt, sodium chloride, on the partitioning of lysozyme (O), yeast alcohol dehydrogenase (●), hemoglobin (□), and bovine serum albumin (Δ) in Pluronic F68/potassium phosphate system at 25 °C. Data extracted from ref [47].

shows that the partitioning coefficients of four different proteins increased with increasing concentration of added NaCl. This was explained by the salting-out effect of the proteins [47, 48].

When Pluronic block copolymers are used as one of the phase forming components the temperature plays an important role in the phase behaviour, since the onset of micelle-formation is very temperature-dependent. Svensson et al. studied the phase diagrams of three Pluronic copolymers (F68, P105, and L64) together with dextran T500, as the other phase-forming component [49]. They found that the binodals as well as the tielines changed dramatically with temperature (see Figure 12).

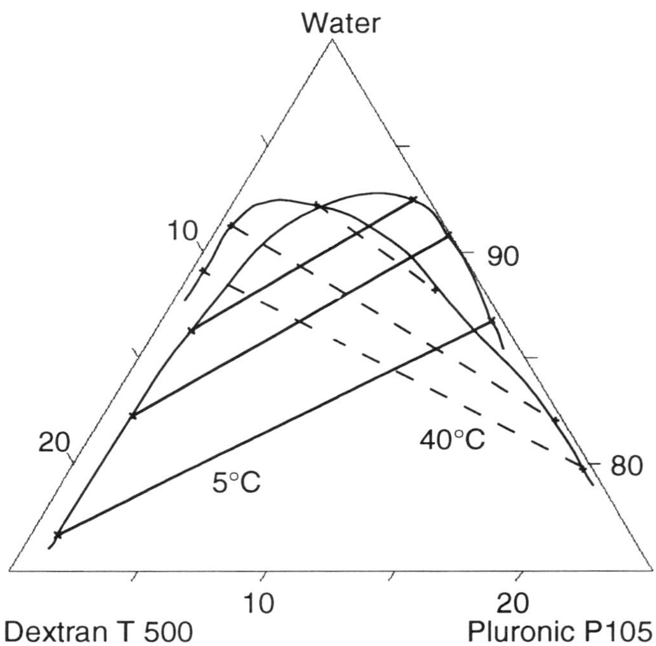

Figure 12. Phase diagram for the Pluronic P105/dextran T500/water system at 5 and 40°C with tie lines. Concentrations in weight percent. From ref [49].

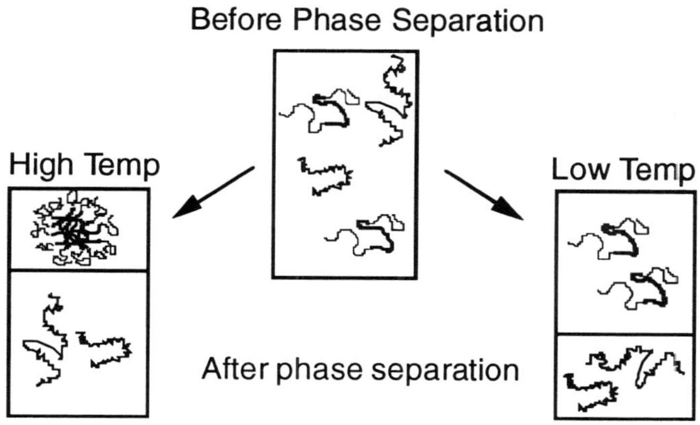

Figure 13. Illustration of temperature dependent two-phase systems containing micelle-forming block copolymer.

When the temperature is elevated, water is transferred from the Pluronic rich phase to the dextran rich phase, caused by the temperature-triggered micellization of the Pluronic molecules. In principal, it is possible, at a fix starting concentration, to form a two-phase system with block copolymer micelles and solubilized unimer at high and low temperature, respectively. This situation is illustrated in Figure 13. As a consequence the relative difference in hydrophobicity between the phases increases at higher temperature. Partitioning of hydrophobic amino acids and oligopeptides in such a situation has been performed [50]. It was found that the partitioning coefficients (K) increased when the temperature was elevated from 5 to 40 °C at a fix point in the P105/dextran T500/water-system. An example of this is shown in Figure 14 where the logarithm of K is plotted against the number of tryptophan residues. This increase can originate from two effects; (i) the top phase (Pluronic rich) is depleted of water at higher temperatures, causing a more concentrated phase, (ii) micelle-formation in the top phase.

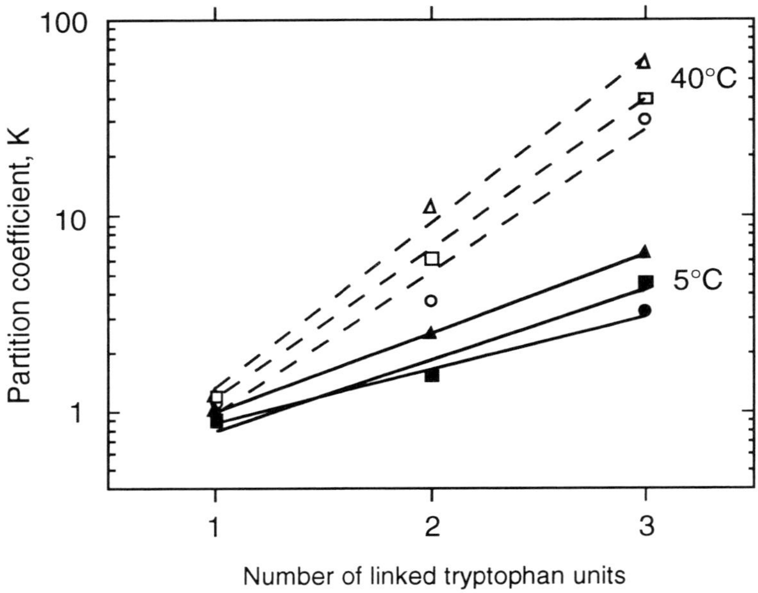

Figure 14. Partition coefficient of tryptophan, di-tryptophan, and tri-tryptophan versus the number of linked tryptophan units in sodium acetate (NaAc) at 40 °C (dashed lines and open symbols) and 5 °C (solid lines and filled symbols), and at different pH regimes: low (circles), neutral (squares), and high (triangles). Ionic strength: 110 mM. From ref [50].

One way to evaluate this is by a plot of ln K against the concentration difference of Pluronic between the two phases at the two temperatures examined. Such a plot is shown in Figure 15. As seen a fairly linear relationship was obtained for the three partitioned components and it was interpreted that micelle formation in this case did not play any major role. Additional work is needed to clarify the potential applications of block copolymers in aqueous two-phase separations. Nevertheless, the use of block copolymers in this separation technique for bio--molecules has been demonstrated. In the case of proteins and amino acids the results resemble those obtained in the more traditional systems PEO/dextran and PEO/salt-systems. The fact that many block copolymers exhibit self-aggregating behaviour will introduce more parameters (temperature, nature and composition of the blocks), to control the partition behaviour.

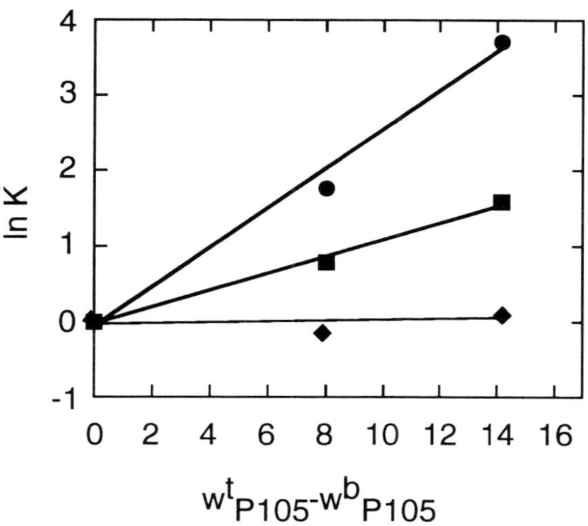

Figure 15. The natural logarithm of partition coefficient of tryptophan (◆), di-tryptophan(■), tri-tryptophan(●) versus the concentration difference of Pluronic P105 between the top and bottom phases (w^t_{P105}-w^b_{P105}) in NaOH/$NaClO_4$ solution at pH 9. Ionic strength: 110 mM. Points at 8 and 14.2 wt% difference were performed at 5 and 40 °C, respectively. From ref [50].

3.5. Polyampholytes in two-phase systems.

A polyampholyte is a polymer which contains acidic and basic groups. A recent review concerning general solution properties of synthetic polyampholytes is given by Bekturov et al. [51]. The polyampholyte contains charged groups at all pH. However, at the isoelectric point (pI) the net charge of the polyampholyte is zero. At pI the solubility of the polyampholyte is usually very low. This property has been utilised in order to recover the polyampholyte from a water solution by adjusting the pH to the pI of the polyampholyte. The polyampholyte may also contain uncharged groups which makes the polymer less hydrophilic. With sufficient amount of uncharged groups the polyampholyte may have a low solubility in the whole pH interval. Polyampholytes based on copolymers of polyacrylic derivatives have been studied as phase forming polymers in aqueous two phase systems, for separation of proteins [52-53]. Examples of polyacrylic residues are given in Fig. 16 . Acidic groups are acrylic acid (AA), methacrylic acid (MA) and ethylacrylic acid (EA). Basic groups are aminoethylmethacrylic acid (AEMA) and dimethylaminoethylmethacrylic acid (DMAEMA). Uncharged groups are methylmethacrylic acid (MMA) and hydroxyethylmethacrylic acid (HEMA).

Acrylic monomers

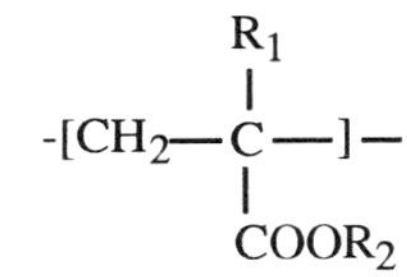

Name	Abbreviation	Composition	
acrylic acid	AA	R_1= H	R_2= H
methacrylic acid	MA	R_1= CH_3	R_2=H
methylmethacrylic acid	MMA	R_1= CH_3	R_2= CH_3
ethylacrylic acid	EA	R_1=C_2H_5	R_2=H
hydroxyethylmethacrylic acid	HEMA	R_1= CH_3	R_2= C_2H_4OH
aminoethylmethacrylic acid	AEMA	R_1= CH_3	R_2= $C_2H_4NH_2$
dimethylaminoethylmethacrylic acid	DMAEMA	R_1= CH_3	R_2=$C_2H_4N(CH_3)_2$

Fig. 16. Composition of acrylic monomers

Hughes et al. investigated the partitioning behaviour of human serum albumin and trypsin (from pancreatic extract) in two phase systems, composed of either two different polyampholytes or one polyampholyte and the uncharged polymer polyvinyl alcohol (PVA) [52]. The polyampholytes had different pI (see Table 2). With two polyampholytes (Amph.1 and Amph.2) with almost similar net charge a two phase system could be obtained at pH 6.6, where the more negatively charged Amph. 2 was enriched in the bottom phase. In this system the total protein and positively charged trypsin were partitioned to the bottom phase ($K_{protein}$=0.34, $K_{trypsin}$=0.05). Upon addition of 1 M NaCl the protein was partitioned to the top phase ($K_{protein}$=2.4) and trypsin remained in the bottom phase ($K_{trypsin}$=0.5).

Two types of PVA/polyampholyte (PVA in the top phase) two phase systems were also studied at pH 6.0: the PVA/(negatively charged Amph. 2) and the PVA/(positively charged Amph. 4) systems. These systems formed one phase systems for salt concentration lower than 50 mM. In the PVA/Amph.2 system trypsin and protein were partitioned to the bottom phase. The preference for the bottom phase decreased for trypsin and increased for protein with increasing salt concentration. In the PVA/Amph.4 system the proteins were partitioned to the bottom phase while trypsin was almost evenly distributed between the phases. With increasing salt concentration the partitioning of proteins became more even while the trypsin partitioned more to the bottom phase.

Table 2 Acrylic polyampholytes (data extracted from ref. [52])

Polyampholyte	composition	monomer ratio	pI	Mw
Amph. 1	AA,EA, HEMA, AEMA, DMAEMA	2.5:2.0:0.1:0.1:1.0	4.8	44 000
Amph. 2	AA, MA, DMAEMA	4:1:1	4.1	64 000
Amph. 3	AA, MA, DMAEMA	3.5:1:1	4.4	50 000
Amph. 4	AA, MMA, DMAEMA	1:1:1	6.5	not available

Effect of pH was also studied. HSA and plasma protein were partitioned in a PVA/amph.3 system. At pH 5 and 6 HSA and proteins partitioned almost quantitatively to the amph. 3 phase while at higher pH the HSA and the protein partitioned to the PVA rich phase. With the addition of salt (NaCl) at high pH the proteins and HSA were again strongly partitioned to the amph. 3 phase.

In the two type of systems: Amph.1/Amph.2 and PVA/Amph.2 the polyampholytes could be precipitated out of the solution by adjusting the pH to pI. The proteins remained in the solution. However, proteins can be precipitated together with an oppositely charged polyampholyte [54-55]. This has been explained as an isoelectric complex between protein and polyampholyte [51, 54-55]

A more recent study of similar systems has been performed by Patrickios et al. [53]. Data for the synthesised polyampholytes used in their study are given in Table 3. They found a very complex phase behaviour of the system PVA/Amph.A. At pH< 5.3 they obtained only one phase systems if the ionic strength was less than 1 M (KCl). At higher pH two phase systems were obtained if the ionic strength was higher than 0.1 M. At low salinity and pH less than 7.2 the polyampholyte was found in the bottom phase. At pH 7.2 and 0.7 M KCl phase inversion was observed with the polyampholyte in the top phase. Qualitatively similar results to the study of Hughes et al. [52] was observed for the partitioning behaviour of proteins. At low salinity (about 0.1 M) positively charged chymotrypsinogen was partitioned to the negatively charged polyampholyte phase, while negatively charged ovalbumin was partitioned to the PVA phase.

Effect of cation on protein partitioning was also studied in PVA/Amph.B system at pH 4.6 where Amph.B is negatively charged. Positively charged chymotrypsinogen and ovalbumin were partitioned to the Amph.B phase, in the presence of alkali chloride salts. Changing the type of alkali cation had no significant influence on the partitioning of the proteins. However, changing from monovalent to divalent cation by using $MgCl_2$ or $SrCl_2$ instead of alkali salts, raised the K values substantially. In this case ovalbumin was partitioned to the PVA phase.

Table 3 Acrylic polyampholytes (data extracted from ref. [53])

Polyampholyte	composition	monomer ratio	pI	Mw
Amph. A	AA, MMA, DMAEMA	0.9:1:1	6.3	88 500
Amph. B	AA, MMA, DMAEMA	3.7:1:1	3.8	43 000

4. CHROMATOGRAPHY

4.1 Displacement and shielding chromatography

In affinity chromatography a ligand is covalently attached to a column matrix. When a sample of proteins containing a target protein is added to the column, a high purification can be obtained if only the target protein binds to the ligand. After washing the column, which removes unbound proteins, the target protein is eluted from the column by adding a soluble ligand to the column (specific elution) or by increasing ionic strength (non-specific elution). Strong binding between the ligand and target protein is desirable and this is can be obtained by choosing ligands with specific affinity to the target protein. However, specific ligands are usually expensive. The relatively inexpensive triazine dyes have been much used as ligands in affinity chromatography [56]. Nucleotide binding proteins have a strong affinity for these dyes. However, many proteins may also bind non-specifically to the ligand with an intermediate binding strength, which leads to a poor purification [57].

Shielding chromatography is a method to avoid that proteins with a weak or intermediate affinity to the ligand will bind to the ligand in the column. This is accomplished by preloading a column with a polymer, which binds to the ligand. When a sample of proteins is loaded on the column only the proteins with a very strong affinity to the ligand are bound, since the polymer is a competitor to the proteins for binding the ligands. The target protein which has a strong affinity to the ligand can compete with the polymer and will locally displace a polymer segment from binding to a ligand. This will decrease the total binding force between polymer and ligand but the polymer will normally not be eluted from the column since it can bind several ligands at the same time. The principle is shown in Figure 17.

Shielding chromatography has been studied and developed by Galaev et al [57-58]. A strategy for the design of polymer shielding dye-affinity chromatography has been presented in ref. [59]. The chosen polymer should form a strong polymer-ligand complex, otherwise the proteins, target protein or added soluble ligands will displace the polymer, leading to polymer leaching.

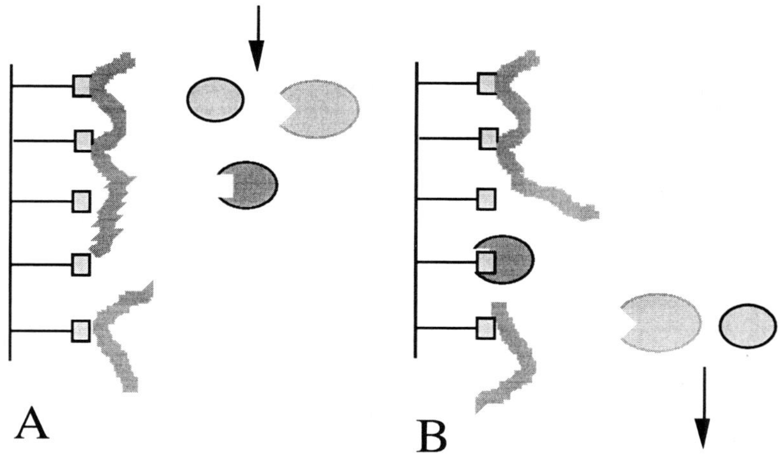

Fig. 17 Shielding chromatography. (A) Loading the polymer shielded column with proteins. (B) Proteins with insufficient affinity for the ligand are washed out of the column. The polymer remains non-covalently bound to the column.

However, the polymer-ligand complex should not be too strong, which could prevent target proteins from binding. The strength of the polymer ligand complex can be modulated by changing polymer size. Larger polymers will bind stronger due to the numerous multi-point contacts with the ligand [59]. The shielding chromatography has been applied to the purification of lactate dehydrogenase (LDH) from porcine muscle. Polyvinylpyrrolidone (PVP-10) with molecular weight of 10 000 was used as shielding polymer in a Procion Blue HERD coupled sepharose column. Shielding was obtained by first percolating the column with 1% solution of PVP. LDH was eluted from the column with 1.5 M KCl. Finally, PVP was removed from the column by addition of a ligand, Cibacron Blue. Another method was to displace PVP with another polymer, polyethyleneimine (PEI). The LDH was purified 30-fold and with 94% recovery [59].

Displacement chromatography is similar to shielding chromatography [60]. The difference is that in the former, a polymer with much higher affinity for the ligand than the target protein, is added to the column after the target protein has been bound to the column. This leads to an effective elution of the target protein. The cationic PEI has been used to displace LDH from the negatively charged Cibacron Blue 3GA ligand (CB). In order to regenerate the column, PEI was removed by increasing the pH to 12. The reduced electrostatic attraction between PEI and CB, due to decreased charge of PEI, was further reduced with high ionic strength. This method was insufficient for a system with Procion Red HE-3B (PR) as ligand. In this system PEI was removed from the PR-bound column by including polyacrylic acid (PAA) in the high pH-high salt buffer. The highly charged PAA formed a soluble polyelectrolyte complex with the weakly charged PEI, which was thus eluted [60].

A new method which combines shielding and displacement effects has been developed by Galaev et al. [61]. In this method a thermoprecipitating polymer, polyvinylcaprolactam (PVCL) was used as a shielding polymer. This polymer thermoprecipitates at 38 °C (the cloud point temperature). In this system the target protein, LDH was bound to the column when the temperature was 40 °C (and ionic strength 0.1 M KCl). Upon decreasing the temperature below the cloud point, the LDH was eluted from the column at the same ionic strength, 0.1 M KCl. This effect was explained as follows: above the cloud point the polymer adopts a compact structure. This leads to a reduced strength of the polymer-ligand complex due to few multi-point contacts between the polymer and the ligands. This reduced affinity allows the LDH to compete with the polymer for ligand binding. Below the cloud point the polymer adopts a coil structure with more ligand binding as consequence. Thus the target protein cannot compete with the polymer and is eluted from the column.

5. PRECIPITATION

5.1. Thermoprecipitating copolymers.

Polymers which display a LCST and phase separates above the cloud point as a solid (flocculation) are called thermoprecipitating polymers. Proteins are natural thermoprecipitating polymers [62]. A list of many synthetic thermoprecipitating polymers and their application is given by Galaev et al. [21]. Many solution properties of these polymers are discussed by Taylor et al. [63]. Examples of homopolymers which display thermoprecipitation are poly (N-isopropylacrylamide) (=PNIPAAM) and polyvinylcaprolactam (PVCL) [21] The LCST of a thermoprecipitating polymer can be regulated by copolymerization of hydrophobic with hydrophilic monomers [63-64]. If the monomers contain titratable groups the LCST will depend on the pH [63-64]. Salts usually decrease the cloud point of a thermoprecipitating polymer while ionic surfactants increase the cloud point [21]. Thermoprecipitating polymers have recently been used in bioseparation processes [21]. Ligands coupled to thermoprecipitating polymers are used in affinity thermoprecipitation [21, 65-67]. Cross linked thermo-precipitating polymers form gels which have been used to concentrate macromolecules. The collapsed gels are added to a protein solution. The gel starts to swell when water molecules diffuse into the gel structure. The proteins are excluded due to the small pore size of the gel. The swollen gel is removed and transformed into a collapsed gel by heating the gel above the cloud point [68].

5.2 Affinity precipitation.

The principle of affinity thermoprecipitation is shown in Figure 18. A target protein is complexed with a polymer linked ligand. The protein-polymer complex is then precipitated by raising the temperature above the cloud point. The precipitated complex is dissolved in a water solution at lower temperature. The target protein can be separated from the ligand-polymer by re-precipitating the polymer in a solution where the ligand-protein complex is weak (for instance at low pH, high ionic strength). The great advantage of affinity precipitation is the large reduction in working volumes after the precipitation step. The strength of the binding constant (K_b) between the target protein and ligand must be very strong (K_b between 10^{-9}-10^{-11}) since coupling of ligand to polymer usually strongly decreases the binding strength between target protein and ligand [100-1000-fold decrease) [67]. Furthermore, the precipitate forms a hydrophobic environment which is energetically unfavourable for the protein. Multi-point attachment between the ligands on the polymer facilitates precipitation of the protein-polymer complex.

NIPAAM and PVCL are unreactive thermoprecipitating polymers. In order to bind a ligand to these polymers reactive groups have been introduced by two ways; copolymerization with a reactive monomer e.g. NIPAAM and glycidyl

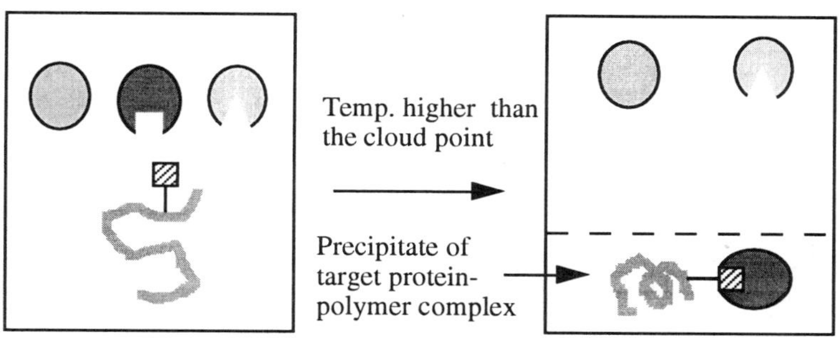

Figure 18. Affinity thermoprecipitation

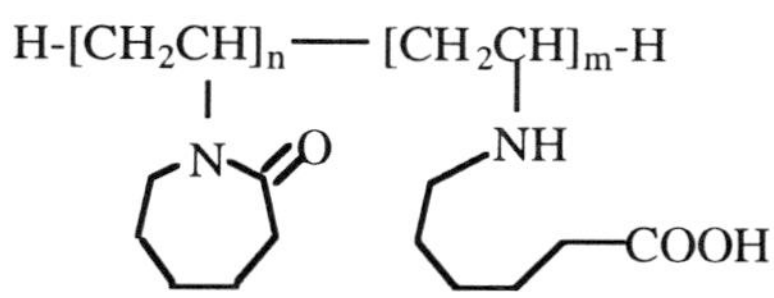

Figure 19. Partially hydrolysed PVCL

methacrylate [65], or transformation of some monomers to reactive groups e.g. partial hydrolysis of PVCL [66] (see Figure 19 above). Examples of coupled ligands are p-aminobenzamidine (trypsin binding) [65], soybean trypsin inhibitor (trypsin binding) [66] and IgG (binds protein A-alkaline phosphatase complex) [65]. Dissociation of trypsin from ligand has been done by dissolving the precipitate in a pH 2 solution with 0.1 M KCl [65-66].

Affinity precipitation can also be performed with charged copolymers which can be precipitated by a pH change. One polymer used was Eudragit S 100, which is a copolymer of methacrylic acid and methylmethacrylic acid, the ratio of the monomers was 1:2, respectively. This polymer has been used for affinity precipitation by decreasing the pH below 4.7 [69-72]. In a new application, the polymer could be precipitated at pH 7.5 by adding 50 mM Ca^{2+} and raising the temperature (40°C) [73-74]. A triazine dye, Cibacron blue 3GA (Fig. 18) was coupled to the polymer. With this ligand-polymer it was possible to extract lactate dehydrogenase (LDH) and puryvate kinase (PK) from muscle extract. Both enzymes were bound to the polymer-ligand and precipitated at 40°C. These enzymes could be separated from each other by suspending the precipitate in buffers with different ionic strength. PK was dissociated from the polymer-ligand at 0.1 M KCl. LDH was dissociated at 0.15 M KCl. The yields and purification factors were (70%, 4,4 fold) and (73%, 7.5 fold), respectively [69]. Another method has been to add Zn^{2+} ions to a solution of the same polymer-ligand. The polymer-ligand forms a strong binding complex with Zn^{2+} and alcoholdehydrogenase [ADH]. This complex was thermoprecipitated at 40°C in 50 mM Ca^{2+} yeast extract solution. ADH was dissociated from the polymer ligand by adding iminodiacetic acid, which formed a strong complex with Zn^{2+}. The yield and purification in this process were 66% and 20-fold, respectively [70].

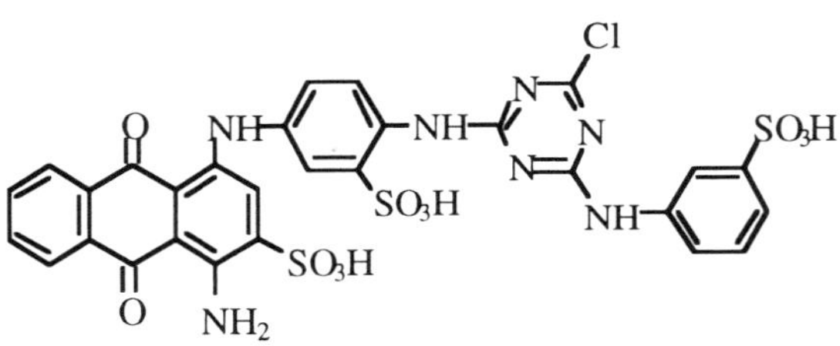

Figure 20. Cibacron blue 3GA

REFERENCES

1. P. N. Hurter, P. Alexandridis and T. A. Hatton, in S. D. Christian and J. F. Scamehorn eds., Solubilization in Surfactant Aggregates, Marcel Dekker, Inc., New York, 1995, p. 191.
2. R. Nagarajan, Current Opinion in Colloid & Interface Science, 1 (1996) 391
3. A. Tontisakis , R. Hilfiker and B. Chu, J. Colloid Interface Sci., 135 (1990) 427.
4. G. J. McFann, K. P. Johnston, P. N. Hurter and A. T. Hatton, Ind. Eng. Chem. Res., 32 (1993) 2336.

5. P. N. Hurter and T. A. Hatton, Langmuir, 8 (1992) 1291.
6. M. Nowakowska , B. White and J. E. Guillet, Macromolecules, 23 (1990) 3375.
7. R. Nagarajan , M. Barry and E. Ruckenstein , Langmuir, 2 (1986) 210.
8. A. V. Kabanov , I. R. Nazarova, I. V. Astafieva, E. V. Batrakova, V. Y. Alakhov, A. A. Yaroslavov and V. A. Kabanov, Macromolecules, 28 (1995) 2303.
9. S-H. Lin and Y. Kawashima, Pharm. Acta Helv., 60 (1985) 339.
10. K. Nakashima, Y. Fujimoto and T. Anzai, Photochemistry and Photobiology, 61 (1995) 592.
11. K. Pandya , P. Badahur, T. N. Nagar and A. Badahur, Colloids Surfaces A: Physicochem. Eng. Aspects, 70 (1993) 219.
12. D. Kiserow , K. Prochazka , C. Ramireddy, Z. Tuzar, P. Munk and S. E. Webber, Macromolecules, 25 (1992) 461.
13. M. Tian , E. Arca, Z. Tuzar , S. E. Webber and P. Munk, J. Polym. Sci. B., 33 (1995) 1713.
14. W. R. Haulbrook, J. L. Feerer, T. A. Hatton and J. W. Tester, Environ. Sci. Technol., 27 (1993) 2783.
15. P. L. Valint and J. Bock , Macromolecules, 21 (1988) 175.
16. A. A. Al-Saden , T. L. Whateley and A. T. Florence, J. Coll. Interface Sci., 90 (1982) 303.
17. M. Nowakowska and J. E. Guillet, Macromolecules, 24 (1991) 474.
18 T. L. Calvert, R. J. Phillips and S. R. Dungan, AIChE J., 40 (1994) 1449.
19 F. Gadelle, W. J. Koros and R. S. Schechter, Macromolecules, 28 (1995) 4883.
20. S. Saeki, N. Kuwahara, M. Nakata and M. Kaneko, Polymer 17 (1976) 685.
21. I. Galaev and B. Mattiasson, Enzyme Microb. Technol. 15 (1993) 354.
22. F. E. Bailey and R. W. Callard, J. Appl. Polym. Sci. 1 (1959),56.
23. A. Louai, D. Sarazin, G. Pollet, J. François and F. Moreaux, Polymer 32 (1991) 703.
24. H.-O. Johansson, G. Karlström and F. Tjerneld, Macromolecules 26 (1993) 4478.
25. K. P. Ananthapadmanabhan and E. D. Goddard, Colloid. Surf. 25 (1987) 393.
26. H.-O. Johansson, G. Karlström and F. Tjerneld, Colloid Polym. Sci., in press.
27. Å. Sjöberg, G. Karlström and F. Tjerneld, Macromolecules 22 (1989) 4512.
28. A. Louai, D. Sarazin, G. Pollet, J. François and F. Moreaux, Polymer 32 (1991) 713.
29. A. Carlsson, G. Karlström, B. Lindman and O. Stenberg, Colloid. Polym. Sci. 266 (1988) 1031.
30. P. A. Alred, R. F. Modlin and F. Tjerneld, J. Chromatogr. 628 (1993) 205.
31. P. A. Harris, G. Karlström and F. Tjerneld, Bioseparation 2 (1991) 237.
32. P. A. Alred, A. Kozlowski, J. M. Harris and F. Tjerneld, J. Chromatogr. 659 (1994) 289.
33. K. Berggren, H.-O. Johansson and F. Tjerneld, J. Chromatogr. A 718 (1995) 67.
34. H.-O. Johansson, G. Lundh, G. Karlström and F. Tjerneld, Biochim. Biophys. Acta 1290 (1996) 290.

35. H.-O. Johansson, G. Karlström, B. Mattiasson and F. Tjerneld Bioseparation 5 (1995) 269
36. H.-O. Johansson, G. Karlström and F. Tjerneld, Biochim. Biophys. Acta, in press.
37. H. Walter and G. Johansson (editors), Methods in Enzymology Vol. 228, 1994, Academic Press, New York.
38. A. L. Nguyen and J. H. T. Luong, Enzyme Microb. Technol. 12 (1990) 663.
39. P. A. Alred, F. Tjerneld, A. Kozlowski and J. M. Harris, Bioseparation 2 (1992) 363.
40. N. Garg , I. Y. Galaev and B. Mattiasson, Biotechnol. Appl. Biochem. 20 (1994) 199.
41. C. Bordier, J. Biol. Chem., 256 (1981) 1604.
42. T. Minuth, J. Thömmes and M.-R. Kula, J. Biotechnol., 38 (1995) 151.
43. D. R. Skuse, R. Norris-Jones, M. Yalpani and D. E. Brooks, Enzyme Microb. Technol., 14 (1992) 785.
44. D. R. Skuse, R. Norris-Jones, D. E. Brooks and M. Yalpani, in M. Yalpani, (edit.), Industrial Polysaccharides, Elsevier Science Publishers, Amsterdam, 1987, p. 377.
45. G. Johansson and M. Joelsson , Enzyme Microb. Technol., 7 (1985) 629.
46. G. Johansson, G. Kopperschläger and P.-Å. Albertsson, Eur. J. Biochem., 131 (1983) 589.
47. T. Kitahara, M. Kamihira and H. Takeuchi , J. Chem. Eng. Jpn., 26 (1993) 183.
48. K. D. Collins and M. W. Washabaugh, Quarterly Review of Biophysics, 18 (1985) 323.
49. M. Svensson, P. Linse and F. Tjerneld, Macromolecules, 28 (1995) 3597.
50. M. Svensson, F. Joabsson, P. Linse and F. Tjerneld, J. Chromatography A., in press.
51. E. A. Bekturov, S. E. Kudaibergenov and S. R. Rafikov, J. Macromol. Sci. C30 (1990) 233.
52. P. Hughes and C. R. Lowe, Enzyme Microb. Technol. 10 (1988) 115.
53. C. S. Patrickios, N. L. Abbott, R. P. Foss and T. A. Hatton, AIChE Symp. Ser., 290 (1992) 80.
54. H. Morawetz and W. L. Hughes, J. Phys. Chem. 56 (1952) 64.
55. D. Guoqiang, R. Kaul and B. Mattiasson, Bioseparation 3 (1993) 333.
56. R. K. Scopes, J. Chromatogr. 376 (1986) 131.
57. I. Y. Galaev and B. Mattiasson, J. Chromatogr. A 662 (1994) 27.
58. I. Y. Galaev, N. Garg and B. Mattiasson J. Chromatogr. A 684 (1994) 45.
59. N. Garg, I. Y. Galaev and B. Mattiasson, J. Molec. Recogn., in press.
60. I. Y. Galaev, P. Arvidsson and B. Mattiasson, J. Chromatogr. A 710 (1995) 259.
61. I. Y. Galaev, C. Warrol and B. Mattiasson, J. Chromatogr. A 684 (1994) 37.
62. F. Macritchie, J. Colloid Inter. Sci. 45 (1973) 235.
63. L. D. Taylor and L. D. Cerankowski J. Polym. Sci. 13 (1975) 2551
64. G. Chen and A. S. Hoffman, Nature 373 (1995) 49.

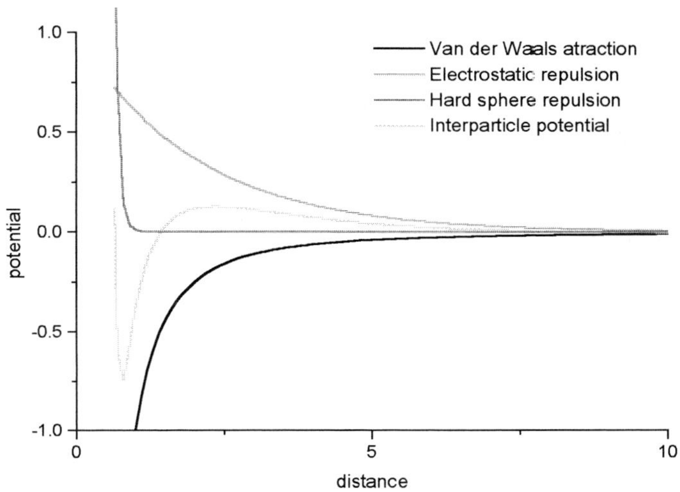

Figure 2a. Schema of electrostatically stabilized interparticle potential.

force created by surface charges can be modulated by the presence of electrolytes that modify the form of the interparticle interaction. Steric repulsion can produce a barrier against flocculation as well. Here the form of the potential is different from that for electric repulsion and a secondary minimum can be present due to the finite length of the steric repulsive force (fig. 2b). The distinction is usually made between weak and strong flocculation. In the former, flocculation is produced in the secondary minimum while in the later the depth of the potential well is much higher and the interparticle distance reaches the primary minimum. If the droplets have a strong tendency to flocculate the aggregates will grow until all emulsion droplets are involved and the continuous phase will be expelled from the floc. However, if the emulsion is weakly flocculated, the interparticle distance is relatively high and can occupy the whole emulsion without expelling continuous phase. Weakly flocculated emulsions are shear thinning due to rupture of the structure by shear [17].

Surfactants modify flocculation by their effect on the interparticle interaction. Charged surfactants produce surface charges. However, emulsions stabilized only by electrical charges are very sensitive to the presence of electrolytes. Indeed, if the c.m.c. (critical micellar concentration) of the surfactant is high, the same surfactant ions can reduce the extension of the electrical double layer and produce flocculation. Nonionic surfactants prevent flocculation by steric repulsion. Hydrophilic nonionic surfactants such as alcohol ethoxylates can be considered as low molecular weight polymeric surfactants, the longer the length of the hydrophilic part the better the stabilization against flocculation of o/w emulsions. However, increasing the length of the hydrophilic part alone can produce a reduction in the adsorption of surfactant and, therefore, a decrease in the steric repulsion.

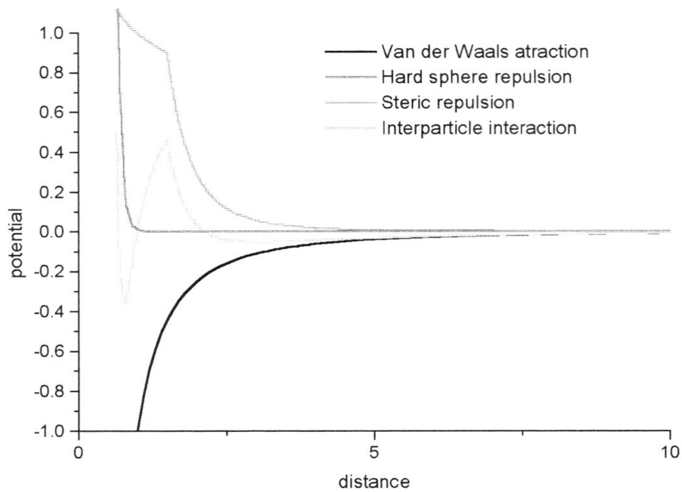

Figure 2b. Schema of sterically stabilized interparticle potential

The macromolecular nature of polymeric surfactants can clearly improve the stabilizing effect due to steric repulsion. In a polymeric surfactant both the hydrophilic and hydrophobic part of the molecule can have a high molecular weight, providing steric repulsion without reducing its adsorption capacity. Concerning ionic polymeric surfactants, similar conclusions can be drawn with the added stabilizing effect of an electrical double layer.

An important aspect of flocculation is that of flocculation induced by depletion flocculation [18]. Differences in osmotic pressure can be produced by non-adsorbed polymers or micelles. The situation is depicted in figure 3. When approaching two droplets, the presence of smaller particles in between produces an additional difficulty for the approach. However, once the distance is of the order of the size of the smaller particles, an attractive force appears due to the difference in osmotic pressure between the bulk and the region between the droplets. At this point the solvent tends to be squeezed out. At higher polymer concentrations a re-estabilization occurs. This type of flocculation is usually weak and can be used to stabilize dispersed systems against sedimentation.

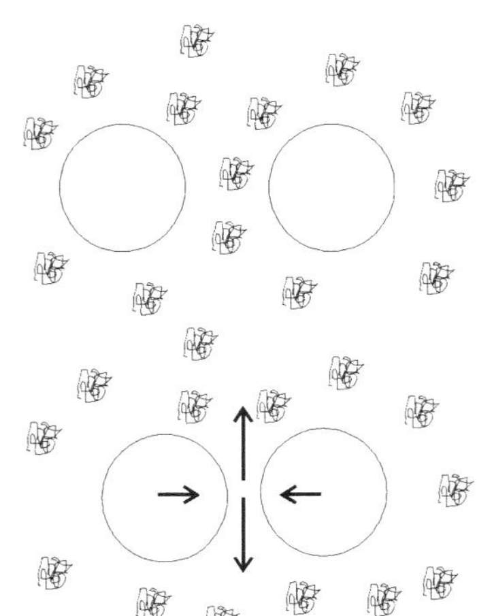

Figure 3. Depletion flocculation mechanism.

The use of polymeric surfactants can result in bridging flocculation. Bridging flocculation can be produced by a too large molecular weight of the stabilizing blocks [19] or because of having more than one anchoring block, these can adsorb at different droplets creating an almost permanent link.

2.3 Coalescence

Coalescence is the process by which two or more droplets fusion to form a bigger droplet. The driving force of this process is the reduction in interfacial area, this is going to be more important the higher the interfacial tension and the lower the droplet size. For two droplets to coalesce they have to come in contact, this contact could be instantaneous (droplet collision) or almost permanent (as in a floc or a sedimented layer). In any situation, for the droplets to coalesce, a thin film is present between the droplets before they can coalesce. Therefore, the stability of an emulsion against coalescence is mainly governed by thin film stability.

The stability of thin films separating two portions of the same phase has been studied and related with emulsion stability [20]. In those studies the coalescence of a droplet on a liquid-liquid interface was studied. Although it could be questionable that macroscopic droplets coalescing against a planar interface would behave in the same way as in an emulsion, good correlation of coalescing times with emulsion stability could be found.

The theories for droplet coalescence are several and difficult to test, it is expected that the preparation of model emulsions with low polydispersity and well characterized will lead to a better testing of the existing theories. Hole nucleation is probably the theory that gives a clearer picture of the process; see for instance the paper by Kabalnov and Wennerström on the subject [4]. In this theory the thin liquid film between two droplets can break due to the formation of holes caused by thermal fluctuations. The energy barrier depends on the natural curvature of the interface and is greatly different if the interface is w/o/w or o/w/o. An alternative view of hole nucleation is kinetic rather than thermodynamic. The growth of the holes to produce complete coalescence depends on the rate at which the surfactant molecules can move to cover the newly formed interface. If this rate is slow, the hole will not grow and eventually can disappear. If the surfactant can cover the new interface in short time, the hole is stabilized and can grow further. Whether the holes grow further or just allow for the passage of the dispersed phase to form a single droplets is of not importance for the overall picture. Related to the hole nucleation would be Gibbs-Marangoni elasticity. The higher the Gibbs-Marangoni elasticity the fewer chances for a hole to form and grow. Gibbs-Marangoni elasticity is enhanced by low diffusivities [21-22] of the surfactant and therefore by the molecular weight.

2.4 Ostwald Ripening

This process produces the growth of the bigger droplets at the expense of the small droplets that, eventually, will disappear. The driving force of this process is the term that takes into account the dimensions of the droplet in the chemical potential. For small droplets the chemical potential should be described with a term in which the intensive property is the Laplace pressure and the extensive property the volume of the droplet. As the Laplace pressure is inversely proportional to the radius of the droplets, this term increases as the radius decreases. The rate at which mater is transferred between the droplets in this process depends on the droplet size and droplet size distribution and also on the solubility of the dispersed phase in the continuous phase. The final stage would be the formation of a single droplet. The rate of growth, according to the Lifshitz-Slezov theory, is: [23]

$$\frac{d}{dt}a^3 = \frac{8\sigma V_m c D}{9RT}\gamma(\varphi) \qquad \textbf{(2)}$$

where a is the mean number average radius, σ the interfacial tension, V_m the molecular volume of the substance, c its solubility in the external phase, D its diffusivity, R the gas constant, T the absolute temperature and $\gamma(\varphi)$ an acceleration factor related to the volume fraction.

This equation shows that the growth rate is constant in terms of volume or that the liner growth of the radius is proportional to the inverse of their surface. Therefore, smaller mean droplet sizes grow at a faster rate. This process can be slowed by addition of small amounts of a third component that partitions favorably in the dispersed phase, that is, a component which solubility in the continuous phase is lower than that of the main dispersed phase [24].

Surfactants have influence mainly on the interfacial tension and solubility. On the one hand, the surfactant reduces the interfacial tension but, increases the solubility of the dispersed phase in surfactant micelles. Micellar solubilization seems to be responsible for increased ripening of emulsions when increasing surfactant concentration [25]. However, predicting the rates using micellar solubilization results in high overestimates. These results support the view that micelle mediated transport between emulsion droplets is a slow process [23]. Polymeric surfactants could be advantageous with respect to normal surfactants if they can effectively reduce the interfacial tension while producing reduced micellar solubilization.

2.5 Inversion

This process is the transformation of the dispersed phase of the emulsion in continuous and vice versa. This process is intimately related to the ability of the surfactant to stabilize one or the other type of emulsion and therefore to coalescence. Parameters that can change this ability will produce phase inversion, this is the fundament of the EIP method [11]. Phase inversion has been treated from the viewpoint of catastrophe theory [26]. This process can be understood as very fast coalescence of all droplets. The former continuous phase would form droplets because, at the volume fractions at which the inversion takes place, most of the continuous phase is already in the Plateau borders. This phase inversion has been used as a method of preparation of low droplet size and polydispersity emulsions [27]. Inversion takes place at a certain volume fraction for a particular system (internal and external phases composition and surfactant), although a strong influence of the stirring conditions is also found. There is some hysteresis if the inversion takes place in one direction or in the other. The stability of the emulsions that undergo phase inversion is usually very different. If the surfactant is mainly soluble in the dispersed phase, phase inversion takes place at a relative low volume fraction and the inverted emulsion is usually much more stable against coalescence than the original emulsion. The surfactant plays an important role in the volume fraction at which inversion takes place and on the extend of the hysteresis. Polymeric surfactants probably lead to higher hysteresis. This would probably be the case of preparation of highly concentrated o/w emulsions with a relatively hydrophobic surfactant as observed by Pons et al. [9]. Strong adsorption and slow diffusivity of the surfactant would prevent inversion to a large extend.

Nonionic surfactants with polyoxiethylene as hydrophilic chain change their affinity for water and oil with changing the temperature. This can produce phase inversion of the emulsions, indeed, a method of emulsion preparation is based on this property of nonionic surfactants (see below). This type of phase inversion is reversible and is fundamentally different of the phase inversion produced by the increase of volume fraction. In this process the affinity of the surfactant for the continuous and dispersed phases is changed by the temperature change while in the previous process the surfactant does not change its partition between the phases.

3 EMULSION FORMATION

3.1 Methods of preparation

Because of their metastable nature, emulsions are highly dependent on the method of preparation. Emulsions have traditionally been prepared by adding the dispersed phase to the continuous phase with vigorous agitation. The surfactant is dissolved in the continuous phase and by adding the dispersed phase with agitation this is disrupted in droplets. The droplet size in this method is usually found to depend strongly on the intensity of agitation. A good review concerning emulsification machinery and emulsification in general can be found in the chapter by Walstra [28]. The processes involved in stirring emulsification are not yet well understood. During emulsification droplets are created by disruption of one of the phases. In the conditions in which disruption is possible one may expect an enhanced rate for coalescence as well. From a dynamic point of view the emulsion formed by shearing or shaking the system will be that in which the rate of coalescence is the lowest for the two types possible. Many parameters influence the properties of the final emulsion: the way the droplets are disrupted, the viscosity of both phases, the temperature, the velocity of addition of the internal phase, the energy input, etc. In addition the process of emulsification can be done in many different ways, for instance, mixing the phase in their final proportion or by additions, placing the surfactant in the would be continuous or dispersed phase, using phase inversion, etc. One may say that the methods of preparation are so many as emulsions can be prepared. All this makes emulsification as one of the worse understood processes related to emulsions. Even preparing an emulsion twice by using apparently the same process often results in very different products. Due to the difficulty of the problem, only very specific processes have been studied in detail and, still those, are poorly understood. The reader is referred to the excellent although a bit old review by Walstra [28]. Only two processes not covered by Walstra review will be discussed in more detail: preparation of monodisperse emulsions by fractioning and preparation of emulsions by thermal phase inversion.

Bibette and coworkers devised a method to prepare monodisperse emulsions by fractioning of polydisperse emulsions. The method is based on the differences in floculation and sedimentation due to droplet size. An initial emulsion is partially flocculated by the addition of a surfactant solution of relatively high concentration. The flocculation is induced by the presence of micelles and the osmotic pressure induced for approaching droplets. This upper and lower fractions are further fractioned by changing the surfactant concentration in the continuous phase. Emulsions with a very narrow droplet size distribution are obtained after a few fractioning steps. The polydispersity is so low as to allow for quasicrystaline packing. Emulsions thus produced are good models for studying emulsion properties such as stability and rheology. These newly developed systems will probably help in our understanding of emulsions.

Small droplet size emulsions with small droplet size and reduced polydispersity can be obtained by the phase inversion technique [27]. Oil, water and an appropriate surfactant are mixed at a temperature close to the PIT of the system. The mixture is rapidly heated or cooled (depending on the emulsion to be w/o or o/w). At a temperature near the PIT the interfacial tension is very low. This helps the formation of very small droplets. However, at these temperatures the emulsions are very unstable and completely phase separate in short time. The surfactant is near the balance and it does not favour either o/w or w/o emulsions. A rapid change of temperature changes the preferred curvature of the surfactant to either o/w or w/o according to the direction of the temperature change. If the temperature change is fast enough, not much growth will occur. The low polydispersity of the system could be related to fast Ostwald ripening occurring while the system is still near the PIT [29] or be due to the mechanism of droplet formation and fast coalescence model applies to the phenomena [4]. This method has also been successfully used to prepare highly concentrated emulsions with evidence of low polydispersity [29].

3.2 Droplet size and droplet size distribution

Droplet size and droplet size distribution are two of the properties most often used to characterize an emulsion. Most, if not all, other properties are related to the droplet size and therefore, its control is of fundamental importance when using or studying emulsions. Several techniques are used to measure droplet size; the most used are: microscopy, laser diffraction and scattering, counting methods, sedimentation, turbidity etcetera. Besides mean droplet size it is important to characterize the droplet size distribution. Two emulsions with the same mean droplet size but different droplet size distributions can have very different properties. For instance a truly monodisperse emulsion would not change due to Ostwald ripening while a wide droplet size distribution will evolve with the disappearance of the small droplets. Rheology is highly influenced by polydispersity too. Usually small droplet sizes are preferred because the overall stability normally increases by decreasing the mean droplet size. However, below a certain limit, Ostwald ripening starts being fast unless the dispersed phase is very insoluble in the continuous phase. Decreasing the droplet size leads to an increase in emulsion viscosity that could sometimes be undesirable. The effect of polymeric surfactants on droplet size is, of course, mediated by their effect on the process of emulsification and on the stability. Factors that increase the stability of emulsions can affect the droplet size.

3.3 Structure

Emulsion structure can refer to two aspects: how the components of the emulsion are distributed and structured between the phases and alternatively how the phases themselves are distributed (that is, droplet size and droplet size distribution).

How the components of the emulsion are distributed between the phases is not a trivial question. The simple picture of an emulsion as consisting of an oil dispersed in an aqueous phase (or the contrary) with the droplets covered by the emulsifier is overall correct. However, realizing that in any emulsion the structure is more complex is important. On the one hand, the excess surfactant that is not covering the emulsion droplets may form micelles in the continuous phase with some dispersed phase solubilized in them. This changes the properties of the continous phase with respect to the pure component and can affect stability, for instance inducing flocculation by the mechanism of depletion flocculation or by increasing the solubility of the dispersed phase in the continuous phase leading to faster Ostwald ripening. The formation of liquid crystalline structures in the continuous phase could have a stabilizing effect due to the increased viscosity of the medium. Coverage of emulsion droplets by lamellar liquid crystalline phases has been claimed to be responsible of the increased stability against coalescence of some emulsions [30]. As mentioned before, polymeric surfactants may form structures at low concentration. However, the almost irreversible adsorption of polymeric surfactants allows for the use of low surfactant concentration and these effects could be prevented. When weak flocculation is desirable, the use of polymeric surfactants could be useful because the same molecule that stabilizes the interfacial film could be used to induce depletion flocculation. Excess polymeric surfactant could produce as well a thickening effect that would slow down sedimentation.

4 Polymeric Surfactants in Emulsion Polymerization

Although polymeric surfactants have been around for some time and have widespread application in emulsion preparation is perhaps in the field of emulsion polymerization where they have been most used and where most of the systematic studies can be found [19]. The reason of this use can probably be found in the interest of using a same surfactant for emulsion polymerization and for the stabilization of the particles formed during the polymerization. This means an interest in looking for surfactants that produce steric (sometimes with combined electric charges) stabilization while having hydrophobic blocs compatible with the particles produced by polymerization. The surfactants used in emulsion polymerization have a wide range of monomer compositions but the common characteristic is the overall structure that is of the block type. Linear diblock and triblock are common as well as grafted copolymers. Random and multiple blocks higher that three are probably less efficient and no data is found in the literature. Triblock structure with the anchor block as the middle blocks is deemed to be better emulsion stabilizers than the corresponding diblocks [31]. Triblock copolymers in which the anchor block is the middle block produce better stabilization than terminal anchoring groups. This could be due to the possibility of strong bridging flocculation.

Initially the main interest of using polymeric surfactants was as stabilizers for nonaqueous continuous phase emulsions. Classic ionic surfactants had been good stabilizers for aqueous emulsion polymerization. However, in nonaqueous systems stabilization is mainly of a steric type. The choice of the blocks is dictated by the nature of continuous and dispersed phases. The block compatible with the continuous phase should be of a large enough molecular weight to produce effective steric stabilization. The block not compatible with the continuous phase should have a strong affinity for the dispersed phase in order to be strongly adsorbed. With respect to flocculation emulsion and suspension requirements for the stabilizing blocks are the same, that is, to prevent flocculation the continuous phase has to be a good solvent. As the solvency can be changed by changing the temperature or the composition of the solvent, flocculation can be induced in these systems. Flocculation can usually be reversed by returning to conditions of good solvency. Too long stabilizing molecular weights can produce instability either by bridging flocculation or by turning the surfactant too soluble in the continuous phase and thus producing low surfactant adsorption.

Most of the studies in emulsion polymerization focus on the produced polymer particles and the emulsion stability is not particularly studied. In these studies long term stability of the emulsion is not a requirement because the polymerization is usually preformed under conditions of good agitation. In addition the polymerization takes only a few hours or days. In emulsion polymerization the reaction takes place in the continuous phase and the role of the emulsion droplets is that of a reservoir of monomer. During emulsion polymerization the content of dispersed phase is continuously reduced and therefore the emulsion as such disappears of the system. An interesting point is that of the intermediate stages of emulsion polymerization in which the system corresponds to a mixture of polymer and emulsion particles. Those systems, termed suspoemulsions, have not received much interest in the past although this kind of situation can be found in many practical systems. It is of interest to know how the emulsion droplets interact with the newly formed particles in the system. Polystyrene particles stabilized by grafted polyoxiethylene chains mixed with hydrocarbon emulsion droplets stabilized with a pluronic surfactant were studied by Pons et al. [32] in some detail. The rheological behaviour of the system showed the difference between deformable particles (emulsion droplets) and solid particles (polystyrene) but, no specific interaction was observed.

The polymeric surfactants used in aqueous emulsion polymerization should conform to the same requirements to those used for nonaqueous systems: good adsorption, long enough stabilizing blocks but not too long to induce bridging flocculation.

5 Other Studies on Polymeric Surfactants Emulsions

Naturally occurring polymeric surfactants (in particular proteins) have been successfully used as stabilizers in food emulsions. However due to the complexity of these molecules it is difficult to extract general and quantitative information from the available results, see for instance the concerned chapters in refs. [33-34]. Studies on emulsions using synthetic polymeric surfactants have been scarce. One reason for this is probably the difficulty of obtaining well characterized polymeric surfactants. Although polymeric surfactants of the Pluronic family have been around since the fifties [13] and some work on their solution and micellar behaviour has appeared over the years, recently increasing attention has been given to their behaviour in more complex mixtures. Only recently, some studies on ternary phase behaviour and on emulsion behaviour have been published [9,35-36]. Systematic studies on a family of diblock and triblock copolymers in which the hydrophilic block is polyethyleneoxyde and the hydrophobic block is polybuthyleneoxyde with arrangements for the blocks both A-B-A and B-A-B were undertaken by J. Szymanovski and coworkers [37-38]. With respect to the emulsion behaviour they studied the influence of block length on the EIP and on the PIT. Both the molecular weight of the hydrophobic part and the proportion of the hydrophilic block affect the EIP. Emulsions invert from w/o to o/w at a higher water content as the hydrophobic molecular mass increases but, interestingly, there is a minimum in the water content that produces inversion at intermediate values of ethylene oxide content. They found that the PIT mainly depended on the content of ethylene oxide on the total molecular weight. The molecular weight of the polyoxybuthylene had a minor influence on this parameter (a 10% of the changes attributable to the changes in the hydrophobic part).

The rheology of emulsions prepared with surfactants of the pluronic family has been studied in some detail. Faers and Luckham [35] compared the rheology of latices and emulsions both stabilized by pluronic surfactants at medium and high dispersed phase volume fractions. The general behaviour is the same, except that at high volume fractions the emulsion droplets can deform, this was also found by Pons, Rossi and Tadros [32] for latices stabilized by grafted polyethyleneoxyde and emulsion droplets stabilized by pluronics. The elastic behaviour of the systems is approximately the same below a volume fraction at which the droplets start touching. Above this, the elastic modulus of the latices continue to increase in an exponential way while for the emulsions the increase inelastic modulus is linear with volume fraction. In the latices the elastic modulus increase because of the compression of the steric layer while for the emulsions, above a threshold, the deformation of the droplets is a lower energy path to deform the system. Highly concentrated o/w emulsions could be prepared using a pluronic surfactant with only a 20% content of ethylene oxide in its composition. This composition corresponds to two blocks of 9 ethylene oxide blocks joined by a propylene oxide block with 50 units. The stability of these o/w emulsions with so short stabilizing chains was attributed to the fact that hydrated propylene oxide although

is not hydrophilic it is not oleophillic either and will strongly adsorb at the interface with low penetration in the oil droplets [9].
Cárdenas and coworkers [39-40] synthesized and studied graft copolymers in which the backbone is poly methyl methacrylate and the stabilizing blocks are grafted methoxy-polyethylene oxide chains. They varied the molecular weight of the backbone and the molecular weight and number of polyethylene oxide branches. Studying the stability of o/w emulsions, they found that the dispersed phase should be a bad solvent both for the stabilizing block and the anchoring block to produce stable emulsions. The authors also conclude that an optimum ratio exists between the molecular weight of the stabilizing and anchoring blocks. This would ensure strong adsorption and a high density of droplet coverage.

Polymeric surfactants have also been studied with relation to the stabilization of multiple emulsions [41-42]. Multiple emulsions in which both the primary and secondary emulsions were stabilized by natural and synthetic polymeric surfactants produced more stable systems than those previously obtained with low molecular weight surfactants. The release of marker molecules from these emulsions was slower than in previously studied multiple emulsions, this could be related to denser and thicker adsorbed layers produced by the polymeric surfactants.

REFERENCES

1. Th. F. Tadros and B. Vincent in P. Becher (Ed.) Encyclopedia of Emulsion Technology, Vol. 1, Marcel Deker Inc., New York, 1983, pg. 129,272.
2. P. Lo Nostro, Adv. Colloid and Int. Sci., 56 (1995) 245.
3. W.D. Bancroft, J. Phys. Chem., 19 (1915) 275.
4. A. Kabalnov and H. Wennerström, Langmuir, 12 (1996), 276.
5. J.T. Davies, in J.F. Danielli, K.G.A. Pankhurst and A.C. Riddiford (Eds.), Progress in Surface Science, Vol. 2, Academic Press, New York, 1964.
6. B.P. Binks, Colloids Surf. A, 71 (1993) 167.
7. R. Pons, I. Llamas, G. Calderó, L. Pérez, M.J. García-Celma, N. Azemar, I. Carrera and C. Solans, Progress in Colloid and Polymer Science, (submited).
8. W.C. Griffin, J. Soc. Cosmet. Chem., 1 (1949) 311.
9. R. Pons, C. Solans and Th. F. Tadros, Langmuir, 11 (1995) 1966.
10. a)K. Shinoda and H. Arai, J. Phys. Chem., 68 (1964) 3485. b) K. Shinoda and H. Saito J. Colloid Interface Sci., 26 (1968) 70.
11. L. Marzall, in M.J. Schick (Ed.) Nonionic Surfactants: Physical Chemistry, Marcel Dekker, Inc.,New York, 1987, 493.
12. G.G. Stokes, Phil. Mag., 1 (1851) 337.
13. I.R. Schmolka and A.J. Raymond, J.A.O.C.S. 42 (1965) 1088.
14. H.M. Princen and A.D. Kiss, Langmuir, 3 (1987) 36.
15. B.V. Derjaguin and L. Landau, Acta Phsicochem. USSR 14 (1941) 633.
16. E.J.W. Verwey and J. Th. G. Overbeek, Theory of the Stability of Lyophobic Colloids, Elsevier, Amsterdam, 1948.
17. Th. F. Tadros Colloids Surf. A, 91 (1994) 39.
18. D.H. Napper, Polymeric Stabilization of Colloidal Dispersions, Academic Press, London, 1983.
19. I. Piirma in Polymeric Surfactants Marcel Dekker Inc., New York, 1992.
20. G.V. Jeffreys and G.A. Davies, Recent Advances in Liquid-Liquid Extraction, Oxford, 1971, pg. 495.

21. E.H. Lucassen-Reynders and K.A. Kuijpers, Colloids Surf. A, 65 (1992) 175
22. E.L. Lucassen -Reynders in P. Becher (Ed.) Encyclopedia of Emulsion Technology, Marcel Dekker, New Yorok, 1996.
23. A.S. Kabalnov, A.V. Pertzov and E.D. Shchukin, J. Colloid and Interface Sci., 118 (1987) 590.
24. W.I. Higuchi and J. Misra J. Pharm. Sci., 51 (1952) 459.
25. P. Taylor, Colloid Surf. A, 99 (1995) 175.
26. J. L. Salager, In P. Becher (Ed.) Encyclopedia of Emulsion Technology, Marcel Dekker, New York, Vol. 3, 1988.
27. a) K. Shinoda, J. Colloid Interface Science, 24 (1967) 4. b) H. Sagitani in S.E. Friberg and B. Lindman Eds. Organized Solutions, Marcel Dekker, Inc., New York, 1992, pg. 259
28. P. Walstra in P. Becher (Ed.) Encyclopedia of Emulsion Technology, Vol. 1, Marcel Deker Inc., New York, 1983, pg. 57.
29. R. Pons, I. Carrera, P. Erra, H. Kunieda, and C. Solans, Colloids Surf. A, 91 (1995) 259.
30. S.E. Friberg in J. Sjöblom (Ed.) Emulsions- A Fundamental and Practical Approach, Kluwer Academic Publishers, Amsterdam, pg 1.
31 J.L. Mura and G. Riess, Polymers for Advanced Tehcnologies, 6 (1995) 497.
32. R. Pons, P. Rossi and Th. F. Tadros, J. Phys. Chem., 99 (1995) 12624.
33. E. Dickinson (Ed.) Food Emulsions and Foams, Royal society of Chemistry, London, 1986.
34. J. Sjöblom (Ed.) Emulsions- A Fundamental and Practical Approach, Kluwer Academic Publishers, Amsterdam.
35. M.A. Faers and P.F. Luckham, Colloids Surf. A, 86 (1994) 317.
36. P. Alexandridis and T. A. Hatton, Colloids Surf. A, 96 (1995) 1.
37. J. Szymanovski and K. Prochaska, Fette Seifen Anstrichmittel, 3 (1981) 172.
38. J. Szymanovski, E. Pietrzak, M. Wisniewski and K. Prochaska, Tenside Detergents, 20 (1983) 4.
39. A.E. Cárdenas-Valera, A.I. Bailey and A. Doroszkowski, Colloids Surf. A, 96 (1995) 53.
40. A.E. Cárdenas-Valera and A.I. Bailey, Colloids Surf. A, 97 (1995) 1.
41. N. Garti and A. Aserin Adv. Colloid and Interface Science 65 (1996) 37
42. M.L. Cole and T.L. Whateley, J. Colloid Interface Science 175 (1995) 281.

Subject Index

A

B

C

D

E

F

G

H

L

M

N

O

P

R

S

T

U

V

W